D0850587

Real Analysis

A First Course

Second Edition

Russell A. Gordon
Whitman College

Boston San Francisco New York
London Toronto Sydney Tokyo Singapore Madrid
Mexico City Munich Paris Cape Town Hong Kong Montreal

Publisher: *Greg Tobin*
Managing Editor: *Karen Guardino*
Assistant Editor: *RoseAnne Johnson*
Associate Production Supervisor: *Julie LaChance*
Marketing Manager: *Michael Boezi*
Cover Designer: *Dardani Gasc Design*
Senior Designer: *Barbara Atkinson*
Manufacturing Buyer: *Caroline Fell*
Manufacturing Manager: *Evelyn Beaton*

Cover Art: "Composition" by Varvara Fedorovna Stepanova.
Tretyakov Gallery/Superstock. Licensed by Vaga, New York, NY.

This text is in the Addison-Wesley Higher Mathematics Series. For more information about Addison-Wesley mathematics books, access our World Wide Web site at *http://www.aw.com/he/* .

Library of Congress Cataloging-in-Publication Data

Gordon, Russell A., 1955–
Real analysis : a first course / Russell A. Gordon.— 2nd ed.
p. cm.
Includes bibliographical references and index.
ISBN 0-201-43727-9
1. Mathematical analysis. I. Title.

QA300 .G593 2001
515—dc21

2001022753

ISBN 0-201-43727-9

1 2 3 4 5 6 7 8 9 10 RRD 04 03 02 01

In memory of

Mary Norton Gordon,

dedicated mother and teacher

Preface

My goal in writing this edition is the same as the first: to create a carefully worded narrative that presents the ideas of elementary real analysis while keeping the perspective of a student in mind. The order and flow of topics has been preserved, but the sections have been reorganized somewhat so that related ideas are grouped together better. A few additional topics have been added; most notably, functions of bounded variation, convex functions, numerical methods of integration, and metric spaces. The biggest change is in the number of exercises; there are now more than 1600 exercises in the text. In fact, this edition is longer than the first because there are roughly 120 pages of exercises. Rather than give a chapter by chapter discussion of contents and changes, I will focus on three topics: integration, topology, and the organization of the exercises.

As most people teaching analysis know, there are two distinct ways to define the Riemann integral: as a limit of Riemann sums or as the common value of the infimum and supremum of upper and lower sums, respectively. Each approach has pedagogical and theoretical merit, but each one has some disadvantages as well. Many instructors feel rather strongly about the approach they prefer. I prefer Riemann sums because the definition of the integral is fairly easy to write down, the ideas are familiar (although poorly understood) to students from calculus, and certain properties of the integral (such as linearity) are more evident. In any event, at some point it is necessary to include a tedious proof that the two approaches are equivalent. In this edition, I have once again used Riemann sums to define the integral, but I have proved the Cauchy criterion for the integral early in the integration chapter and (in essence) shown that both methods of defining the integral are equivalent. Consequently, two different ways of establishing the existence of an integral are available; most results in elementary integration theory can be proved relatively easily using one of them. For those who are interested, the upper and lower sum approach is outlined in the supplementary exercises of the integration chapter. If desired, it is possible to follow this approach and skim Sections 5.1 and 5.2. The rest of the chapter should not be affected very much. The material on numerical methods of integration (new to this edition) is included because the proofs of the error formulas present some nice applications of the properties of derivatives and integrals.

Another area of dispute among teachers of analysis is the role topology should play in the course. In my experience, many students have a great deal of difficulty with the concepts in topology. (I think that many of us with advanced degrees in mathematics have forgotten how foreign these ideas are to the novice.) Introducing topology early and using these ideas in the statements of theorems in real analysis tends to cloud the meaning of the analysis results. Since topology is not necessary to state or prove these results, I prefer to leave topology until the end of the course or for a second semester. This does not mean that topology is not important or that students should be sheltered from abstract ideas. I just think that students are in a better position to absorb and appreciate topological ideas after getting a handle on the theory behind calculus. Thus, a discussion of point-set topology does not appear until the last chapter of this book. However, with some care in the assignment of exercises, the first three sections of the topology chapter can be covered concurrently with the rest of the book. The last section of the topology chapter introduces the concept of a metric space. It presents many results about metric spaces, but leaves the proofs as exercises. This section and its exercises are essentially a guided self-study that I have used when teaching a second semester of analysis.

Over the years, I have compiled a list of problems to assign to students in real analysis. Most of these problems have found their way into this edition of the text. The exercises range in difficulty from routine to challenging. There are exercises at the end of each section and, except for the first and last chapters, there are supplementary exercises at the end of each chapter. The exercises for a section appear in the same order as the topics in that section. (The order is implicit; there are no special headings.) This makes it easy to find the exercises related to a particular topic if only part of a section is covered. As the level of difficulty of an exercise (which is a highly subjective observation) is not noted in the text, some care must be taken in assigning exercises. For the most part, the exercises following a section range from easy to moderate in degree of difficulty. The supplementary exercises are usually more difficult for students simply due to the fact that it is not clear which theorems or results will be needed to solve the problems. Some of these exercises are fairly easy, but most of them range from moderate to challenging. They include additional material not discussed in the main body of the text (such as absolute continuity, Newton's method, and arc length); a number of these can be used as projects for students. Although it makes for a lot of reading, I recommend that the instructor skim through all the exercises to find those exercises that can be best assigned for the direction of the course they would like to teach. As with the first edition, no explicit hints are given for the exercises (the timing and nature of a hint is thus left to the discretion of the teacher) and solutions to the exercises are not included in the text.

This textbook includes almost all of the basic ideas in elementary real analysis, along with some additional topics that I find interesting and accessible to students at this level. For the most part, the extra material has been placed at the end of sections and/or chapters or in the exercises, making it easy to skip for those who choose to do so. One such topic is a study of convex functions. This topic is included because, as some reviewers pointed out, the chapter on differentiation in the first edition included very little material that was not already found in standard calculus books. Hence, beginning with the familiar notion of a concave up graph, students using this edition can get a glimpse of one of the more advanced topics in differentiation theory.

Many students taking a course in real analysis have either had an introduction to higher mathematics course and/or an abstract algebra course. For those students who have not had an exposure to higher mathematics, real analysis can present quite a challenge. There is some material in the appendices that can help these students and provide a review for those who are already somewhat familiar with it. It is probably best to visit this material as it is needed rather than just plow directly through it. However, the material in the appendices is not designed to be a short course introducing the concepts of abstract mathematics; there are no exercises and the number of examples is limited.

At Whitman College, the typical real analysis class has less than a dozen students. When I teach this class, I do very little lecturing. For each class period, the students are expected to read several pages of the book and to work on some of the problems at the end of the section. My goal is that the students understand every sentence of the text. If any statement is unclear, they are expected to ask questions. On occasion, small gaps (usually unstated) are left in the proofs in the text. The intention here is that the student should identify the gap and pause long enough to fill in the details. This activity is one of my attempts to teach students to become critical readers. I usually designate one or two problems for the students to write up carefully and turn in to be graded. Class time is spent discussing the material in the section and the remaining exercises that were assigned. Using student questions and my own questions, we clarify the concepts and proofs in the section and create examples to further illustrate the concepts. For the problems that are not to be turned in, I sometimes have students write their solutions on the board and give the rest of the class the responsibility of critiquing the proofs. This is a good opportunity to discover what the class knows and does not know. The mistakes that occur open the door to a discussion of logic and foundations, as well as the specific topic of real analysis. However, the main goal is to get the students to communicate mathematics with each other. On occasion, I will give a short lecture on some topic or develop the solution to an exercise in full detail. For these exercises, we begin with muddled attempts, dead ends, and sketchy outlines, then move to a rough solution, and finally write up a polished proof. It is important for students to see this entire process. Mathematicians do not copy down a problem, write the word "proof", then proceed to solve the problem. Observing the entire solution process helps students learn how to solve problems.

A Note to the Student

Real analysis is the study of real numbers, sets of real numbers, and functions defined on sets of real numbers. The adjective "real" makes some students wonder if there is a "fake" or "pretend" analysis competing for equal time. The word "real" is used here to distinguish this type of analysis from another branch of mathematics known as complex analysis. The adjectives "real" and "complex" refer to the real numbers and the complex numbers, respectively. Since every real number is a complex number whose imaginary part is zero, the study of real numbers is essentially included in the study of complex numbers. While it is true that complex analysis is a rich field which offers much insight into the inner workings of real analysis, a thorough grounding in real analysis is quite helpful before tackling complex analysis. The purpose of this textbook is to present some of the basic ideas and concepts of real analysis. In a nutshell, this book presents the topics of a first-year calculus course, with all of the proofs and without the applications.

The fact that the focus of this book is on proofs may discourage you; perhaps you tend to avoid proofs. It is unfortunate that the word "proof" conjures up negative emotional energy. Proofs are the mode of communication of pure mathematics. Even in applied mathematics, it is necessary to convince others that your method and solution are valid. In fact, proofs or (to use a more common term) arguments are used in every discipline. An argument presents reasons why some statement should be accepted as true. In mathematics, as opposed to other disciplines, these arguments are more formal and there is much more agreement on what statements may be accepted as true. To make progress in this book you must confront formal proofs head on. I assume that you have some familiarity with the notion of proof and know some of the basic ideas of logic. You should have picked up some of these ideas in a calculus class; even more so if you have had a course in linear algebra or abstract algebra. Appendix A discusses some of the essential ideas of mathematical logic. As you proceed through this book, your knowledge of logic will deepen and your ability to write proofs will improve. There may be some frustration at first, but after getting past the initial hurdle of writing proofs, it can become a very rewarding experience.

In reading a proof (whether one in this book, one that your classmate has written, or one that you have just written yourself), you should strive to understand every word. You should be able to justify each and every statement. Some of the proofs in this textbook have small gaps in them, that is, some effort is required to understand why a certain statement is true. The purpose of these gaps is to assist you in learning to carefully analyze proofs. My intention is that you notice these gaps while reading the book and that you take the time to fill in the missing details. The missing details may involve some algebra or the use of a previous theorem. If you are unable to supply the missing details, you should ask for assistance. Struggling with ideas, thinking about concepts, working with examples, and discussing the material with others is the best way to learn the contents of this book. You need to become an active reader and a critical thinker.

There are exercises after each section and (with the exception of the first and last chapters) supplementary exercises at the end of each chapter. Most of the exercises following a section refer to the material discussed in that section. This often provides a hint as to how to start the exercises. For example, an exercise in the section following the Intermediate Value Theorem probably requires the Intermediate Value Theorem in its solution. In your solutions to the exercises, you should only use material in the book up to that point. It may happen that a quick and easy solution to a problem involves the use of a theorem in a later section, but that is not the intended way to solve the problem at the given point. The phrases "use the definition" and "prove directly" mean to use only the definition of the concept rather than theorems concerning the given concept. The purpose of such exercises is to provide practice working with the definitions. The supplementary exercises at the end of a chapter are in no particular order and involve the ideas in the corresponding chapter in a less direct way. These exercises are more challenging because it may be harder to get started, but learning how to get started on a problem is an important skill to acquire. Many of the exercises in this textbook are either routine applications of definitions and theorems or exercises of moderate difficulty. However, there are some quite challenging exercises scattered throughout the book. Being warned in advance that an exercise is difficult affects (usually in a negative way) a student's attitude about that exercise. For this reason, the challenging exercises are not marked or noted in any way.

I have worked hard to make the book readable, but I do not pretend that the material in this book is easy. Engaging in mathematics is both frustrating and fun, challenging and aggravating. This is true at all levels of mathematics. It is important to not allow yourself to become discouraged. Keep working, studying, thinking, and asking questions. With discipline and patience, you will get a sense of the beauty of this subject.

Acknowledgments

As with the first edition, two of my colleagues at Whitman College have provided input for this text. Bob Fontenot and I often had spontaneous discussions concerning content and pedagogy in the teaching of real analysis. Certain aspects of this text reflect these discussions. David Guichard once again provided technical support for the preparation of the manuscript. His efforts and expertise made it possible for me to focus on writing rather than on implementing various TEX macros. A grant from the Louis B. Perry summer research scholarship fund allowed Derek Garton, a student at Whitman, to spend a summer reading and commenting on the manuscript. His thoughtful comments helped to smooth out a number of rough spots in the text and to eliminate some unclear statements. The first edition was formally reviewed by several mathematicians. I appreciate the time and effort they invested in their reviews; many of their comments have been incorporated into this edition. My wife, Brenda, provided emotional support and was tolerant of my frequent excursions into "mathland". Although her contributions are intangible, her love and support are evident in the pages of this book. In spite of all this assistance, there are bound to be some problems with the text. Feel free to contact me at the mathematics department of Whitman College, Walla Walla, WA 99362 or at gordon@whitman.edu with questions and comments.

Some Comments on Notation

There are some standard conventions for the use of letters as symbols in real analysis. None of the choices is mandatory and there are often exceptions to the conventions, but it is helpful to be aware of the following conventions.

The letters a, b, c, d generally represent constants.

The letters s, t, u, v, w, x, y, z generally represent variables.

The letters i, j, k, m, n, p, q generally represent integers.

The letters f, g, h generally represent functions.

Capital letters are also used, but their meanings follow less definite patterns. Because the number of letters is limited, mathematicians often use Greek letters to represent various quantities as well. For reference, the Greek alphabet is displayed below; the lower case symbol is followed by the upper case symbol for each letter.

α	A	alpha
β	B	beta
γ	Γ	gamma
δ	Δ	delta
ϵ	E	epsilon
ζ	Z	zeta
η	H	eta
θ	Θ	theta
ι	I	iota
κ	K	kappa
λ	Λ	lambda
μ	M	mu
ν	N	nu
ξ	Ξ	xi
o	O	omicron
π	Π	pi
ρ	P	rho
σ	Σ	sigma
τ	T	tau
υ	Υ	upsilon
ϕ	Φ	phi
χ	X	chi
ψ	Ψ	psi
ω	Ω	omega

Contents

3

Limits and Continuity 81

4

Differentiation 129

5

Integration 163

6

Infinite Series 209

7

Sequences and Series of Functions 241

8

Point-Set Topology 291

A

Mathematical Logic 341

B

Sets and Functions 359

1

Real Numbers

Where should this book begin? Every author faces this question as he or she sits before an empty piece of paper or a blank computer screen. For a novel, it is important to catch the reader's attention and create a desire to continue reading. For a mathematics textbook, it is generally assumed that the reader already has some motivation to study the given subject—perhaps an intrinsic interest or a graduation requirement. The more relevant question is what background to assume. What mathematical knowledge does the reader possess as he or she opens this book for the first time? A novel that is boring will collect dust on a shelf; a mathematics textbook that is confusing will experience a similar fate.

The assumed background for a reader of this textbook is a complete calculus sequence and some degree of mathematical sophistication. While this last phrase is rather ambiguous, it essentially means that the reader understands the need for proofs in mathematics, is willing to attempt to read and understand proofs, and is able to think abstractly. In addition, the reader needs to have some informal knowledge of the following topics:

1. sets and operations on sets;
2. functions and properties of functions;
3. mathematical induction;
4. some basic proof strategies such as proof by contradiction.

Because many of these more advanced concepts are more thoroughly learned in the context of some specific content, the best suggestion is to start reading this book and to ask questions about ideas that are not clear. There is some material in the appendices that may also be helpful. However, even with all of this prerequisite knowledge, the reader should not expect to understand the material in this book without a struggle. Abstract mathematics requires patience, concentration, and discipline.

Since real analysis is the study of real numbers, sets of real numbers, and functions defined on sets of real numbers, it is probably a good idea to begin with the definition of a real number. This may seem like an unnecessary place to start since anyone reading this book has been working with real numbers for at least several years. Although the question, "What is a real number?", may appear to have an obvious answer, an attempt to give a precise answer to this question often leads to circular definitions or an appeal to common sense. (For an interesting and entertaining discussion concerning the foundation for a course in real analysis, see Pourciau [20].) As with many objects in this remarkable world, close scrutiny leads to amazing and aggravating puzzles. Atoms, cells, and galaxies are much richer than anyone ever could have imagined, and the same is true of the set of real numbers. In this chapter, we will examine the concept of a real number and set the stage for the rest of the material in this book.

1.1 What Is a Real Number?

The **natural numbers** (or the counting numbers) arise, well, quite naturally. These are the numbers

$$1, 2, 3, 4, 5, 6, 7, 8, 9, 10, \ldots,$$

which are used for counting objects. The natural numbers can also be associated with the passage of time; one day follows another just as 2 follows 1, 3 follows 2, and so on. Some people will argue that these numbers are hard-wired into the human brain while others claim the natural numbers are a property of the universe in which we live, but we will not enter the realms of psychology or philosophy here. The **integers** are the numbers

$$\ldots, -5, -4, -3, -2, -1, 0, 1, 2, 3, 4, 5, \ldots,$$

which extend indefinitely in either direction. In this context, the natural numbers are referred to as the **positive integers**. The negative integers are introduced to solve problems, such as $x + 5 = 2$, for which there are no solutions in the set of positive integers. We will use the symbol $\mathbb{Z}$ to represent the set of integers and the symbol $\mathbb{Z}^+$ to denote the set of positive integers. The symbol $\mathbb{Z}$ comes from the German word for number, *Zählen*.

When numbers are used for measurement, it is sometimes necessary to consider parts of a whole. This need for parts or fractions of an integer leads to the concept of a rational number. A **rational number** is a number of the form p/q, where p and q are integers with $q \neq 0$. In other words, a rational number is a ratio of two integers. The standard symbol for the set of rational numbers is the symbol $\mathbb{Q}$, where the use of the letter Q follows from the fact that rational numbers are quotients. The number $3/5$ can be interpreted as follows: divide a stick into 5 equal pieces and take 3 of the pieces. The rational numbers include the integers since an integer n can also be represented as $n/1$. Since the rational numbers are the familiar numbers

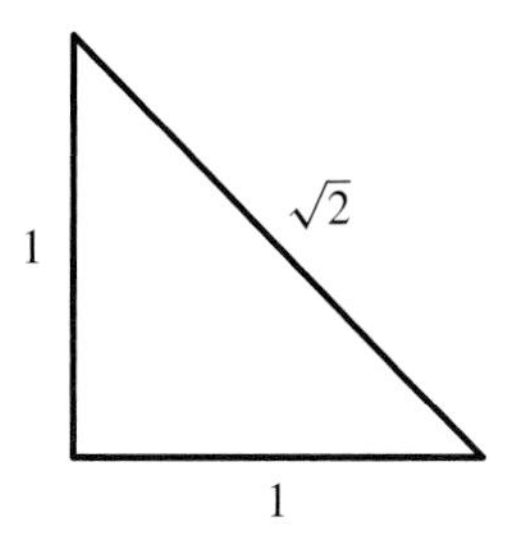

Figure 1.1 The length of the hypotenuse is not a rational number

that arise in everyday life, a great deal of elementary school mathematics is devoted to a study of these numbers.

Since there are an infinite number of rational numbers and (more importantly) since there is another rational number between any two distinct rational numbers, it would appear that the rational numbers are suitable for all possible measurements. This "optimistic" view of the rational numbers is easily dispelled. By the Pythagorean Theorem, the hypotenuse of a right triangle with sides of length 1 has a length of $\sqrt{2}$ (see Figure 1.1). The ancient Greek mathematicians knew that this length could not be described by dividing a stick into equal parts and taking some number of the parts. In other words, the number $\sqrt{2}$ is not rational. Although some readers may be familiar with a proof that $\sqrt{2}$ is not rational, a proof is given in the next paragraph.

The proof is an example of a proof by contradiction. Suppose that $\sqrt{2}$ is a rational number. Then $\sqrt{2} = p/q$, where p and q are positive integers that have no common divisors greater than 1. It follows that $p^2 = 2q^2$, which indicates that p^2 is an even integer. The only way for p^2 to be even is for p to be even. Let $p = 2r$, where r is a positive integer, and compute

$$2q^2 = p^2 = (2r)^2 = 4r^2.$$

This implies that $q^2 = 2r^2$, so q is also an even integer. But p and q cannot both be even integers since they have no common divisors greater than 1. This contradiction proves that $\sqrt{2}$ is not a rational number.

There are many other "lengths" that cannot be represented as the ratio of two integers. It follows that the set of rational numbers is not big enough for the purpose of exact measurement. Consequently, it is necessary to introduce a larger set of numbers known as the set of **real numbers** and denoted by the symbol $\mathbb{R}$. Every rational number is a real number, and the real numbers that are not rational numbers are called **irrational numbers**. In a nutshell, the real numbers contain all of the numbers needed in the development of calculus. The rational numbers have a simple description as the ratio of two integers but, as we will see, there is not an equally simple description of the set of real numbers.

Two of the familiar ways to describe a real number are the following.

1. A real number is a decimal expansion.
2. A real number represents a point on a number line.

The problem with these two interpretations of a real number is that they each have some serious shortcomings. We will take a few moments to consider these shortcomings, beginning with those of decimal expansions. The first problem to note is that a **decimal expansion** is actually an infinite sum. For instance, the decimal expansion

$$0.143143143143\ldots$$

represents the sum

$$\frac{1}{10}+\frac{4}{10^2}+\frac{3}{10^3}+\frac{1}{10^4}+\frac{4}{10^5}+\frac{3}{10^6}+\cdots.$$

Thus, an elementary knowledge of infinite series is required to truly understand real numbers expressed in this way. A second problem is that the integer 10 plays a key role in decimal expansions, but it is certainly possible to represent a real number using powers of integers other than 10; common examples are 2, 16, and 60. However, a real number is independent of the base (the integer whose powers are used) in which it is represented. Even with decimal expansions, the representation of a real number need not be unique. For example,

$$0.2000000\ldots = 0.1999999\ldots.$$

To see this, let $x = 0.1999999\ldots$ and compute

$$9x = 10x - x = 1.9999999\ldots - 0.1999999\ldots = 1.8.$$

It follows that $x = 0.2$. These difficulties do not indicate that decimal expansions are invalid or useless, but they do indicate that decimal expansions are not suitable for a mathematical definition of a real number.

In terms of decimal expansions, there is a simple distinction between rational numbers and irrational numbers: a real number is a rational number if and only if its decimal expansion has a repeating pattern. Rather than give a general proof of this result, we will illustrate it with two examples. Consider the rational number $15/101$. To convert this number to a decimal, we must perform long division:

$$\begin{array}{r|l}
 & .148514\cdots \\ \hline
101 & 15.000000\cdots \\
 & \underline{10\ 1} \\
 & 4\ 90 \\
 & \underline{4\ 04} \\
 & 860 \\
 & \underline{808} \\
 & 520 \\
 & \underline{505} \\
 & 150 \\
 & \underline{101} \\
 & 490 \\
 & \vdots
\end{array}$$

It follows that

$$\frac{15}{101} = 0.1485148514851485\ldots.$$

Since the only possible remainders in this case are 0, 1, 2, ..., 100, eventually one of them will appear again (rather quickly in this particular case) and start a repeating pattern. The same thing happens when any rational number is converted to a decimal. Now consider the number represented by the decimal

$$0.143143143143143\ldots.$$

Call this number x and compute

$$1000x - x = 143.143143143\ldots - 0.143143143\ldots = 143,$$

which indicates that $x = 143/999$. A similar computation is possible for any decimal expansion that develops a repeating pattern. The number

$$0.101001000100001000001000001\ldots$$

is an irrational number since its decimal expansion has no repeating pattern (the number of 0's between 1's continues to increase). Note that the key here is the word *repeating*. The preceding decimal expansion has a pattern, but it is not a repeating pattern. The fact that decimal expansions of irrational numbers have no repeating pattern reveals another problem. In terms of decimal expansions, what is the number $\sqrt{2}$? Since this number is irrational, its decimal expansion has no repeating pattern. Consequently, there are real numbers whose complete decimal expansions are unknown.

The second interpretation of a real number is geometrical in nature. In this view, we mark off two points on a straight line and call the left point 0 and the right point 1. All other real numbers then represent points on the line. The number 1/2 is the point midway between 0 and 1, the number 2 is twice as far to the right of 0 as 1, and the number $\sqrt{2}$ represents the length of the hypotenuse of a right triangle with sides of length 1. Negative numbers appear to the left of 0 using the same unit of measurement. A portion of the **number line** appears below.

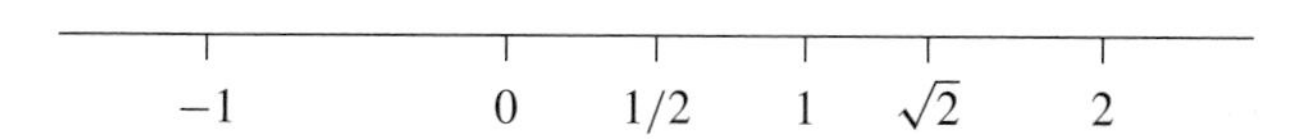

One of the obvious problems with this interpretation of a real number is that operations such as multiplication and division are difficult to perform. In addition, the point on the line that represents a real number varies with the choice of 0 and 1, but the concept of a real number should not depend on a given spatial configuration. In other words, real numbers should be independent of geometry. Finally, there is a philosophical problem with the number line. A point has no length, a line has length, and a line is made up of points. How can you combine "things" that do not have a property and end up with something that suddenly has that property? You may recall that some of Zeno's paradoxes (see Katz [11] for a discussion of these) are concerned with this problem. (Zeno of Elea was a Greek philosopher who lived in the fifth century B.C.E.)

The last few paragraphs are not intended to convince the reader to stop thinking of a real number as a decimal expansion or as a point on a line. These are very useful interpretations of a real number, but they are not suitable for a rigorous definition of a real number. What then is the mathematical definition of a real number? The definition of a new term in mathematics must involve previously defined terms, and, in this case, the rational numbers are the "known" quantities. A real number is defined to be some sort of object involving the rational numbers: either a set of rational numbers (known as Dedekind cuts) or as equivalence classes of Cauchy sequences of rational numbers (see the next chapter for the definition of a Cauchy sequence). The operations of addition and multiplication are then defined on these objects and all of the usual properties of the real numbers follow as theorems. This process is not all that difficult, but it is tedious and requires some mathematical sophistication. Depending on your personality, it is either a fascinating process or a pointless exercise. In any case, we are not going to include this development here. The reader can consult Rudin [22] for a development of the real numbers using Dedekind cuts; a thorough account of both definitions can be found in Hobson [10].

In between the useful but problematic familiar interpretations of a real number and the rigorous but very abstract interpretations of a real number is an interpretation that we will use for this book. Consider the following formula, which illustrates the distributive property of the real numbers:

$$n(x + a) = nx + na.$$

While working with a formula of this type, it is unlikely that the symbols n, x, and a are considered to be decimal expansions, points on a line, or some type of set composed of rational numbers. The symbols are just objects—namely, real numbers—that satisfy the given property. In other words, in most manipulations involving real numbers, the focus is on the properties of the real numbers rather than on some particular interpretation of the real numbers. It is possible to give a short list of properties of the real numbers from which all of the others follow. A real number is then considered to be a member of a set of objects that has all of these properties. The purpose of the next few paragraphs is to list these properties.

If the reader is encountering abstract mathematics for the first time, then the next paragraphs will probably seem rather mysterious. If you are in this position, then as you read through the following material, think about the usual properties of numbers that you have used over the years and note how each numbered item in the various definitions lists one of those properties. Mathematicians seek a firm logical foundation for the terms and symbols they use; the concept of an ordered field is such a foundation for the set of real numbers. In order to work through this text, it is important that the reader be familiar with the properties of the set of real numbers. However, a complete understanding of the formal definition of the set of real numbers is not crucial for the remainder of the text. (By the way, the ideas presented here will make more sense as your level of mathematical maturity increases.)

DEFINITION 1.1 A **field** is a nonempty set F of objects that has two operations defined on it. These operations are called addition and multiplication and are denoted in the usual way. Addition and multiplication satisfy the following properties:

1. $x + y \in F$ for all $x, y \in F$.
2. $x + y = y + x$ for all $x, y \in F$.

3. $(x + y) + z = x + (y + z)$ for all $x, y, z \in F$.
4. F contains an element 0 such that $x + 0 = x$ for all $x \in F$.
5. For each $x \in F$ there exists $y \in F$ such that $x + y = 0$.
6. $xy \in F$ for all $x, y \in F$.
7. $xy = yx$ for all $x, y \in F$.
8. $(xy)z = x(yz)$ for all $x, y, z \in F$.
9. F contains an element 1 such that $x \cdot 1 = x$ for all $x \in F$.
10. For each $x \in F$ such that $x \neq 0$ there exists $y \in F$ such that $xy = 1$.
11. $x(y + z) = xy + xz$ for all $x, y, z \in F$.

The field properties are very familiar properties of the real numbers and are used correctly by most people most of the time with very little conscious thought. Properties (1) and (6) assert that the set F is closed under addition and multiplication, that is, the sum and product of two elements in the field are elements of the field. Properties (2), (3), (7), and (8) are the commutative and associative properties of addition and multiplication, respectively. Properties (4) and (9) establish the existence of additive and multiplicative identities, while properties (5) and (10) establish the existence of additive and multiplicative inverses. Finally, property (11) is the distributive property.

It is worth noting that our definition of a field makes a number of assumptions. For instance, it is assumed that the reader knows what an operation is, is familiar with the notation for addition and multiplication, and understands the order in which operations are to be performed. To list and define all of these assumptions would take us too far afield, so we will leave such a discussion to another book. The interested reader can consult the abstract algebra texts by Birkhoff and MacLane [2] or Gallian [7]. The set of rational numbers is one example of a field, but there are other examples as well; most texts on abstract algebra will contain many examples of fields.

DEFINITION 1.2 An **order** $<$ on a set S is a relation that satisfies the following two properties:

1. If $x, y \in S$, then exactly one of $x < y$, $x = y$, or $y < x$ is true.
2. For all $x, y, z \in S$, if $x < y$ and $y < z$, then $x < z$.

An **ordered set** is a set with an order defined on it.

Although the term "relation" has a precise mathematical definition, it will not be given here. It is sufficient to think of a relation as a relationship between two numbers. For this informal presentation, we are assuming that the reader is familiar with comparing the sizes of real numbers. For real numbers x and y, the inequality $x < y$ means that the number x is less than the number y. As an example, the set of integers with the symbol $m < n$ meaning (as usual) that m is less than n is an ordered set.

DEFINITION 1.3 An **ordered field** is a field F that is an ordered set with the following additional properties:

1. If $x > 0$ and $y > 0$, then $x + y > 0$.
2. If $x > 0$ and $y > 0$, then $xy > 0$.
3. $x < y$ if and only if $y - x > 0$.

For the record, there are other ways to define an ordered field, and even within the approach that has been adopted here, there are variations. In fact, Definition 1.3 involves a degree of redundancy: it is possible to prove that property (3) implies property (1). However, the three properties listed in Definition 1.3 are very familiar properties: the sum of two positive numbers is positive, the product of two positive numbers is positive, and x is less than y if and only if $y - x$ is positive.

All of the familiar properties of equalities and inequalities are valid in an ordered field. As a typical example, the property,

$$\text{if } x < y \text{ and } z > 0, \text{ then } xz < yz,$$

follows from the properties of an ordered field. We will neither state nor prove these results here and simply assume that the reader is familiar with these properties of real numbers. In any case, proving these properties is not the purpose of this textbook. Once again, the interested reader is referred to a book on abstract algebra.

The two most familiar examples of ordered fields are the set of rational numbers and the set of real numbers. (As an aside, the set of complex numbers is a field that is not an ordered field.) In other words, the field properties and the order properties do not distinguish between the sets $\mathbb{Q}$ and $\mathbb{R}$; both sets satisfy all of the properties that have been listed thus far. Since the sets $\mathbb{Q}$ and $\mathbb{R}$ have some differences, the real numbers must possess some additional property that the rational numbers do not possess. A discussion of the distinction between these two sets of numbers will be the topic of Section 1.3.

Exercises

1. It was stated in the text that the rational numbers form a field. In particular, the rational numbers are closed under addition and multiplication. Prove this fact by showing that the sum of two rational numbers is a rational number and the product of two rational numbers is a rational number.
2. Prove that there is a rational number between any two distinct rational numbers.
3. Convert each of the rational numbers into a repeating decimal.
   a) $8/27$ **b)** $4/21$ **c)** $5/19$
4. Convert each repeating decimal into a rational number of the form p/q, where p and q are positive integers with no common divisors.
   a) $0.357357357\ldots$ **b)** $0.327272727\ldots$ **c)** $0.211538461538 46\ldots$
5. Find the millionth digit in the decimal expansion of $2/7$.
6. Prove that the reciprocal of an irrational number is an irrational number.
7. Prove that the sum of a rational number and an irrational number is irrational.
8. Prove that the product of a nonzero rational number and an irrational number is an irrational number.

9. Let x be an irrational number. Prove that there exists an irrational number y such that xy is a rational number.

10. Let x be a real number. Prove that at least one of the numbers $\sqrt{2} - x$ or $\sqrt{2} + x$ is irrational.

11. Let n be a positive integer that is not a perfect square. Prove that $\sqrt{n}$ is irrational.

12. Prove that $\sqrt{2} + \sqrt{3}$ is irrational.

13. Prove that $\sqrt{n-1} + \sqrt{n+1}$ is irrational for every positive integer n.

14. Prove that there is no rational number r such that $2^r = 3$.

15. Let x and y be irrational numbers such that $x - y$ is also irrational. Define sets A and B by $A = \{x + r : r \in \mathbb{Q}\}$ and $B = \{y + r : r \in \mathbb{Q}\}$. Prove that the sets A and B have no elements in common.

16. Let A be the set of all numbers of the form $a + b\sqrt{2}$, where a and b are arbitrary rational numbers. Let addition and multiplication be defined on A in the same way they are defined for real numbers. Prove that the set A is a field.

17. Let B be the set of all irrational numbers together with the numbers 0, 1, and -1. Let addition and multiplication be defined on B in the same way they are defined for real numbers. Determine the field properties that are satisfied by B. Is B a field?

18. Let S be the set of all ordered pairs of positive integers and adopt the convention that $(a, b) = (c, d)$ if and only if $a = c$ and $b = d$. For each given relation, determine whether or not the relation satisfies the properties listed in Definition 1.2.

a) Define a relation $<$ on S by $(a, b) < (c, d)$ if and only if $ab < cd$.

b) Define a relation $<$ on S by $(a, b) < (c, d)$ if and only if $ad < bc$.

19. Referring to Definition 1.3, prove that property (3) implies property (1). Be certain to use only properties that have been proved or already assumed to be true.

20. Use the properties of an ordered field to prove the following: if $x < y$ and $z > 0$, then $xz < yz$.

1.2 Absolute Value, Intervals, and Inequalities

This section contains a discussion of the absolute value function, the concept of an interval, a formula for geometric sums, and a brief exploration of two interesting inequalities. The properties of the absolute value function will be used throughout the text, primarily as a measure of the distance between two numbers. Almost all the theorems discussed in this book are stated for functions that are defined on an interval, so it is important to understand exactly what is meant by the term "interval". The formula for a geometric sum is quite simple and appears now and again in proofs and examples. Although the last two inequalities discussed in this section will not be used in this text, they are important in other areas of real analysis and are related to ideas that have been considered thus far.

In analysis, it is often necessary to measure the distance between two points. Since the points considered in analysis may represent numbers, vectors, functions, or sets, the notion of distance between points and how to compute it can become rather abstract. It is possible to study this notion in a more general setting (see the discussion of metric spaces in Section 8.5), but at this stage of the game, we will remain in the familiar territory of the real numbers. The absolute value function can

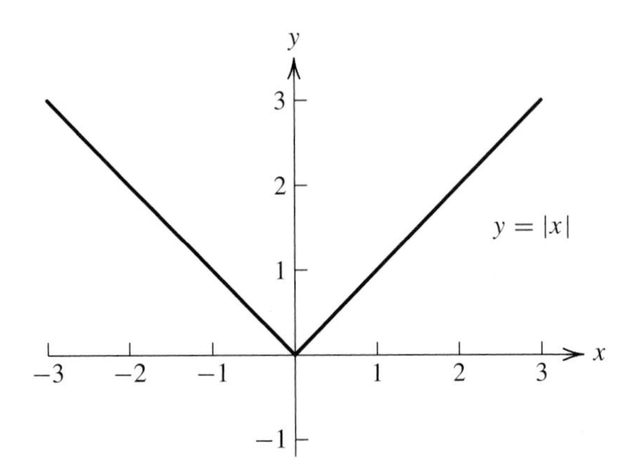

Figure 1.2 A graph of the absolute value function

be used to measure the distance between two real numbers. This function and its properties will be considered in the next few paragraphs.

Given a real number x, the **absolute value** of x, denoted by $|x|$, is defined by

$$|x| = \begin{cases} x, & \text{if } x \geq 0; \\ -x, & \text{if } x < 0. \end{cases}$$

Note that $|x| \geq 0$ for all x and that $|x| = 0$ if and only if $x = 0$. The number $|x|$ can be interpreted as the distance on the number line from the point x to the point 0. A graph of the absolute value function is shown in Figure 1.2. The following theorem records some simple properties of the absolute value function; the proof will be left as an exercise.

THEOREM 1.4 The absolute value function has the following properties.

a) $-|a| \leq a \leq |a|$ for every real number a.

b) $|ab| = |a|\,|b|$ for all real numbers a and b.

c) $|-a| = |a|$ for every real number a.

d) Let $c > 0$. Then $|a| < c$ if and only if $-c < a < c$ and $|a| \leq c$ if and only if $-c \leq a \leq c$. ■

The limit concept is fundamental in the theory of calculus and, as the reader may recall, this concept depends on a measure of the distance between two real numbers. If x and y are real numbers, then $|x - y|$ is the **distance** between x and y. Consequently, the absolute value function will be seen frequently in this text. Several important inequalities involving the absolute value function are recorded in the next two theorems; the first is by far the most commonly used inequality in analysis.

THEOREM 1.5 Triangle Inequality The inequality $|a + b| \leq |a| + |b|$ is valid for all real numbers a and b.

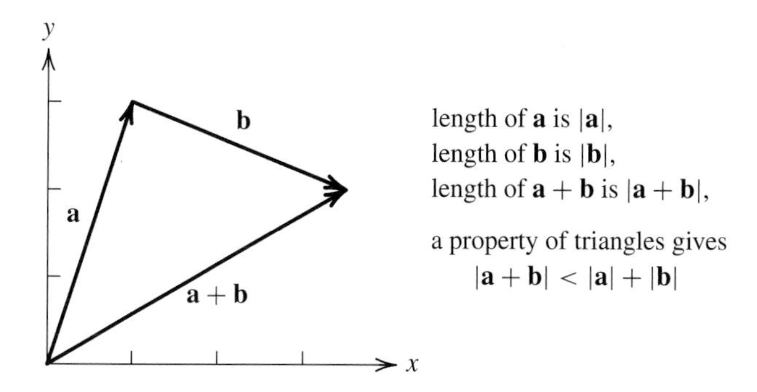

Figure 1.3 A geometric interpretation of the Triangle Inequality

Proof. Let a and b be real numbers. Adding the inequalities $-|a| \le a \le |a|$ and $-|b| \le b \le |b|$ yields

$$-|a| - |b| \le a + b \le |a| + |b|.$$

Part (d) of Theorem 1.4 then shows that $|a + b| \le |a| + |b|$. ■

THEOREM 1.6 Reverse Triangle Inequality The inequality

$$\big||a| - |b|\big| \le |a - b|$$

is valid for all real numbers a and b.

Proof. Let a and b be distinct real numbers. (If $a = b$, the conclusion is trivial.) By the Triangle Inequality,

$$|a| = |a - b + b| \le |a - b| + |b| \quad \text{and} \quad |b| = |b - a + a| \le |b - a| + |a|.$$

These two inequalities can be written as

$$|a| - |b| \le |a - b| \quad \text{and} \quad |b| - |a| \le |a - b|,$$

where part (c) of Theorem 1.4 has been used in the second one. Writing these inequalities as one inequality yields

$$-|a - b| \le |a| - |b| \le |a - b|.$$

The Reverse Triangle Inequality now follows from part (d) of Theorem 1.4. This completes the proof. ■

Since we are working in one dimension, the use of the term "triangle" in the Triangle Inequality may be puzzling. The name for this inequality has its origins in vector geometry. If **a** and **b** are vectors in $\mathbb{R}^2$, then the vector sum **a** + **b** can be represented using a triangle, as shown in Figure 1.3. In this context, the Triangle Inequality simply states that the sum of the lengths of two sides of a triangle is greater than the length of the third side. For the record, this geometric observation does not constitute a proof of the Triangle Inequality.

The Triangle Inequality will be used frequently in this text. Since it is such a basic property of the real numbers, in most of the cases in which it is used, it will not even be mentioned. The most common use of the Triangle Inequality in analysis is in a slightly different form than the one stated in Theorem 1.5. Letting $a = x - y$ and $b = y - z$, the Triangle Inequality becomes

$$|x - z| \leq |x - y| + |y - z|.$$

Inequalities such as this will occur often; an appropriate y term will simply appear with little or no explanation. For example, the inequality

$$|x - 3| \leq |x + 4| + 7$$

is valid for all real numbers x. The reader should pause long enough to determine the extra term that was used in this case. Since inequalities of this type will be used many times in the text, it is important to understand them. As one last example, let S be a set of real numbers and let L be a real number. Suppose that $|x - L| < 1$ for all $x \in S$. Then

$$|x| = |x - L + L| \leq |x - L| + |L| < 1 + |L|$$

for all $x \in S$.

The next result essentially states that the only nonnegative number that is less than every positive number is 0. This is, of course, a simple observation, but its use in proofs can be a sticking point for some people. Let a and b be two real numbers. To show that $a = b$, it is sufficient to prove that the inequality $|a - b| < \epsilon$ holds for each positive number ϵ. In other words, if the distance from a to b is less than any positive number, then a and b are the same number.

THEOREM 1.7 Let x be a real number. If $|x| < \epsilon$ for each $\epsilon > 0$, then $x = 0$.

Proof. The proof will be a proof by contradiction. Suppose that $x \neq 0$. It follows that $|x|/2 > 0$. Since the hypothesis is valid for every positive number, we find that $|x| < |x|/2$, a contradiction. It follows that $x = 0$. ■

Let $n \geq 2$ be a positive integer and let $a_1, a_2, \ldots, a_n$ be real numbers. The **maximum** (or largest) and **minimum** (or smallest) numbers in this list will be denoted by

$$\max\{a_1, a_2, \ldots, a_n\} \quad \text{and} \quad \min\{a_1, a_2, \ldots, a_n\},$$

respectively. For the case in which $n = 2$, the symbols $x \vee y$ and $x \wedge y$ are sometimes used to represent the numbers $\max\{x, y\}$ and $\min\{x, y\}$, respectively. The absolute value function can be used to provide an algebraic formula for $x \vee y$ and $x \wedge y$. It will be left as an exercise to verify the formulas

$$x \vee y = \frac{x + y + |x - y|}{2} \quad \text{and} \quad x \wedge y = \frac{x + y - |x - y|}{2}.$$

These formulas are sometimes useful when proving facts that involve the maximum or minimum of two quantities.

In a study of real analysis, it is necessary to consider sets of real numbers. Although an arbitrary set of real numbers can be quite complicated (an exposure to more advanced ideas, such as those in point-set topology discussed in Chapter 8, is required to truly appreciate this fact), most of the sets considered in this text will be intervals. The reader is certainly familiar with intervals from calculus, but a formal definition of an interval is required when proving theorems in analysis.

DEFINITION 1.8 A set S of real numbers is an **interval** if and only if S contains at least two points and for any two points $x, y \in S$, every real number between x and y belongs to S as well.

The next theorem records the familiar possible forms for an interval. Recall that the symbol ∞ represents a quantity that is larger than any real number. Similarly, the symbol $-\infty$ represents a quantity that is smaller than any real number. Neither ∞ nor $-\infty$ is a real number; they are simply convenient symbols to represent a certain concept.

THEOREM 1.9 An interval has one of the following nine forms.

1. $(a, b) = \{x \in \mathbb{R} : a < x < b\}$, a bounded, open interval;
2. $[a, b) = \{x \in \mathbb{R} : a \leq x < b\}$, a bounded, half-open interval;
3. $(a, b] = \{x \in \mathbb{R} : a < x \leq b\}$, a bounded, half-open interval;
4. $[a, b] = \{x \in \mathbb{R} : a \leq x \leq b\}$, a bounded, closed interval;
5. $(-\infty, a) = \{x \in \mathbb{R} : x < a\}$, an unbounded, open interval;
6. $(-\infty, a] = \{x \in \mathbb{R} : x \leq a\}$, an unbounded, closed interval;
7. $(b, \infty) = \{x \in \mathbb{R} : x > b\}$, an unbounded, open interval;
8. $[b, \infty) = \{x \in \mathbb{R} : x \geq b\}$, an unbounded, closed interval;
9. $(-\infty, \infty) = \mathbb{R}$, an unbounded interval, both open and closed.

Proof. It is easy to verify that each of the nine sets listed in the theorem satisfies the definition of an interval. A proof that every interval has one of the nine forms essentially amounts to finding the largest and smallest element in the set. A rigorous proof requires the Completeness Axiom, so it will be postponed until the next section (see Exercise 14 in Section 1.3). ■

The terms describing the intervals listed in Theorem 1.9 may warrant some explanation. As they are used here, the adjectives bounded and unbounded are probably self-explanatory; an interval is bounded if both of its endpoints are real numbers. (A formal definition of a bounded set will be presented in the next section.) As indicated in the theorem, there are four types of bounded intervals and five types of unbounded intervals. When writing a bounded interval, it is always assumed that the first number is strictly less than the second number. For example, the notation $[u, v]$ implies that $u < v$. The convention, which should be familiar to the reader from earlier mathematics courses, that a bounded, open interval does not include its endpoints and a bounded, closed interval does include its endpoints is standard (so a half-open interval contains just one of its endpoints), but the terms open and closed also have rigorous mathematical definitions. Exercise 18 at the end of this section will help explain why intervals (6) and (8) are closed. For the record, the set $\mathbb{R}$ is considered to be an interval that is both open and closed. The reader may consult Section 8.1 for a further study of open and closed sets.

A **geometric sum** is a sum of the form

$$a + ar + ar^2 + ar^3 + \cdots + ar^n,$$

where a and r are real numbers and n is a positive integer. The cases in which $a = 0$ or $r = 1$ are trivial and generally excluded from consideration. The defining characteristic of a geometric sum is the fact that each term in the sum is the previous term multiplied by a fixed constant; in other words, the ratio of consecutive terms is constant. This type of sum appears quite often both in applications and in proofs of results in analysis. As shown in the following theorem, the sum of all the terms in a geometric sum is easy to express without all of the plus signs.

THEOREM 1.10 Geometric Sum If $a \neq 0$ and $r \neq 1$ are real numbers, then

$$a + ar + ar^2 + ar^3 + \cdots + ar^n = a \cdot \frac{1 - r^{n+1}}{1 - r}.$$

Proof. Let S be the sum of the given geometric sum. Writing out the quantities S and rS, then subtracting, yields

$$\begin{aligned} S &= a + ar + ar^2 + ar^3 + \cdots + ar^n; \\ rS &= ar + ar^2 + ar^3 + ar^4 + \cdots + ar^{n+1}; \\ (1 - r)S &= a - ar^{n+1}. \end{aligned}$$

The result follows easily from the last equation. ∎

It is highly recommended that the method for finding the sum of a geometric sum be remembered. In some ways, the method of this proof is more useful than the formula itself because geometric sums often appear in forms not quite like the form stated in the theorem. To use the formula in these situations, it is necessary to manipulate the sum into the desired form—this is sometimes more work than simply computing $S - rS$ for the particular sum. For example, one way to find the sum $r^2 + r^3 + \cdots + r^n$ is to write S for the sum and note that $S - rS = r^2 - r^{n+1}$. A value for S follows easily. Another way to find S is to first rewrite the sum, then use the formula in the theorem:

$$r^2 + r^3 + \cdots + r^n = r^2\big(1 + r + \cdots + r^{n-2}\big) = r^2 \cdot \frac{1 - r^{n-1}}{1 - r}.$$

The reader can decide which of these two methods appears to work most efficiently.

We conclude this section with two interesting and important inequalities. The first inequality concerns the relationship between the arithmetic mean and geometric mean of a set of real numbers. Let n be a positive integer and let $a_1, a_2, \ldots, a_n$ be nonnegative real numbers. Then the **arithmetic mean** and the **geometric mean** of this set of numbers are defined by

$$\frac{a_1 + a_2 + \cdots + a_n}{n} \quad \text{and} \quad \big(a_1 a_2 \cdots a_n\big)^{1/n},$$

respectively. For the record, the arithmetic mean of a set of numbers, which is sometimes called the average of the numbers, can be defined even if the numbers are not nonnegative. The arithmetic mean of two numbers represents the number that is halfway between the two numbers. For two positive real numbers x and y, their geometric mean $\sqrt{xy}$ represents the length of the side of a square whose area is the same as the area of a rectangle with sides of lengths x and y. It is easy to verify that the geometric mean of two positive numbers is less than or equal to the arithmetic mean of the numbers (see the exercises). This result is true for every set of n nonnegative numbers, but a proof for values of $n > 2$ is more difficult. The proof presented here begins with a lemma. (As discussed in Appendix A, a lemma is a result that is primarily useful in proving some other result.)

LEMMA 1.11 Let $n \geq 2$ be an integer. Suppose that $b_1, b_2, \ldots, b_n$ are positive real numbers that are not all equal. If $b_1 b_2 \cdots b_n = 1$, then $b_1 + b_2 + \cdots + b_n > n$.

Proof. The proof will use the Principle of Mathematical Induction. For the case $n = 2$, we know that $b_1 \neq b_2$ and $b_1 b_2 = 1$. It follows that

$$0 < \left(\sqrt{b_1} - \sqrt{b_2}\right)^2 = b_1 - 2\sqrt{b_1 b_2} + b_2 = b_1 - 2 + b_2,$$

so the result is true when $n = 2$. Now suppose the result is valid for some positive integer $p \geq 2$. Let $b_1, b_2, \ldots, b_p, b_{p+1}$ be positive real numbers that are not all equal and satisfy $b_1 b_2 \cdots b_p b_{p+1} = 1$. Without loss of generality, we may assume that the numbers are in increasing order: $b_1 \leq b_2 \leq \cdots \leq b_p \leq b_{p+1}$. By the assumptions on these numbers, we must have $b_1 < 1 < b_{p+1}$. Since the conclusion of the lemma is assumed to be true when $n = p$ and since $(b_1 b_{p+1}) b_2 \cdots b_p = 1$,

$$b_1 b_{p+1} + b_2 + \cdots b_p \geq p.$$

(Equality is possible in this case since each of the numbers in the sum could be 1.) Since $(b_{p+1} - 1)(1 - b_1)$ is the product of two positive numbers,

$$\begin{aligned} b_1 + b_2 + \cdots + b_{p+1} &= (b_1 b_{p+1} + b_2 + \cdots + b_p) + 1 + (b_{p+1} - 1)(1 - b_1) \\ &\geq p + 1 + (b_{p+1} - 1)(1 - b_1) \\ &> p + 1. \end{aligned}$$

This shows that the result holds when $n = p + 1$. By the Principle of Mathematical Induction, the inequality is valid for all integers $n \geq 2$. ■

THEOREM 1.12 Arithmetic Mean/Geometric Mean Inequality Let n be a positive integer. If $a_1, a_2, \ldots, a_n$ are nonnegative real numbers, then

$$\left(a_1 a_2 \cdots a_n\right)^{1/n} \leq \frac{a_1 + a_2 + \cdots + a_n}{n}.$$

Equality occurs if and only if $a_1 = a_2 = \cdots = a_n$.

Proof. Equality certainly occurs if $a_1 = a_2 = \cdots = a_n$. In addition, the result is trivial if $n = 1$ or if one of the a_k's is 0. Suppose that $n \geq 2$, that all of the a_k's are positive, and that the a_k's are not all equal. Let $r = \left(a_1 a_2 \cdots a_n\right)^{1/n}$ and note that

$$\frac{a_1}{r} \cdot \frac{a_2}{r} \cdots \frac{a_n}{r} = \frac{a_1 a_2 \cdots a_n}{r^n} = 1.$$

By the previous lemma,

$$\frac{a_1}{r} + \frac{a_2}{r} + \cdots + \frac{a_n}{r} > n,$$

which is equivalent to

$$\frac{a_1 + a_2 + \cdots + a_n}{n} > r = \left(a_1 a_2 \cdots a_n\right)^{1/n}.$$

This completes the proof. ■

To see one application of this inequality, consider the following optimization problem: find the minimum surface area of an open top rectangular box having a square base and a fixed volume V. To solve this problem, let x be the length and width of the base of the box and let h be the height of the box. Then the volume V and surface area S of the box are given by $V = x^2h$ and $S = x^2 + 4xh$. The Arithmetic Mean/Geometric Mean Inequality yields

$$S = x^2 + 4xh = x^2 + 2xh + 2xh \geq 3\sqrt[3]{x^2 \cdot 2xh \cdot 2xh} = 3\sqrt[3]{4}\, V^{2/3}.$$

Hence, the minimum surface area of the box is $3\sqrt[3]{4}\, V^{2/3}$, and it occurs when $x^2 = 2xh$ or $x = 2h$. Note that the minimum surface area occurs when the area of the base of the box is the same as the area of two opposite sides of the box. The crucial step in this particular solution to the problem is writing the expression for S in such a way that the product of all the terms gives an expression for V. In practice, it may take some trial and error to find the right combination. By the way, make sure you see why writing $S = x^2 + xh + 3xh$ will not work. If you are intrigued by this solution of a "calculus" problem without calculus, see Niven [18].

The last inequality considered in this section is known as the Cauchy-Schwarz Inequality. This inequality plays an important role in the study of the vector space $\mathbb{R}^p$. However, rather than express the result using inner products and lengths of vectors (see the supplementary exercises of Chapter 2), we will simply phrase it in terms of sums.

THEOREM 1.13 Cauchy-Schwarz Inequality Let n be a positive integer. If $a_1, a_2, \ldots, a_n$ and $b_1, b_2, \ldots, b_n$ are real numbers, then

$$\left(\sum_{k=1}^{n} a_k b_k\right)^2 \leq \left(\sum_{k=1}^{n} a_k^2\right)\left(\sum_{k=1}^{n} b_k^2\right).$$

Equality occurs if and only if there is a constant c such that $a_k = cb_k$ for all integers $k = 1, 2, \ldots, n$.

Proof. If $\sum_{k=1}^{n} b_k^2 = 0$, then all of the b_k's are 0 and both sides of the inequality are 0. Suppose that this sum is positive. For any constant z,

$$0 \leq \sum_{k=1}^{n} \left(a_k - zb_k\right)^2 = \sum_{k=1}^{n} a_k^2 - 2z \sum_{k=1}^{n} a_k b_k + z^2 \sum_{k=1}^{n} b_k^2.$$

For an appropriate choice of the constant z (see the exercises),

$$\sum_{k=1}^{n} a_k^2 - 2z\sum_{k=1}^{n} a_k b_k + z^2\sum_{k=1}^{n} b_k^2 = \sum_{k=1}^{n} a_k^2 - \frac{\left(\sum_{k=1}^{n} a_k b_k\right)^2}{\sum_{k=1}^{n} b_k^2}.$$

The Cauchy-Schwarz Inequality follows by noting that the right side of this equation is nonnegative. The only way in which equality can occur is if there is a value of z for which $a_k - zb_k = 0$ for all k. This completes the proof. ■

Exercises

1. Sketch the graphs of the equations $y = 3|x| + |x - 2|$ and $y = 3 - |2x - 5|$.
2. Sketch the graph of the equation $y = |x^2 - 2x|$.
3. Given a real number x, what is the value of $\sqrt{x^2}$?
4. Prove Theorem 1.4.
5. Under what conditions on the numbers a and b does equality occur in the Triangle Inequality? Answer the same question for the Reverse Triangle Inequality.
6. Let a, x, and y be real numbers and let $\epsilon > 0$. Suppose that $|x-a| < \epsilon$ and $|y-a| < \epsilon$. Use the Triangle Inequality to find an estimate for the magnitude of $|x - y|$.
7. Let A and B be two nonempty sets of real numbers and let L be a real number. Suppose that there exist positive real numbers x and y such that $|a - L| < x$ for all $a \in A$ and that $|2b - L| < y$ for all $b \in B$. Use the Triangle Inequality to prove that $|a - 4b| < x + 2y + |L|$ for all $a \in A$ and $b \in B$.
8. Let $n > 1$ be a positive integer and let $a_1, a_2, \ldots, a_n$ be real numbers.

   a) Prove that $\left|\sum_{k=1}^{n} a_k\right| \le \sum_{k=1}^{n} |a_k|$.

   b) Prove that $\left|\sum_{k=1}^{n} a_k\right| \ge |a_1| - \sum_{k=2}^{n} |a_k|$.
9. Let S be a nonempty set of real numbers and let a be a nonzero real number. Suppose that $|x - a| < |a|/2$ for all $x \in S$. Prove that $|x| > |a|/2$ for all $x \in S$.
10. Give a two dimensional vector interpretation of the Reverse Triangle Inequality.
11. Write the following statement using mathematical symbols, then verify that it is valid. The distance between the absolute values of two numbers is less than or equal to the distance between the numbers.
12. Let x and a be real numbers. Suppose that $x < a + \epsilon$ for all positive numbers ϵ. Prove that $x \le a$.
13. Given real numbers x and y, prove that

$$x \vee y = \frac{x + y + |x - y|}{2} \quad \text{and} \quad x \wedge y = \frac{x + y - |x - y|}{2}.$$

14. Find an example of an interval that satisfies the given condition.

    a) a bounded, open interval that contains no integers

    b) a bounded, closed interval that contains exactly two integers

c) an unbounded, closed interval that contains no negative numbers

d) an unbounded, open interval that contains all the positive integers, but no other integers

15. Write each of the following sets as intervals.

a) $\{x : x^2 < 6\}$

b) $\{x : x^3 \geq 8\}$

c) $\{x : |x| \leq 2 \text{ and } |x - 2| > 1\}$

d) $\{x : 2x + |x| \leq 3\}$

e) $\{x : x/(x + 2) < 0\}$

16. Let I and J be two intervals and suppose that $I \cap J \neq \emptyset$.

a) Prove that $I \cap J$ is an interval if it contains more than one point.

b) Give an example to show that $I \cap J$ may consist of a single point.

c) Suppose that I and J are bounded, open intervals. Prove that $I \cap J$ is a bounded, open interval.

d) Give an example in which I and J are bounded, half-open intervals and $I \cap J$ is a closed interval.

17. Let a and b be real numbers with $a < b$, and let x be a real number. Suppose that for each $\epsilon > 0$, the number x belongs to the open interval $(a - \epsilon, b + \epsilon)$. Prove that x belongs to the interval $[a, b]$.

18. A set S of real numbers is defined to be an open set if it has the following property: for each $x \in S$, there exists a positive number r such that $(x - r, x + r) \subseteq S$. The set S is closed if the set $\mathbb{R} \setminus S$ is open.

a) Prove that the interval (a, b) is an open set.

b) Prove that the interval (a, ∞) is an open set.

c) Prove that $\emptyset$ and $\mathbb{R}$ are open sets.

d) Prove that the union and intersection of two open sets is an open set.

e) Prove that the interval $[a, b]$ is a closed set.

f) Use the previous results to justify the use of the adjectives open and closed in the statement of Theorem 1.9.

19. Find each of the following geometric sums. Assume that $r \neq 1$ and $n \geq 4$.

a) $2 + 6 + 18 + 54 + \cdots + 1458$

b) $4 - 2 + 1 - \dfrac{1}{2} + \cdots - \dfrac{1}{512}$

c) $3r + 3r^2 + 3r^3 + \cdots + 3r^n$

d) $r^3 - r^4 + r^5 - \cdots + r^{2n-1}$

20. A certain ball has a bounce coefficient of 0.85. This means that the ball bounces to a height $0.85h$ when dropped from a height h. Suppose that this ball is dropped from a height of ten feet. How far has it traveled when it hits the floor for the twelfth time?

21. An **arithmetic sum** is a sum of the form

$$a + (a + d) + (a + 2d) + (a + 3d) + \cdots + (a + nd),$$

where a and d are real numbers and n is a positive integer. The defining characteristic of an arithmetic sum is the fact that the difference of consecutive terms is a fixed constant. Find a formula for the sum of an arithmetic sum that does not involve all of the plus signs.

22. Formulas for the sums of the powers of the first n positive integers are also interesting to find. These sums look like

$$\begin{aligned}&1+2+3+\cdots+n,\\&1^2+2^2+3^2+\cdots+n^2,\\&1^3+2^3+3^3+\cdots+n^3,\\&1^4+2^4+3^4+\cdots+n^4,\\&1^5+2^5+3^5+\cdots+n^5,\end{aligned}$$

and so on, where n is a positive integer. It is possible to write down the first few sums in each case, search for a pattern, make a conjecture for the sum, and prove that the conjecture is correct using mathematical induction. However, the formulation of a conjecture becomes increasingly difficult as the power increases. This exercise presents another way to find these sums. The key is recognizing that a certain sum can be expressed in two different ways. For example,

$$(n+1)^2-1^2=\sum_{i=1}^{n}\left((i+1)^2-i^2\right)=2\sum_{i=1}^{n}i+\sum_{i=1}^{n}1.$$

It follows that

$$\sum_{i=1}^{n}i=\frac{1}{2}\left((n+1)^2-1^2-\sum_{i=1}^{n}1\right).$$

Since $\sum_{i=1}^{n}1=n$, a formula for $\sum_{i=1}^{n}i$, which simplifies to $n(n+1)/2$, has been found. (There is an easier way to find this particular sum; the reader can probably discover it.) The rest of the sums are found in the same way. For the sum of the squares, write

$$(n+1)^3-1^3=\sum_{i=1}^{n}\left((i+1)^3-i^3\right)=3\sum_{i=1}^{n}i^2+3\sum_{i=1}^{n}i+\sum_{i=1}^{n}1.$$

Since two of the sums are known, the third can be found. Use this process to find the formulas for the sums of the powers listed at the beginning of the exercise.

23. Let x and y be two positive numbers. Prove that $\sqrt{xy}\le(x+y)/2$. This is the Arithmetic Mean/Geometric Mean Inequality for the special case in which $n=2$. You should be able to give a simple proof for this case.

24. Let n be a positive integer and let $a_1, a_2, \ldots, a_n$ be nonnegative real numbers. Prove that the arithmetic mean and the geometric mean of this set of numbers lie in the closed interval $[m, M]$, where $m=\min\{a_1,a_2,\ldots,a_n\}$ and $M=\max\{a_1,a_2,\ldots,a_n\}$.

25. Let a and b be fixed positive numbers. Find the value of $x>0$ that minimizes the given expression and determine the minimum value of the expression.

a) $ax+\dfrac{b}{x}$ **b)** $ax^2+\dfrac{b}{x}$ **c)** $ax+\dfrac{b}{x^2}$

26. Let x and y be positive numbers. For each of the following conditions on x and y, find the maximum value for the product xy.

a) $4x+9y=36$ **b)** $4x^2+9y^2=36$ **c)** $4x^2+9y=36$

27. Let x and y be positive numbers. Find the minimum value for the sum $4x+9y$, subject to the condition $x^2y^3=100$.

28. Let x, y, and z be positive numbers. Find the minimum value for the sum $x^2+y^2+z^2$, subject to the condition $xy^2z = 12$.

29. Find the maximum volume for a right circular cylinder if its surface area is a constant value S. Consider the case in which the cylinder has a bottom but no top as well as the case in which it has a top and a bottom.

30. Let n be a positive integer and let $a_1, a_2, \ldots, a_n$ be positive numbers. The **harmonic mean** of these numbers is the reciprocal of the arithmetic mean of the reciprocals of the numbers. Prove that the harmonic mean of a set of positive numbers is less than or equal to the geometric mean. When does equality occur?

31. Let n be a positive integer. Suppose that $a_1, a_2, \ldots, a_n$ and $b_1, b_2, \ldots, b_n$ are real numbers and that at least one of the b_k's is not zero. For each real number t, let $P(t) = \sum_{k=1}^{n} (a_k - tb_k)^2$.

a) Show that P is a polynomial in t of degree 2.

b) By completing the square, find the value of t that minimizes P.

c) Show that the value of t from part (b) is the appropriate choice for the constant z that is needed in the proof of the Cauchy-Schwarz Inequality.

32. Prove the Cauchy-Schwarz Inequality for the case $n = 2$ by writing out both sides of the inequality, then multiplying and rearranging terms until a familiar inequality is obtained. Make certain that the steps are reversible!

33. Rephrase the Cauchy-Schwarz Inequality in the language of the vector space $\mathbb{R}^n$.

34. Let $a_1, a_2, \ldots, a_n$ be real numbers. Prove that $\sum_{k=1}^{n} |a_k| \le \sqrt{n}\sqrt{\sum_{k=1}^{n} a_k^2}$.

35. Let r be a fixed positive real number. Suppose that a, b, and c are real numbers that satisfy $a^2+b^2+c^2 = r^2$. Find the maximum value of $|a|+|b|+|c|$ and the values of a, b, and c that generate the maximum value.

36. Suppose that $a_1, a_2, \ldots, a_n$ are positive real numbers. Find the minimum value of the expression $\left(\sum_{k=1}^{n} a_k\right)\left(\sum_{k=1}^{n} \frac{1}{a_k}\right)$.

37. Suppose that $x > -1$ and that $x \neq 0$. Prove that $(1+x)^n > 1+nx$ for each positive integer $n > 1$. This result is known as **Bernouilli's Inequality**.

1.3 The Completeness Axiom

The rational numbers are closed under addition and multiplication, but the rational numbers are not closed under the process of finding roots. This was illustrated in Section 1.1 with a proof that $\sqrt{2}$ is not a rational number. The rational numbers are also not closed under the limit process since there are convergent sequences of rational numbers that do not converge to a rational number. (We assume that the reader has had a little exposure to the limit process.) For example, the sequence

$$.101,\ .101001,\ .1010010001,\ .101001000100001,\ .101001000100001000001,\ \ldots$$

is a sequence of rational numbers that converges to an irrational number (the limit is a number whose decimal expansion has no repeating pattern). Since the limit process is central to the development of calculus, this deficiency of the rational numbers is a serious problem. The set of real numbers is an ordered field with one extra property that overcomes both deficiencies (roots and limits) of the ordered field of rational numbers. This extra property of the real numbers can be formulated in various equivalent ways; we will adopt one of these formulations as an axiom. (An axiom is a statement that is assumed to be true; see Appendix A). Since the form of the axiom adopted here does not mention roots or limits explicitly, it will not be immediately apparent how this axiom will resolve the deficiencies of the rational numbers. However, after spending some time working with the axiom, its usefulness will become clear.

Before presenting this important axiom, it is necessary to introduce some new terminology. This is accomplished in the next two definitions.

DEFINITION 1.14 Let S be a nonempty set of real numbers.

a) The set S is **bounded above** if there is a number M such that $x \leq M$ for all $x \in S$. The number M is called an **upper bound** of S.

b) The set S is **bounded below** if there is a number m such that $x \geq m$ for all $x \in S$. The number m is called a **lower bound** of S.

c) The set S is **bounded** if there is a number M such that $|x| \leq M$ for all $x \in S$. The number M is called a **bound** for S.

The straightforward proof that a set is bounded if and only if it is bounded above and bounded below will be left as an exercise. If a set has an upper bound, then it has many upper bounds: any number greater than an upper bound is another upper bound. Analogous statements are true for lower bounds and bounds. Here are a few simple examples to illustrate the definition.

1. The set $\mathbb{Z}^+$ is bounded below; any negative number is a lower bound.
2. The set $\{x/(x+1) : x \in \mathbb{Q} \text{ and } x > 0\}$ is bounded; any number $M \geq 1$ is a bound for this set.
3. The set $\{x \in \mathbb{R} : x < 0\}$ is bounded above, but not bounded.

A set that is not bounded is said to be unbounded. To understand what it means for a set to be unbounded, it is necessary to negate the definition of a bounded set. Referring to Definition 1.14, a set S is **unbounded** if for each number M there is a point $x \in S$ such that $|x| > M$.

DEFINITION 1.15 Let S be a nonempty set of real numbers.

a) Suppose that S is bounded above. A number β is the **supremum** of S if β is an upper bound of S and any number less than β is not an upper bound of S. We will write $\beta = \sup S$.

b) Suppose that S is bounded below. A number α is the **infimum** of S if α is a lower bound of S and any number greater than α is not a lower bound of S. We will write $\alpha = \inf S$.

It should be clear that a set can have at most one supremum and one infimum; a proof of this fact will be left as an exercise. The supremum of a set is sometimes called the **least upper bound** of the set since any upper bound of the set must be greater than or equal to the supremum. Similarly, the infimum of a set is sometimes called the **greatest lower bound** of the set. The set $\mathbb{Z}^+$ is bounded below, and it is easy to see that $\inf \mathbb{Z}^+ = 1$. The set $\{x \in \mathbb{R} : x^2 \leq 6.25\}$ is a bounded set, and its supremum and infimum are 2.5 and -2.5, respectively. It is tempting to think of the supremum of a set as the largest element in the set and the infimum of a set as the smallest element in the set (as in the two examples just considered), but this is incorrect. The open interval $(0, 1)$ does not contain any upper or any lower bounds of itself. Consequently, the interval $(0, 1)$ does not contain its infimum 0 or its supremum 1. In general, do not assume that a bounded set contains its infimum or its supremum. However, a nonempty finite set (a set with n elements for some positive integer n) always contains its infimum and supremum. A proof of this fact will be left as an exercise.

Does every set that is bounded above have a supremum? From our knowledge of the number line, it certainly appears that the answer is yes. However, this property of the real numbers does not follow from the algebraic or order properties of the real numbers. In fact, the rational numbers, which have the same algebraic and order properties as the real numbers, do not satisfy this property. This should come as no surprise since we have already mentioned some deficiencies of the rational numbers. An example of this particular deficiency of the rational numbers is probably helpful.

Let S be the set of rational numbers defined by

$$S = \{x \in \mathbb{Q} : x > 0 \text{ and } x^2 < 2\}.$$

The set S is a nonempty set of rational numbers that is bounded above: any positive rational number y such that $y^2 > 2$ is an upper bound of S. In addition, any upper bound of S must be greater than 1. We will show that S does not have a supremum in the set of rational numbers. Let $p \in \mathbb{Q}$ be any upper bound of S and define

$$q = p - \frac{p^2 - 2}{p + 2}.$$

Note that q is a rational number. Performing some simple algebra yields

$$q^2 - 2 = \left(\frac{2p + 2}{p + 2}\right)^2 - 2 = \frac{2(p^2 - 2)}{(p + 2)^2}.$$

Using the equations for q and $q^2 - 2$ reveals the following:

$$\text{if } p^2 < 2, \text{ then } q > p \text{ and } q^2 < 2;$$
$$\text{if } p^2 > 2, \text{ then } q < p \text{ and } q^2 > 2.$$

The first statement contradicts the fact that p is an upper bound of S, and the second statement shows that p is not the supremum of S. (Note that $p^2 = 2$ is not possible since $\sqrt{2}$ is not a rational number.) Since p was an arbitrary rational upper bound, we have shown that the set S has no supremum in the set of rational numbers. Once again, the rational numbers have been shown to be incomplete. For the record, in the set of real numbers, we find that $\inf S = 0$ and $\sup S = \sqrt{2}$.

The following axiom guarantees that every nonempty set of real numbers that is bounded above has a supremum. As we will see, it forms the basis for all the

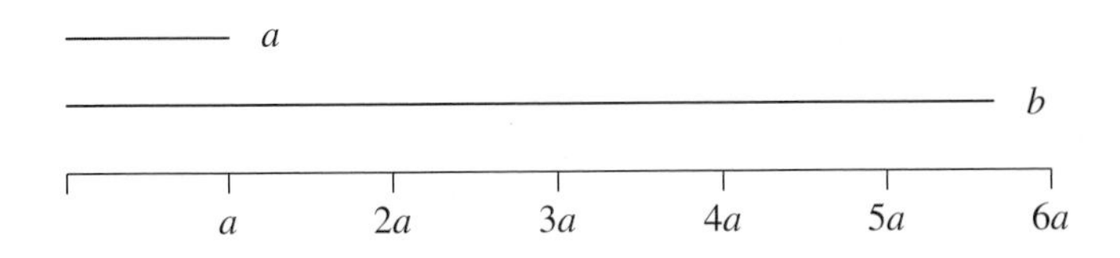

Figure 1.4 Six copies of a are longer than b

limit operations of calculus. In the development of the set of real numbers using Dedekind cuts or Cauchy sequences, it is possible to derive this statement as a theorem from more basic principles. For our purposes, it will suffice to accept it as an axiom.

Completeness Axiom: Each nonempty set of real numbers that is bounded above has a supremum.

As shown by the previous example involving the set S, the set of rational numbers does not satisfy the Completeness Axiom. The Completeness Axiom separates the set of real numbers from the set of rational numbers. To summarize Section 1.1 and the discussion in this section, the foundation for this text on real analysis is the fact that the set of real numbers is an ordered field that satisfies the Completeness Axiom.

The Completeness Axiom asserts that something exists—namely, the supremum of a nonempty set of real numbers that is bounded above. It does not indicate how the supremum is found, but only states that such a number exists. This axiom lies behind most of the important existence results in real analysis, such as the Intermediate Value Theorem and the Mean Value Theorem. Once again, the Completeness Axiom guarantees the existence of some real number even if it is not clear what that number is or how to go about finding it.

The next few results explore some of the consequences of the Completeness Axiom. The conclusion of the first result is intuitively clear: given positive numbers a and b, there is an integer n such that $na > b$. In terms of measurement, this means that given two sticks of unequal length, it is always possible to put down enough copies of the shorter stick end to end to obtain a length longer than the other stick (see Figure 1.4). This result is referred to as the Archimedean Property of the real numbers, in honor of Archimedes of Syracuse (a famous Greek scientist who lived in the third century B.C.E.).

THEOREM 1.16 Archimedean Property of the Real Numbers If a and b are positive real numbers, then there exists a positive integer n such that $na > b$.

Proof. The result is trivial if $a \geq b > 0$, so assume that $0 < a < b$. The proof will be by contradiction. Suppose that there is no positive integer n such that $na > b$. Then $na \leq b$ for every positive integer n, so the set $S = \{na : n \in \mathbb{Z}^+\}$ is bounded above by b. By the Completeness Axiom, the set S has a supremum β. Now $\beta - a < \beta$ since $a > 0$ and, by the definition of supremum, the number $\beta - a$ is not an upper bound of S. Consequently, there exists a positive integer p such that $pa > \beta - a$. It follows that

$$\beta < pa + a = (p + 1)a.$$

Since $(p + 1)a \in S$, this contradicts the fact that β is an upper bound of S. Hence, there exists a positive integer n such that $na > b$. ■

The next theorem asserts that four statements are equivalent. This means that if one of the statements is assumed to be true, then the other three statements are also true; the four results essentially say the same thing. Since statement (1) is Theorem 1.16, all four statements are valid. In the remainder of the text, any one of these statements will be referred to as the Archimedean Property of the real numbers.

Two comments concerning the proof of the following theorem should make the proof easier to follow. The first comment is an explanation of the mathematical symbol $\Rightarrow$ for implication. The statement (1) $\Rightarrow$ (2), which is read as "one implies two", means "if (1), then (2)"; the first statement is the hypothesis and the second is the conclusion. The second comment concerns the Well-Ordering Property of the positive integers. This property states that every nonempty set of positive integers contains a least integer. That is, if A is a nonempty set of positive integers, then there exists $n \in A$ such that $n \le a$ for all $a \in A$. The Well-Ordering Property of the positive integers is equivalent to the Principle of Mathematical Induction. (Additional information on this topic can be found in Appendix C).

THEOREM 1.17 The following statements are equivalent.

1. If a and b are positive real numbers, then there exists a positive integer n such that $na > b$.
2. The set of positive integers is not bounded above.
3. For each real number x, there exists an integer n such that $n \le x < n + 1$.
4. For each positive real number x, there exists a positive integer n such that $1/n < x$.

Proof. In order to prove a theorem of this form, it is sufficient to prove that (1) $\Rightarrow$ (2) $\Rightarrow$ (3) $\Rightarrow$ (4) $\Rightarrow$ (1). We will prove (2) $\Rightarrow$ (3) and (4) $\Rightarrow$ (1), leaving the proofs of the other two implications as exercises.

Assume that condition (2) is satisfied. Since (3) is obvious if $x \in (-1, 1)$ or if x is an integer, suppose that x is not an integer and $x > 1$. By hypothesis, the set $S_x = \{p \in \mathbb{Z}^+ : p > x\}$ is nonempty. By the Well-Ordering Property of the positive integers, the set S_x contains a least integer q. Since $q - 1 < x < q$, the integer $n = q - 1$ has the desired property. Now suppose that x is not an integer and $x < -1$. Since $-x > 1$, the first part of the proof shows that there exists an integer m such that $m - 1 < -x < m$. It follows that $-m < x < -m + 1$. This shows that (2) $\Rightarrow$ (3).

Now suppose that condition (4) holds. As in the proof of the previous theorem, it is only necessary to consider the case in which $0 < a < b$. Since a/b is a positive number, condition (4) asserts that there exists a positive integer n such that $1/n < a/b$. This inequality is equivalent to $na > b$. Thus (4) $\Rightarrow$ (1). ■

The four statements recorded in Theorem 1.17 are properties of the real numbers that are almost obvious to anyone who has completed an elementary algebra course. The only surprising fact about these properties is that they require any proof at all. However, if one thinks of the set of real numbers as the set of rational numbers with some extra numbers tossed in for completeness (so that the deficiencies of the rational numbers are overcome), then the Archimedean Property of the real numbers shows that the extra numbers fill in the gaps left by the rational numbers as opposed to being located somewhere beyond all of the integers (see statements (2) and (3) as well as Theorem 1.18). It is important to note that the Completeness Axiom is needed to prove that any one of these four statements is true (see the proof of Theorem 1.16).

It is easy to prove that there is a rational number between any two distinct rational numbers. The next result shows that between any two distinct real numbers there is both an irrational number and a rational number. This fact is another result that follows from the Completeness Axiom.

THEOREM 1.18 Between any two distinct real numbers there is a rational number and an irrational number.

Proof. Suppose that x and y are distinct real numbers with $x < y$. By the Archimedean Property of the real numbers, there exists a positive integer q such that $q(y - x) > 1$. Since $qy - qx > 1$, there exists an integer p such that $qx < p < qy$. (A proof of this statement will be left as an exercise.) The number p/q is rational and $x < p/q < y$. A similar argument establishes the existence of a rational number $r \neq 0$ such that $\sqrt{2}x < r < \sqrt{2}y$. The number $r/\sqrt{2}$ is an irrational number between x and y. This completes the proof. ■

Since there is both a rational number and an irrational number between any two distinct real numbers, it appears that the rational numbers and the irrational numbers are evenly distributed on the real line. It turns out, however, that the sizes of these two sets are dramatically different. A discussion of this important distinction between the set of rational numbers and the set of irrational numbers will be presented in the next section.

The Completeness Axiom separates the ordered field of the rational numbers from the ordered field of the real numbers. As a result of this axiom, the set of real numbers overcomes the deficiencies of the set of rational numbers. The fact that the real numbers overcome the deficiency concerning limits will become apparent in the next two chapters. The other deficiency is the fact that a root of a positive rational number may not be a rational number. However, every root of a positive real number is a real number. The proof of the next theorem indicates how the Completeness Axiom guarantees the existence of roots. It is not a general result since, as stated previously, it is not one of the goals of this book to rigorously prove all of the properties of the real numbers. However, the proof of the existence of the specific root in the following theorem illustrates the power of the Completeness Axiom.

THEOREM 1.19 There exists a real number x such that $x^3 = 5$.

Proof. Let A be the set of real numbers defined by

$$A = \{x \in \mathbb{R} : x > 0 \text{ and } x^3 < 5\}.$$

The set A is a nonempty set of real numbers that is bounded above. For example, the number 1.1 belongs to the set A and 2 is an upper bound of A. By the Completeness Axiom, the set A has a supremum. Let $\beta = \sup A$ and note that $1 < \beta < 2$. We will prove that $\beta^3 = 5$ by showing that both of the assumptions $\beta^3 < 5$ and $\beta^3 > 5$ lead to contradictions.

Suppose first that $\beta^3 < 5$. We claim that there exists a real number h such that $0 < h < 1$ and $(\beta + h)^3 < 5$. To find the number h, use the facts $1 < \beta < 2$ and $0 < h < 1$ to compute

$$(\beta + h)^3 = \beta^3 + 3\beta^2 h + 3\beta h^2 + h^3 < \beta^3 + 12h + 6h + h = \beta^3 + 19h.$$

The number $h = (5 - \beta^3)/19$ has the desired properties. Since $\beta < \beta + h$ and $\beta + h \in A$, the existence of h contradicts the fact that β is an upper bound of A.

Now suppose that $\beta^3 > 5$ and let $h = (\beta^3 - 5)/13$. (The algebra used to find h has been omitted.) Note that $0 < h < 1$; the second inequality follows from the fact that $\beta < 2$. Since

$$(\beta - h)^3 = \beta^3 - 3\beta^2 h + 3\beta h^2 - h^3 > \beta^3 - 12h + 0 - h = \beta^3 - 13h = 5,$$

it follows that $\beta - h < \beta$ and that $\beta - h$ is an upper bound of A. (A proof that a number b is an upper bound of A if $b^3 > 5$ will be left as an exercise.) This is a contradiction to the fact that $\beta = \sup A$.

Since both of the assumptions $\beta^3 < 5$ and $\beta^3 > 5$ lead to contradictions, it follows that $\beta^3 = 5$. This completes the proof. ∎

There will be occasions in this text when the interpretation of a real number as a point on a number line or as a decimal expansion will be useful. The use of the number line to visualize real numbers follows from geometric intuition and the concept of magnitude. The number line provides a nice visual interpretation for some concepts; it can therefore provide insight into potential theorems and proof strategies. However, visual insight does not constitute a proof; a rigorous proof using only the properties of the real numbers is still required. As an example of the use of this visualization, see the proof of Theorem 1.29 and Figure 1.5.

The correspondence between decimal expansions and real numbers is not as intuitive as the number line. As you might expect, the fact that each real number has a decimal expansion is a consequence of the Completeness Axiom. A proof that positive real numbers have a decimal expansion (the result for negative numbers follows easily) is given in the following theorem. A proof that every decimal expansion represents a real number will be left as an exercise.

THEOREM 1.20 Each positive real number has a decimal expansion.

Proof. Let x be a positive real number. By the Archimedean Property of the real numbers, there exists an integer d_0 such that $d_0 \le x < d_0 + 1$. Let d_1 be the largest integer such that

$$d_0 + \frac{d_1}{10} \le x.$$

Note that $0 \le d_1 \le 9$. Let d_2 be the largest integer such that

$$d_0 + \frac{d_1}{10} + \frac{d_2}{10^2} \le x$$

and note that $0 \le d_2 \le 9$. Suppose that, for some integer $n \ge 2$, integers $d_0, d_1, \ldots, d_{n-1}$ have been chosen in this way. Let d_n be the largest integer such that

$$d_0 + \frac{d_1}{10} + \cdots + \frac{d_{n-1}}{10^{n-1}} + \frac{d_n}{10^n} \le x$$

and note that $0 \le d_n \le 9$. This defines d_n for all positive integers n (using the Principle of Mathematical Induction). Let

$$D = \left\{ d_0 + \frac{d_1}{10} + \cdots + \frac{d_n}{10^n} : n \in \mathbb{Z}^+ \right\}.$$

The set D is nonempty and x is an upper bound of D. By the Completeness Axiom, the number $\beta = \sup D$ exists. Note that $\beta \le x$ since x is an upper bound of D. Suppose that $\beta < x$. According to the Archimedean Property of the real numbers, there exists a positive integer p such that $1/p < x - \beta$. Using the definitions of d_p and β, we find that

$$x < d_0 + \frac{d_1}{10} + \cdots + \frac{d_p}{10^p} + \frac{1}{10^p} \le \beta + \frac{1}{10^p} < \beta + \frac{1}{p} < x,$$

a contradiction. We conclude that $\beta = x$. In other words,

$$d_0.d_1d_2d_3d_4\ldots = d_0 + \frac{d_1}{10} + \frac{d_2}{10^2} + \frac{d_3}{10^3} + \cdots$$

is a decimal expansion of x. This completes the proof. ∎

Exercises

1. Prove that a nonempty set is bounded if and only if it is both bounded above and bounded below. Note that this proves that a bounded interval is a bounded set.
2. Prove that the set $\{x \in \mathbb{R} : 10\sqrt{x} - x > 0\}$ is bounded.
3. Prove that the set $\{x \in \mathbb{R} : x^2 - 25x > 0\}$ is unbounded.
4. Prove that every finite set is bounded.
5. Prove that a nonempty set is bounded if and only if it is contained in a bounded interval.
6. Prove that the union of two bounded sets is a bounded set.
7. Consider the set S of rational numbers discussed prior to the statement of the Completeness Axiom, as well as the numbers p and q defined there. Prove each of the following.
   a) If y is a rational number such that $y^2 > 2$, then y is an upper bound of S.
   b) Every rational number that is an upper bound of S is greater than 1.
   c) The number q is rational.

8. Prove that a nonempty set that is bounded above has only one supremum.

9. The Completeness Axiom only asserts something about sets that are bounded above. Use the Completeness Axiom to prove that every nonempty set of real numbers that is bounded below has an infimum.

10. Prove that the infimum and supremum of the interval (a, b) are a and b, respectively.

11. Prove that a nonempty finite set contains its infimum.

12. Let S be a nonempty set of real numbers that is bounded above and let $\beta = \sup S$. Prove that for each $\epsilon > 0$ there exists a point $x \in S$ such that $x > \beta - \epsilon$.

13. Find the supremum of the set $\{x : 3x^2 + 3 < 10x\}$.

14. Use the Completeness Axiom to finish the proof of Theorem 1.9, that is, to prove that an interval has one of nine possible forms. There are a number of cases to consider: the set is bounded, the set is bounded above but not below, etc.

15. Let a be a positive number. Prove that for each real number x there is an integer n such that $na \leq x < (n + 1)a$.

16. Referring to Theorem 1.17, prove (1) $\Rightarrow$ (2).

17. Referring to Theorem 1.17, prove (3) $\Rightarrow$ (4).

18. Prove each of the following results—give a direct proof of each one—without use of the Completeness Axiom. These results could be called the Archimedean Property of the rational numbers.

a) If a and b are positive rational numbers, then there exists a positive integer n such that $na > b$.

b) For each positive integer n, there exists a rational number r such that $r > n$.

c) For each rational number x, there exists an integer n such that $n \leq x < n + 1$.

d) For each positive rational number x, there is a positive integer n such that $1/n < x$.

19. Let s and t be real numbers such that $t - s > 1$. Prove that there exists an integer p such that $s < p < t$.

20. Let x be a real number. Prove the following statement: for each $\epsilon > 0$, there exists a rational number r such that $0 < |x - r| < \epsilon$.

21. A real number of the form $p/2^n$, where p is an integer and n is a nonnegative integer, is known as a **dyadic rational number**. Prove that there is a dyadic rational number between any two distinct real numbers.

22. Consider the set A defined in the proof of Theorem 1.19. Prove that b is an upper bound of A if $b^3 > 5$.

23. Use an argument similar to the one in the proof of Theorem 1.19 to prove

a) there is a real number x such that $x^2 = 2$;

b) there is a real number y such that $y^3 = 7$.

24. Prove that every decimal expansion represents a real number.

Remark. The rest of the exercises in this section are in no particular order.

25. Prove that each real number $x \in [0, 1]$ has a **binary expansion**. That is, for each $x \in [0, 1]$, prove that there is a sequence $b_{x1}, b_{x2}, b_{x3}, \ldots$ of numbers such that b_{xk} is either 0 or 1, and

$$x = \frac{b_{x1}}{2} + \frac{b_{x2}}{2^2} + \frac{b_{x3}}{2^3} + \frac{b_{x4}}{2^4} + \frac{b_{x5}}{2^5} + \cdots.$$

26. Let A be a nonempty bounded set and suppose that S is a nonempty subset of A. Prove that $\inf A \le \inf S \le \sup S \le \sup A$.

27. Let A be a nonempty bounded set. The maximum value of A is a number $x \in A$ such that $a \le x$ for all $a \in A$. Prove that a nonempty bounded set has a maximum value if and only if it contains its supremum.

28. Let A and B be nonempty bounded sets. Prove that

$$\inf(A \cup B) = \min\{\inf A, \inf B\} \quad \text{and} \quad \sup(A \cup B) = \max\{\sup A, \sup B\}.$$

Is there an analogous result for $A \cap B$, assuming that this set is nonempty? Provide either a proof or a counterexample.

29. Let S be a nonempty bounded set of real numbers. Prove that $\inf S = -\sup(-S)$, where $-S = \{-x : x \in S\}$.

30. Let S be a nonempty bounded set of real numbers and let k be a real number. Let kS be the set defined by $kS = \{kx : x \in S\}$. How are the numbers $\inf(kS)$ and $\sup(kS)$ related to the numbers $\inf S$ and $\sup S$? Be sure to consider both of the cases $k > 0$ and $k < 0$ and to provide proofs of your conjectures.

31. Let A and B be nonempty bounded sets of positive numbers. Define a set C by $C = \{ab : a \in A, b \in B\}$. Prove that $\sup C = \sup A \sup B$. Does this result remain valid if either A or B contain negative numbers?

1.4 Countable and Uncountable Sets

The Completeness Axiom is the property that distinguishes the set of real numbers from the set of rational numbers. Consequently, every property that is valid for $\mathbb{R}$ but not $\mathbb{Q}$ (such as the existence of roots and limits) has its origins in the Completeness Axiom. This axiom also leads to another interesting distinction between the set of rational numbers and the set of real numbers: the set of real numbers is "larger" than the set of rational numbers. On the surface, this seems like a simplistic observation. Since every rational number is a real number and there are real numbers that are not rational numbers, it is clear that the set of real numbers contains more elements than the set of rational numbers. Another valid statement, "there are more irrational numbers than there are rational numbers", is not as easy to dismiss. These two sets have no elements in common and both sets are infinite. What does it mean to say that one set is "more infinite" than another set? To answer this question, we must first agree on a definition of size for infinite sets. Since the definition of the size of an infinite set relies heavily on the notions of sets and functions, it might be helpful to review the material on sets and functions found in Appendix B.

A good way to motivate the definition of size for infinite sets is with a thought experiment. Consider a large auditorium that is filled with people. The fire code states that each person must have a seat, while the management wants every seat full in order to maximize revenue. How can you determine if both conditions are met? One method is to actually count the number of people and to count the number of seats. If there are 9850 people and 9850 seats, then everyone is satisfied. This would be a tedious task, and errors in counting could easily occur. A much more efficient method is to have everyone take a seat. If each person has a seat and if no seat is empty, then there are the same number of people as seats. This is a true statement even if you do not know either the number of people or the number of seats.

It is thus possible to show that two sets have the same size without knowing the number of elements in each set: simply pair off the elements of each set. This method easily extends to infinite sets. Two infinite sets have the same size if their elements can be put into a one-to-one correspondence. More formally, two nonempty sets A and B have the **same size** if there exists a function $f: A \to B$ that is both one-to-one and onto. (Two sets that can be put into a one-to-one correspondence are sometimes said to have the same **cardinality**, but we will not use this term.) The difficulty, if one can call it that, is that this definition leads to intuitively bizarre results. For example, the set of positive integers and the set of even positive integers have the same size. This follows from the pairing

$$\begin{array}{cccccccccccc} 1 & 2 & 3 & 4 & 5 & 6 & 7 & 8 & 9 & 10 & \ldots \\ 2 & 4 & 6 & 8 & 10 & 12 & 14 & 16 & 18 & 20 & \ldots \end{array}$$

which establishes a one-to-one correspondence between the two sets. The fact that this pairing is a one-to-one correspondence is clear, but it seems just as clear that there are more positive integers than there are even positive integers. If a definition leads to contradictions (in the logical sense), it must be discarded; if it leads to results that seem to violate common sense, then the definition can either be left aside or intuition can rise to the occasion. In this case, it is intuition that must rise to the occasion. In other words, we will accept this definition and see where it leads.

The discussion in this section will be somewhat informal, that is, we will accept some statements as true without offering a rigorous proof. Although this violates the spirit of the book, there are a couple of reasons for doing so. The first is that this topic is somewhat peripheral to a study of basic real analysis. The notion of a countably infinite set (defined below) will appear several times in the text, but it does not play a crucial role in any of the major theorems or definitions of introductory real analysis. A second reason is that there are a number of "obvious" results in this area that have tedious proofs and, more importantly for us, the proofs do not really add to an understanding of the concepts. In general, of course, it is necessary to prove facts (obvious or not) from definitions, axioms, and previous results. However, since this particular topic is peripheral to a study of basic real analysis, we will leave some of the more tedious and unenlightening proofs to another text.

The next definition formalizes the previous discussion concerning the relative sizes of arbitrary sets. It also introduces some adjectives that describe the various sizes of sets that will be considered in this text.

DEFINITION 1.21 Let A be an arbitrary set.

a) The set A is **finite** if it is empty or if its elements can be put in a one-to-one correspondence with the set $\{1, 2, \ldots, n\}$ for some positive integer n.

b) The set A is **infinite** if it is not finite.

c) The set A is **countably infinite** if its elements can be put in a one-to-one correspondence with the set of positive integers.

d) The set A is **countable** if it is either finite or countably infinite.

e) The set A is **uncountable** if it is not countable.

The distinction between finite sets and infinite sets is generally easy to grasp: a finite set is eventually exhausted when you start listing out its elements, whereas an infinite set is not. For instance, the number of license plates using three letters (from the alphabet) and three single-digit numbers is finite. There are many of them, but in theory if you start writing down all of the possibilities, eventually the list would end. The set of positive integers is infinite because a list of positive integers never ends. (Actually, the fact that the set of positive integers is an infinite set or, more generally, that a countably infinite set is an infinite set requires a proof. You must show that part (a) of Definition 1.21 is not satisfied for any positive integer n. This is one of those proofs that we are leaving out of this discussion.) The set of rational numbers is an infinite set, as is the set of real numbers.

A set A is countably infinite if there exists a function $f: A \to \mathbb{Z}^+$ (or a function $g: \mathbb{Z}^+ \to A$) that is both one-to-one and onto. As indicated earlier, the set of even positive integers is countably infinite. Letting E^+ be the set of even positive integers, the function $f: \mathbb{Z}^+ \to E^+$ defined by $f(n) = 2n$ is one-to-one and onto. The set of all integers greater than -1000 is also countably infinite; the function g defined on $\mathbb{Z}^+$ by $g(n) = n - 1000$ provides a one-to-one correspondence between these two sets. Since it is often difficult to write down a correspondence between two sets as a function, a pairing of the elements of two sets is sometimes just written down as a pattern, with the assumption that the pattern continues. For example, the pairing

$$\begin{array}{ccccccccccc} 1 & 2 & 3 & 4 & 5 & 6 & 7 & 8 & 9 & 10 & \ldots \\ 0 & 1 & -1 & 2 & -2 & 3 & -3 & 4 & -4 & 5 & \ldots \end{array}$$

shows that the set $\mathbb{Z}$ is countably infinite.

The next theorem lists some results that are easy to believe based upon the definitions of the concepts. We will accept these results without writing out the details of the proofs. However, the remaining results in this section will be proved.

THEOREM 1.22 The following results on finite and infinite sets are valid.

1. A subset of a finite set is finite.
2. The union of two finite sets is a finite set.
3. A set that contains an infinite subset is infinite.
4. A set that contains an uncountable subset is uncountable. ■

The result stated in the next theorem is just as easy to believe as parts (3) and (4) of Theorem 1.22. However, to present some idea of the nature of the proofs in this area of mathematics, the details of the proof will be included. The Well-Ordering Property of the positive integers (see Appendix C) will be used in the proof. The corollary to the theorem follows immediately from the theorem and the definition of a countable set.

THEOREM 1.23 A subset of a countably infinite set is countable.

Proof. Since any countably infinite set can be put in a one-to-one correspondence with the set of positive integers, it is sufficient to prove that every subset of positive integers is countable. Let A be a subset of the positive integers. If A is finite, then A is countable and the proof is complete. Suppose that A is an infinite set. By the Well-Ordering Property of the positive integers, the set A contains a least element. Let $f(1)$ be the smallest integer in A. Similarly, let $f(2)$ be the smallest integer in $A \setminus \{f(1)\}$. In general, let $f(n+1)$ be the smallest integer in $A \setminus \{f(1), \ldots, f(n)\}$. Since the set A is infinite, the function f is defined for each positive integer n, that is, the function f maps $\mathbb{Z}^+$ into A. It is clear from the definition of f that f is a one-to-one function. To show that f is onto, let $a \in A$. The number of integers in A that are less than a is finite (there are at most $a - 1$ such integers). Let p be the number of elements in the set A that are less than a. By the definition of the function f, we find that a is the smallest integer in the set $A \setminus \{f(1), f(2), \ldots, f(p)\}$. Then $f(p+1) = a$, and it follows that f is onto. Therefore, the function f establishes a one-to-one correspondence between $\mathbb{Z}^+$ and A. This shows that A is countably infinite. ■

COROLLARY 1.24 An infinite subset of a countably infinite set is countably infinite. ■

The assertion that there are more irrational numbers than there are rational numbers can now be stated precisely as follows: the set of rational numbers is countably infinite and the set of irrational numbers is uncountable. This fact, which was first published by Georg Cantor (1845–1918), came as a surprise to mathematicians of the time. As a first step toward a proof, we will prove that the union of a countable number of countable sets is a countable set. (The formation of a set of this type is explained in the proof of the theorem.)

THEOREM 1.25 A countable union of countable sets is countable.

Proof. It is sufficient to prove that a countably infinite union of disjoint countably infinite sets is countably infinite (see the exercises). In order to have a countably infinite number of sets, there must be one set corresponding to each positive integer n. Let $\{A_n : n \in \mathbb{Z}^+\}$ be a countably infinite collection of sets. Suppose that each A_n is a countably infinite set and that none of the sets have any elements in common. We must prove that the set $A = \bigcup_{n=1}^{\infty} A_n$ is countably infinite. Since each A_n is countably infinite, its elements can be put into a one-to-one correspondence with the set of positive integers. For each n, let $A_n = \{x_{nk} : k = 1, 2, \ldots\}$, that is,

$$\begin{aligned} A_1 &= \{x_{11}, x_{12}, x_{13}, x_{14}, \ldots\}, \\ A_2 &= \{x_{21}, x_{22}, x_{23}, x_{24}, \ldots\}, \\ A_3 &= \{x_{31}, x_{32}, x_{33}, x_{34}, \ldots\}, \end{aligned}$$

and so on. By the Fundamental Theorem of Arithmetic, which states that the factorization of positive integers into products of primes is unique (see Appendix C), the pairing $x_{nk} \longleftrightarrow 2^n 3^k$ is a one-to-one correspondence between A and a subset of the positive integers. By Corollary 1.24, the set A is countably infinite. ■

The previous theorem is often used to prove that an infinite set is countably infinite. If it is possible to decompose the set into a countably infinite number of subsets, each of which is countable, then the set is countably infinite. The advantage of this method for proving that a set is countably infinite is that there is no need to find a formula of correspondence or even to illustrate how the elements of the set can be paired with the positive integers. To illustrate the use of Theorem 1.25, we will use it to prove that the set of rational numbers is countably infinite. The method of proof, which is summarized in the next few sentences, is quite typical. Let A be a set. For each positive integer n, define a subset A_n of A. The sets A_n must be defined in such a way that $A = \bigcup_{n=1}^{\infty} A_n$ and that a method for proving that each A_n is a countable set is apparent. It is this step of the proof that may require some creativity. The conclusion that A is countable then follows from Theorem 1.25. For the record, the ratio p/q of two integers is said to be in **reduced form** if p and q have no common divisors.

THEOREM 1.26 The set of rational numbers is countably infinite.

Proof. Let A_1 be the set of all rational numbers that in reduced form have a denominator of 1. The set A_1 is actually the set of integers and is thus countably infinite. Let A_2 be the set of all rational numbers that in reduced form have a denominator of 2. Since the only possible choices for the numerators are odd integers, the set A_2 is also countably infinite. In general, for each positive integer n, let A_n be the set of all rational numbers that in reduced form have a denominator of n. For instance,

$$A_2 = \left\{\ldots, -\frac{5}{2}, -\frac{3}{2}, -\frac{1}{2}, \frac{1}{2}, \frac{3}{2}, \frac{5}{2}, \ldots\right\};$$

$$A_3 = \left\{\ldots, -\frac{4}{3}, -\frac{2}{3}, -\frac{1}{3}, \frac{1}{3}, \frac{2}{3}, \frac{4}{3}, \ldots\right\};$$

$$A_4 = \left\{\ldots, -\frac{5}{4}, -\frac{3}{4}, -\frac{1}{4}, \frac{1}{4}, \frac{3}{4}, \frac{5}{4}, \ldots\right\}.$$

It should be clear that $\mathbb{Q} = \bigcup_{n=1}^{\infty} A_n$ and that each of the sets A_n is countable. By Theorem 1.25, the set $\mathbb{Q}$ is countably infinite. ∎

Since a countably infinite union of countably infinite sets is still countably infinite, it is difficult to imagine a set that is uncountable. However, such sets do exist. To prove that a set is uncountable, it is necessary to verify that there is no one-to-one correspondence between the given set and the set of positive integers. The typical proof begins by supposing that there is such a one-to-one correspondence and proceeds to show that this assumption leads to a contradiction. This is the technique used in the proof of the next theorem. The proofs of the two lemmas preceding the theorem will be left as exercises.

LEMMA 1.27 If $[c, d]$ is a closed interval and x is a real number, then there exists a closed interval $[u, v]$ such that $[u, v] \subseteq [c, d]$ and $x \notin [u, v]$. ∎

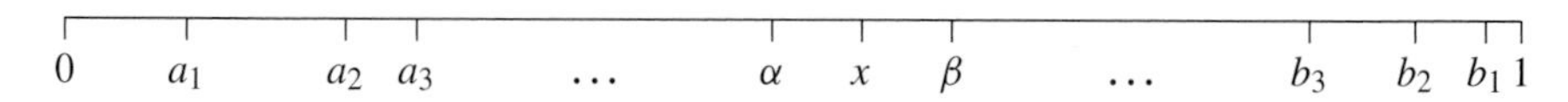

Figure 1.5 A possible representation for the numbers in the proof of Theorem 1.29

LEMMA 1.28 For each positive integer n, let $[a_n, b_n]$ be a closed interval. If $[a_n, b_n] \subseteq [a_{n-1}, b_{n-1}]$ for each integer $n > 1$, then the sets $\{a_n : n \in \mathbb{Z}^+\}$ and $\{b_n : n \in \mathbb{Z}^+\}$ are bounded and $\sup\{a_n : n \in \mathbb{Z}^+\} \leq \inf\{b_n : n \in \mathbb{Z}^+\}$. ■

THEOREM 1.29 The set of real numbers is uncountable.

Proof. By part (4) of Theorem 1.22, it is sufficient to prove that the closed interval $[0, 1]$ is uncountable. To do this, we will assume that the interval $[0, 1]$ is countably infinite and reach a contradiction. By the assumption that $[0, 1]$ is countably infinite, there exists a function $f: \mathbb{Z}^+ \to [0, 1]$ that is a one-to-one correspondence between $\mathbb{Z}^+$ and $[0, 1]$. By Lemma 1.27,

there exists an interval $[a_1, b_1] \subseteq [0, 1]$ such that $f(1) \notin [a_1, b_1]$,

there exists an interval $[a_2, b_2] \subseteq [a_1, b_1]$ such that $f(2) \notin [a_2, b_2]$,

there exists an interval $[a_3, b_3] \subseteq [a_2, b_2]$ such that $f(3) \notin [a_3, b_3]$,

and so on. Since this process can be repeated indefinitely (the Principle of Mathematical Induction is implicitly used here), for each positive integer n, there exists an interval $[a_n, b_n]$, and these intervals have the following properties: $f(n) \notin [a_n, b_n]$ for all n and $[a_n, b_n] \subseteq [a_{n-1}, b_{n-1}]$ for all $n > 1$. By Lemma 1.28, the numbers α and β defined by $\alpha = \sup\{a_n : n \in \mathbb{Z}^+\}$ and $\beta = \inf\{b_n : n \in \mathbb{Z}^+\}$ satisfy $\alpha \leq \beta$. It should also be clear that $\alpha, \beta \in [0, 1]$. Suppose that x is any number that satisfies $\alpha \leq x \leq \beta$ (see Figure 1.5). By the definitions of α and β, it follows that $a_n \leq x \leq b_n$ for every positive integer n. Since $f(n) \notin [a_n, b_n]$ for all n, we find that $f(n) \neq x$ for all $n \in \mathbb{Z}^+$. This means that x is an element of $[0, 1]$ that is not in the range of f, a contradiction to the fact that f is a one-to-one correspondence between $\mathbb{Z}^+$ and $[0, 1]$. Therefore, the interval $[0, 1]$ is uncountable. ■

COROLLARY 1.30 The set of irrational numbers is uncountable.

Proof. This will be another proof by contradiction. Let K be the set of irrational numbers and suppose that K is countably infinite. Since the rational numbers $\mathbb{Q}$ are countably infinite and $\mathbb{R} = \mathbb{Q} \cup K$, it follows from Theorem 1.25 that the set of real numbers is countably infinite. As this is a contradiction to Theorem 1.29, the set of irrational numbers is uncountable. ■

Between any two distinct rational numbers there is an irrational number and between any two distinct irrational numbers there is a rational number. Yet the set of irrational numbers is uncountable and the set of rational numbers is countably infinite. It is difficult to make sense of these two statements at the same time. Nevertheless, both statements are valid and both statements are consequences of the Completeness Axiom. For a further comparison of the sizes of these two sets of real numbers, see Exercise 47 in Section 5.6.

Exercises

1. For each pair of sets, find an explicit one-to-one correspondence between them. That is, find a one-to-one function that maps one set onto the other.

a) the positive integers and the odd positive integers

b) the even integers and the odd integers

c) the positive integers and the integers

d) the interval $(1, 2)$ and the interval $(1, 6)$

e) the interval $(0, 1)$ and the interval $(4, 50)$

f) the set of real numbers and the interval $(0, 1)$

g) the interval $(0, 1)$ and the interval $[0, 1]$

2. Let I be a bounded interval. Prove that $I \cap Z$ is a finite set.

3. Use Definition 1.21 to prove that the set of odd integers is countably infinite.

4. Let S be a nonempty set of real numbers that is bounded above and let $\beta = \sup S$. Suppose that $\beta \notin S$. Prove that for each $\epsilon > 0$, the set $\{x \in S : x > \beta - \epsilon\}$ is infinite.

5. Give an induction argument to prove that the union of a finite number of finite sets is a finite set. Assume that the base case is true as indicated in part (2) of Theorem 1.22.

6. For each rational number x, write $x = p/q$, where p and q are integers with no common factors and $q \geq 1$, and let $d(x) = q$. (Note that $d(n) = 1$ for each integer n.) Let $[a, b]$ be a closed interval and let r be a positive real number. Prove that $\{x \in [a, b] \cap \mathbb{Q} : d(x) \leq r\}$ is a finite set.

7. Prove that the union of two disjoint countably infinite sets is countably infinite by finding a one-to-one correspondence between the union of the sets and the set $\mathbb{Z}^+$.

8. Let A and B be two sets.

a) Suppose that A and B are both countably infinite sets. Prove that there is a one-to-one correspondence between A and B.

b) Suppose that A is countably infinite and that there is a one-to-one correspondence between A and B. Prove that B is countably infinite.

9. Prove that the union of a countable set and an uncountable set is uncountable.

10. Justify the first sentence of the proof of Theorem 1.25.

11. Prove that the set of all ordered pairs of positive integers is countably infinite.

12. Consider the set $\mathbb{Z}^+$ of positive integers.

a) Prove that $\mathbb{Z}^+$ can be expressed as a countably infinite union of disjoint countably infinite sets.

b) Use part (a) to give a different proof of Theorem 1.25.

13. Adopting the notation of the proof of Theorem 1.26, prove that each set A_n is countably infinite.

14. Interpret the set of all ordered pairs of positive integers as a grid of dots in the first quadrant of the xy-plane. Consider the "path" that traverses these dots in the following order:

$$(1, 1),\ (2, 1),\ (1, 2),\ (3, 1),\ (2, 2),\ (1, 3),\ (4, 1),\ (3, 2),\ (2, 3),\ (1, 4),\ (5, 1),\ \ldots.$$

a) Show that this constitutes a proof that the set of all ordered pairs of positive integers is countably infinite.

b) Use the idea in part (a) to determine an explicit one-to-one correspondence between $\mathbb{Z}^+$ and the set of all ordered pairs of positive integers.

c) Use part (a) to give a different proof of Theorem 1.26.

15. Prove Lemma 1.27.

16. Prove Lemma 1.28.

17. Suppose that a and b are distinct real numbers such that $a < b$.

a) Prove that the set $\{x \in \mathbb{R} : a < x < b\}$ is uncountable.

b) Prove that the set $\{x \in \mathbb{Q} : a < x < b\}$ is countably infinite.

Remark. The rest of the exercises in this section are in no particular order.

18. Let A be the collection of all sequences of 0's and 1's. One example of an element of the set A is the sequence

$$1, 0, 1, 0, 1, 0, 1, 0, 1, 0, 1, 0, 1, 0, 1, 0, \ldots.$$

Let B be the subset of A that consists of all sequences of 0's and 1's for which the number of 1's is finite. One example of an element of the set B is the sequence

$$1, 0, 1, 0, 1, 0, 0, 0, 0, 0, 0, 0, 0, 0, 0, 0, \ldots.$$

a) Use Theorem 1.25 to show that the set B is countably infinite.

b) Prove that the set A is uncountable. (The standard proof of this result is known as **Cantor's diagonal process**. Assume that the set A is countably infinite, that is, each positive integer corresponds to one sequence in A. Write out the sequences underneath each other, a first sequence, a second sequence, and so on. The resulting figure looks like a matrix composed of 0's and 1's. Use the diagonal of this matrix to generate a sequence in A that is not in the list.)

c) Prove that the collection of all subsets of positive integers is uncountable by establishing a one-to-one correspondence between this set and the set A.

d) Use Exercise 25 in Section 1.3 and parts (a) and (b) to prove that the interval $[0, 1]$ is uncountable.

19. Use the fact that every real number has a unique decimal expansion that does not end in all 9's to prove that the interval $(0, 1)$ is an uncountable set.

20. Let $\mathcal{F}$ be a collection of disjoint intervals of real numbers. This means that if I and J are two distinct intervals in $\mathcal{F}$, then $I \cap J = \emptyset$. Prove that the set $\mathcal{F}$ is countable.

21. Prove that every infinite set contains a countably infinite subset.

22. Prove that a set is infinite if and only if it can be put into a one-to-one correspondence with one of its proper subsets. (A set A is a proper subset of B if $A \subseteq B$ but $A \neq B$.)

23. The set of real numbers is composed of the rational numbers and the irrational numbers. Another classification of real numbers involves algebraic and transcendental numbers. An **algebraic number** is any number that is a root of a polynomial with integer coefficients. For example, the number $\sqrt{2}$ is algebraic since it is a root of the polynomial $x^2 - 2$. A **transcendental number** is a real number that is not an algebraic number.

a) Prove that every rational number is an algebraic number.

b) Prove that the square root and cube root of every positive integer is an algebraic number.

c) Prove that $2 - \sqrt{3}$ and $\sqrt{4 + \sqrt[3]{3}}$ are algebraic numbers.

d) Prove that the set of all algebraic numbers is countably infinite.

e) Prove that there exist transcendental numbers. It should be pointed out that a proof that a specific number (such as e or π) is transcendental is usually extremely difficult. The interested reader can consult Niven [17].

24. Although there exists a one-to-one correspondence between any two countably infinite sets (see Exercise 8), a similar result is not valid for uncountable sets. That is, if A and B are uncountable sets, it is not necessarily true that there is a one-to-one correspondence between the elements of A and B. To see this, let A be any nonempty set and let $\mathcal{P}(A)$ represent the collection of all subsets of A; this set is known as the **power set** of A.

a) Suppose that A has n elements. Prove that $\mathcal{P}(A)$ has 2^n elements.

b) Suppose that A is a countably infinite set. Prove that $\mathcal{P}(A)$ is uncountable.

c) Suppose that A is uncountable. Prove that there is no one-to-one correspondence between A and $\mathcal{P}(A)$.

25. Let S be an uncountable set and let B be a countably infinite subset of S. Prove that there is a one-to-one correspondence between S and $S \setminus B$.

1.5 Real-Valued Functions

A **real-valued function** is a function whose codomain is the set $\mathbb{R}$, that is, whose output values are real numbers. The domain of almost every function considered in this text will be a set of real numbers. In fact, the domain will usually be an interval. Consequently, most of the functions in this text will be (either explicitly or implicitly) of the form $f: I \to \mathbb{R}$, where I is an interval. When it is stated that a function f is defined on an interval I, this means that f is of the form $f: I \to \mathbb{R}$. In this section, we will discuss various types of real-valued functions and some of the adjectives used to describe such functions. Since anyone who has taken a calculus course has had a great deal of experience with real-valued functions, much of the material in this section will be familiar. It is included here so that it can be used as a reference as the reader works through the text.

Let n be a positive integer. A **polynomial** P of **degree** n is a function of the form

$$P(x) = a_n x^n + a_{n-1} x^{n-1} + \cdots + a_1 x + a_0,$$

where $a_0, a_1, \ldots, a_n$ are fixed real numbers and $a_n \neq 0$. The numbers a_i are known as the **coefficients** of P. A constant function, a function that has the same value at every point, is referred to as a polynomial of degree 0. For the record, a **root** of a polynomial is a real number r such that $P(r) = 0$. A **rational function** is a ratio of two polynomials. The collection of all rational functions includes all polynomials since the constant function 1 is a polynomial. The determination of the value of a rational function at a particular real number involves only the operations of addition, subtraction, multiplication, and division. An **algebraic function** is a function that involves these four operations along with the process of finding roots. For example, the functions f and g defined by

$$f(x) = \sqrt[3]{4x^2 + 5x - 1} \quad \text{and} \quad g(x) = \frac{\sqrt{x} + 2x}{\sqrt[4]{x^3 + 5} - x^2}$$

are algebraic functions. The five operations used to define algebraic functions are all familiar to students who have studied high school algebra, and, if necessary, simple examples of these operations can be performed by students without a calculator.

In addition to the algebraic functions, there are many other functions considered in algebra and calculus. The most familiar ones are the **trigonometric functions**, the **inverse trigonometric functions**, the **exponential functions**, and the **logarithmic functions**. These functions are known as **transcendental functions** because the evaluation of these functions "transcends" the algebraic operations. (You may have seen calculus books that have the phrase "early transcendentals" as part of the title. This means that these functions are introduced early in the text.) Since the transcendental functions transcend the algebraic operations, they are more difficult to define rigorously. The most common ways to do this (assuming, for the moment, some facts from calculus) are with differential equations, integrals, or power series. For example, three ways to define the sine function are presented (somewhat cryptically) in the following list.

1. The sine function is the function f that satisfies $f(0) = 0$, $f'(0) = 1$, and $f''(x) + f(x) = 0$ for all real numbers x.
2. The sine function is the inverse of the function g defined by
$$g(x) = \int_0^x \frac{1}{\sqrt{1-t^2}}\, dt$$
for each $x \in [-1, 1]$. The sine function is then extended periodically so that it is defined for all real numbers x.
3. For each real number x, the value of $\sin x$ is given by the sum of the power series
$$\sum_{k=0}^{\infty} \frac{(-1)^k}{(2k+1)!} x^{2k+1}.$$
(It can be shown that this series converges for each real number x.)

There are three main difficulties with these definitions. The first is that some knowledge of the theory of either differential equations, integrals, or power series is required before the definitions make sense. A second difficulty is that the usual properties of the sine function are not evident from the definition. In fact, these abstract definitions disguise the original motivation behind the development of the sine function. Finally, it is quite helpful in elementary real analysis to have the transcendental functions around as a source of examples. Waiting until enough theory has been developed to define these functions deprives the reader of much of the intuition and motivation for the creation of the theory in the first place.

Providing formal definitions of the transcendental functions is not one of the goals of this course. We will simply use these functions from time to time and assume that the reader is familiar with their properties. (Some discussion on the functions $\ln x$ and e^x can be found in the supplementary exercises of Chapter 5. For a discussion of the trigonometric functions, see Bartle [1], Rudin [22], or Stromberg [24]. Fitzpatrick [6] gives an elementary treatment of the transcendental functions in terms of differential equations.) For example, we will assume that the reader is familiar with the unit circle definition of $\sin x$ and $\cos x$ and the basic properties and identities satisfied by the trigonometric functions. Since the definitions of the inverse trigonometric functions vary slightly from text to text, we record the definitions that will be used in this book. (Texts vary in the definitions of the functions $\operatorname{arcsec} x$ and $\operatorname{arccsc} x$. The choices for the ranges of these functions determine whether or not the absolute value function appears in the derivative formulas; some authors prefer one form for the derivative over another.)

1. For each real number $x \in [-1, 1]$, $\arcsin x$ is the unique real number in the interval $[-\pi/2, \pi/2]$ that satisfies $\sin(\arcsin x) = x$.
2. For each real number $x \in [-1, 1]$, $\arccos x = \dfrac{\pi}{2} - \arcsin x$.
3. For each real number x, $\arctan x = \arcsin\left(\dfrac{x}{\sqrt{x^2+1}}\right)$.
4. For each real number x, $\operatorname{arccot} x = \dfrac{\pi}{2} - \arctan x$.
5. For each real number x that satisfies $|x| \geq 1$, $\operatorname{arccsc} x = \arcsin(1/x)$.
6. For each real number x that satisfies $|x| \geq 1$, $\operatorname{arcsec} x = \dfrac{\pi}{2} - \operatorname{arccsc} x$.

When a real-valued function is defined by a formula, the domain of the function is implicitly assumed to be the set of all real numbers for which the formula is defined. For example, the domain of the function f defined by

$$f(x) = \frac{x^2}{1 + \cos x}$$

is the set $\mathbb{R} \setminus \{(2n+1)\pi : n \in \mathbb{Z}\}$. However, in the statements of definitions and theorems in this text, we will always consider functions that are either defined on an interval or defined on an interval except at one point in that interval. The function f mentioned earlier in this paragraph is defined on the interval $(-\pi, \pi)$ and is defined on the interval $[1, 5]$ except at the point π. It is certainly possible to allow for more general domains, but the purpose here is to focus on the theorems and not get bogged down in the technical details of general sets of real numbers; intervals are familiar and convenient sets of real numbers and almost every function the reader has considered has a domain that contains many intervals. As an example to illustrate a common way that functions will be presented in this text, consider the function $g\colon [0, 1] \to \mathbb{R}$ defined by

$$g(x) = \begin{cases} \sin(1/x), & \text{if } x \neq 0; \\ 0, & \text{if } x = 0. \end{cases}$$

The formula $\sin(1/x)$ that appears in the definition of g is defined for values of x other than those in the interval $(0, 1]$, but for the particular purpose of whatever discussion follows, only those values of x in the interval $[0, 1]$ will be considered when working with the function g.

It is helpful to have some descriptive terminology to describe the behavior of functions. We begin with the notion of a monotone function.

DEFINITION 1.31 Let I be an interval, let $f: I \to \mathbb{R}$, and let J be a subinterval of I.

a) The function f is **increasing** on J if $f(x) \leq f(y)$ for all $x, y \in J$ that satisfy $x < y$ and **strictly increasing** on J if $f(x) < f(y)$ for all $x, y \in J$ that satisfy $x < y$.

b) The function f is **decreasing** on J if $f(x) \geq f(y)$ for all $x, y \in J$ that satisfy $x < y$ and **strictly decreasing** on J if $f(x) > f(y)$ for all $x, y \in J$ that satisfy $x < y$.

c) The function f is **monotone** on J if it is either increasing or decreasing on J and **strictly monotone** on J if it is either strictly increasing or strictly decreasing on J.

As with any new concept, some examples are helpful. The reader should recall that the graph of an increasing function moves up as you look at the graph from left to right. Using this observation, the following examples should be clear.

1. The function $f: [-1, 1] \to \mathbb{R}$ defined by $f(x) = x^2$ is strictly increasing on the interval $[0, 1]$, but not monotone on the interval $[-1, 1]$.

2. The function $g: [-4, 0) \to \mathbb{R}$ defined by $g(x) = 1/x$ is strictly decreasing on the interval $[-4, 0)$.

3. The function $h: \mathbb{R} \to \mathbb{R}$ defined by $h(x) = x^2/(x^2 + 1)$ is increasing on the interval $[0, \infty)$, decreasing on the interval $(-\infty, 0]$, and not monotone on $\mathbb{R}$.

4. According to the definition, a constant function is both increasing and decreasing on $\mathbb{R}$.

Of course, appealing to a graph is not a proof that a function is monotone. Here is one proof that the function F defined by $F(x) = x^3$ is strictly increasing on $\mathbb{R}$. Let x and y be real numbers with $x < y$. There are three cases to consider. Suppose first that $0 \leq x < y$. Since

$$F(y) - F(x) = y^3 - x^3 = (y - x)(y^2 + xy + x^2)$$

is the product of two positive terms, we find that $y^3 > x^3$. The second case is trivial: if $x < 0 \leq y$, then certainly $y^3 > x^3$. Finally, suppose that $x < y \leq 0$. Then $0 \leq -y < -x$, and the first case yields $(-x)^3 > (-y)^3$, which is equivalent to $x^3 < y^3$. Thus, the function F is strictly increasing on $\mathbb{R}$.

Monotone functions will be discussed in greater detail in Section 3.5. We now turn to the concept of a bounded function.

DEFINITION 1.32 Let I be an interval, let $f: I \to \mathbb{R}$, and let J be a subinterval of I.

a) The function f is **bounded above** on J if there exists a number M such that $f(x) \leq M$ for all $x \in J$. The number M is called an **upper bound** of f on J.

b) The function f is **bounded below** on J if there exists a number m such that $f(x) \geq m$ for all $x \in J$. The number m is called a **lower bound** of f on J.

c) The function f is **bounded** on J if there exists a number M such that $|f(x)| \leq M$ for all $x \in J$. The number M is called a **bound** for f on J.

In other words, a function f is bounded above, bounded below, or bounded on an interval J if the set $f(J) = \{f(x) : x \in J\}$ is bounded above, bounded below, or bounded, respectively. As with sets, a function that is not bounded is said to be **unbounded**. Since the graph of a constant function is a horizontal line, the graph of a function that is bounded above lies below some horizontal line, and the graph of a bounded function lies between a pair of horizontal lines. For example, the graph of $y = \sin x$ lies between the graphs of the lines $y = -1$ and $y = 1$. Some further examples illustrating the definition are given below.

1. The function $f: \mathbb{R} \to \mathbb{R}$ defined by $f(x) = x^2$ is bounded on the interval $(0, 4)$, unbounded on the interval $[1, \infty)$, and bounded below on $\mathbb{R}$.

2. The function $g: [-4, 0) \to \mathbb{R}$ defined by $g(x) = 1/x$ is bounded above on $(-1, 0)$, unbounded on $[-4, 0)$, and bounded on $(-3, -1)$.

3. The function $h: \mathbb{R} \to \mathbb{R}$ defined by $h(x) = x^2/(x^2 + 1)$ is bounded on $\mathbb{R}$. The number 1 is a bound for this function.

By the Completeness Axiom, a nonempty bounded set has a supremum, but it does not follow that such a set has a maximum value. The **maximum value** of a nonempty bounded set A is a number $x \in A$ such that $a \leq x$ for all $a \in A$. The key point here is that the maximum value of A must belong to the set A. Hence, a nonempty bounded set has a maximum value if and only if it contains its supremum (see Exercise 27 in Section 1.3). Similarly, a bounded function may or may not have a maximum value (or maximum output): a bounded function $f: I \to \mathbb{R}$ has a maximum value if and only if the set $f(I)$ contains its supremum. The usual way of expressing this concept, as well as the analogous concept of a minimum value, is presented in the following definition.

DEFINITION 1.33 Let I be an interval, let $f: I \to \mathbb{R}$, and let $c \in I$.

a) The function f has a **maximum value** at c if $f(x) \leq f(c)$ for all $x \in I$.

b) The function f has a **minimum value** at c if $f(x) \geq f(c)$ for all $x \in I$.

c) The function f has an **extreme value** at c if it has either a maximum value or a minimum value at c.

d) The function f has a **relative maximum value** at c if there exists $\delta > 0$ such that $f(x) \leq f(c)$ for all $x \in I$ that satisfy $|x - c| < \delta$.

e) The function f has a **relative minimum value** at c if there exists $\delta > 0$ such that $f(x) \geq f(c)$ for all $x \in I$ that satisfy $|x - c| < \delta$.

f) The function f has a **relative extreme value** at c if it has either a relative maximum value or a relative minimum value at c.

It should be clear from the definition that a function has a relative extreme value at a point for which it has an extreme value, but that a function can have a relative extreme value at a point for which it does not have an extreme value. The distinction between a relative extreme value and an extreme value lies in the domain of the function: a relative extreme value is an extreme value if the domain of the function is restricted to a small enough interval. It is important to note that the restricted interval is an open interval unless the point c happens to be an endpoint of I. As indicated in the following three examples, it is often easy to spot points at which a function has an extreme value.

1. The function $f: \mathbb{R} \to \mathbb{R}$ defined by $f(x) = x^2$ has a minimum value of 0 at 0.
2. The function $g: \mathbb{R} \to \mathbb{R}$ defined by $g(x) = \cos x$ has a maximum value of 1 at each even multiple of π.
3. The function $h: [-2, 2] \to \mathbb{R}$ defined by $h(x) = x^2 - x^4$ has a relative minimum value of 0 at 0. To see this, note that
$$h(x) = x^2 - x^4 = x^2(1 - x^2) > 0 = h(0)$$
for all x that satisfy $0 < |x| < 1$. Consequently, we may take $\delta = 1$ in part (e) of Definition 1.33. The function h does not have a minimum value at 0 since h assumes negative values on the interval $[-2, 2]$.

The concept of a relative extreme value of a function normally conjures up an image of a graph with a hump in it. This is certainly true in many cases, especially for the types of functions that appear in calculus. However, a constant function has a minimum value and a maximum value at every point in its domain. This might seem a little bizarre, but it is consistent with the definition. It is important to keep such "extreme" examples in mind when making general statements about these concepts. (This last statement is true of every definition; be careful not to add conditions to the definition by restricting yourself to a simple mental picture.)

Here is a quick comment on terminology. Suppose that $f(c) \le f(x) \le f(d)$ for all $x \in [a, b]$. Then the maximum value of f on $[a, b]$ is $f(d)$ and the minimum value of f on $[a, b]$ is $f(c)$. A common error is to assert that the point $(d, f(d))$ is the maximum value of f. This is not correct because the values of the function f are real numbers, not ordered pairs of real numbers. The maximum value of f occurs at d, but the maximum value of f is $f(d)$.

Since the algebraic and transcendental functions are well-known, the reader has some ideas about the behavior of real-valued functions. However, compared to an arbitrary real-valued function, these functions are extremely well-behaved, that is, they are continuous and differentiable at most points, their graphs are easy to visualize, and they can be expressed in terms of power series. A typical real-valued function has none of these properties. As we will see, there are functions that are nowhere continuous, there are continuous functions that are nowhere differentiable, and there are increasing functions that have a countably infinite number of discontinuities in every bounded interval. In one sense, functions such as these are somewhat unusual. Although this text is not directed toward a study of strange and unusual functions, some familiarity with unusual functions is needed to avoid errors

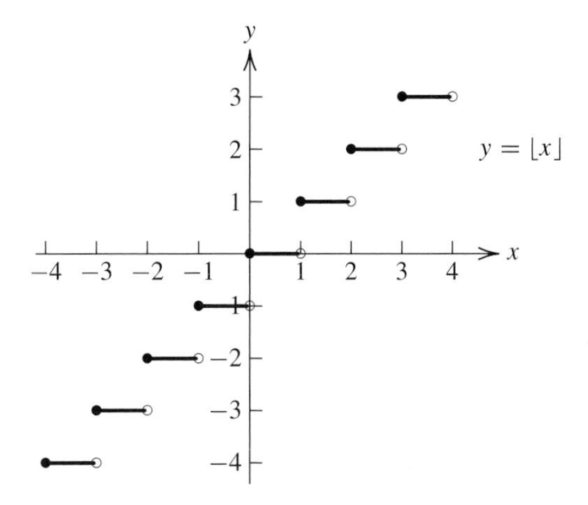

Figure 1.6 A graph of the greatest integer function

in thinking about real-valued functions. For example, if you think of the graph of a continuous function as a nice smooth curve with a few humps in it, you may be led to make some incorrect statements or write an invalid proof. To increase your awareness of the range of possibilities for a real-valued function, a number of exercises at the end of this section consider some unfamiliar functions. The **greatest integer function**, which may be familiar, will be referred to regularly in the text. The symbol $\lfloor x \rfloor$ represents the greatest integer that is less than or equal to x. A graph of this function can be found in Figure 1.6.

As a final comment, it is important to note that the symbols f and $f(x)$ are not interchangeable. The symbol f represents a function, whereas the symbol $f(x)$ represents the value of f at x. A good way to illustrate this is with a calculator. The symbol x represents the number that is entered into the calculator, the symbol f represents the function key that is pressed (the squaring key, the reciprocal key, etc.), and $f(x)$ represents the displayed output. There is a clear distinction between the function keys of the calculator and the displayed outputs. To define a function $f: I \to \mathbb{R}$, you must give a rule for determining an output value for each input value chosen from the interval I. The function f is said to be **well-defined** when there is a unique output for each input. For example, the function $h: [-1, 1] \to \mathbb{R}$ defined by $h(x) = \sqrt{x}$ is not well-defined.

Given two functions f and g defined on an interval I, we can define a new function $f + g$ on I by

$$(f + g)(x) = f(x) + g(x);$$

that is, the value of the function $f + g$ at x is the sum of the values of f and g at x. Technically speaking, the $+$ in $f + g$ is different from the $+$ in $f(x) + g(x)$; the first refers to the addition of functions and the second refers to the addition of numbers. However, this distinction is rarely noted and does not have much of an impact. The functions $f - g$, fg, and f/g are defined similarly. These algebraic combinations of functions will appear frequently in the next few chapters.

Exercises

1. Show that the sum of two polynomials of degree n may not be of degree n.

2. What is the degree of the product of a polynomial of degree n with itself? Justify your answer.

3. Use the technique of completing the square to derive the quadratic formula for the roots of an arbitrary polynomial of degree 2.

4. Find all of the roots of the polynomial $x^3 + 4x^2 - 2x - 3$.

5. Give an example of a polynomial P of degree 4 such that P has no real roots and all of the coefficients of P are nonzero.

6. Give an example of a polynomial Q of degree 2 such that Q has two distinct rational roots and all of the coefficients of Q are irrational.

7. Consider the polynomial $P\colon \mathbb{R} \to \mathbb{R}$ defined by $P(x) = x^4 + x^3 + x^2 + x + 1$. Using only basic algebra, prove that P is neither one-to-one nor onto.

8. Let n be a positive integer and consider the polynomial P defined by

$$P(x) = x^n + a_{n-1}x^{n-1} + \cdots + a_1 x + a_0,$$

where the coefficients a_i are integers and $a_0 \neq 0$. Suppose that r is a rational root of P. Prove that r is an integer that divides a_0. You may find the Fundamental Theorem of Arithmetic (see Appendix C) useful for this problem.

9. For each part, give an example of a rational function with the indicated properties.

a) The graph of the function has three x-intercepts and two vertical asymptotes.

b) Let n be a positive integer. The domain of the function is $\mathbb{R} \setminus A$, where A is a set of real numbers with n elements.

c) The degree of the numerator is 3, the degree of the denominator is 4, and the domain of the function is all real numbers.

10. Consider the rational function $f\colon [0, \infty) \to \mathbb{R}$ defined by $f(x) = (x - 2)/(x + 3)$. Prove that f is one-to-one.

11. Sketch the graphs of the curves $y = \sin(1/x)$, $y = x\sin(1/x)$, $y = x^2\sin(1/x)$, and $y = (1/x)\sin(1/x)$.

12. Determine the range of each of the six inverse trigonometric functions.

13. Find the domain of the given function.

a) $f(x) = \sqrt{x^2 - x - 2}$ **b)** $g(x) = \dfrac{x^2}{1 + \sin x}$ **c)** $h(x) = \dfrac{x}{\sin x + \cos x}$

14. Let a be a positive number and assume the "usual" properties of the functions e^x and $\ln x$. This exercise indicates that all other exponential functions and logarithmic functions can be defined in terms of these two functions.

a) Prove that $a^x = e^{x \ln a}$ for all real numbers x.

b) Prove that $\log_a x = \ln x / \ln a$ for all positive numbers x.

15. For each real number x, define $\sinh x = (e^x - e^{-x})/2$ and $\cosh x = (e^x + e^{-x})/2$. These functions are known as the **hyperbolic sine** and **hyperbolic cosine**, respectively.

a) Prove that $\cosh^2 x - \sinh^2 x = 1$ for all x. (This is why these functions are known as hyperbolic trigonometric functions; they can be thought of as ordered pairs on the hyperbola $x^2 - y^2 = 1$ rather than as ordered pairs on the circle $x^2 + y^2 = 1$.)

b) Sketch the graphs of $y = \sinh x$ and $y = \cosh x$.

c) Find and prove identities for $\sinh 2x$ and $\cosh 2x$.

d) Find the inverse of $\sinh x$ for $x \in \mathbb{R}$ and the inverse of $\cosh x$ for $x \geq 0$.

16. Consider the function g defined by $g(x) = \cot(\pi x)$.

a) Find a closed interval I such that g is defined on I.

b) Find an open interval J such that $4 \in J$ and g is defined on J except at 4.

c) Find the largest open interval K containing $\sqrt{2}$ such that g is defined on K.

17. Consider the function f defined by $f(x) = \sqrt{\sin(1/x^2)}$. Show that there is no open interval I containing 0 such that f is defined on I except at the point 0.

18. Let n be an odd positive integer. Prove that the function x^n is strictly increasing on $\mathbb{R}$.

19. Let n be an even positive integer. Prove that the function x^n is strictly increasing on $[0, \infty)$ and strictly decreasing on $(-\infty, 0]$.

20. Define a function g on $\mathbb{R}$ by $g(x) = 1/x$ if $x < 0$ and $g(x) = -x$ if $x \geq 0$. Prove that g is strictly decreasing on the intervals $(-\infty, 0)$ and $[0, \infty)$, but that g is not decreasing on $\mathbb{R}$.

21. Suppose that f is an increasing function on an interval I. Prove that $-f$ is a decreasing function on I.

22. Suppose that f is a positive increasing function on an interval I. Prove that $1/f$ is decreasing on I.

23. Prove that the sum of two increasing functions is an increasing function.

24. Give an example to show that the difference of two monotone functions may not be a monotone function.

25. Show that the product of two increasing functions may not be increasing. Find a condition that guarantees that the product of two increasing functions is an increasing function.

26. Let I be an interval, let f be a positive increasing function defined on I, and let g be a positive decreasing function defined on I. Give an example to show that fg may be strictly increasing on I and one to show that fg may be strictly decreasing on I.

27. Prove that the composition of two increasing functions is an increasing function. A solution to this exercise must include appropriate hypotheses so that the composition is defined.

28. Give an example of a function $f\colon [a, b] \to \mathbb{R}$ such that f is increasing on (a, b) but not increasing on $[a, b]$.

29. Let f and g be two functions defined on an interval I. Define functions $f \wedge g$ and $f \vee g$ on I by

$$(f \wedge g)(x) = \max\{f(x), g(x)\} \quad \text{and} \quad (f \vee g)(x) = \min\{f(x), g(x)\}.$$

(See the discussion following Theorem 1.7 in Section 1.2.) Suppose that f and g are increasing on I. Determine whether or not the functions $f \wedge g$ and $f \vee g$ are increasing on I.

30. Let $f\colon [a, b] \to \mathbb{R}$ be a function and suppose that for each $\epsilon > 0$, the function g defined by $g(x) = f(x) + \epsilon x$ is increasing on $[a, b]$. Prove that f is increasing on $[a, b]$.

31. Let f be the function defined by $f(x) = 1/x$. Prove that f is bounded on the interval $[a, 1]$ for every real number a in the interval $(0, 1)$, but that f is not bounded on $(0, 1)$.

32. Prove that the sum and product of two bounded functions is a bounded function.

33. Find general conditions on a bounded function f that guarantee the function $1/f$ is also a bounded function.

34. Suppose that $f: [a, b] \to \mathbb{R}$ is monotone on $[a, b]$. Prove that f is bounded on $[a, b]$.

35. Suppose that $f: (a, b) \to \mathbb{R}$ is monotone on (a, b). Give an example to show that f may not be bounded on (a, b).

36. Give an example of a function $f: (0, 2) \to \mathbb{R}$ such that f has a relative maximum value at 1, but f is not bounded above on $(0, 2)$.

37. Consider the function $h: (0, 1) \to \mathbb{R}$ defined by $h(x) = \cos(\pi/x)$. Determine the set of all x such that h has a relative extreme value at x.

38. Let I be an interval, let $f: I \to \mathbb{R}$, and let $c \in I$. Consider the following statement: the function f has a relative maximum value at c if there exists an interval $J \subseteq I$ such that $c \in J$ and $f: J \to \mathbb{R}$ has a maximum value at c. Show that this statement is false, then determine how to modify the statement (a slight modification should be enough) so that it is true.

39. Sketch the graphs of $y = 3\lfloor x/2 \rfloor$, $y = \lfloor x^2 \rfloor$, and $y = \lfloor x \rfloor^2$.

40. Let f be the function defined by $f(x) = x - \lfloor x \rfloor$. Describe the function f in words and sketch its graph.

41. Prove that the function f defined by $f(x) = 1$ if x is rational and $f(x) = 0$ if x is irrational assumes the values 0 and 1 infinitely often in every interval.

42. Prove that the function f defined by $f(x) = x$ if x is rational and $f(x) = -x$ if x is irrational is not monotone on any interval.

43. Consider the function g defined by $g(x) = \min\{x - \lfloor x \rfloor, 1 + \lfloor x \rfloor - x\}$ for each real number x. Sketch the graph of this function, then write down a simple descriptive definition of this function.

Remark. The rest of the exercises in this section are in no particular order.

44. Let a, b, c, and d be real numbers with $ac \neq 0$ and consider the function f defined by $f(x) = (ax + b)/(cx + d)$.

a) Prove that f is one-to-one.

b) Find a formula for $f_{\text{inv}}(x)$.

c) Find conditions on a, b, c, and d so that $f = f_{\text{inv}}$.

45. Consider the function h defined by

$$h(x) = \begin{cases} 0, & \text{if } x \notin \mathbb{Q}; \\ 1/q, & \text{if } x = p/q. \end{cases}$$

Here it is assumed that the rational number p/q is in reduced form and that $q > 0$.

a) Find $h(n)$ for each integer n.

b) Find three solutions to the equation $h(x) = 1/3$.

c) Find all of the solutions to the equation $h(x) = 1/7$ that lie in the interval $(3, 4)$.

d) Prove that the set of all solutions to the equation $h(x) = 1/5$ is countably infinite.

e) Let (a, b) be any interval and let $\epsilon > 0$. Prove that $\{x \in (a, b) : h(x) \geq \epsilon\}$ is a finite set.

46. This exercise introduces a function known as the **Cantor ternary function**. It assumes a little knowledge of infinite series and sequences, primarily in a form related to geometric series.

a) Show that for each real number $x \in [0, 1]$, there exists a sequence of integers $t_{x1}, t_{x2}, t_{x3}, \ldots$ such that t_{xk} is either 0, 1, or 2, and

$$x = \frac{t_{x1}}{3} + \frac{t_{x2}}{3^2} + \frac{t_{x3}}{3^3} + \frac{t_{x4}}{3^4} + \frac{t_{x5}}{3^5} + \cdots.$$

In other words, show that the number x has a ternary expansion (the base is 3 rather than 10).

b) Define a function $f\colon [0, 1] \to \mathbb{R}$ as follows. If the digit 1 does not appear in the ternary expansion of x, define

$$f(x) = \sum_{k=1}^{\infty} \frac{t_{xk}/2}{2^k}.$$

If the digit 1 does appear in the ternary expansion of x, let $j_x = \min\{k : t_{xk} = 1\}$ and define

$$f(x) = \sum_{k=1}^{j_x - 1} \frac{t_{xk}/2}{2^k} + \frac{1}{2^{j_x}}.$$

Note that the expression for $f(x)$ is a binary expansion for some real number (see Exercise 25 in Section 1.3). Show that f is well-defined (that is, show that if x has two different ternary expansions, the value $f(x)$ is the same no matter which representation is used) and that $0 \le f(x) \le 1$ for all $x \in [0, 1]$.

c) Find $f(1/3)$, $f(2/3)$, $f(4/27)$, $f(80/81)$, and $f(1/2)$.

d) Prove that the function f is increasing on $[0, 1]$.

e) Determine all the intervals on which f is constant.

f) Find the sum of the lengths of all the intervals from part (e).

47. Let f and g be bounded functions defined on $[a, b]$.

a) Give a proof of the inequality

$$\sup\{f(x) + g(x) : x \in [a, b]\} \le \sup\{f(x) : x \in [a, b]\} + \sup\{g(x) : x \in [a, b]\}.$$

Provide examples to show that either equality or strict inequality may occur.

b) Establish an inequality similar to the one in part (a) that is valid for the infima of these sets. Provide examples to show that either equality or strict inequality may occur.

c) Suppose further that f and g are nonnegative. Find and prove inequalities involving the numbers

$$\sup\{f(x)g(x) : x \in [a, b]\} \quad \text{and} \quad \inf\{f(x)g(x) : x \in [a, b]\}.$$

48. A function $f\colon \mathbb{R} \to \mathbb{R}$ has a **proper relative maximum value** at c if there exists $\delta > 0$ such that $f(x) < f(c)$ for all x that satisfy $0 < |x - c| < \delta$. Prove that the set of points at which f has a proper relative maximum value is countable.

49. Let $\mathcal{F}$ be the collection of all functions $f\colon \mathbb{R} \to \mathbb{R}$. This exercise shows that the set $\mathcal{F}$ cannot be put into a one-to-one correspondence with $\mathbb{R}$. Suppose that $h\colon \mathbb{R} \to \mathcal{F}$ is a one-to-one function. For each real number x, let $h(x) = f_x$, where f_x is a function mapping $\mathbb{R}$ into $\mathbb{R}$. Show that the function $f\colon \mathbb{R} \to \mathbb{R}$ defined by $f(x) = f_x(x) + 1$ is not in the range of h. Therefore, the function h is not onto.

2

Sequences

Since an introduction to sequences of real numbers is usually included in calculus, the reader should have some familiarity with sequences. In a typical calculus course, the definition of a sequence and a discussion of the properties of sequences are covered quickly, often without proof, just prior to a study of infinite series. This is both understandable and unfortunate—understandable because the proofs of some of the key results on sequences require the Completeness Axiom and unfortunate because a knowledge of sequences is critical in the study of real analysis and analysis in general. In this chapter, we will consider sequences of real numbers and prove a number of important facts about them. The limit process appears in its simplest form in the context of sequences, and the use of the Completeness Axiom to guarantee the existence of limits is readily seen.

2.1 Convergent Sequences

The term "sequence" occurs in ordinary conversation; the phrase "an unfortunate sequence of events" is one example. In this case, we envision a first event causing a second event which in turn causes a third event and so on. A small amount of mathematical insight reveals that the positive integers are somehow connected with sequences. The first event corresponds with 1, the second event with 2, and the third event with 3. A sequence of real numbers represents one number followed by another number; a first number, a second number, a third number, and so on indefinitely. There is not necessarily a connection between the numbers, but there often is. The sequence of real numbers

$$1, \frac{1}{2}, \frac{1}{3}, \frac{1}{4}, \frac{1}{5}, \frac{1}{6}, \frac{1}{7}, \ldots$$

is usually written as $\{1/n\}_{n=1}^{\infty}$, or more simply as $\{1/n\}$. In general, the sequence $\{x_n\}$ represents $x_1, x_2, x_3, \ldots$, where it is understood that the index n runs over the set of positive integers. If we think of x_1 as $x(1)$, x_2 as $x(2)$, and so on, it becomes evident that a sequence is simply a special type of function.

DEFINITION 2.1 A **sequence** is a function whose domain is the set of positive integers. A **sequence of real numbers** is a sequence whose codomain is the set $\mathbb{R}$. Although a sequence is a function, the standard notation for a sequence of real numbers is $\{x_n\}$, which represents $x_1, x_2, x_3, \ldots$. The subscript n is known as the **index** of the sequence and is assumed to represent a positive integer. Sometimes the notation $\{x_n\}_{n=1}^{\infty}$ is used to emphasize that the index n begins with 1. The numbers $x_1, x_2, \ldots$ are called the **terms** of the sequence.

A little more generally, a sequence is a function whose domain is the set of all integers $n \geq i$, where i is an integer; a sequence of this type is denoted by $\{x_n\}_{n=i}^{\infty}$. The most common choices for i are 0 and 1. The terms of a sequence can be the elements of any set, but the focus in this chapter will be sequences of real numbers. Later in the text, we will consider sequences of intervals, sequences of functions, and sequences of sets. The type of object represented by the terms of a sequence should be clear from the context.

The most common ways to describe a sequence of real numbers are explicit formulas, implicitly assumed patterns, and recursive definitions. To describe a sequence $\{x_n\}$ by an explicit formula is easy; just take a function of n that is defined for all positive integers n. For example,

$$\left\{\frac{2^n - 1}{2^n}\right\} = \frac{1}{2}, \frac{3}{4}, \frac{7}{8}, \frac{15}{16}, \frac{31}{32}, \ldots \quad \text{and} \quad \left\{\frac{(-1)^n}{2n}\right\} = -\frac{1}{2}, \frac{1}{4}, -\frac{1}{6}, \frac{1}{8}, -\frac{1}{10}, \ldots.$$

Sometimes it is easier to write out enough terms to establish a pattern; the assumption is that the pattern continues indefinitely. The sequences

$$0, 1, 0, 1, 0, 1, 0, 1, \ldots \quad \text{and} \quad 1, -1, 1, -2, 1, -3, 1, -4, \ldots$$

are examples of sequences of this type. To define a sequence **recursively** means that the first few terms of a sequence are given and a rule to generate successive terms of the sequence is provided. As an example, define a sequence $\{x_n\}$ by letting $x_1 = 1$, $x_2 = 1$, and $x_n = x_{n-2} + x_{n-1}$ for all $n > 2$. The fact that x_n is defined for every positive integer n follows from the Principle of Mathematical Induction. The first few terms of this sequence are

$$1, 1, 2, 3, 5, 8, 13, 21, 34, 55, 89, \ldots,$$

and the reader may recognize this sequence as the sequence of **Fibonacci numbers**.

It is important to make a distinction between a sequence $\{x_n\}$ and its range $\{x_n : n \in \mathbb{Z}^+\}$. For example, let $x_n = \sin(n\pi/2)$ for each positive integer n. The sequence $\{x_n\}$ represents all of the terms in the order imposed on them by the positive integers:

$$1, 0, -1, 0, 1, 0, -1, 0, 1, 0, -1, 0, \ldots.$$

The range $\{x_n : n \in \mathbb{Z}^+\}$ of this sequence is simply the set $\{-1, 0, 1\}$. Every sequence has a countably infinite number of terms, but in this case the range is a finite set. It is a little unfortunate that the notation $\{x_n\}_{n=1}^{\infty}$ for a sequence and $\{x_n : n \in \mathbb{Z}^+\}$ for a set is so similar, especially when $\{x_n\}_{n=1}^{\infty}$ is usually abbreviated as $\{x_n\}$. However, the context should make the notation clear, and it is important to remember the distinction between a sequence and its range.

To begin a discussion of the properties of sequences of real numbers, we introduce some descriptive terminology. There is nothing fancy going on here, just the clarification of terms. Note the similarities between the definitions of bounded sequences and bounded sets.

DEFINITION 2.2 Let $\{x_n\}$ be a sequence of real numbers.

a) The sequence $\{x_n\}$ is **bounded above** if there exists a number M such that $x_n \leq M$ for all n. The number M is called an **upper bound** of $\{x_n\}$.

b) The sequence $\{x_n\}$ is **bounded below** if there exists a number m such that $x_n \geq m$ for all n. The number m is called a **lower bound** of $\{x_n\}$.

c) The sequence $\{x_n\}$ is **bounded** if there is a number M such that $|x_n| \leq M$ for all n. The number M is called a **bound** for $\{x_n\}$.

d) The sequence $\{x_n\}$ is **increasing** if $x_n \leq x_{n+1}$ for all n and **strictly increasing** if $x_n < x_{n+1}$ for all n.

e) The sequence $\{x_n\}$ is **decreasing** if $x_n \geq x_{n+1}$ for all n and **strictly decreasing** if $x_n > x_{n+1}$ for all n.

f) The sequence $\{x_n\}$ is **monotone** if it is increasing or decreasing and **strictly monotone** if it is strictly increasing or strictly decreasing.

As with any new definition, it is important to find some simple examples to illustrate the terms. The reader should verify the following statements:

1. The sequence $\{n^2\}$ is bounded below and strictly increasing.
2. The sequence $\{(-1)^n\}$ is bounded, but not monotone.
3. The sequence $1, 1, \frac{1}{2}, \frac{1}{2}, \frac{1}{3}, \frac{1}{3}, \ldots$ is decreasing and bounded.
4. A sequence $\{x_n\}$ of positive numbers is increasing if and only if the inequality $x_{n+1}/x_n \geq 1$ is valid for all n.
5. If $\{x_n\}$ is a decreasing sequence, then $\{-x_n\}$ is an increasing sequence.

Every strictly increasing sequence is increasing but an increasing sequence may not be strictly increasing. There are rare occasions for which the distinction between increasing and strictly increasing (or decreasing and strictly decreasing) is important, but for the most part these sequences have identical properties.

One way to think of a sequence of real numbers is as a set of stepping stones on the real line. The stones must be traversed in the order imposed upon them by the positive integers, that is, in the order $1, 2, 3, \ldots$. If this "path" along the number line leads to some fixed point, the sequence is said to converge. For example, the

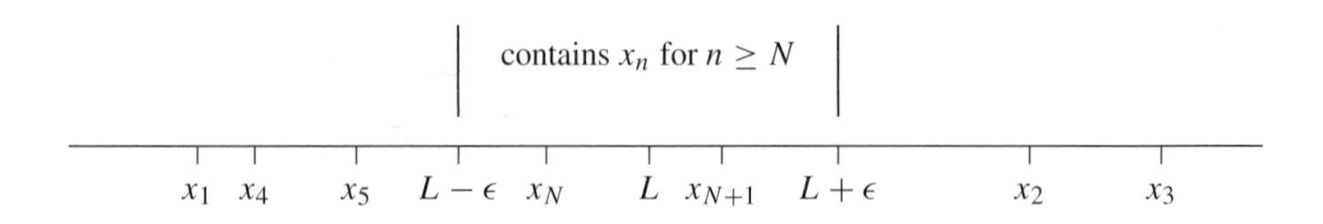

Figure 2.1 The sequence $\{x_n\}$ converges to L

sequence

$$1, \frac{1}{2}, \frac{1}{3}, \frac{1}{4}, \frac{1}{5}, \frac{1}{6}, \frac{1}{7}, \ldots$$

seems to be leading to the point 0. The mathematical definition of convergence for a sequence of real numbers is given in the following definition.

DEFINITION 2.3 A sequence $\{x_n\}$ **converges to a number** L if for each $\epsilon > 0$ there exists a positive integer N such that $|x_n - L| < \epsilon$ for all $n \geq N$. The number L is called the **limit** of the sequence and the notation $\lim_{n\to\infty} x_n = L$ or $x_n \to L$ will sometimes be used. The sequence $\{x_n\}$ is **convergent** (or converges) if there exists a number L such that $\{x_n\}$ converges to L. The sequence $\{x_n\}$ is **divergent** (or diverges) if it does not converge.

What exactly does this definition mean? An informal explanation of convergence goes like this: as you go out farther and farther in the sequence, the terms get closer and closer to L. The problem with this definition is that "farther" and "closer" are rather vague terms. The mathematical definition carefully quantifies these measures. Given an error tolerance $\epsilon > 0$, it must be possible to find a positive integer N such that every term in the sequence past the Nth term is within ϵ of L. This means that the set $\{x_n : n \geq N\}$ is contained in the interval $(L-\epsilon, L+\epsilon)$. In the notation of the definition, the positive number ϵ measures how close the terms are to L and the positive integer N records how far out in the sequence you must go in order to guarantee that the terms are within ϵ of L. The crucial point is that ϵ can be any positive number. No matter what positive number ϵ is, it is always possible to find a positive integer N with the given property. It is sometimes helpful (although we will not do this) to write N as $N(\epsilon)$ since the choice of N usually depends on ϵ. Intuitively, as ϵ gets smaller, N gets bigger. It is worth pointing out that $\epsilon > 0$ comes first, then an appropriate N is sought. A number line view of a convergent sequence can be found in Figure 2.1.

As an example of a convergent sequence, we will prove that the sequence $\{7n/(4n+5)\}$ converges to $7/4$. The solution to this problem begins with the inequality

$$\left|\frac{7n}{4n+5} - \frac{7}{4}\right| < \epsilon.$$

Remember that $\epsilon > 0$ is given first, then N must be chosen so that the above inequality is valid for all $n \geq N$. For each $n \geq N$, a little algebra yields

$$\left|\frac{7n}{4n+5} - \frac{7}{4}\right| = \frac{35}{4(4n+5)} < \frac{35}{4(4n)} < \frac{3}{n} \leq \frac{3}{N}.$$

We thus need to choose a positive integer N so that $3/N < \epsilon$. The fact that such an integer exists follows from the Archimedean property of the real numbers (which in turn follows from the Completeness Axiom). The formal proof reads as follows.

Let $\epsilon > 0$. By the Archimedean property of the real numbers, there exists a positive integer N such that $3/N < \epsilon$. Then for all $n \geq N$,

$$\left|\frac{7n}{4n+5} - \frac{7}{4}\right| = \frac{35}{4(4n+5)} < \frac{35}{4(4n)} < \frac{3}{n} \leq \frac{3}{N} < \epsilon.$$

Hence, the sequence $\{7n/(4n+5)\}$ converges to $7/4$.

It is a good idea to make several comments about this proof. The existence of the integer N does follow from the Archimedean property of the real numbers, but this is such a simple property of $\mathbb{R}$ that we will soon cease referring to it by name. For the first few proofs, it may be beneficial to explicitly mention where the Archimedean property is used. Note that the computations involved in the scratch work and those made in the proof are almost in reverse order. This often occurs in proofs of this type and, in general, a proof does not usually reveal the thought process of the author. The final point is that overkill estimates can simplify the algebra a great deal. In this convergence proof, we could have used the equation

$$\frac{35}{4(4N+5)} < \epsilon$$

to solve for N. Although this would give a more accurate value for N—that is, a better idea of when the terms of the sequence are first within ϵ of $7/4$—the extra work is seldom worth the effort.

It probably seems obvious that a sequence can have at most one limit, but this statement does require proof. Intuition often provides the idea for a proof, then it is necessary to write out the details using only definitions and theorems. In this case, if all of the terms of a sequence are eventually close to a and also close to b, then a and b are close to each other. Since the measure of closeness is arbitrary, the numbers a and b are the same. There are two things to note in the following proof. The first is the use of the Triangle Inequality and the "trick" of adding zero in a creative way; this type of argument will appear frequently in this book. The second is the use of $\epsilon/2$ rather than ϵ. The definition of convergence states that there is a positive integer N that "works" for each positive number (which just happens to be called ϵ). If $\epsilon > 0$, then $\epsilon/2$ is a positive number as well.

THEOREM 2.4 The limit of a convergent sequence is unique.

Proof. Let $\{x_n\}$ be a convergent sequence and suppose that $\{x_n\}$ converges to both the numbers a and b. Let $\epsilon > 0$. By definition, there exists a positive integer N_1 such that $|x_n - a| < \epsilon/2$ for all $n \geq N_1$. Similarly, there exists a positive integer

N_2 such that $|x_n - b| < \epsilon/2$ for all $n \geq N_2$. Now for $N = \max\{N_1, N_2\}$,

$$|a - b| = |a - x_N + x_N - b| \leq |a - x_N| + |x_N - b| < \frac{\epsilon}{2} + \frac{\epsilon}{2} = \epsilon.$$

Since $\epsilon > 0$ was arbitrary, it follows that $a = b$ (see Theorem 1.7). Hence, the sequence $\{x_n\}$ has a unique limit. ■

A sequence $\{x_n\}$ does not converge if there is no number L such that $x_n \to L$. What does it mean for the sequence $\{x_n\}$ not to converge to L? To answer this question, it is necessary to negate the definition of convergence. In order for a sequence $\{x_n\}$ not to converge to L, there must be at least one positive number ϵ for which no integer N can be found with the property that $|x_n - L| < \epsilon$ for all $n \geq N$. All it takes is one value of $n \geq N$ for which $|x_n - L| \geq \epsilon$ to make this happen. In words, the sequence $\{x_n\}$ does not converge to L if there exists an $\epsilon > 0$ such that for every positive integer N there exists an integer $n \geq N$ for which $|x_n - L| \geq \epsilon$. (If you find this discussion confusing, Appendix A may be helpful.) As an illustration, consider the sequence

$$t_n = \begin{cases} 1, & \text{if } n \text{ is not a power of 2;} \\ 2, & \text{if } n \text{ is a power of 2.} \end{cases}$$

The first few terms of this sequence are

$$2, 2, 1, 2, 1, 1, 1, 2, 1, 1, 1, 1, 1, 1, 1, 2, 1, 1 \ldots.$$

This sequence does not converge to 1 since $|t_{2^n} - 1| = 1$ for all n. Referring to the negated definition of convergence, we can take $\epsilon = 1$ and $n = 2^N > N$. A similar argument shows that $\{t_n\}$ does not converge to L for any real number L. It follows that $\{t_n\}$ does not converge. We will develop more efficient methods for proving that a sequence does not converge later in this chapter.

It is useful to note that the definition of a convergent sequence has several equivalent forms. For example, the inequality $|x_n - L| < \epsilon$ can be replaced by $|x_n - L| \leq \epsilon$ or $|x_n - L| < c\epsilon$, where c is some fixed positive constant. The inequality $n \geq N$ can be replaced with $n > N$. It is not necessary that N be a positive integer, but we will always assume that it is. Finally, it is sufficient to only consider the case in which $0 < \epsilon < 1$ or some other small positive value other than 1. These alternate definitions sometimes make a proof or the notation of a proof a little simpler. The proof that all of these definitions are equivalent is left to the reader.

The next theorem states that a convergent sequence is bounded. The intuition behind the proof of this theorem is that the terms of a convergent sequence "pile up" near the limit (see Figure 2.2). This leaves a finite number of strays to account for in computing the bound, and a finite set is always bounded. The number 1 used in the proof is arbitrary; any positive number will work.

THEOREM 2.5 A convergent sequence is bounded.

Proof. Let $\{x_n\}$ be a convergent sequence and let L be the limit of this sequence. Corresponding to $\epsilon = 1$, there exists a positive integer N such that $|x_n - L| < 1$ for all $n > N$. The number $M = \max\{|x_1|, |x_2|, \ldots, |x_N|, |L| + 1\}$ is a bound for

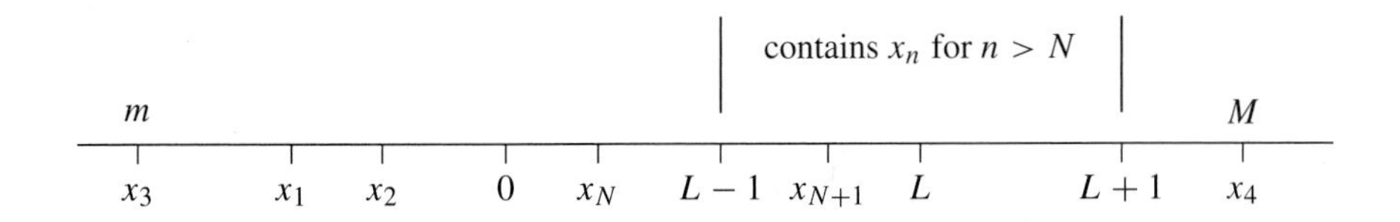

Figure 2.2 A convergent sequence is bounded above and below

the sequence $\{x_n\}$. To see this, first note that $|x_n| \leq M$ for all $n \leq N$ simply by the definition of M. For $n > N$, use the Triangle Inequality to compute

$$|x_n| \leq |x_n - L| + |L| < 1 + |L| \leq M.$$

It follows that $|x_n| \leq M$ for all n. Therefore, the sequence $\{x_n\}$ is bounded. ■

In the more traditional "if, then" form, this theorem states the following:

If $\{x_n\}$ converges, then $\{x_n\}$ is bounded.

The **contrapositive** of the statement "if P, then Q" is the statement "if not Q, then not P". The contrapositive is a true statement precisely when the original statement is true. Thus the following statement is valid:

If $\{x_n\}$ is not bounded, then $\{x_n\}$ does not converge.

For example, the sequence $\{2^n\}$ does not converge since it is not bounded. The **converse** of the statement "if P, then Q" is the statement "if Q, then P". The converse may or may not be true when the original statement is true. In this case, the converse states the following:

If $\{x_n\}$ is bounded, then $\{x_n\}$ converges.

This statement is false because there are bounded sequences that do not converge. The sequence $\{(-1)^n\}$ is one simple example. Although discussions such as this of the implications of a theorem will not appear often in the text, a similar analysis should be carried out for most of the theorems in this book.

An unbounded sequence does not converge, but it is sometimes useful to consider sequences that "converge to infinity". Each of the sequences

$$1, 2, 3, 4, 5, 6, 7, 8, \ldots \quad \text{and} \quad 0, 1, 0, 2, 0, 3, 0, 4, \ldots$$

is unbounded, but it can be said that the terms of the first sequence are leading to ∞, whereas the terms of the second are not. This concept is made precise in the following definition.

DEFINITION 2.6 A sequence $\{x_n\}$ of real numbers **converges to** ∞ if for each $M > 0$ there exists a positive integer N such that $x_n > M$ for all $n \geq N$. We will write $\lim_{n\to\infty} x_n = \infty$ or $x_n \to \infty$ when $\{x_n\}$ converges to ∞. The sequence $\{x_n\}$ converges to $-\infty$ if $\{-x_n\}$ converges to ∞.

To illustrate this concept, consider the following examples.

1. The sequence $\{2^n\}$ converges to ∞.
2. The sequence $\{-n^2\}$ converges to $-\infty$.
3. The sequence $\{(-1)^n n\}$ is unbounded but does not converge to ∞ or $-\infty$.

As indicated by the last example, an unbounded sequence may not converge to ∞ or $-\infty$. It is worth reiterating that a sequence $\{x_n\}$ for which $x_n \to \infty$ does not converge since ∞ is not a real number.

A comment on notation for sequences may be helpful. For starters, letters other than x and n may be used to denote sequences. The subscript in a sequence is a **dummy variable**—its name does not matter. The sequences $\{x_n\}$ and $\{x_k\}$ both represent $x_1, x_2, x_3, x_4, x_5, \ldots$. Since the subscript represents an integer, the usual letters that appear in the subscript are i, j, k, m, and n. Almost any letter may be used in place of x. For example, $\{a_k\}$, $\{b_j\}$, and $\{y_n\}$ all represent sequences. In order to keep track of sequences and their limits, we often let a denote the limit of $\{a_k\}$ and let b denote the limit of $\{b_j\}$. None of this notation is mandatory, but an effort should be made to avoid notation that can cause confusion.

We next consider the relationship between limits of sequences and algebraic combinations of sequences. The results listed in the next theorem are not surprising, but it takes some effort to understand the proofs if this is your first exposure to ϵ-N arguments. It is good practice, both for improving your ability to write proofs and for understanding the definition of a convergent sequence, to go through each of the proofs in detail. Since this is best done while actually writing a proof, all of the proofs will be left as exercises except for a proof for the product of two sequences. This result, expressed informally, states that $a_n b_n$ is near ab when a_n is near a and b_n is near b. To show this, it is necessary to express $a_n b_n - ab$ in terms of $a_n - a$ and $b_n - b$. This relationship can be found in the last displayed equation in the proof of the next theorem. (Another relationship between these terms is discussed in the exercises.) It might be helpful to start with this equation and see how the pieces of the proof are put together in order to make the quantity $|a_n b_n - ab|$ less than ϵ.

THEOREM 2.7 Suppose that $\{a_n\}$ converges to a and $\{b_n\}$ converges to b. Then

a) the sequence $\{ca_n\}$ converges to ca, where c is any constant;

b) the sequence $\{a_n + b_n\}$ converges to $a + b$;

c) the sequence $\{a_n - b_n\}$ converges to $a - b$;

d) the sequence $\{a_n b_n\}$ converges to ab;

e) the sequence $\{a_n/b_n\}$ converges to a/b, provided that $b_n \neq 0$ for all n and $b \neq 0$.

Proof. We will prove part (d) only and leave the proofs of the remaining parts as exercises. By Theorem 2.5, the sequence $\{a_n\}$ is bounded. Let $M > 1$ be a bound for the sequence $\{a_n\}$ and let $\epsilon > 0$. By definition, there exist positive integers N_1 and N_2 such that

$$|a_n - a| < \frac{\epsilon}{2(1 + |b|)} \quad \text{for all } n \geq N_1 \text{ and } |b_n - b| < \frac{\epsilon}{2M} \quad \text{for all } n \geq N_2.$$

(We use $1 + |b|$ in the denominator instead of $|b|$ to cover the case in which $b = 0$.) Let $N = \max\{N_1, N_2\}$. For all $n \geq N$,

$$\begin{aligned} |a_n b_n - ab| &\leq |a_n b_n - a_n b| + |a_n b - ab| \\ &= |a_n|\,|b_n - b| + |b|\,|a_n - a| \\ &< M\frac{\epsilon}{2M} + |b|\frac{\epsilon}{2(1+|b|)} < \frac{\epsilon}{2} + \frac{\epsilon}{2} = \epsilon. \end{aligned}$$

Hence, the sequence $\{a_n b_n\}$ converges to ab. ■

The next two theorems relate inequalities and sequences. These two results are immediately apparent when sequences are thought of as points on a number line, but the proofs only involve the definition of convergent sequence and the properties of inequalities. For reasons that should be clear, the first theorem is often called the squeeze theorem for sequences.

THEOREM 2.8 Squeeze Theorem for Sequences Let $\{a_n\}$ and $\{b_n\}$ be convergent sequences and suppose that $\{x_n\}$ is a sequence such that $a_n \leq x_n \leq b_n$ for all n. If the sequences $\{a_n\}$ and $\{b_n\}$ both converge to L, then the sequence $\{x_n\}$ converges to L.

Proof. Let $\epsilon > 0$. By the definition of convergent sequence, there exists a positive integer N such that $|a_n - L| < \epsilon$ and $|b_n - L| < \epsilon$ for all $n \geq N$. For each $n \geq N$,

$$L - \epsilon < a_n \leq x_n \leq b_n < L + \epsilon,$$

which is equivalent to $|x_n - L| < \epsilon$. This shows that $\{x_n\}$ converges to L. ■

As an example of this theorem, consider the sequence $\{n/10^n\}$. Since $n^2 < 10^n$ for all positive integers n (this can be proved using induction), we find that

$$0 < \frac{n}{10^n} < \frac{n}{n^2} = \frac{1}{n}$$

for all positive integers n. Since the sequences $\{0\}$ (a constant sequence) and $\{1/n\}$ both converge to 0, the Squeeze Theorem asserts that the sequence $\{n/10^n\}$ converges to 0.

Let A be a set of real numbers. A sequence $\{x_n\}$ is said to be in A if each term x_n of the sequence belongs to the set A. In particular, a sequence $\{x_n\}$ is in $[a, b]$ if $a \leq x_n \leq b$ for each positive integer n.

THEOREM 2.9 Let $\{x_n\}$ be a sequence in a closed interval $[a, b]$. If $\{x_n\}$ converges to x, then $x \in [a, b]$.

Proof. We first prove the following result: if $x_n \geq 0$ for all n, then $x \geq 0$. Let $\epsilon > 0$. By definition, there exists a positive integer p such that $|x_n - x| < \epsilon$ for all $n \geq p$. In particular, the inequality $x_p - x < \epsilon$ is valid. This inequality, together with $x_p \geq 0$, yields $x > x_p - \epsilon \geq -\epsilon$. Since $\epsilon > 0$ was arbitrary, it follows that $x \geq 0$.

Now suppose that $a \leq x_n \leq b$ for all n. All of the terms of the sequence $\{x_n - a\}$ are nonnegative and, by Theorem 2.7, this sequence converges to $x - a$.

By the result in the first paragraph, it follows that $x \geq a$. Similarly, the limit $b - x$ of the sequence $\{b - x_n\}$ is nonnegative, so $x \leq b$. It follows that $x \in [a, b]$. ■

It is important to note that strict inequalities are not preserved by limit operations. In particular, suppose that $\{x_n\}$ is a sequence in an open interval (a, b) and suppose that $\{x_n\}$ converges to x. It then follows that $x \in [a, b]$, but it does not follow that $x \in (a, b)$. Thus, a convergent sequence in (a, b) may converge to a limit that is not in (a, b), whereas a convergent sequence in $[a, b]$ always converges to a limit in $[a, b]$. This is an important distinction between open and closed intervals.

Exercises

1. Let $\{a_n\}$ be the sequence defined by $a_n = \big(1 \cdot 3 \cdot 5 \cdot \cdots \cdot (2n + 1)\big)/\big((2n)!\big)$. Find the first five terms of this sequence and a simple formula for the ratio a_{n+1}/a_n.

2. Let $\{a_n\}$ be the sequence defined by $a_1 = 3$ and $a_n = a_{n-1} + n$ for all $n \geq 2$. Find the first five terms of this sequence and a formula for a_n that does not depend on knowing the previous term.

3. Determine the range of the sequence $\{2\cos(n\pi/4)\}$.

4. Prove that a sequence is bounded if and only if it is bounded above and bounded below.

5. Suppose that the sequences $\{x_n\}$ and $\{y_n\}$ are bounded. Prove that the sequences $\{x_n + y_n\}$, $\{x_n - y_n\}$, and $\{x_n y_n\}$ are bounded.

6. Let $\{a_n\}$ be the sequence defined by $a_1 = 1$ and $a_n = 3 - \big(1/a_{n-1}\big)$ for all $n \geq 2$. Prove that $\{a_n\}$ is a bounded increasing sequence.

7. Find an example of a sequence with the given property.

a) The sequence is monotone but not bounded.

b) The sequence is bounded but not monotone.

c) The sequence is bounded and strictly increasing.

d) The sequence is convergent but not monotone.

e) The sequence is strictly decreasing but not convergent.

f) The sequence is neither bounded nor monotone.

g) The sequence is bounded below, not bounded above, and contains an infinite number of negative terms.

8. Prove that each of the sequences is monotone.

a) $\{2^n/n!\}$ **b)** $\{(n + 3)/(6n - 1)\}$ **c)** $\{\sqrt[n]{4}\}$

9. Use the definition of convergence to prove that the sequence converges.

a) $\{(5n + 17)/(2n)\}$ **b)** $\{2n/(3n + 2)\}$ **c)** $\{n/(2n - 51)\}$

10. Prove that the sequence $\{(-1)^n\}$ does not converge.

11. Let $\{x_n\}$ be a sequence of real numbers.

a) Suppose that $\{x_n\}$ converges to L. Prove that $\{|x_n|\}$ converges to $|L|$.

b) Suppose that $\{|x_n|\}$ converges. Give an example to show that $\{x_n\}$ may not converge.

c) Suppose that $\{|x_n|\}$ converges to 0. Prove that $\{x_n\}$ converges to 0.

12. Let $\{a_n\}$ and $\{b_n\}$ be two sequences and suppose that the set $\{n : a_n \neq b_n\}$ is finite. Prove that the sequences either both converge to the same limit or both diverge.

13. Let $\{a_n\}$ be a sequence that converges to L and let p be a fixed positive integer. Prove that the sequence $\{a_{n+p}\}$ converges to L.

14. Let $\{b_k\}$ be a sequence of nonnegative real numbers that converges to b and suppose that $b > 0$. Prove that $\{\sqrt{b_k}\}$ converges to $\sqrt{b}$. Look over your proof and determine if it is valid for $b = 0$. If not, give a separate proof that is valid for $b = 0$.

15. Find an example of a sequence with the given property.

a) $x_n \to 3$ and $x_n \neq 3$ for all n.

b) $y_n \to 3$ and the sets $\{n : y_n = 3\}$ and $\{n : y_n \neq 3\}$ are both infinite.

c) $z_n \to 3$ and $\{z_n\}$ is strictly increasing.

16. Suppose that a sequence has a finite range. Under what conditions will the sequence converge?

17. Let $\{x_k\}$ and $\{y_k\}$ be two sequences and let $\{r_k\}$ be a sequence of positive numbers that converges to 0. Suppose that $0 < |y_k - x_k| < r_k$ for each positive integer k.

a) Give an example to show that the sequences $\{x_k\}$ and $\{y_k\}$ may not converge.

b) Suppose that $\{x_k\}$ converges to L. Prove that the sequence $\{y_k\}$ converges to L.

18. Suppose that $\{a_n\}$ converges to a and that $a > 0$. Prove that there exists a positive number m and a positive integer q such that $a_n > m$ for all $n \geq q$.

19. Suppose that $\{x_n\}$ is a sequence of nonzero numbers that converges to ∞. Prove that $\{1/x_n\}$ converges to 0.

20. Let $\{x_n\}$ be a sequence of real numbers that converges to ∞.

a) Prove that $\{x_n\}$ is bounded below.

b) Suppose that $a_n \geq x_n$ for all n. Prove that $\{a_n\}$ converges to ∞.

c) Suppose that $\{b_n\}$ converges. Prove that $\{x_n + b_n\}$ converges to ∞.

d) Suppose that c is a nonzero number. What can be said about the sequence $\{cx_n\}$? Prove your conjectures.

21. Suppose that $x_n \to \infty$ and $y_n \to -\infty$. Find examples of such sequences with the given property.

a) The sequence $\{x_n + y_n\}$ converges to ∞.

b) The sequence $\{x_n + y_n\}$ converges to $-\infty$.

c) The sequence $\{x_n + y_n\}$ converges to a, where a is a real number.

d) The sequence $\{x_n + y_n\}$ is bounded but does not converge.

22. Let $\{y_n\}$ be an unbounded sequence of positive terms. Does it necessarily follow that $\lim_{n\to\infty} y_n = \infty$?

23. Prove part (a) of Theorem 2.7.

24. Prove part (b) of Theorem 2.7.

25. Use parts (a) and (b) to prove part (c) of Theorem 2.7.

26. Use the fact that $a_n = a + (a_n - a)$ and $b_n = b + (b_n - b)$ to establish the equality

$$a_n b_n - ab = (a_n - a)(b_n - b) + b(a_n - a) + a(b_n - b).$$

Then use this equality to give a different proof of part (d) of Theorem 2.7. Is there any advantage to this proof over the one given in the text?

27. Complete the following steps to prove part (e) of Theorem 2.7. Suppose that $\{b_n\}$ converges to b, that $b_n \neq 0$ for all n, and that $b \neq 0$.

a) Prove that the sequence $\{|b_n|\}$ is bounded below by a positive number by first showing that $|b_n| > |b|/2$ for all sufficiently large n.

b) Prove that the sequence $\{1/b_n\}$ is bounded.

c) Prove that $\{1/b_n\}$ converges to $1/b$ by noting that there exists a positive constant M such that $|1/b_n - 1/b| \leq M|b_n - b|$ for all n.

d) Finish the proof by writing $a_n/b_n = a_n(1/b_n)$ and using the result for the product of two convergent sequences.

28. Use some algebra and Theorem 2.7 to find the limit of the sequence $\left\{\dfrac{2n^2 - 5n + 1}{6 - n^2}\right\}$. Be sure to explain all of your steps and note all of the theorems that are used.

29. Let P and Q be polynomials of the same degree and suppose that $Q(n) \neq 0$ for all positive integers n. Express the limit of the sequence $\{P(n)/Q(n)\}$ in terms of the coefficients of P and Q.

30. Let P be a polynomial. Prove that the sequence $\{P(1/n)\}$ converges to $P(0)$.

31. Suppose that $\{a_n\}$ converges to a and $\{b_n\}$ converges to b. Find the limit of the sequence $\{(2a_n + 3nb_n)/(4n + 1)\}$.

32. Show that the sum of a convergent sequence and a divergent sequence must be a divergent sequence. What can you say about the sum of two divergent sequences?

33. Use the Squeeze Theorem to prove that each of the following sequences converges.

a) $\{\sin n/n\}$ **b)** $\{n^3/2^n\}$ **c)** $\{3^n/n!\}$

34. Prove Theorem 2.9 using a proof by contradiction.

35. Give an example of a convergent sequence of negative numbers whose limit is not negative.

36. Let $\{a_n\}$ and $\{b_n\}$ be two convergent sequences with limits a and b, respectively, and suppose that $a_n \leq b_n$ for all n. Prove that $a \leq b$.

Remark. The rest of the exercises in this section are in no particular order.

37. A sequence $\{x_n\}$ is said to be **eventually bounded** if there exist a positive integer N and a positive number M such that $|x_n| < M$ for all $n > N$. Prove that a sequence is bounded if and only if it is eventually bounded.

38. Referring to the previous exercise, explain what it means for a sequence to be **eventually monotone**. Give an example of a sequence that is not eventually monotone and an example of a sequence that is eventually increasing but not increasing.

39. Suppose that $\{x_n\}$ is a bounded sequence and for each positive integer n, let

$$b_n = \sup\{x_n, x_{n+1}, x_{n+2}, x_{n+3}, x_{n+4}, \ldots\}.$$

Prove that $\{b_n\}$ is bounded and decreasing.

40. Use the greatest integer function (see Section 1.5) to find a formula for the terms of the sequence $1, 1, \frac{1}{2}, \frac{1}{2}, \frac{1}{3}, \frac{1}{3}, \ldots$.

41. Prove that the sequence $\{\lfloor \sin 2n \rfloor\}$ does not converge.

42. Find a sequence $\{x_n\}$ such that $\{x_n\}$ converges to 0 and $\{\sin(1/x_n)\}$ converges to $1/2$.

43. Determine (with proof) whether or not the sequence $\{n!/n^n\}$ converges.

44. Prove that the sequence of Fibonacci numbers is strictly increasing and unbounded.

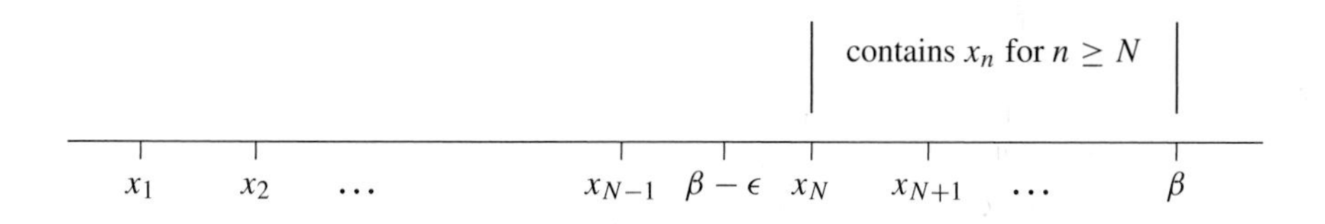

Figure 2.3 A bounded increasing sequence converges

2.2 Monotone Sequences and Cauchy Sequences

By definition, a sequence $\{x_n\}$ of real numbers converges if there is a real number L such that $x_n \to L$. Thus, in order to use the definition to prove that a sequence converges, it is first necessary to know the exact value of the limit of the sequence. However, there are many situations for which it is either difficult or impossible to find the exact value of the limit. Furthermore, especially in the theory behind analysis, it is sometimes more important to know that a sequence converges than it is to find the actual limit of the sequence. In this section, we will consider two ways to prove that a sequence of real numbers converges without knowing the value of its limit.

Consider an increasing sequence $\{x_n\}$ and think of the terms of the sequence as points on a number line. As n increases, the points x_n move to the right and there are only two possibilities; either the points move off indefinitely far to the right (as does the sequence $\{n\}$) or they pile up in front of some blocking value (as does the sequence $\{n/(n+1)\}$). This observation leads to the following theorem.

THEOREM 2.10 A monotone sequence converges if and only if it is bounded.

Proof. Since every convergent sequence is bounded (see Theorem 2.5), it remains to prove that every bounded monotone sequence converges. Let $\{x_n\}$ be an increasing sequence that is bounded. Since the sequence $\{x_n\}$ is bounded, the Completeness Axiom guarantees that the number $\beta = \sup\{x_n : n \in \mathbb{Z}^+\}$ exists. Let $\epsilon > 0$. Since $\beta - \epsilon$ is not an upper bound of the set $\{x_n : n \in \mathbb{Z}^+\}$, there exists a positive integer N such that $x_N > \beta - \epsilon$. Now use the fact that $\{x_n\}$ is increasing to conclude that

$$\beta - \epsilon < x_N \leq x_n \leq \beta$$

for all $n \geq N$ (see Figure 2.3). It follows that $|x_n - \beta| < \epsilon$ for all $n \geq N$. Therefore, the sequence $\{x_n\}$ converges to β. A proof that bounded decreasing sequences converge will be left as an exercise. This completes the proof. ■

To illustrate how this theorem is used to prove that a sequence converges, consider the sequence $\{a_n\}$, where $a_n = \sum_{k=1}^{n} 1/k2^k$ for each positive integer n. The first six terms of this sequence are

$$\frac{1}{2}, \frac{5}{8}, \frac{2}{3}, \frac{131}{192}, \frac{661}{960}, \text{ and } \frac{1327}{1920}.$$

Since a_{n+1} is obtained from a_n by adding a positive number, it is clear that the sequence is increasing. In addition, all of the terms are positive and

$$a_n = \sum_{k=1}^{n} \frac{1}{k2^k} \leq \sum_{k=1}^{n} \frac{1}{2^k} = 1 - \frac{1}{2^n} < 1$$

for all n. (The second equality uses the formula for a geometric sum, see Theorem 1.10.) Hence, the sequence $\{a_n\}$ is bounded by 1. Since $\{a_n\}$ is bounded and increasing, it is convergent by Theorem 2.10.

As a second example, consider the sequence $\{b_n\}$ defined recursively by $b_1 = 3$ and

$$b_{n+1} = \frac{b_n}{2} + \frac{3}{b_n}$$

for all $n \geq 1$. The first four terms of this sequence are 3, 2.5, 2.45, and 2.4495. It should be clear from the recursive definition that all of the terms of this sequence are positive. Based upon the first few terms, it appears that the sequence is strictly decreasing. To verify this, we must show that $b_{n+1} - b_n < 0$ for all n. Since

$$b_{n+1} - b_n = \frac{b_n}{2} + \frac{3}{b_n} - b_n = \frac{6 - b_n^2}{2b_n},$$

the desired inequality is equivalent to $b_n^2 > 6$ for all n. This statement can be proved using mathematical induction. It is clear that $b_1^2 > 6$. Now suppose that $b_p^2 > 6$ for some positive integer p and compute

$$b_{p+1}^2 - 6 = \left(\frac{b_p}{2} + \frac{3}{b_p}\right)^2 - 6 = \left(\frac{b_p}{2} - \frac{3}{b_p}\right)^2 > 0.$$

It follows that $b_n^2 > 6$ for all n by the Principle of Mathematical Induction. This shows that the sequence $\{b_n\}$ is bounded below by $\sqrt{6}$ and is a decreasing sequence. By Theorem 2.10, the sequence $\{b_n\}$ converges.

Although it has been shown that the sequences $\{a_n\}$ and $\{b_n\}$ converge, we have not found the limit of either sequence. It is always possible to approximate the limit of a convergent sequence $\{x_n\}$ by computing x_N for a large value of N or computing many terms of the sequence and looking for a pattern in the decimal expansions. Since the limit is unknown, it is difficult to check the accuracy of these approximations. Various methods for determining the accuracy of an approximation are addressed in a branch of mathematics known as numerical analysis, but these methods will not be discussed here. However, there are some techniques that can be used to determine the exact value of the limit of a sequence. One such technique is useful in finding the limit of recursively defined sequences. As an example, we will find the limit of the sequence $\{b_n\}$. Let L be the limit of $\{b_n\}$ and note that $L \geq \sqrt{6}$. By Theorem 2.7, the sequence $\{b_n/2\}$ converges to $L/2$ and the sequence $\{3/b_n\}$ converges to $3/L$. The sequence $\{b_{n+1}\}$ also converges to L since it is just the sequence $\{b_n\}$ without the first term. Hence, the equality

$$b_{n+1} = \frac{b_n}{2} + \frac{3}{b_n} \quad \text{yields} \quad L = \frac{L}{2} + \frac{3}{L} \quad \text{or} \quad \frac{L}{2} = \frac{3}{L}.$$

Since L must be positive, it follows that $L = \sqrt{6}$. In other words, the sequence $\{b_n\}$ converges to $\sqrt{6}$.

Thus, one way to prove that a sequence converges without finding its limit is to show that the sequence is bounded and monotone. Since convergent sequences are not necessarily monotone, it would be nice to have a condition for convergence that does not require the sequence to be monotone. Fortunately, there is such a condition, and sequences that satisfy this condition are known as Cauchy sequences in honor of Augustin-Louis Cauchy (1789–1867). Cauchy sequences play an important role in analysis and will appear throughout this text. The definition of a Cauchy sequence is similar to the definition of a convergent sequence, but rather than stating that the terms are eventually near some number L, it states that the terms are eventually near each other.

DEFINITION 2.11 A sequence $\{x_n\}$ is a **Cauchy sequence** if for each $\epsilon > 0$ there exists a positive integer N such that $|x_m - x_n| < \epsilon$ for all $m, n \geq N$.

To illustrate the definition, consider the sequence $\{c_n\}$, where $c_n = \sum_{k=1}^{n} 1/k^2$ for each positive integer n. Let $\epsilon > 0$ and choose a positive integer N such that $1/N < \epsilon$. If $m, n \geq N$ with $n > m$, then

$$\begin{aligned} |c_n - c_m| &= \sum_{k=m+1}^{n} \frac{1}{k^2} < \sum_{k=m+1}^{n} \frac{1}{k(k-1)} = \sum_{k=m+1}^{n} \left(\frac{1}{k-1} - \frac{1}{k} \right) \\ &= \frac{1}{m} - \frac{1}{n} < \frac{1}{N} < \epsilon. \end{aligned}$$

Hence, the sequence $\{c_n\}$ is a Cauchy sequence.

It is important to note that the inequality $|x_m - x_n| < \epsilon$ must be valid for all integers m and n that satisfy $m, n \geq N$. In particular, a sequence $\{x_n\}$ with the property that for each $\epsilon > 0$ there exists a positive integer N such that $|x_{n+1} - x_n| < \epsilon$ for all $n \geq N$ may not be a Cauchy sequence. An example of such a sequence is requested in Exercise 20.

For a convergent sequence, the terms are eventually close to the limit, whereas for a Cauchy sequence, the terms are eventually close to each other. If these two ideas appear to be the same, it is because they are—at least for sequences of real numbers. A proof of this fact will be given in a moment. The essence of the proof resides in the need to establish the existence of a limit for a Cauchy sequence. This is where the Completeness Axiom comes into play.

It is a simple consequence of the definitions that every convergent sequence is a Cauchy sequence (see the exercises), but the fact that every Cauchy sequence of real numbers converges requires more work to prove. The proof, as you should expect, requires the Completeness Axiom. It also uses the fact that a Cauchy sequence is bounded. A proof of this result, which is similar to the proof that convergent sequences are bounded, will be left as an exercise.

THEOREM 2.12 Every Cauchy sequence of real numbers is bounded. ■

THEOREM 2.13 A sequence of real numbers converges if and only if it is a Cauchy sequence.

Proof. Let $\{x_n\}$ be a Cauchy sequence. Since $\{x_n\}$ is bounded by Theorem 2.12, for each positive integer n, the number

$$a_n = \inf\{x_n, x_{n+1}, x_{n+2}, x_{n+3}, \ldots\}$$

exists by the Completeness Axiom. The sequence $\{a_n\}$ is increasing and bounded (see the exercises) and therefore converges by Theorem 2.10. Let a be the limit of the sequence $\{a_n\}$. We will prove that $\{x_n\}$ converges to a. Let $\epsilon > 0$. By the definition of a Cauchy sequence, there exists a positive integer N such that $|x_n - x_m| < \epsilon$ for all $m, n \geq N$. It follows that both of the sequences $\{x_n\}_{n=N}^{\infty}$ and $\{a_n\}_{n=N}^{\infty}$ are contained in the interval $[x_N - \epsilon, x_N + \epsilon]$. By Theorem 2.9, the limit a is in the interval $[x_N - \epsilon, x_N + \epsilon]$ and, since x_n is also in this interval for all $n \geq N$, it follows that $|x_n - a| \leq 2\epsilon$ for all $n \geq N$. Hence, the sequence $\{x_n\}$ converges to a. This shows that every Cauchy sequence of real numbers converges. The proof of the converse will be left as an exercise. ■

The fact that every Cauchy sequence of real numbers converges to some real number is a very important property of the set of real numbers. It distinguishes the real numbers from the rational numbers since, as we have mentioned, a Cauchy sequence of rational numbers may not converge to a rational number. The advantage of Cauchy sequences lies in the fact that one can establish the convergence of a sequence without finding a value for the limit or needing the sequence to be monotone. The definition of a Cauchy sequence only involves the distance between points in the sequence; it is an internal property of sequences.

The next theorem provides a list of commonly occurring convergent sequences. Most of these results follow from a combination of the Binomial Theorem (see Appendix C for a statement of this theorem as well as the definition of binomial coefficients) and the Squeeze Theorem for sequences. Some of the proofs will be left as exercises, along with a proof that the sequences in (7) and (8) have the same limit.

THEOREM 2.14

1. If $r > 1$, then the sequence $\{r^n\}$ converges to ∞.
2. If $|r| < 1$, then the sequence $\{r^n\}$ converges to 0.
3. If $|r| < 1$, then the sequence $\left\{\sum_{k=0}^{n} r^k\right\}$ converges to $1/(1-r)$.
4. If $|r| < 1$ and $p \in \mathbb{Z}^+$, then the sequence $\{n^p r^n\}$ converges to 0.
5. If $a > 0$, then the sequence $\{\sqrt[n]{a}\}$ converges to 1.
6. The sequence $\{\sqrt[n]{n}\}$ converges to 1.
7. The sequence $\{(1 + (1/n))^n\}$ converges.
8. The sequence $\left\{\sum_{k=0}^{n} 1/k!\right\}$ converges.

Proof. Suppose that $r > 1$. By the Binomial Theorem,

$$r^n = \big(1 + (r-1)\big)^n = \sum_{k=0}^{n} \binom{n}{k} (r-1)^k \geq 1 + n(r-1)$$

for all positive integers n. Since the sequence $\{n(r-1)\}$ converges to ∞ (see the exercises), so does the sequence $\{r^n\}$. This proves (1). Part (2) follows from part (1) by considering reciprocals, and part (3) follows from part (2) and the formula for a finite geometric sum.

To prove part (6), let $t_n = \sqrt[n]{n} - 1$ for each n. It is sufficient to prove that $\{t_n\}$ converges to 0. Use the Binomial Theorem to obtain

$$n = (1 + t_n)^n = \sum_{k=0}^{n} \binom{n}{k} t_n^k > \binom{n}{2} t_n^2 = \frac{n(n-1)}{2} t_n^2$$

for all $n \geq 2$. Rearranging this inequality and using the fact that t_n is positive yields $0 < t_n < \sqrt{2/(n-1)}$ for all $n \geq 2$. Hence, the sequence $\{t_n\}$ converges to 0 by the Squeeze Theorem.

To prove part (7), we first note that

$$\binom{n+1}{i} \frac{1}{(n+1)^i} \geq \binom{n}{i} \frac{1}{n^i} \quad \text{and} \quad \binom{n}{i} \frac{1}{n^i} \leq \frac{1}{i!}$$

for all $0 \leq i \leq n$ (see the exercises). Using the first result, we obtain

$$\left(1 + \frac{1}{n+1}\right)^{n+1} = \sum_{i=0}^{n+1} \binom{n+1}{i} \left(\frac{1}{n+1}\right)^i > \sum_{i=0}^{n} \binom{n}{i} \left(\frac{1}{n}\right)^i = \left(1 + \frac{1}{n}\right)^n,$$

which shows that the sequence $\{(1+(1/n))^n\}$ is increasing. The second result, along with the inequality $i! \geq 2^{i-1}$ for all $i \geq 1$ and the formula for a finite geometric sum, yields

$$\left(1 + \frac{1}{n}\right)^n = \sum_{i=0}^{n} \binom{n}{i} \left(\frac{1}{n}\right)^i \leq \sum_{i=0}^{n} \frac{1}{i!} \leq 1 + \sum_{i=1}^{n} \frac{1}{2^{i-1}} < 3$$

for all n, so the sequence $\{(1+(1/n))^n\}$ is bounded. By Theorem 2.10, the sequence converges.

The proofs of the remaining parts will be left as exercises. ∎

We conclude this section with a theorem that is sometimes useful for proving results in elementary real analysis. A **nested sequence of intervals** is a sequence $\{I_n\}$ of intervals with the property $I_{n+1} \subseteq I_n$ for all n. In other words, each interval in the sequence is contained in the previous one. The Nested Intervals Theorem states that given any nested sequence of closed and bounded intervals, there exists a point that belongs to all of the intervals.

THEOREM 2.15 Nested Intervals Theorem If $\{[a_n, b_n]\}$ is a nested sequence of closed and bounded intervals, then there exists a point z that belongs to all of the intervals $[a_n, b_n]$. Furthermore, if $\lim_{n\to\infty} (b_n - a_n) = 0$, then the point z is unique.

Proof. We will provide a sketch of the proof and leave the details as an exercise. The sequences $\{a_n\}$ and $\{b_n\}$ are bounded and monotone and therefore convergent. Let a be the limit of the sequence $\{a_n\}$ and let b be the limit of the sequence $\{b_n\}$. Then $a \leq b$ and any number z that satisfies the inequality $a \leq z \leq b$ belongs to all of the intervals $[a_n, b_n]$. This completes the proof. ■

Exercises

1. This exercise shows how to complete the proof of Theorem 2.10 in two different ways. Let $\{x_n\}$ be a bounded decreasing sequence.

a) Prove that $\{x_n\}$ converges by modifying the proof for increasing sequences that is presented in the text.

b) Prove that $\{x_n\}$ converges by considering the sequence $\{-x_n\}$ and using the result for increasing sequences.

c) Which of the two methods do you prefer? Give a reason for your answer.

2. For each positive integer n, let

$$x_n = \frac{1}{n+1} + \frac{1}{n+2} + \cdots + \frac{1}{2n}.$$

Prove that the sequence $\{x_n\}$ converges.

3. Let $\{x_n\}$ be the sequence defined recursively by $x_1 = 2$ and

$$x_{n+1} = \frac{x_n}{2} + \frac{5}{x_n}$$

for all $n \geq 1$. Prove that $\{x_n\}$ converges and find the limit of the sequence.

4. Let $p \in \mathbb{Z}^+$ and let $\{x_n\}$ be the sequence defined recursively by $x_1 = p$ and

$$x_{n+1} = \frac{x_n}{2} + \frac{p}{x_n}$$

for all $n \geq 1$. Prove that $\{x_n\}$ converges and find the limit of the sequence.

5. Prove that $\{x_n\}$ converges, where $x_n = \sum_{k=1}^{n} \frac{2^k}{k^2 3^k}$ for each positive integer n.

6. Prove that $\{x_n\}$ converges, where $x_n = \sum_{k=1}^{n} k^2 2^{-k}$ for each positive integer n.

7. Suppose that $x_1 = 0.4$ and that $x_{n+1} = (x_n^3 + 2)/3$ for each $n \geq 1$. Prove that the sequence $\{x_n\}$ converges. What is the limit of this sequence?

8. For each positive integer n, let $x_n = \sum_{k=1}^{n} 1/\sqrt{k}$. Prove that $\{x_n\}$ is increasing and unbounded.

9. For each positive integer n, let

$$a_n = 1 + \frac{1}{\sqrt{2}} + \frac{1}{\sqrt{3}} + \frac{1}{\sqrt{4}} + \cdots + \frac{1}{\sqrt{n}} - 2\sqrt{n}.$$

Prove that the sequence $\{a_n\}$ converges.

10. Let $\{a_n\}$ be the sequence defined by $a_1 = 1$ and $a_{n+1} = 3 - (1/a_n)$ for all $n \geq 1$. Prove that $\{a_n\}$ converges and find the limit of the sequence.

11. Consider the sequence that is generated in the following way: start with any positive number and find the square root of twice this number, then take the resulting number and find the square root of twice this number, and continue this process indefinitely. Express this sequence as a recursively defined sequence, then prove that the sequence converges and find its limit.

12. Use the definition to prove that $\{n/(n+3)\}$ is a Cauchy sequence.

13. Let $x_n = \sum_{k=1}^{n} k^{-k}$ for each positive integer n. Use the definition to prove that $\{x_n\}$ is a Cauchy sequence.

14. Use the definition to prove that $\{(-1)^n\}$ is not a Cauchy sequence.

15. Prove Theorem 2.12.

16. Prove that every convergent sequence is a Cauchy sequence.

17. Adopting the notation of the proof of Theorem 2.13, prove that the sequence $\{a_n\}$ is increasing and bounded.

18. Prove that the given sequence converges.

a) $\left\{\sum_{k=1}^{n} \frac{\sin k}{k^2}\right\}$ **b)** $\left\{\sum_{k=1}^{n} \frac{(-1)^k}{k^2}\right\}$ **c)** $\left\{\sum_{k=1}^{n} \frac{(-1)^k}{k}\right\}$

19. Prove that the sequence $\left\{\sin(n\pi/3)\right\}$ does not converge.

20. Let $\{x_n\}$ be a sequence and suppose that the sequence $\{x_{n+1} - x_n\}$ converges to 0. Give an example to show that the sequence $\{x_n\}$ may not converge. Hence, the condition that $|x_n - x_m| < \epsilon$ for all $m, n \geq N$ is crucial in the definition of a Cauchy sequence.

21. In the proof of part (1) of Theorem 2.14, it is stated that the sequence $\{n(r-1)\}$ converges to ∞. Give a careful proof of this statement, including the names of all the results you use.

22. Fill in the details for the proofs of parts (2) and (3) of Theorem 2.14.

23. Prove part (2) of Theorem 2.14 in the following alternate way. Suppose that $|r| < 1$.

a) Show that $\{|r|^n\}$ is a decreasing sequence of positive numbers and conclude that $\{|r|^n\}$ converges.

b) Find the limit of $\{|r|^n\}$ by treating it as a recursive sequence defined by $x_1 = |r|$ and $x_n = |r|x_{n-1}$ for all $n > 1$.

24. Let $\{x_n\}$ be a sequence and let r be a real number that satisfies $0 < r < 1$. Suppose that $|x_{n+1} - x_n| \leq r|x_n - x_{n-1}|$ for all $n > 1$. Prove that $\{x_n\}$ is a Cauchy sequence and hence convergent.

25. Let a_0 and a_1 be distinct real numbers. For each $n > 1$, let $a_n = (a_{n-1} + a_{n-2})/2$. Use the previous exercise to show that the sequence $\{a_n\}$ is a Cauchy sequence. Express the limit of the sequence in terms of a_0 and a_1.

26. Given that $0 < r < 1$, find the limit of each sequence.

a) $\left\{\sum_{k=1}^{n} r^{k+1}\right\}$ **b)** $\left\{\sum_{k=1}^{n} r^{2k}\right\}$ **c)** $\left\{\sum_{k=0}^{n} (1-r)^k\right\}$

27. Use the following steps to prove part (4) of Theorem 2.14.

a) Explain why it is sufficient to consider the case in which $0 < r < 1$.

b) Suppose that k and n are positive integers with $n > 2k$. Prove that $\binom{n}{k} > \frac{n^k}{2^k k!}$.

c) Choose a positive number a so that $1 + a = 1/r$ and explain why

$$\frac{1}{r^n} > \frac{n^{p+1}}{2^{p+1}(p+1)!}\, a^{p+1}$$

for all $n > 2p + 2$.

d) Show that $n^p r^n < 2^{p+1}(p+1)!/(n\, a^{p+1})$ for all sufficiently large n and complete the proof of part (4).

28. Let p be any positive integer and suppose that $0 < r < 1$. Prove that the sequence $\{n^p r^n\}$ is eventually decreasing.

29. Let q be any rational number and suppose that $0 < r < 1$. Prove that the sequence $\{n^q r^n\}$ converges to 0.

30. Prove part (5) of Theorem 2.14 in each of the following ways.

a) Use the Binomial Theorem.

b) Use the result from part (6) of Theorem 2.14.

31. Establish, with proof, the limit of the given sequence.

a) $\{\sqrt[n]{4n}\}$ **b)** $\{\sqrt[n]{n^3}\}$ **c)** $\{10^{1/n^2}\}$

32. Suppose that $r > 1$. Prove that the sequence $\{\sqrt[n]{1 + r^n}\}$ converges and find its limit. Generalize this result to the sequence $\{\sqrt[n]{a + r^n}\}$, where a is any positive number.

33. Let n be a positive integer greater than 2 and prove that

$$\binom{n}{i}\frac{1}{n^i} = \frac{1}{i!}\left(1 - \frac{1}{n}\right)\left(1 - \frac{2}{n}\right)\cdots\left(1 - \frac{i-1}{n}\right)$$

for all integers $2 \le i \le n$. Use this result to prove the inequalities needed in the proof of part (7) of Theorem 2.14.

34. Prove part (8) of Theorem 2.14.

35. This exercise outlines a proof that the sequences in parts (7) and (8) of Theorem 2.14 have the same limit. (The common value is the familiar number e.)

a) Prove that $\left(1 + \frac{1}{n}\right)^n \le \sum_{i=0}^{n} \frac{1}{i!}$ for all n. Show that this result indicates that one limit is greater than or equal to the other.

b) To establish the reverse inequality, suppose that $m \ge 2$ is a positive integer and prove that

$$\left(1 + \frac{1}{n}\right)^n \ge 1 + 1 + \sum_{i=2}^{m} \frac{1}{i!}\left(1 - \frac{1}{n}\right)\left(1 - \frac{2}{n}\right)\cdots\left(1 - \frac{i-1}{n}\right)$$

for all $n > m$. Then observe what happens to this inequality as $n \to \infty$ and m remains fixed.

36. Give an example of a nested sequence of open bounded intervals such that there is no point that belongs to all of the intervals.

37. Give an example of a nested sequence of closed intervals such that there is no point that belongs to all of the intervals.

38. Fill in the details of the proof of the Nested Intervals Theorem.

39. Let $[a, b]$ be an interval, let $A \subseteq [a, b]$, and suppose that A is an infinite set.

a) Prove that there exists a nested sequence $\{[a_n, b_n]\}$ of intervals in $[a, b]$ such that $A \cap [a_n, b_n]$ is infinite for each n and $\lim_{n\to\infty}(b_n - a_n) = 0$.

b) Let z be the unique point that belongs to all of the intervals $[a_n, b_n]$. Show that if I is any open interval that contains z, then $A \cap I$ is infinite.

c) Suppose that A is uncountable. Prove that there exists a point $z \in [a, b]$ such that $A \cap I$ is uncountable for every open interval I that contains z.

2.3 SUBSEQUENCES

A set A is a subset of another set B if each element of A is an element of B. A sequence $\{x_n\}$ is a subsequence of another sequence $\{y_n\}$ if each term of $\{x_n\}$ is a term of $\{y_n\}$ and the terms of $\{x_n\}$ appear in the same relative order that they appear in $\{y_n\}$. For example, the sequence $\{1/n^2\}$ is a subsequence of $\{1/n\}$ and the sequence $\{1\}$ is a subsequence of $\{(-1)^n\}$. Since the notion of order of terms is a bit vague, especially when many of the terms have the same value, a more precise definition of subsequence is needed.

DEFINITION 2.16 Let $\{x_n\}$ be a sequence and let $\{p_n\}$ be a strictly increasing sequence of positive integers. The sequence $\{x_{p_n}\}$ is called a **subsequence** of $\{x_n\}$.

Since there are infinitely many strictly increasing sequences of positive integers, a sequence has an infinite number of subsequences. The sequences $\{x_{2n}\}$, $\{x_{5n+2}\}$, and $\{x_{n^2}\}$ are all subsequences of $\{x_n\}$. In addition, a sequence is a subsequence of itself; simply let $p_n = n$ for all positive integers n. For the general case, note that $p_n \geq n$ for all values of n. In keeping with our stepping stone analogy for sequences (see the discussion following Definition 2.3), a subsequence is obtained from a sequence by removing some of the stones and leaving the rest in their original order. As indicated in the second part of the next theorem, subsequences can provide a simple way to show that a sequence does not converge.

THEOREM 2.17 Let $\{x_n\}$ be a sequence of real numbers.

a) If $\{x_n\}$ converges to L, then every subsequence of $\{x_n\}$ converges to L.

b) If $\{x_n\}$ has two subsequences that converge to different limits, then $\{x_n\}$ does not converge.

Proof. Suppose that $\{x_n\}$ converges to L and let $\{x_{p_n}\}$ be any subsequence of $\{x_n\}$. Given $\epsilon > 0$, there exists a positive integer N such that $|x_n - L| < \epsilon$ for all $n \geq N$. Since $p_n \geq n$ for all n, it follows that $|x_{p_n} - L| < \epsilon$ for all $n \geq N$. Hence, the sequence $\{x_{p_n}\}$ converges to L. This proves part (a); a proof of part (b) will be left as an exercise. ■

As an illustration of part (b) of this theorem, consider the sequence $\{(-1)^n\}$. This sequence has subsequences that converge to different limits; the subsequence $\{(-1)^{2n}\}$ converges to 1 and the subsequence $\{(-1)^{2n+1}\}$ converges to -1. Therefore, the sequence $\{(-1)^n\}$ does not converge.

An interesting property of sequences of real numbers is the fact that every sequence of real numbers has a monotone subsequence. For sequences given by formulas or patterns, it is usually easy to spot a monotone subsequence. The sequence

$$1,\ 3.9,\ 2.1,\ -1,\ 1,\ 3.99,\ 2.01,\ -1,\ 1,\ 3.999,\ 2.001,\ -1, \ldots$$

has at least four monotone subsequences; the constant sequences $\{1\}$ and $\{-1\}$ and the sequences

$$2.1,\ 2.01,\ 2.001,\ 2.0001, \ldots \quad \text{and} \quad 3.9,\ 3.99,\ 3.999,\ 3.9999, \ldots$$

which are decreasing and increasing, respectively. However, it is more difficult to write down a representation for a monotone subsequence of an arbitrary sequence of real numbers. The proof of the following theorem provides a quick and clever way to represent a generic monotone subsequence.

THEOREM 2.18 Every sequence of real numbers has a monotone subsequence.

Proof. Let $\{a_n\}$ be an arbitrary sequence of real numbers. Let S be the set of all integers n such that a_n is a lower bound for the set $\{a_{n+1}, a_{n+2}, a_{n+3}, \ldots\}$. If S is infinite, then S can be expressed as a strictly increasing sequence $\{p_n\}$ and the subsequence $\{a_{p_n}\}$ is increasing. If S is finite, then there exists an integer N larger than every element in S. Let p_1 be any integer greater than N. Since $p_1 \notin S$, it is not a lower bound of the set $\{a_{p_1+1}, a_{p_1+2}, a_{p_1+3}, \ldots\}$, so there exists an integer $p_2 > p_1$ such that $a_{p_2} < a_{p_1}$. Similarly, there exists an integer $p_3 > p_2$ such that $a_{p_3} < a_{p_2}$. Continuing this process yields a decreasing sequence $\{a_{p_n}\}$. This completes the proof. ∎

Since bounded monotone sequences converge, a monotone subsequence of a bounded sequence must converge. Hence, the previous theorem shows that every bounded sequence has a convergent subsequence. This is a very important property of sequences and will be used in the proofs of several key results in later chapters. The names of two mathematicians, Bernhard Bolzano (1781–1848) and Karl Weierstrass (1815–1897), are usually associated with this theorem. The special case of the theorem that will be used most often in this book is included in the statement of the theorem; it follows easily from Theorem 2.9.

THEOREM 2.19 **Bolzano-Weierstrass Theorem** Every bounded sequence has a convergent subsequence. In particular, if $\{x_n\}$ is a sequence in $[a, b]$, then $\{x_n\}$ has a subsequence $\{x_{p_n}\}$ that converges to a number in $[a, b]$. ∎

The Bolzano-Weierstrass Theorem states that every bounded sequence of real numbers has a convergent subsequence, but a bounded sequence can certainly have more than one convergent subsequence. In fact, a bounded sequence that does not converge always has at least two subsequences that converge to different limits; a proof of this statement will be left as an exercise. It is sometimes useful to know the largest and smallest limits that a subsequence of a given sequence can have. This idea leads to the notion of the limit superior and limit inferior of a sequence.

Let $\{x_n\}$ be a bounded sequence of real numbers. For each positive integer n, define

$$a_n = \inf\{x_n, x_{n+1}, x_{n+2}, x_{n+3}, \ldots\} \quad \text{and} \quad b_n = \sup\{x_n, x_{n+1}, x_{n+2}, x_{n+3}, \ldots\};$$

the existence of these numbers is guaranteed by the Completeness Axiom. Note that $\{a_n\}$ is a bounded increasing sequence and that $\{b_n\}$ is a bounded decreasing sequence. By Theorem 2.10, the sequences $\{a_n\}$ and $\{b_n\}$ converge; the limits of these sequences are known as the limit inferior and limit superior of the sequence $\{x_n\}$, respectively. This idea is incorporated into the following definition, which also includes the possibility of unbounded sequences.

DEFINITION 2.20 Let $\{x_n\}$ be a sequence of real numbers.

a) If $\{x_n\}$ is bounded below, then the **limit inferior** of $\{x_n\}$ is defined by

$$\liminf_{n\to\infty} x_n = \lim_{n\to\infty} \left(\inf\{x_k : k \geq n\}\right).$$

If $\{x_n\}$ is not bounded below, then the limit inferior of $\{x_n\}$ is $-\infty$.

b) If $\{x_n\}$ is bounded above, then the **limit superior** of $\{x_n\}$ is defined by

$$\limsup_{n\to\infty} x_n = \lim_{n\to\infty} \left(\sup\{x_k : k \geq n\}\right).$$

If $\{x_n\}$ is not bounded above, then the limit superior of $\{x_n\}$ is ∞.

The limits in this definition are to be considered in the extended sense, that is, it is possible for $\liminf_{n\to\infty} x_n = \infty$ and $\limsup_{n\to\infty} x_n = -\infty$.

Note that the sequences $\inf\{x_k : k \geq n\}_{n=1}^{\infty}$ and $\sup\{x_k : k \geq n\}_{n=1}^{\infty}$, which appear in Definition 2.20, are increasing and decreasing, respectively. It follows that these sequences either converge to a real number or to either ∞ or $-\infty$. Since $\inf\{x_k : k \geq n\} \leq \sup\{x_k : k \geq n\}$ for each positive integer n, it is clear that $\liminf_{n\to\infty} x_n \leq \limsup_{n\to\infty} x_n$. Although this definition requires some time and effort to become familiar, the limit inferior and limit superior of a specific sequence is often easy to compute. For the sequence $\{(-1)^n\}$,

$$\inf\{(-1)^k : k \geq n\} = -1 \quad \text{and} \quad \sup\{(-1)^k : k \geq n\} = 1$$

for all positive integers n; hence $\liminf_{n\to\infty}(-1)^n = -1$ and $\limsup_{n\to\infty}(-1)^n = 1$. For the sequence $\{x_n\}$, whose terms are

$$4,\ 3.1,\ 5.9,\ 4,\ 3.01,\ 5.99,\ 4,\ 3.001,\ 5.999, \ldots,$$

it should be clear that $\liminf_{n\to\infty} x_n = 3$ and $\limsup_{n\to\infty} x_n = 6$. As far as unbounded sequences are concerned, note that $\liminf_{n\to\infty} n^2 = \infty$ and $\liminf_{n\to\infty}(-1)^n n = -\infty$. The next theorem lists some of the basic properties of the limit inferior and limit superior of a bounded sequence.

THEOREM 2.21 Let $\{x_n\}$ be a bounded sequence of real numbers.

a) If $m \leq x_n \leq M$ for all $n \geq N$, then $\liminf_{n\to\infty} x_n \geq m$ and $\limsup_{n\to\infty} x_n \leq M$.

b) If $\beta > \limsup_{n\to\infty} x_n$, then there exists a positive integer N such that $x_n < \beta$ for all $n \geq N$.

c) If $\alpha < \liminf_{n\to\infty} x_n$, then there exists a positive integer N such that $x_n > \alpha$ for all $n \geq N$.

d) The sequence $\{x_n\}$ converges if and only if $\liminf_{n\to\infty} x_n = \limsup_{n\to\infty} x_n$.

e) Both $\liminf_{n\to\infty}(c+x_n) = c+\liminf_{n\to\infty} x_n$ and $\limsup_{n\to\infty}(c+x_n) = c+\limsup_{n\to\infty} x_n$ are valid for any real number c.

f) If $c > 0$, then $\liminf_{n\to\infty}(cx_n) = c\liminf_{n\to\infty} x_n$ and $\limsup_{n\to\infty}(cx_n) = c\limsup_{n\to\infty} x_n$.

g) If $c < 0$, then $\liminf_{n\to\infty}(cx_n) = c\limsup_{n\to\infty} x_n$ and $\limsup_{n\to\infty}(cx_n) = c\liminf_{n\to\infty} x_n$.

h) If $\{x_{p_n}\}$ is any subsequence of $\{x_n\}$, then

$$\liminf_{n\to\infty} x_n \leq \liminf_{n\to\infty} x_{p_n} \leq \limsup_{n\to\infty} x_{p_n} \leq \limsup_{n\to\infty} x_n.$$

i) For each $\epsilon > 0$, both of the sets of integers $\{n : x_n < \liminf_{n\to\infty} x_n + \epsilon\}$ and $\{n : x_n > \limsup_{n\to\infty} x_n - \epsilon\}$ are infinite.

Proof. These results all follow from the definitions of limit inferior and limit superior and the properties of the infimum and supremum of sets of real numbers (see Section 1.3). We will prove parts (c), (d), and (i) only and leave the proofs of the other parts as exercises.

Suppose that $\alpha < \liminf_{n\to\infty} x_n$. Since the sequence $\{\inf\{x_k : k \geq n\}\}_{n=1}^{\infty}$ is increasing and converges to $\liminf_{n\to\infty} x_n$, there exists a positive integer N such that $\inf\{x_k : k \geq N\} > \alpha$. It follows that $x_n > \alpha$ for all $n \geq N$. This proves part (c).

To prove part (d), suppose first that $\{x_n\}$ converges to x and let $\epsilon > 0$. By definition, there exists a positive integer N such that $x - \epsilon < x_n < x + \epsilon$ for all $n \geq N$. Part (a) of the theorem then yields

$$x - \epsilon \leq \liminf_{n\to\infty} x_n \leq \limsup_{n\to\infty} x_n \leq x + \epsilon.$$

Since $\epsilon > 0$ was arbitrary, it follows that $\liminf_{n\to\infty} x_n = \limsup_{n\to\infty} x_n$. Now suppose that the limit inferior and limit superior of the sequence are equal and let x be the common value. Given $\epsilon > 0$, use the results of parts (b) and (c) to choose a positive integer N such that $x - \epsilon < x_n < x + \epsilon$ for all $n \geq N$. This shows that the sequence $\{x_n\}$ converges to x.

We will prove part (i) by contradiction. Suppose that for some $\epsilon > 0$, the set $\{n : x_n < \liminf_{n\to\infty} x_n + \epsilon\}$ is finite. Then there exists a positive integer N such that $x_n \geq \liminf_{n\to\infty} x_n + \epsilon$ for all $n \geq N$. Part (a) of the theorem then yields

$$\liminf_{n\to\infty} x_n \geq \liminf_{n\to\infty} x_n + \epsilon,$$

a contradiction. The proof that the other set is infinite is similar. ■

Many of the results recorded in Theorem 2.21 can be extended to unbounded sequences. For example, it is not difficult to show that

$$\liminf_{n\to\infty} x_n = -\limsup_{n\to\infty}(-x_n)$$

for any sequence $\{x_n\}$. It is often possible to use this connection between the limit inferior and limit superior of a sequence to obtain results for the limit inferior from the corresponding result for the limit superior. We will leave the statements and proofs of the results for unbounded sequences as an exercise for the interested reader. It is important to realize that some simple properties of limits of sequences are not valid for the limit superior and the limit inferior of a sequence. For instance, the equation

$$\limsup_{n\to\infty}(x_n + y_n) = \limsup_{n\to\infty} x_n + \limsup_{n\to\infty} y_n$$

may not be valid, see Exercise 32.

Let $\{x_n\}$ be a bounded sequence of real numbers. A **subsequential limit** of $\{x_n\}$ is any real number x such that there exists a subsequence of $\{x_n\}$ that converges to x. The next theorem justifies the terminology that we have been using: the limit inferior of a sequence is the smallest of the subsequential limits of a sequence and the limit superior of a sequence is the largest of the subsequential limits of a sequence.

THEOREM 2.22 Let $\{x_n\}$ be a bounded sequence and let S be the set of all subsequential limits of $\{x_n\}$. Then the set S contains its infimum and supremum and

$$\liminf_{n\to\infty} x_n = \inf S \quad \text{and} \quad \limsup_{n\to\infty} x_n = \sup S.$$

Proof. The set S is bounded since $\{x_n\}$ is bounded; hence the numbers $\inf S$ and $\sup S$ exist by the Completeness Axiom. We will prove that $\sup S \in S$ and that $\limsup_{n\to\infty} x_n = \sup S$ and leave the rest of the proof as an exercise.

Let $s \in S$ and let $\{x_{q_n}\}$ be a subsequence of $\{x_n\}$ that converges to s. By parts (d) and (h) of Theorem 2.21,

$$s = \lim_{n\to\infty} x_{q_n} = \limsup_{n\to\infty} x_{q_n} \le \limsup_{n\to\infty} x_n.$$

This shows that $\sup S \le \limsup_{n\to\infty} x_n$. To complete the proof, it is sufficient to find a subsequence of $\{x_n\}$ that converges to $\limsup_{n\to\infty} x_n$. Let $z = \limsup_{n\to\infty} x_n$. By parts (b) and (i) of Theorem 2.21, the set $\{n : z - 1 < x_n < z + 1\}$ is infinite. Choose an integer p_1 from this set. Similarly, the set $\{n : z - 1/2 < x_n < z + 1/2\}$ is infinite so there exists an integer $p_2 > p_1$ that belongs to this set. Continuing in this manner, we obtain a subsequence $\{x_{p_n}\}$ with the property that $z - 1/n < x_{p_n} < z + 1/n$ for all n. It follows that $\{x_{p_n}\}$ converges to z. ■

Exercises

1. Give an example of sequences $\{x_n\}$ and $\{y_n\}$ such that $\{x_n : n \in \mathbb{Z}^+\} \subseteq \{y_n : n \in \mathbb{Z}^+\}$ but $\{x_n\}$ is not a subsequence of $\{y_n\}$.
2. For the sequence $\{x_n\} = \{(-1)^n n/(2n+3)\}$, find explicit formulas for $\{x_{2n}\}$ and $\{x_{2n+1}\}$ and find the limits of each of these subsequences.
3. Provide an example of a sequence with the given property.
   a) a sequence that has subsequences that converge to 1, 2, and 3
   b) a sequence that has subsequences that converge to ∞ and $-\infty$
   c) a sequence that has a strictly increasing subsequence, a strictly decreasing subsequence, and a constant subsequence
   d) an unbounded sequence which has a convergent subsequence
   e) a sequence that has no convergent subsequence
4. Find at least six convergent subsequences of $\{\sin(n\pi/4)\}$.
5. Prove part (b) of Theorem 2.17.
6. Let $\{a_n\}$ be a sequence of real numbers. Suppose that the subsequences $\{a_{2n}\}$ and $\{a_{2n-1}\}$ converge to the same number. Prove that $\{a_n\}$ converges.
7. State and prove a generalization of the previous exercise.
8. Let $x_1 = 1$ and let $x_n = (1 + x_{n-1})^{-1}$ for all $n \geq 2$.
   a) For each n, let $a_n = x_{2n}$. Find a recursive definition for $\{a_n\}$, then use it to prove that $a_n^2 + a_n < 1$ for all n and that $\{a_n\}$ is increasing.
   b) For each n, let $b_n = x_{2n-1}$. Prove that $\{b_n\}$ is decreasing.
   c) Prove that the sequence $\{x_n\}$ converges and find the limit of the sequence.
9. Let $\{a_n\}$ be an unbounded sequence. Prove that there exists a subsequence $\{a_{p_n}\}$ of $\{a_n\}$ such that $\{1/a_{p_n}\}$ converges to 0.
10. For each of the following sequences, find the monotone subsequence that is generated by the method of the proof of Theorem 2.18. Can you find monotone subsequences other than the one generated by the method of the proof?
    a) $0, 0, 0, 1, 2, 3, 1, 2, 3, 1, 2, 3, 1, 2, 3, 1, 2, 3, \ldots$
    b) $1, -1, \frac{1}{2}, -2, \frac{1}{3}, -3, \frac{1}{4}, -4, \frac{1}{5}, -5, \frac{1}{6}, -6, \ldots$
    c) $1, 1, 0, 2, \frac{3}{2}, \frac{1}{2}, 3, \frac{5}{3}, \frac{3}{4}, 4, \frac{7}{4}, \frac{7}{8}, 5, \frac{9}{5}, \frac{15}{16}, 6, \frac{11}{6}, \frac{31}{32}, \ldots$
11. Referring to the proof of Theorem 2.18, give an example of a sequence for which $S = \{2, 4, 7, 10\}$ and an example of a sequence for which S is the set of all positive multiples of 5.
12. It is possible to use the Nested Intervals Theorem to prove the Bolzano-Weierstrass Theorem. Let $\{x_n\}$ be a bounded sequence of real numbers.
    a) Let $S = \{x_n : n \in \mathbb{Z}^+\}$. Show that the result is trivial if S is finite.
    b) Suppose that S is infinite. Show that there exists a nested sequence of intervals $\{[a_n, b_n]\}$ such that $[a_n, b_n] \cap S$ is infinite for all n and $\{b_n - a_n\}$ converges to 0.
    c) Let c be the point that lies in all of the intervals $[a_n, b_n]$ and show that there is a subsequence of $\{x_n\}$ that converges to c.

13. Let $\{x_n\}$ be a bounded sequence and suppose that every convergent subsequence of $\{x_n\}$ converges to L. (This statement does not guarantee that every subsequence converges!) Prove that $\{x_n\}$ converges to L.

14. Let $\{x_n\}$ be a sequence that does not converge and let L be any real number. Prove that there exist $\epsilon > 0$ and a subsequence $\{x_{p_n}\}$ of $\{x_n\}$ such that $|x_{p_n} - L| > \epsilon$ for all n.

15. Let $\{x_n\}$ be a bounded sequence that does not converge. Prove that $\{x_n\}$ has at least two subsequences that converge to different limits.

16. Prove the following statement about Cauchy sequences without using the fact that a Cauchy sequence of real numbers converges: if a subsequence of a Cauchy sequence converges, then the Cauchy sequence converges.

17. Find the limit inferior and limit superior of the given sequence.
a) $\{(n/3) - \lfloor n/3 \rfloor\}$ **b)** $\{(-1)^n(1 + 1/n)\}$ **c)** $\{n \sin(n\pi/3)\}$

18. Find the limit inferior and limit superior of the sequence $\{\lfloor 5 \sin n \rfloor\}$.

19. Let $\{r_n\}$ be any listing of the rational numbers in the interval (a, b). Establish, with proof, $\liminf_{n\to\infty} r_n$ and $\limsup_{n\to\infty} r_n$.

20. Prove part (a) of Theorem 2.21.

21. Prove part (b) of Theorem 2.21.

22. Prove part (e) of Theorem 2.21.

23. Prove parts (f) and (g) of Theorem 2.21.

24. Prove part (h) of Theorem 2.21.

25. Determine which of the results stated in Theorem 2.21 can be extended to unbounded sequences. This is a rather vague question, but you should give examples for results that cannot be extended and proofs for those that can. If a result cannot be extended as stated, determine if it is possible to modify the statement so that it can be extended.

26. Find the set of all subsequential limits for the sequence $\{\cos(n\pi/3)\}$.

27. Find the set of all subsequential limits for the given sequence.
a) $\{4 + 5(-1)^{\lfloor n/2 \rfloor}\}$ **b)** $\{3(-1)^n + (-1)^{\lfloor n/2 \rfloor}\}$ **c)** $\{(-1)^{\lfloor n/2 \rfloor} + 2(-1)^{\lfloor n/3 \rfloor}\}$

28. Explain why there is no sequence whose set of subsequential limits is $\{1/n : n \in \mathbb{Z}^+\}$.

29. Give an example of a sequence whose set of subsequential limits is the point 0 along with the set $\{1/n : n \in \mathbb{Z}^+\}$.

30. Let $\{x_n\}$ be a sequence for which $\limsup_{n\to\infty} x_n = \infty$. Prove that there exists a subsequence $\{x_{p_n}\}$ of $\{x_n\}$ such that $x_{p_n} \to \infty$.

31. Let $\{x_n\}$ be a sequence of real numbers and let $a_n = (x_1 + \cdots + x_n)/n$.

a) Prove that $\liminf_{n\to\infty} x_n \le \liminf_{n\to\infty} a_n \le \limsup_{n\to\infty} a_n \le \limsup_{n\to\infty} x_n$.

b) Find an example for which all of the limits in part (a) are finite and all of the inequalities are strict.

c) Find an example for which some of the limits are infinite and others are finite.

32. Consider the sequences $\{x_n\} = \{(-1)^n\}$ and $\{y_n\} = \{(-1)^{n+1}\}$. Compare the numbers

$$\limsup_{n\to\infty}(x_n + y_n) \quad \text{and} \quad \limsup_{n\to\infty} x_n + \limsup_{n\to\infty} y_n,$$

as well as the numbers

$$\limsup_{n\to\infty}(x_n y_n) \quad \text{and} \quad \limsup_{n\to\infty} x_n \cdot \limsup_{n\to\infty} y_n.$$

33. Let $\{x_n\}$ and $\{y_n\}$ be two bounded sequences.

a) Suppose that $x_n \le y_n$ for all $n \ge N$. Prove that

$$\liminf_{n\to\infty} x_n \le \liminf_{n\to\infty} y_n \quad \text{and} \quad \limsup_{n\to\infty} x_n \le \limsup_{n\to\infty} y_n.$$

b) Suppose that $\{x_n\}$ converges to x. Prove that

$$\liminf_{n\to\infty}(x_n + y_n) = x + \liminf_{n\to\infty} y_n \quad \text{and} \quad \limsup_{n\to\infty}(x_n + y_n) = x + \limsup_{n\to\infty} y_n.$$

c) Prove that

$$\liminf_{n\to\infty} x_n + \liminf_{n\to\infty} y_n \le \liminf_{n\to\infty}(x_n + y_n)$$
$$\le \limsup_{n\to\infty}(x_n + y_n) \le \limsup_{n\to\infty} x_n + \limsup_{n\to\infty} y_n.$$

Give an example for which all of the inequalities are strict.

d) Suppose that $\{x_n\}$ converges to $x > 0$ and $y_n \ge 0$ for all n. Prove that

$$\liminf_{n\to\infty}(x_n y_n) = x \liminf_{n\to\infty} y_n \quad \text{and} \quad \limsup_{n\to\infty}(x_n y_n) = x \limsup_{n\to\infty} y_n.$$

What happens if $x = 0$?

e) Suppose that $x_n, y_n \ge 0$ for all n. Prove that

$$\liminf_{n\to\infty} x_n \cdot \liminf_{n\to\infty} y_n \le \liminf_{n\to\infty}(x_n y_n)$$
$$\le \limsup_{n\to\infty}(x_n y_n) \le \limsup_{n\to\infty} x_n \cdot \limsup_{n\to\infty} y_n.$$

Give an example for which all of the inequalities are strict.

34. Let $\{x_n\}$ and $\{y_n\}$ be sequences of nonnegative numbers such that $\{x_n\}$ converges to $x > 0$ and $\limsup_{n\to\infty} y_n = \infty$. Prove that $\limsup_{n\to\infty}(x_n y_n) = \infty$.

2.4 Supplementary Exercises

1. Let $\{x_n\}$ be the sequence defined by $x_1 = 2$ and $x_n = x_{n-1} + n(n+1)$ for all $n \ge 2$. Find the first five terms of this sequence and a formula for x_n that does not depend on knowing the previous term.

2. Let $\{x_n\}$ be a sequence and let $a_n = (x_1 + x_2 + \cdots + x_n)/n$ for each positive integer n.

a) Suppose that $\{x_n\}$ converges to x. Prove that $\{a_n\}$ converges to x.

b) Give an example for which $\{a_n\}$ converges but $\{x_n\}$ does not.

3. Let $\{x_n\}$ be a sequence of positive numbers and let $g_n = \sqrt[n]{x_1 x_2 \cdots x_n}$ for each positive integer n.

a) Suppose that $\{x_n\}$ converges to x. Prove that $\{g_n\}$ converges to x.

b) Give an example for which $\{g_n\}$ converges but $\{x_n\}$ does not.

4. Suppose that $\{x_n\}$ and $\{y_n\}$ are convergent sequences and define sequences $\{m_n\}$ and $\{M_n\}$ by $m_n = \min\{x_n, y_n\}$ and $M_n = \max\{x_n, y_n\}$, respectively, for each n. Prove that the sequences $\{m_n\}$ and $\{M_n\}$ converge.

5. Let x be a real number.

a) Prove that there exists a strictly increasing sequence of irrational numbers that converges to x.

b) Prove that there exists a strictly decreasing sequence of rational numbers that converges to x.

6. Suppose that $\{a_n\}$ and $\{b_n\}$ are sequences such that $\{a_n\}$ converges to a nonzero number and $\{a_n b_n\}$ converges. Prove that $\{b_n\}$ converges.

7. Suppose that $\{a_n\}$ converges to 0 and $\{b_n\}$ converges to ∞.

a) Give an example for which $\{a_n b_n\}$ converges to 0.

b) Give an example for which $\{a_n b_n\}$ converges to ∞.

c) Give an example for which $\{a_n b_n\}$ converges to π.

8. Let $\{a_k\}$ be a sequence of positive numbers that converges to a positive number. Prove that the sequence $\{a_k\}$ is bounded below by a positive number.

9. Find a sequence $\{x_n\}$ with the following property: for every positive integer p, there exists a subsequence of $\{x_n\}$ that converges to p.

10. Determine whether or not the given sequence converges.

a) $\left\{\sqrt{n+1}-\sqrt{n}\right\}$ **b)** $\left\{\dfrac{n!}{n^n}\right\}$ **c)** $\left\{\dfrac{1 \cdot 3 \cdot 5 \cdot \cdots \cdot (2n-1)}{n!}\right\}$

11. Determine the limit of the given sequence.

a) $\left\{(\sqrt[n]{n}-1)^n\right\}$ **b)** $\left\{\displaystyle\sum_{k=1}^{n} \frac{1}{\sqrt{n^2+k}}\right\}$ **c)** $\left\{\dfrac{2^n+n^2}{5^n-n}\right\}$

12. Let $\{x_n\}$ be a sequence of positive numbers and suppose that the sequence $\{x_{n+1}/x_n\}$ converges to L.

a) Suppose that $L < 1$. Prove that the sequence $\{x_n\}$ converges to 0.

b) Suppose that $L > 1$. Prove that the sequence $\{x_n\}$ is unbounded.

c) What happens if $L = 1$?

13. Use the previous exercise to prove part (4) of Theorem 2.14.

14. Apply Exercise 12 to each of the following sequences. Assume $|r| < 1$ and $p \in \mathbb{Z}^+$.

a) $\left\{\dfrac{r^n}{n!}\right\}$ **b)** $\left\{\dfrac{n^p}{n!}\right\}$ **c)** $\left\{\dfrac{n^n}{n!}\right\}$

15. Let $x_1 > 1$ and let $x_n = 2 - 1/x_{n-1}$ for each positive integer $n > 1$. Show that $\{x_n\}$ is bounded and monotone. What is the limit of this sequence?

16. Let $p > 1$ be a positive integer. For each positive integer n, let $b_n = \sum_{i=n+1}^{pn} 1/i$. Determine whether or not the sequence $\{b_n\}$ converges.

17. Let a be a positive number and consider the sequence $\left\{\sqrt{n^2+an}-n\right\}$. Does this sequence converge?

18. Let $x_1 = 1$ and $x_{n+1} = x_n + 1/x_n^2$ for each $n \geq 1$. Prove that $\{x_n\}$ is unbounded.

19. Suppose that $0 < a < b$. Prove that the sequence $\{(a^n + b^n)^{1/n}\}$ converges to b.

20. Let z be any real number and consider the sequence $\left\{\lfloor nz \rfloor / n\right\}$. Prove that this sequence converges and find its limit.

21. Let a_0 and a_1 be distinct real numbers and let r and s be positive numbers such that $r + s = 1$. For each $n > 1$, define $a_n = ra_{n-1} + sa_{n-2}$. Prove that the sequence $\{a_n\}$ converges and find an expression for the limit in terms of a_0, a_1, r, and s.

22. Let x_1 be a real number and let $x_{n+1} = x_n + 0.8x_n(1 - x_n) - 0.072$ for $n > 1$.

a) Let $x_1 = 0.6$. Prove that the sequence $\{x_n\}$ converges and find its limit.

b) Find a set A of real numbers so that $\{x_n\}$ converges if $x_1 \in A$.

23. Let $k \in \mathbb{Z}^+$, let $a_0 = k$, and let $a_n = \sqrt{k + a_{n-1}}$ for $n \geq 1$. Prove that the sequence $\{a_n\}$ converges. For what values of k is the limit of this sequence an integer?

24. Let A be a nonempty bounded set. Prove that there exists an increasing sequence $\{x_n\}$ such that $x_n \in A$ for all n and $x_n \to \sup A$. Find a condition that guarantees that the sequence $\{x_n\}$ can be chosen to be strictly increasing.

25. When requested to give an example of a bounded sequence that does not converge, students often propose $\{\sin n\}$. Prove that this sequence has the desired properties.

26. Let $\{x_n\}$ be a sequence of real numbers and let $y_n = \max\{x_1, x_2, \ldots, x_n\}$ for each positive integer n.

a) Suppose that $\{x_n\}$ is bounded. Prove that $\{y_n\}$ converges to $\sup\{x_n : n \in \mathbb{Z}^+\}$.

b) Suppose that $\{x_n\}$ is not bounded above. Prove that $\{y_n\}$ converges to ∞.

c) Give an example of an unbounded sequence $\{x_n\}$ for which $\{y_n\}$ converges.

27. Let $\{I_n\}$ be a sequence of open intervals such that $[a, b] \subseteq \bigcup_{n=1}^{\infty} I_n$. Use the Nested Intervals Theorem to prove that there exists a finite set $\{n_k : 1 \leq k \leq p\}$ of positive integers such that $[a, b] \subseteq \bigcup_{k=1}^{p} I_{n_k}$.

Remark. Given a positive integer p, let $\mathbb{R}^p$ denote the set of all ordered p-tuples of real numbers. We will use bold-faced letters such as $\mathbf{a}$, $\mathbf{b}$, $\mathbf{x}$, and $\mathbf{y}$ to represent elements of $\mathbb{R}^p$ and write $\mathbf{x} = (x_1, x_2, \ldots, x_p)$ when it is necessary to express $\mathbf{x}$ as an ordered p-tuple. It will be assumed that the reader has some familiarity with $\mathbb{R}^p$, at least for $p = 2$ and $p = 3$, from several variables calculus and/or a course in linear algebra. Given $\mathbf{x}, \mathbf{y} \in \mathbb{R}^p$ and $c \in \mathbb{R}$, define

$$\mathbf{x} + \mathbf{y} = (x_1 + y_1, x_2 + y_2, \ldots, x_p + y_p) \quad \text{and} \quad c\mathbf{x} = (cx_1, cx_2, \ldots, cx_p).$$

The symbol $\mathbf{0}$ represents the p-tuple consisting of all 0's. For $\mathbf{x}, \mathbf{y} \in \mathbb{R}^p$, define the **magnitude** $\|\mathbf{x}\|$ of $\mathbf{x}$ and the **dot product** $\mathbf{x} \cdot \mathbf{y}$ of $\mathbf{x}$ and $\mathbf{y}$ by

$$\|\mathbf{x}\|^2 = \sum_{i=1}^{p} x_i^2 \quad \text{and} \quad \mathbf{x} \cdot \mathbf{y} = \sum_{i=1}^{p} x_i y_i.$$

(Note that $\|\mathbf{x}\|^2 = \mathbf{x} \cdot \mathbf{x}$.) The **distance** between $\mathbf{x}$ and $\mathbf{y}$ is then defined by $\|\mathbf{x} - \mathbf{y}\|$. A sequence $\{\mathbf{x}_n\}$ in $\mathbb{R}^p$ is a Cauchy sequence if for each $\epsilon > 0$ there exists a positive integer N such that $\|\mathbf{x}_n - \mathbf{x}_m\| < \epsilon$ for all $m, n \geq N$. The sequence $\{\mathbf{x}_n\}$ converges to $\mathbf{a} \in \mathbb{R}^p$ if for each $\epsilon > 0$ there exists a positive integer N such that $\|\mathbf{x}_n - \mathbf{a}\| < \epsilon$ for all $n \geq N$. (Note that the concept of a monotone sequence in $\mathbb{R}^p$ makes no sense.) The following exercises explore some of the properties of operations in $\mathbb{R}^p$ and of sequences in $\mathbb{R}^p$.

28. Prove the following properties. Part (e) is the Triangle Inequality for $\mathbb{R}^p$.

a) $\|\mathbf{x}\| \geq 0$ for all $x \in \mathbb{R}^p$.

b) $\|\mathbf{x}\| = 0$ if and only if $\mathbf{x} = \mathbf{0}$.

c) $\|c\mathbf{x}\| = |c|\|\mathbf{x}\|$ for all $c \in \mathbb{R}$ and all $\mathbf{x} \in \mathbb{R}^p$.

d) $|\mathbf{x} \cdot \mathbf{y}| \leq \|\mathbf{x}\|\,\|\mathbf{y}\|$ for all $\mathbf{x}, \mathbf{y} \in \mathbb{R}^p$.

e) $\|\mathbf{x} + \mathbf{y}\| \leq \|\mathbf{x}\| + \|\mathbf{y}\|$ for all $\mathbf{x}, \mathbf{y} \in \mathbb{R}^p$.

f) $\|\mathbf{x} - \mathbf{z}\| \leq \|\mathbf{x} - \mathbf{y}\| + \|\mathbf{y} - \mathbf{z}\|$ for all $\mathbf{x}, \mathbf{y}, \mathbf{z} \in \mathbb{R}^p$.

29. Write out part (d) of the previous exercise in terms of the components of $\mathbf{x}$ and $\mathbf{y}$ and note that this result is the Cauchy-Schwarz Inequality.

30. Let $\mathbf{x} \in \mathbb{R}^p$ and write $\mathbf{x} = (x_1, x_2, \ldots, x_p)$. Prove that

$$|x_i| \leq \|\mathbf{x}\| \leq \sqrt{p}\max\{|x_1|, |x_2|, \ldots, |x_p|\}$$

for $1 \leq i \leq p$.

31. Let $\{\mathbf{x}_n\}$ be any sequence in $\mathbb{R}^p$. Although the notation is a bit awkward, write $\mathbf{x}_n = (x_{n1}, x_{n2}, \ldots, x_{np})$ for each positive integer n.

a) Suppose that $\{\mathbf{x}_n\}$ is a convergent sequence in $\mathbb{R}^p$. Prove that the sequence $\{x_{ni}\}_{n=1}^{\infty}$ is a convergent sequence of real numbers for each $1 \leq i \leq p$.

b) Suppose that $\{\mathbf{x}_n\}$ is a Cauchy sequence in $\mathbb{R}^p$. Prove that the sequence $\{x_{ni}\}_{n=1}^{\infty}$ is a Cauchy sequence of real numbers for each $1 \leq i \leq p$.

32. Prove that every Cauchy sequence in $\mathbb{R}^p$ converges to some point in $\mathbb{R}^p$.

33. Suppose that $\{\mathbf{x}_n\}$ converges to $\mathbf{x}$ and $\{\mathbf{y}_n\}$ converges to $\mathbf{y}$ and let c be a real number. Prove each of the following. (You may need to verify some further properties of $\|\mathbf{x}\|$ and $\mathbf{x} \cdot \mathbf{y}$ along the way.)

a) The sequence $\{\mathbf{x}_n + \mathbf{y}_n\}$ converges to $\mathbf{x} + \mathbf{y}$.

b) The sequence $\{c\mathbf{x}_n\}$ converges to $c\mathbf{x}$.

c) The sequence $\{\mathbf{x}_n \cdot \mathbf{y}_n\}$ converges to $\mathbf{x} \cdot \mathbf{y}$.

d) The sequence $\{\|\mathbf{x}_n\|\}$ converges to $\|\mathbf{x}\|$.

Remark. In Chapter 1, it was mentioned that one of the more abstract interpretations of a real number is as an equivalence class of Cauchy sequences of rational numbers. (There is no need to know what an equivalence relation is in order to continue reading this remark. The interested reader can find the definition in almost any text on abstract algebra.) Two Cauchy sequences $\{x_n\}$ and $\{y_n\}$ of rational numbers are equivalent if the sequence $\{x_n - y_n\}$ converges to 0. (In this development of the real numbers, only rational numbers can be used. Hence, in the definition of a convergent sequence, only rational numbers are used for $\epsilon > 0$.) It is not difficult to verify that this defines an equivalence relation on the set of all Cauchy sequences of rational numbers; a real number is defined to be one of these equivalence classes. It is then necessary to define addition and multiplication on this new collection of numbers. Using this definition for the set of real numbers, it is possible to prove that the set of real numbers has the property that every Cauchy sequence of real numbers converges to a real number. What we have called the Completeness Axiom can then be derived as a theorem, that is, the Completeness Axiom is no longer an axiom since it can be derived from other results. Although we have not adopted this approach, it is an interesting exercise to prove that the Completeness Axiom is a consequence of the fact that every Cauchy sequence converges. In fact, several of the properties of sequences discussed in this chapter are equivalent to the Completeness Axiom; these are listed in the next exercise.

34. Prove that all of the following are equivalent.

1) the Completeness Axiom

2) every Cauchy sequence of real numbers converges

3) every bounded monotone sequence of real numbers converges

4) the Bolzano-Weierstrass Theorem

5) the Nested Intervals Theorem

3

Limits and Continuity

The limit concept separates calculus from algebra and forms the basis for the definition of a continuous function, the derivative of a function, and the integral of a function. It is an idea that makes good intuitive sense but, at the same time, is quite difficult to grasp fully. The notion of a limit has been around for centuries, and it was the key idea behind the derivative concept that was developed in the late seventeenth century. However, the current definition of a limit was not formulated until the middle of the nineteenth century. This shows that even mathematicians struggled for many years with the definition of a limit and how best to express it.

The reader has no doubt used the notation $\lim_{x \to c} f(x) = L$ and performed some computations with limits. However, the definition of limit is usually not quite so familiar. When introducing limits to calculus students, it is sometimes helpful to use the concept of motion. As the inputs x move toward c, the outputs $f(x)$ move toward L. This way of thinking about limits leads to phrases such as "$f(x)$ goes to L as x approaches c". Although a visualization such as this may be helpful, it is important to realize that the limit concept is independent of motion and geometry. It is an arithmetic property of the set of real numbers. As we will see, the definition of a limit depends on the properties of the absolute value function, and the existence of limits is guaranteed by the Completeness Axiom.

In this chapter, the limit of a function will be defined and then used as the basis for the definition of a continuous function. The rest of the chapter will be devoted to a study of some of the properties of continuous functions. Several of these properties, such as the Intermediate Value Theorem and the Extreme Value Theorem, will be familiar to the reader from calculus, but reading and understanding their proofs, which require the Completeness Axiom, may be a new experience.

3.1 The Limit of a Function

To begin a discussion of limits, suppose that f is a function that is defined on $\mathbb{R}$ except possibly at the point 0. What does it mean to say that the function f has a limit at 0? The following examples provide some indication of what can happen.

1. For the function f_1 defined by $f_1(x) = x^2$, inputs near 0 generate outputs near 0; the square of a small number is a small number.
2. For the function f_2 defined by $f_2(x) = 1/x^2$, inputs near 0 generate large outputs; the reciprocal of a small positive number is a large positive number.
3. For the function f_3 defined by $f_3(x) = \sin x/x$, inputs near 0 generate outputs near 1; this follows from the fact that $\sin x \approx x$ when x is near 0.
4. For the function f_4 defined by $f_4(x) = \sin(1/x)$, inputs near 0 generate outputs that vary between -1 and 1; the function $\sin(1/x)$ oscillates wildly for values of x close to 0.
5. For the function f_5 defined by $f_5(x) = \lfloor x \rfloor$, positive inputs near 0 generate outputs near 0 and negative inputs near 0 generate outputs near -1.

As these examples indicate, the behavior of a function for values of x near 0 can vary dramatically. Functions f_1 and f_3 have limits at 0; the other functions do not have limits at 0, but each one fails to have a limit for a different reason.

A function f has a limit of L at a point c if the output values of f are near L when the input values are near c. In order to turn this sentence into a mathematical definition, it is necessary to explicitly quantify the phrases "near L" and "near c". The absolute value function can be used to determine when one number is near another since $|a - b|$ represents the distance between the points a and b. If $|x - c|$ and $|f(x) - L|$ are small, then x is near c and $f(x)$ is near L. In order for the limit of f at c to be L, the functional values $f(x)$ must be within a prescribed (small and arbitrary) distance of L for all input values of x close enough to c, but not including c. The value of the function at c does not come into play when finding the limit at c, and the function may not even be defined at c (see functions f_2, f_3, and f_4 discussed in the previous paragraph). A function f has a limit L at c if given any positive number ϵ, it is possible to find another positive number δ so that $|f(x) - L| < \epsilon$ for all x that satisfy $0 < |x - c| < \delta$. That is, if you want the functional values of f to be within ϵ of L, you must choose the input values of f to be within δ of c (see Figure 3.1). If a $\delta > 0$ can be found for each $\epsilon > 0$ (for each $\epsilon > 0$ is the key to the concept; the values of f can be made closer to L than any positive number), then f has a limit L at c.

DEFINITION 3.1 Let I be an open interval that contains the point c and suppose that f is a function that is defined on I except possibly at the point c. The function f has **limit** L at c if for each $\epsilon > 0$ there exists $\delta > 0$ such that $|f(x) - L| < \epsilon$ for all $x \in I$ that satisfy $0 < |x - c| < \delta$. We will write $\lim_{x \to c} f(x) = L$ when f has limit L at c. The function f has a limit at c if there exists a number L such that $\lim_{x \to c} f(x) = L$.

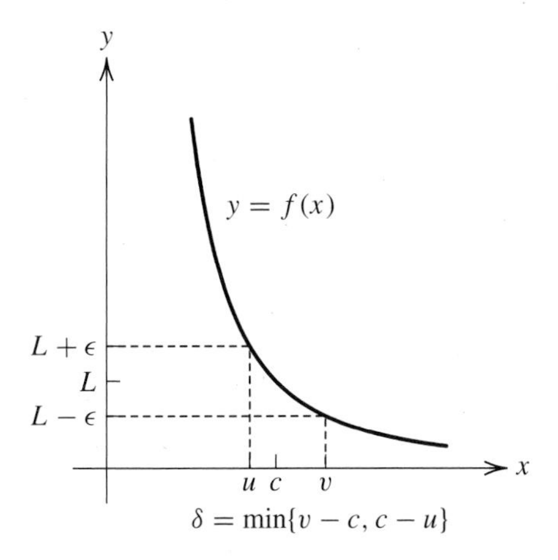

Figure 3.1 A graphical view of the limit of a function

The first sentence of the definition of the limit of a function may seem overly wordy, but it is designed to include functions such as $f_3(x) = \sin x/x$, which are not defined at the point under consideration. Even though this function is not defined at 0, it is still possible to consider the limit of $\sin x/x$ at $x = 0$. Of course, we could just make up a value for this function at 0:

$$f_3(x) = \begin{cases} \sin x/x, & \text{if } x \neq 0; \\ \pi, & \text{if } x = 0; \end{cases}$$

but this is a rather artificial way of avoiding functions that are not defined at certain points, and, in any case, the value of this function at 0 is irrelevant for the determination of the limit at 0. It is best to stick with the definition as stated and become accustomed to seeing a few extra words.

The definition for the limit of a function has some similarities to the definition for the limit of a sequence. For sequences, for each $\epsilon > 0$, one seeks a (probably large) positive integer N such that $|x_n - L| < \epsilon$ for all $n \geq N$, while for functions, for each $\epsilon > 0$, one seeks a (probably small) positive number δ such that $|f(x) - L| < \epsilon$ for all x that satisfy $0 < |x - c| < \delta$. The essential idea is the same; for each positive number ϵ, it is possible to guarantee that all of the output values are within ϵ of L if the inputs are restricted by a certain condition. As with the limit of a sequence, the limit of a function at a point is unique; a proof of this fact will be left as an exercise.

Limit proofs for functions have some similarities to limit proofs for sequences, but there are also important differences. Remember that $\epsilon > 0$ is given first, then a correct $\delta > 0$ must be determined. For many limit problems involving specific functions, it is possible to begin with the expression $|f(x) - L|$ and simplify it so that the expression $|x - c|$ appears. The final step is to choose $\delta > 0$ so that $0 < |x - c| < \delta$ implies $|f(x) - L| < \epsilon$. It is important to note that we have complete freedom in choosing $\delta > 0$; this means that we can control the size of $|x - c|$. As with proofs for sequences, the polished proof does not indicate all of the effort involved in solving the problem. This is evident in the following two examples.

Consider the statement $\lim_{x\to 2}(4x-5)=3$. In this case, the function f under consideration is defined by $f(x)=4x-5$, and it is easy to see that $c=2$ and $L=3$. Given $\epsilon>0$, we need to find $\delta>0$ so that $|f(x)-3|<\epsilon$ for all x that satisfy $|x-2|<\delta$. In order to make the quantity

$$|f(x)-L|=|(4x-5)-3|=4|x-2|$$

less than ϵ, it is clear that $|x-2|$ needs to be less than $\epsilon/4$. In other words, given $\epsilon>0$, we must choose $\delta=\epsilon/4$. The formal proof reads as follows.

Let $\epsilon>0$ and choose $\delta=\epsilon/4$. For all x that satisfy the inequality $0<|x-2|<\delta$, we find that

$$|(4x-5)-3|=|4x-8|=4|x-2|<4\delta=\epsilon.$$

This shows that $\lim_{x\to 2}(4x-5)=3$.

For a second example, we will show that $\lim_{x\to 3}x^2=9$. In this case, $f(x)=x^2$, $c=3$, and $L=9$. We want to make the quantity $|x^2-9|$ less than ϵ for all x that satisfy $0<|x-3|<\delta$ for some choice of δ. It is easy to see how $|x-3|$ is related to $|x^2-9|$; simple factoring yields $|x^2-9|=|x+3|\,|x-3|$. The difference between this problem and the previous one is that $|x-3|$ is multiplied by a factor that contains a variable rather than a constant. We cannot let $\delta=\epsilon/|x+3|$; if δ depends on the variable x, then δ is not a constant. What is needed here is an upper bound for $|x+3|$. This quantity is not bounded above, but remember that we are only interested in values of x near 3; for such values, the term $|x+3|$ is bounded above. Since we can choose any positive number for δ, let's agree to pick $\delta\le 1$ (1 is simply a convenient positive number). Then $|x-3|<1$ implies that

$$|x+3|=|x-3+6|\le|x-3|+6<7.$$

Returning to the expression $|x^2-9|$, we find that

$$|x^2-9|=|x+3|\,|x-3|<7|x-3|<7\delta.$$

To make this less than ϵ, we should choose $\delta=\epsilon/7$. Since it was decided earlier that $\delta\le 1$, the final choice for δ is $\min\{1,\epsilon/7\}$. More likely than not, $\epsilon/7$ will be less than 1, but the definition allows for any positive number ϵ. The proof, as usual, does not include all of this thought process.

Let $\epsilon>0$ and choose $\delta=\min\{1,\epsilon/7\}$. Now $0<|x-3|<\delta$ implies that $|x+3|\le|x-3|+6<7$. It follows that

$$|x^2-9|=|x+3|\,|x-3|<7\,|x-3|<7\delta\le\epsilon$$

for all x that satisfy $0<|x-3|<\delta$. We conclude that $\lim_{x\to 3}x^2=9$.

The next theorem provides an important relationship between limits of functions and limits of sequences. This result is very useful in some cases since it is often easier to work with sequences rather than functions and a number of results involving limits of sequences have been established. The proof of this result involves the negation of the definition of a limit, so the reader should determine how to express $\lim_{x\to c}f(x)\ne L$

mathematically before reading the proof. (It might be helpful to read the discussion that follows Theorem 2.4 or to look at Appendix A.)

THEOREM 3.2 Let I be an open interval that contains the point c and suppose that f is a function that is defined on I except possibly at the point c.

a) The function f has limit L at c if and only if for each sequence $\{x_n\}$ in $I \setminus \{c\}$ that converges to c, the sequence $\{f(x_n)\}$ converges to L.

b) Suppose that there are sequences $\{x_n\}$ and $\{y_n\}$ in $I \setminus \{c\}$ that converge to c such that $\{f(x_n)\}$ converges to L_1 and $\{f(y_n)\}$ converges to L_2. If $L_1 \neq L_2$, then the function f does not have a limit at c.

Proof. Let P be the statement "the function f has limit L at c" and let Q be the statement "for each sequence $\{x_n\}$ in $I \setminus \{c\}$ that converges to c, the sequence $\{f(x_n)\}$ converges to L". A proof that $P \Rightarrow Q$ will be left as an exercise. To prove that $Q \Rightarrow P$, we will prove the contrapositive.

Suppose $\lim_{x \to c} f(x) \neq L$. By the negation of the definition, there exists $\epsilon > 0$ such that for each $\delta > 0$ there is a point $x \in I$ such that

$$0 < |x - c| < \delta \text{ and } |f(x) - L| \geq \epsilon.$$

In particular, for each positive integer n, there exists a point $x_n \in I$ such that

$$0 < |x_n - c| < 1/n \text{ and } |f(x_n) - L| \geq \epsilon.$$

Now $\{x_n\}$ is a sequence in $I \setminus \{c\}$ that converges to c, but the sequence $\{f(x_n)\}$ does not converge to L. This completes the proof of part (a). Part (b) of the theorem follows from part (a); the details will be left as an exercise. ■

The argument found in the proof of part (a) of Theorem 3.2 is a common one. The fact that something is true for each positive number is used to generate a sequence. This is done by taking $1/n$ as the positive number for each positive integer n and finding an appropriate x_n. Since this type of argument will be used several times in this text, it is important to study it carefully.

Part (b) of Theorem 3.2 provides a condition for proving that a limit does not exist that is often much easier to apply than the definition of a limit. As an example, consider the function h defined by

$$h(x) = \begin{cases} x, & \text{if } x \text{ is rational;} \\ -x, & \text{if } x \text{ is irrational;} \end{cases}$$

and let c be any nonzero number. By previous results, there exists a sequence $\{x_n\}$ of rational numbers such that $x_n \to c$ and $x_n \neq c$ for all n, and there exists a sequence $\{y_n\}$ of irrational numbers such that $y_n \to c$ and $y_n \neq c$ for all n. Then

$$\lim_{n\to\infty} h(x_n) = \lim_{n\to\infty} x_n = c \quad \text{and} \quad \lim_{n\to\infty} h(y_n) = \lim_{n\to\infty} (-y_n) = -c.$$

Since the sequences $\{h(x_n)\}$ and $\{h(y_n)\}$ have different limits, the function h does not have a limit at c. It will be left as an exercise to prove that h does have a limit at 0.

The next theorem records the algebraic properties of limits of functions. These properties can be proved by either using the definition of limit of a function or using Theorem 3.2 and the corresponding result for sequences; it is advantageous to be familiar with both methods. The statements in the theorem may appear a little redundant, but there are actually two assertions being made. For example, the theorem states that if f and g have limits at c, then the function $f+g$ has a limit at c. It also shows how the limit of $f+g$ is related to the limits of f and g: the limit of the sum is the sum of the limits.

THEOREM 3.3 Let I be an open interval that contains the point c, let f and g be functions that are defined on I except possibly at the point c, and let k be a constant. Suppose that both of the limits $\lim\limits_{x\to c} f(x)$ and $\lim\limits_{x\to c} g(x)$ exist. Then

a) $f+g$ has a limit at c and $\lim\limits_{x\to c}\big(f(x)+g(x)\big) = \lim\limits_{x\to c} f(x) + \lim\limits_{x\to c} g(x)$;

b) $f-g$ has a limit at c and $\lim\limits_{x\to c}\big(f(x)-g(x)\big) = \lim\limits_{x\to c} f(x) - \lim\limits_{x\to c} g(x)$;

c) kf has a limit at c and $\lim\limits_{x\to c}\big(kf(x)\big) = k\lim\limits_{x\to c} f(x)$;

d) fg has a limit at c and $\lim\limits_{x\to c}\big(f(x)g(x)\big) = \lim\limits_{x\to c} f(x)\cdot\lim\limits_{x\to c} g(x)$.

e) f/g has a limit at c and $\lim\limits_{x\to c}\left(\dfrac{f(x)}{g(x)}\right) = \dfrac{\lim\limits_{x\to c} f(x)}{\lim\limits_{x\to c} g(x)}$, if $\lim\limits_{x\to c} g(x) \neq 0$.

Proof. We will prove the results for sums and quotients and leave the other parts of the proof as exercises. To illustrate the two methods of proof mentioned before the theorem, the definition of limit will be used to prove the result for sums and sequences will be used to prove the result for quotients. To simplify the notation, let $S = \lim\limits_{x\to c} f(x)$ and $T = \lim\limits_{x\to c} g(x)$.

Let $\epsilon > 0$. By Definition 3.1, there exist $\delta_1 > 0$ and $\delta_2 > 0$ such that $|f(x) - S| < \epsilon/2$ for all $x \in I$ that satisfy $0 < |x-c| < \delta_1$ and $|g(x) - T| < \epsilon/2$ for all $x \in I$ that satisfy $0 < |x-c| < \delta_2$. Let $\delta = \min\{\delta_1, \delta_2\}$. For each $x \in I$ that satisfies $0 < |x-c| < \delta$,

$$\big|\big(f(x)+g(x)\big) - (S+T)\big| \le \big|f(x)-S\big| + \big|g(x)-T\big| < \epsilon/2 + \epsilon/2 = \epsilon.$$

This shows that $f+g$ has a limit of $S+T$ at c.

To prove the result for quotients, let $\{x_n\}$ be any sequence in $I\setminus\{c\}$ that converges to c. By Theorem 3.2, the sequence $\{f(x_n)\}$ converges to S and the sequence $\{g(x_n)\}$ converges to T. Since $T \neq 0$, part (e) of Theorem 2.7 indicates that $\{f(x_n)/g(x_n)\}$ converges to S/T. By Theorem 3.2 once again (note that we have used each part of the biconditional statement of Theorem 3.2), the function f/g has limit S/T at c. This completes the proof. ∎

The next two theorems provide relationships between the limit of a function and inequalities satisfied by the function. These results are analogous to Theorems 2.8 and 2.9, which involve similar properties for sequences. A proof of the first result will be left as an exercise.

THEOREM 3.4 Let I be an open interval that contains the point c and suppose that f is a function that is defined on I except possibly at the point c. If $m \le f(x) \le M$ for all x in $I \setminus \{c\}$ and $\lim_{x \to c} f(x) = L$, then $m \le L \le M$. ■

THEOREM 3.5 Squeeze Theorem for Functions Let I be an open interval that contains the point c and suppose that f, g, and h are functions that are defined on I except possibly at the point c. Suppose that $g(x) \le f(x) \le h(x)$ for all x in $I \setminus \{c\}$. If $\lim_{x \to c} g(x) = L = \lim_{x \to c} h(x)$, then f has a limit at c and $\lim_{x \to c} f(x) = L$.

Proof. Let $\{x_n\}$ be any sequence in $I \setminus \{c\}$ that converges to c. By Theorem 3.2, each of the sequences $\{g(x_n)\}$ and $\{h(x_n)\}$ converges to the number L. Since $g(x_n) \le f(x_n) \le h(x_n)$ for all n, the sequence $\{f(x_n)\}$ converges to L by the squeeze theorem for sequences. It follows that $\lim_{x \to c} f(x) = L$ by Theorem 3.2. ■

The definition of the limit of a function f at a point c requires that the function f be defined on both sides of c, but it is possible and sometimes necessary to consider functions that are not defined on both sides of c. For example, the expression $\lim_{x \to 0} \sqrt{x}$ has not been assigned a meaning since $\sqrt{x}$ is not defined for all values of x in an open interval that contains 0. In this case, we would only consider values of x that satisfy $0 < x < \delta$. As a second example, define a function g by

$$g(x) = \begin{cases} x^2, & \text{if } x < 2; \\ x + 2, & \text{if } x \ge 2. \end{cases}$$

To check that the function g has a limit at 2, we need to consider the cases $x > 2$ and $x < 2$ separately since g is defined differently in these two intervals. In general, rather than insisting that $|f(x) - L| < \epsilon$ for all values of x that satisfy $0 < |x - c| < \delta$, we can focus only on values of x in the interval $(c - \delta, c)$ or in the interval $(c, c + \delta)$. Situations such as these lead to the notion of **one-sided limits**.

DEFINITION 3.6 Let I be an open interval that either contains the point c or has c as an endpoint and suppose that f is a function that is defined on I except possibly at the point c.

a) If $c \in I$ or if c is a left endpoint of I, then the function f has **right-hand limit** L at c if for each $\epsilon > 0$ there exists $\delta > 0$ such that $|f(x) - L| < \epsilon$ for all $x \in (c, c + \delta)$. We will write $\lim_{x \to c^+} f(x) = L$ when f has a right-hand limit of L at c and, on occasion, use the notation $L = f(c+)$.

b) If $c \in I$ or if c is a right endpoint of I, then the function f has **left-hand limit** L at c if for each $\epsilon > 0$ there exists $\delta > 0$ such that $|f(x) - L| < \epsilon$ for all $x \in (c - \delta, c)$. We will write $\lim_{x \to c^-} f(x) = L$ when f has a left-hand limit of L at c and, on occasion, use the notation $L = f(c-)$.

In this context, the usual limit is called a **two-sided limit**. It is certainly possible for both one-sided limits to exist even if the two-sided limit does not exist; a graphical example is shown in Figure 3.2. However, a simple relationship between one-sided

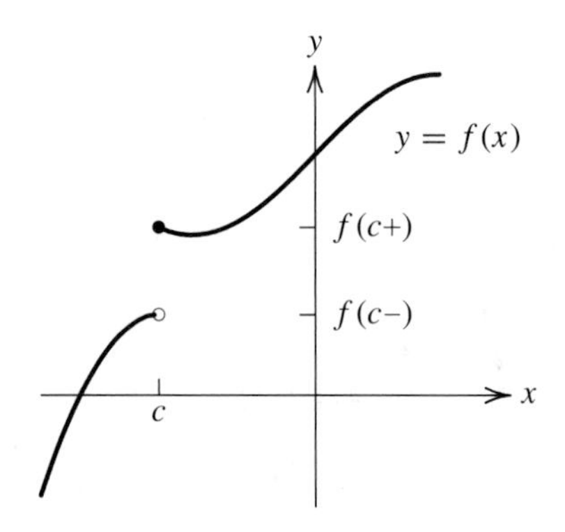

Figure 3.2 One-sided limits exist at c but a two-sided limit does not exist at c

limits and two-sided limits is recorded in the next theorem; the proof of this result will be left as an exercise.

THEOREM 3.7 Let I be an open interval that contains the point c and suppose that f is a function that is defined on I except possibly at the point c. The function f has a limit at c if and only if both one-sided limits exist at c and are equal. ■

As a simple illustration of this theorem, consider the functions f and g defined by

$$f(x) = \begin{cases} x^2, & \text{if } x < 2; \\ 6, & \text{if } x = 2; \\ x+2, & \text{if } x > 2; \end{cases} \quad \text{and} \quad g(x) = \begin{cases} x^2, & \text{if } x < 2; \\ 5, & \text{if } x = 2; \\ x+8, & \text{if } x > 2. \end{cases}$$

For the function f,

$$\lim_{x\to 2^-} f(x) = \lim_{x\to 2^-} x^2 = 4 \quad \text{and} \quad \lim_{x\to 2^+} f(x) = \lim_{x\to 2^+} (x+2) = 4.$$

It follows that $\lim_{x\to 2} f(x)$ exists and has a value of 4. Since $g(2-) = 4$ and $g(2+) = 10$, the function g does not have a limit at 2. Notice that these examples illustrate once again that the value of the function at c does not play any role in the computation of the limit at c.

10/3

Suppose that f is a function defined on a closed interval $[a, b]$. The statement, "f has a limit at each point of $[a, b]$", means that f has a two-sided limit at each point of (a, b), a right-hand limit at a, and a left-hand limit at b. The statement, "f has one-sided limits at each point of $[a, b]$", means that f has both a left-hand limit and a right-hand limit at each point of (a, b), a right-hand limit at a, and a left-hand limit at b. It is important to remember that there are functions that do not have one-sided limits; the function $\sin(1/x)$, which does not have one-sided limits at 0, is one of many examples (see the exercises).

There are other variations on the limit concept; two of these are included in the following definition. The reader is requested to supply definitions for other variations in the exercises.

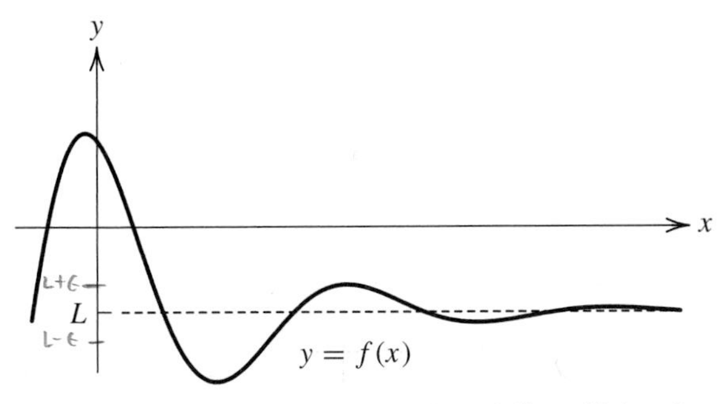

Figure 3.3 A graphical illustration of $\lim_{x\to\infty} f(x) = L$

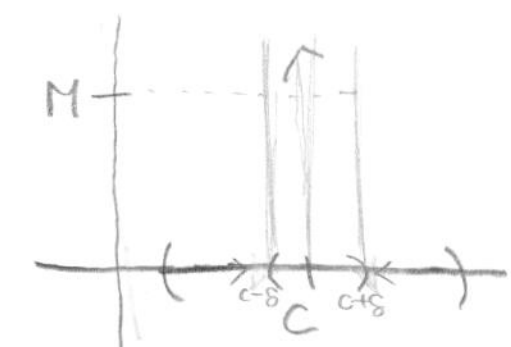

DEFINITION 3.8

a) Let a be a real number and suppose that f is a function that is defined on the interval (a, ∞). The function f has limit L as x tends to ∞ if for each $\epsilon > 0$ there exists $b > a$ such that $|f(x) - L| < \epsilon$ for all $x \geq b$. In this case, we will write $\lim_{x\to\infty} f(x) = L$.

b) Let I be an open interval that contains the point c and suppose that f is a function that is defined on I except possibly at the point c. We will write $\lim_{x\to c} f(x) = \infty$ if for each $M > 0$ there exists $\delta > 0$ such that $f(x) > M$ for all $x \in I$ that satisfy $0 < |x - c| < \delta$.

The limit concepts presented in Definition 3.8 have simple graphical interpretations. If $\lim_{x\to\infty} f(x) = L$, then the graph of f resembles the horizontal line $y = L$ for large values of x (see Figure 3.3). If $\lim_{x\to c} f(x) = \infty$, then the graph of f "blows up" at c, which means that the graph goes off the scale vertically for values of x near c. The graph of the function f defined by $f(x) = 1/x^2$ has this property at 0. As a final comment, note that $\lim_{x\to 0^-}(1/x) = -\infty$ and $\lim_{x\to 0^+}(1/x) = \infty$. (Definitions for these types of limits are similar to part (b) of Definition 3.8; the explicit statements of these will be left to the reader.) It follows that $\lim_{x\to 0}(1/x)$ does not exist even in the sense of part (b) of Definition 3.8.

Since ∞ is not a number, a function f does not have a limit at a point c when $\lim_{x\to c} f(x) = \infty$. The notation does indicate that the output values of the function f are very large for values of $x \neq c$ in an interval containing c. As an example, we will prove that

$$\lim_{x\to 2^-} \frac{x}{\sqrt{4-x^2}} = \infty.$$

Note that for x in the interval $(1, 2)$,

$$\frac{x}{\sqrt{4-x^2}} = \frac{x}{\sqrt{2+x}} \cdot \frac{1}{\sqrt{2-x}} \geq \frac{1}{2} \cdot \frac{1}{\sqrt{2-x}}.$$

Let $M > 1$ and choose $\delta = 1/(2M)^2$. If $2 - \delta < x < 2$, then $x \in (1, 2)$ and

$$\frac{x}{\sqrt{4 - x^2}} \geq \frac{1}{2} \cdot \frac{1}{\sqrt{2 - x}} > \frac{1}{2} \cdot \frac{1}{\sqrt{\delta}} = M.$$

This shows that $\lim_{x \to 2^-} x/\sqrt{4 - x^2} = \infty$.

Many of the limit theorems listed earlier in this section for two-sided limits remain valid for one-sided limits and for other variations of the limit concept. We will not state these results here, but will use them later in the text if the need arises. Exercise 39 at the end of this section explores some of these results.

Exercises

1. As a point of logic, what happens in the definition of limit if the condition that δ is positive is dropped?

2. Prove that the limit of a function is unique.

3. Let f be a function defined by $f(x) = 0$ for $x \leq 0$ and $f(x) = 10^{-9}$ for $x > 0$ and consider the claim that $\lim_{x \to 0} f(x) = 0$. For what values of $\epsilon > 0$ is it possible to find a $\delta > 0$ that satisfies the definition of limit? Does this function have a limit at 0?

4. Use the definition of limit to prove each of the following.

a) $\lim_{x \to 2} (5x - 11) = -1$ **b)** $\lim_{x \to 1} (x^2 + x - 1) = 1$ **c)** $\lim_{x \to -2} (x - 3x^2) = -14$

d) $\lim_{x \to 4} \sqrt{x} = 2$ **e)** $\lim_{x \to -2} x^3 = -8$ **f)** $\lim_{x \to 1} (4/(3x + 2)) = 4/5$

5. Use the definition of limit to prove that $\lim_{x \to c} x^2 = c^2$ for every real number c.

6. Use the definition of limit to prove that $\lim_{x \to 0} (x/|x|) \neq -1$.

7. Finish the proof of part (a) of Theorem 3.2.

8. Prove part (b) of Theorem 3.2.

9. Use Theorem 3.2 to prove that $\lim_{x \to 2} \dfrac{2x + 3}{x^2 + 5x} = \dfrac{1}{2}$.

10. Prove that $\lim_{x \to 0} \sin(1/x)$ does not exist.

11. Let I be an open interval that contains the point c and let f be a function that is defined on I except possibly at the point c.

a) Suppose that $\lim_{x \to c} f(x)$ exists. Prove that $\lim_{x \to c} |f(x)|$ exists.

b) Suppose that $\lim_{x \to c} |f(x)|$ exists. Give an example to show that $\lim_{x \to c} f(x)$ may not exist.

c) Suppose that $\lim_{x \to c} |f(x)| = 0$. Prove that $\lim_{x \to c} f(x) = 0$.

12. Prove that the function h defined after Theorem 3.2 has a limit at 0.

13. Show that the function f defined by $f(x) = 0$ if x is irrational and $f(x) = 1$ if x is rational does not have a limit at any point.

14. Use Theorem 3.2 to finish the proof of Theorem 3.3.

15. Use the definition of limit to prove the kf and $f - g$ results in Theorem 3.3.

16. Use the definition of limit to prove the result in Theorem 3.3 for fg. Begin with the equality

$$f(x)g(x) - ST = \big(S + (f(x) - S)\big)\big(T + (g(x) - T)\big) - ST,$$

where S and T are the limits defined in the proof of Theorem 3.3.

17. This exercise outlines a proof that the limit of a quotient is the quotient of the limits using the definition of the limit of a function. We will use the notation of both the statement and proof of Theorem 3.3.

a) Explain why there is an open interval J containing c such that f/g is defined on J except possibly at the point c.

b) Prove that there exists $\delta_1 > 0$ such that $\left|\dfrac{1}{g(x)}\right| < \dfrac{2}{|T|}$ for all $x \in I$ that satisfy $0 < |x - c| < \delta_1$.

c) Prove that $\displaystyle\lim_{x \to c} \frac{1}{g(x)} = \frac{1}{T}$.

d) Use part (c) and the result for products to complete the proof.

18. Let I be an open interval that contains the point c and let f and g be functions that are defined on I except possibly at the point c. Suppose that $\lim_{x \to c} f(x)$ exists and that $\lim_{x \to c} g(x)$ does not exist. Prove that $\lim_{x \to c} (f(x) + g(x))$ does not exist.

19. Let I be an open interval that contains the point c and let f be a function that is defined on I except possibly at the point c. Suppose that $f(x) > 0$ for all $x \in I \setminus \{c\}$, that $\lim_{x \to c} f(x)$ exists, and that

$$\lim_{x \to c} f(x) = \lim_{x \to c} \big((f(x))^2 + 5f(x) - 3\big).$$

Find the value of $\lim_{x \to c} f(x)$. Explain each step.

20. Complete the following set of results.

a) Use the definition of limit to prove that $\lim_{x \to c} k = k$ for any constant k and any real number c.

b) Use the definition of limit to prove that $\lim_{x \to c} x = c$ for any real number c.

c) Prove that $\lim_{x \to c} x^n = c^n$ for any positive integer n and for any real number c.

d) Let P be a polynomial and let c be a real number. Prove that $\lim_{x \to c} P(x) = P(c)$.

21. Prove Theorem 3.4.

22. Use the definition of limit to prove Theorem 3.5.

23. Use Theorem 3.5 to prove that $\lim_{x \to 0} x^2 \cos(1/x) = 0$. In addition, give a proof of this result without using Theorem 3.5.

24. Find, with proof, the value of $\lim_{x \to 0^+} \sqrt{x} \sin(1/x)$.

25. Prove Theorem 3.7.

26. Prove that the function $\sin(1/x)$ does not have one-sided limits at 0.

27. Give an example of a bounded function that does not have a limit at any point.

28. Give an example of a function that has a limit at each integer but does not have a limit at any other real number.

29. Suppose that f has a one-sided limit at each point of $[a, b]$. Prove that f is bounded on $[a, b]$.

30. Consider the function f defined by $f(x) = x - \lfloor x \rfloor$. Find the one-sided limits of this function at each integer.

31. Find all of the points at which the function f defined by $f(x) = 1/x - \lfloor 1/x \rfloor$ does not have a two-sided limit. Does the function have one-sided limits at these points?

32. Consider the function f defined by

$$f(x) = \begin{cases} -x^2, & \text{if } x < 0; \\ 1, & \text{if } x = 0; \\ 2, & \text{if } 0 < x < 2; \\ x + 3, & \text{if } x \geq 2. \end{cases}$$

Find $f(-1+)$, $f(0-)$, $f(0+)$, $f(2-)$, $f(2+)$, and $f(3-)$.

33. Find constants a and b so that the function f defined by

$$f(x) = \begin{cases} 3ax^2 + 1, & \text{if } 0 \leq x \leq 1; \\ ax + b, & \text{if } 1 < x < 2; \\ 2bx + a, & \text{if } 2 \leq x \leq 4; \end{cases}$$

has a limit at each point of $[0, 4]$.

34. Provide definitions for each of the following. Give an example, both numerical and graphical, for each limit.

a) $\lim_{x \to c^+} f(x) = -\infty$ **b)** $\lim_{x \to c^-} f(x) = \infty$ **c)** $\lim_{x \to \infty} f(x) = \infty$

d) $\lim_{x \to -\infty} f(x) = \infty$ **e)** $\lim_{x \to -\infty} f(x) = L$

35. Find, with proof, the value of each limit.

a) $\lim_{x \to 2^+} \dfrac{1}{x^2 - 2x}$ **b)** $\lim_{x \to \infty} \dfrac{x^2 - 1}{5x - 2x^2}$ **c)** $\lim_{x \to \infty} \dfrac{\sin x}{x}$

36. Let $f\colon [1, \infty) \to \mathbb{R}$ and define $g\colon (0, 1] \to \mathbb{R}$ by $g(x) = f(1/x)$. Prove that $\lim_{x \to \infty} f(x) = L$ if and only if $\lim_{x \to 0^+} g(x) = L$.

37. Let a be a real number and let f, g be functions defined on $[a, \infty)$. Suppose that $\lim_{x \to \infty} f(x) = L$ and $\lim_{x \to \infty} g(x) = \infty$. Prove that $\lim_{x \to \infty} f(g(x)) = L$. Give an example of functions f and g that illustrate this fact.

38. Let a be a real number. Evaluate $\lim_{x \to \infty} \left(\sqrt{x^2 + ax} - x \right)$.

39. For each result, state and prove a similar result that is valid for (i) a one-sided limit, (ii) a limit as $x \to \infty$, and (iii) a limit of the form $\lim_{x \to c} f(x) = \infty$.

a) Theorem 3.2.

b) The limit of a sum is the sum of the limits.

c) The limit of a product is the product of the limits.

d) Theorem 3.5.

40. Let P be a polynomial of degree at least 1 and assume that the leading coefficient of P is positive.

a) Prove that $\lim_{x \to \infty} P(x) = \infty$.

b) Prove that $\lim_{x \to -\infty} P(x) = \infty$ if the degree of P is even.

c) Prove that $\lim_{x \to -\infty} P(x) = -\infty$ if the degree of P is odd.

41. Let P and Q be polynomials and consider the limit of $P(x)/Q(x)$ as $x \to \infty$. State and prove a conjecture for the value of this limit that depends on the degrees of P and Q and their leading coefficients.

3.2 Continuous Functions

What does it mean for a function f to be continuous? The most common informal definition of this concept states that a function f is continuous if you can sketch its graph without raising your pencil. In other words, the graph has no breaks or gaps in it. This is a global view of continuity—look at the graph of f as a whole and see if there are any breaks in it. However, if a break does occur in the graph, then this break will occur at some point. It is thus possible to interpret continuity as a local phenomenon by saying that a function is continuous if it is continuous at each point. It should be clear that this visual representation of continuity is not suitable as a mathematical definition. To drive home this point, remember that a graph, even a computer-generated graph, is drawn by plotting some points and sketching a curve through those points. A sketch created in this fashion assumes in advance that the function is continuous.

Although the notion of a graph without breaks cannot be used as a definition of a continuous function, it does provide the intuition for a mathematical definition of continuity. In order for the graph of a function f to not have a break at a point c, inputs near c must generate outputs near $f(c)$. This statement is made precise in the following definition.

DEFINITION 3.9 Let I be an interval, let $f: I \to \mathbb{R}$, and let $c \in I$. The function f is **continuous** at c if for each $\epsilon > 0$ there exists $\delta > 0$ such that $|f(x) - f(c)| < \epsilon$ for all $x \in I$ that satisfy $|x - c| < \delta$. The function f is continuous on I if f is continuous at each point of I.

The definition of continuity at a point c is quite similar to the definition of a limit at c, but there are three important differences. First, the inequality $0 < |x - c| < \delta$ has been replaced by $|x - c| < \delta$, which means that the point c is now an allowed input for the function. Second, the number L has been replaced by $f(c)$—that is, the value of the limit is $f(c)$. Thus, continuity brings the point c and the value of f at c back into play. Continuous functions are predictable in the sense that inputs near c generate outputs near $f(c)$. Finally, it is no longer assumed that the interval I is open. If c happens to be an endpoint of I, then only values of x on one side of c are considered. In this case, the definition of continuity resembles the definition of one-sided limits.

Definition 3.9 considers a function f defined on an interval I. The definition can still be applied even if the domain of f is larger than the set I, but the values of f for $x \notin I$ are ignored. This may sound awkward, but an example should clarify any confusion. The greatest integer function $\lfloor x \rfloor$ is defined for all real numbers and it should be clear that $\lfloor x \rfloor$ is not continuous at any integer if we let $I = \mathbb{R}$ in Definition 3.9. However, all of the following statements are valid for the function f whose formula is $f(x) = \lfloor x \rfloor$:

1. The function $f : [0, 1) \to \mathbb{R}$ is continuous at 0.

2. The function $f : (-1, 1) \to \mathbb{R}$ is not continuous at 0.
3. The function $f : [0, 1) \to \mathbb{R}$ is continuous on $[0, 1)$.
4. The function $f : [0, 1] \to \mathbb{R}$ is not continuous on $[0, 1]$.

The bottom line here is that the domain of the function is important, so we leave the reader with the warning to pay attention to the domain of a given function.

The observation that the definition of continuity at c and the definition of the existence of a limit at c are almost identical leads to the following theorem. It provides some useful equivalent ways to view continuity.

THEOREM 3.10 Let I be an interval, let $f: I \to \mathbb{R}$, and let $c \in I$. The following are equivalent:

1. The function f is continuous at c.
2. The function f has a limit at c and $\lim_{x \to c} f(x) = f(c)$. (If c is an endpoint of I, then an appropriate one-sided limit is used here.)
3. The sequence $\{f(x_n)\}$ converges to $f(c)$ for each sequence $\{x_n\}$ in I that converges to c.

Proof. The fact that statements (1) and (2) are equivalent follows from the definitions of the two concepts. The equivalence of statements (2) and (3) is essentially Theorem 3.2. The details will be left as an exercise. ■

Continuity is preserved by the usual algebraic operations. This should come as no surprise since continuity is equivalent to the existence of a certain limit and limits obey the usual rules of algebra. The results for continuity at a point are given in the next theorem. The corollary to the theorem simply extends the results from continuity at a point to continuity on an interval.

THEOREM 3.11 Let f and g be functions defined on an interval I, let $c \in I$, and let k be a constant. If f and g are continuous at c, then the functions $f+g$, $f-g$, kf, and fg are continuous at c. If $g(c) \neq 0$, then f/g is continuous at c.

Proof. We will prove the result for quotients only; the other parts are just as easy. Since f and g are continuous at c, it follows that $\lim_{x \to c} f(x) = f(c)$ and $\lim_{x \to c} g(x) = g(c)$. (If c is an endpoint of I, then an appropriate one-sided limit must be used.) Since $g(c) \neq 0$, Theorem 3.3 yields

$$\lim_{x \to c} \left(\frac{f(x)}{g(x)} \right) = \frac{\lim_{x \to c} f(x)}{\lim_{x \to c} g(x)} = \frac{f(c)}{g(c)}.$$

Hence, the function f/g is continuous at c. ■

COROLLARY 3.12 Let f and g be continuous functions defined on an interval I and let k be a constant. Then the functions $f+g$, $f-g$, kf, and fg are continuous on I. If $g(x) \neq 0$ for all $x \in I$, then f/g is continuous on I. ■

In addition to the algebraic operations, the composition of two functions is another way to combine functions. The following result indicates that the composition of two continuous functions is a continuous function. As discussed in Appendix B, if g is defined on an interval I, then $g(I)$ represents the set $\{g(x) : x \in I\}$.

THEOREM 3.13 Let I be an interval, let $g: I \to \mathbb{R}$, let $c \in I$, and let f be a function that is defined on an interval J that contains $g(I)$. If g is continuous at c and f is continuous at $g(c)$, then $f \circ g$ is continuous at c. Hence, if g is continuous on I and f is continuous on J, then $f \circ g$ is continuous on I.

Proof. The hypotheses guarantee that the function $f \circ g$ is defined on the interval I. Let $\{x_n\}$ be a sequence in I that converges to c. Since g is continuous at c, the sequence $\{g(x_n)\}$ converges to $g(c)$. Since f is continuous at $g(c)$, the sequence $\{f(g(x_n))\}$ converges to $f(g(c))$. By Theorem 3.10, the function $f \circ g$ is continuous at c. ■

The last two theorems guarantee that any function that is an algebraic combination and/or a composition of continuous functions is a continuous function. In order to apply these results, we first need a collection of continuous functions. Most familiar functions are continuous at almost all of the points in their domain; some of these functions are listed in the next theorem.

THEOREM 3.14

a) A polynomial is continuous at every real number.

b) A rational function is continuous at every real number for which it is defined.

c) The six trigonometric functions are continuous at every real number for which they are defined.

Proof. It is trivial to verify that constant functions are continuous at every real number and almost as easy to show that the function $f(x) = x$ is continuous at every real number. These two facts, when combined with Theorem 3.11 and mathematical induction, show that a polynomial is continuous at every real number. The details will be left as an exercise. The result for rational functions then follows from the quotient property of continuous functions. A proof that the six trigonometric functions are continuous at every real number for which they are defined is outlined in the exercises. ■

According to Theorem 3.10, a function f is continuous at c if and only if $\lim_{x \to c} f(x) = f(c)$. This formulation of continuity can be divided into three parts; the function must be defined at c, the function must have a limit at c, and these two values must be the same. A function can always be assigned a value at c so the first condition can be satisfied even if the formula for the function is not defined at c. There are several ways for a function not to have a limit at c; it can be unbounded, it can oscillate, or it can have different one-sided limits. For any of these situations, the function cannot be continuous at c. In a calculus course, it rarely happens that the limit of a function f at c exists and has a value different from $f(c)$, but there

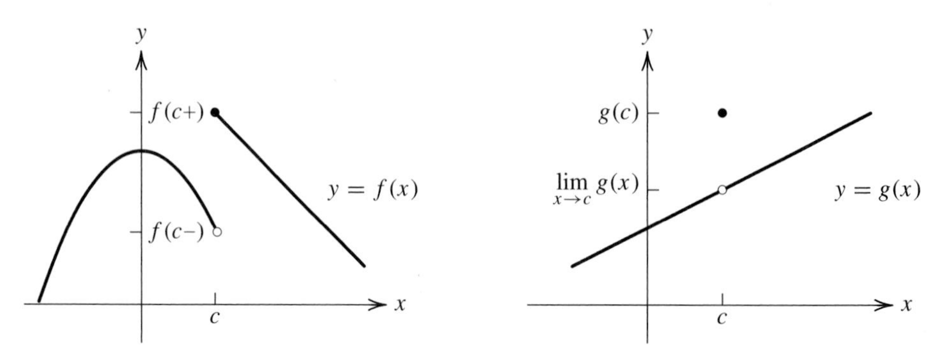

Figure 3.4 A jump discontinuity at c (left) and a removable discontinuity at c

are functions with this property. The function f defined by $f(x) = \lfloor x \rfloor + \lfloor -x \rfloor$ has a limit at every integer n, but this limit is not $f(n)$ (see the exercises). Of the many ways in which a function may fail to be continuous at a point, two are given a special name.

DEFINITION 3.15 Let I be an open interval, let $c \in I$, and let f be a function defined on I except possibly at the point c.

a) The function f has a **jump discontinuity** at c if the one-sided limits of f exist at c but are not equal.

b) The function f has a **removable discontinuity** at c if $\lim_{x \to c} f(x)$ exists, but $f(c)$ is either not defined or has a value different from the limit.

These two types of discontinuities are illustrated in Figure 3.4. A function f has a jump discontinuity at c when both $\lim_{x \to c^-} f(x)$ and $\lim_{x \to c^+} f(x)$ exist but are not equal. The name is indicative of what the graph looks like at the point of discontinuity since the graph "jumps" from $f(c-)$ to $f(c+)$ as x increases through c. As a specific example, the function g defined by $g(x) = x/|x|$ has a jump discontinuity at 0 since the left-hand limit is -1 and the right-hand limit is 1. A removable discontinuity can be removed by giving the function the value of the limit at the given point. The function h defined by $h(x) = x \sin(1/x)$ has a removable discontinuity at 0; assigning $h(0) = 0$ makes h a continuous function at 0.

For the record, there are discontinuities that are neither removable discontinuities nor jump discontinuities. An example is the discontinuity of the function $\sin(1/x)$ at 0. This function does not have one-sided limits at 0 since its graph oscillates wildly near the origin. As x decreases from 1 to 0, the term $1/x$ increases from 1 to ∞. In essence, the graph of $\sin(1/x)$ on $(0, 1)$ is the graph of $\sin x$ on $(1, \infty)$ compressed against the y-axis. Since the graph goes up and down an infinite number of times on the interval $(0, 1)$, it is not possible to sketch all of this graph. A computer screen will give either a black blob near the origin or offer a jagged graph due to a lack of resolution. In Figure 3.5, a small portion of the graph of $y = \sin(1/x)$ is shown. The oscillations from -1 to 1 continue to become more frequent as x approaches 0. Of course, this does not represent the entire graph, but

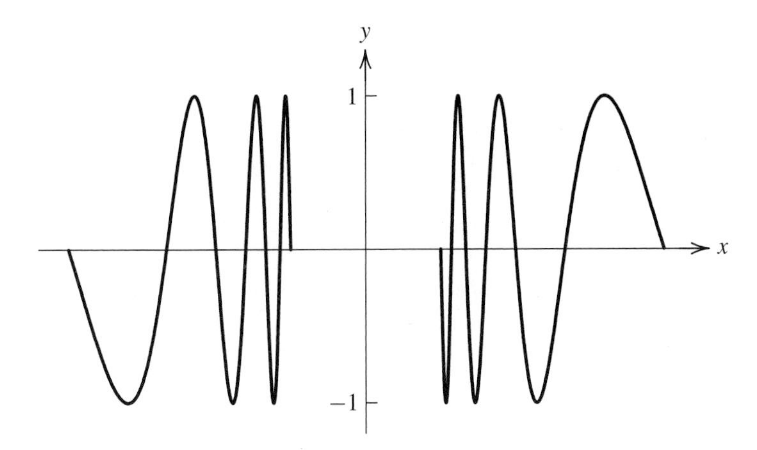

Figure 3.5 A portion of the graph of $y = \sin(1/x)$

it should give a good idea of the behavior of $\sin(1/x)$ near 0. This function and functions similar to it are important to consider when making general statements about the properties of functions. They offer an antidote for simplistic thinking and are the basis for a number of counterexamples that occur later in the text.

Exercises

1. Use the definition of continuity to prove that the given function is continuous at every real number.
 a) $f(x) = 6x - 11$ **b)** $g(x) = |x|$ **c)** $h(x) = x^2$

2. Let a and b be real numbers with $a \neq 0$. Use the definition of continuity to prove that the function f defined by $f(x) = ax + b$ is continuous at every real number.

3. Use the definition of continuity to prove that the function f defined by $f(x) = \sqrt{x}$ is continuous at every nonnegative real number.

4. Let $f: [a, b] \to \mathbb{R}$ be continuous at $c \in [a, b]$ and suppose that $f(c) > 0$. Prove that there exist a positive number m and an interval $[u, v] \subseteq [a, b]$ such that $c \in [u, v]$ and $f(x) \geq m$ for all $x \in [u, v]$.

5. Write out the details for a proof of Theorem 3.10.

6. Let f be defined by $f(x) = x - \lfloor x \rfloor$. Find a closed interval I such that $2 \in I$ and $f: I \to \mathbb{R}$ is continuous on I.

7. Let $f: [a, b] \to \mathbb{R}$ be continuous on $[a, b]$ and suppose that $f(x) = 0$ for each rational number x in $[a, b]$. Prove that $f(x) = 0$ for all x in $[a, b]$.

8. Prove that the function f defined by $f(x) = x$ if x is rational and $f(x) = -x$ if x is irrational is continuous at 0 only.

9. This exercise concerns the proof of Theorem 3.11.

 a) Use the definition of continuity to prove that the sum of two continuous functions is continuous.

 b) Prove that the product of two continuous functions is continuous.

10. Use the definition of continuity to prove Theorem 3.13.

11. Suppose that f is a continuous function defined on an interval I.

a) Prove that $|f|$ is continuous on I.

b) Suppose that f is nonnegative on I. Prove that $\sqrt{f}$ is continuous on I.

12. Let f and g be continuous functions defined on an interval I. Prove that $f \vee g$ and $f \wedge g$ are continuous on I. (See Exercise 29 in Section 1.5 for the definitions of these functions.)

13. Fill in the details for the proof that every polynomial is continuous on $\mathbb{R}$.

14. Let f be defined by $f(x) = x/(x^2 - 20)$. Find the largest interval I such that $4 \in I$ and $f: I \to \mathbb{R}$ is continuous on I.

15. The purpose of this exercise is to verify that $|\sin a - \sin b| \leq |a - b|$ for all real numbers a and b.

a) Explain geometrically why $\sin x \leq x$ for all $x \in [0, \pi/2]$. (This inequality can be proved more rigorously using the power series for $\sin x$.)

b) Use the properties of $\sin x$ to show that $|\sin x| \leq |x|$ for all real numbers x.

c) Use the identities for $\sin(x + y)$ and $\sin(x - y)$ to obtain the identity

$$\sin(x + y) - \sin(x - y) = 2 \sin y \cos x,$$

which is valid for all real numbers x and y.

d) Find values for x and y so that $x + y = a$ and $x - y = b$, then use the information in parts (b) and (c) to prove the assertion.

16. Use the inequality $|\sin a - \sin b| \leq |a - b|$, which is valid for all real numbers a and b, to prove that $\sin x$ is continuous on $\mathbb{R}$.

17. Show how the continuity of the other five trigonometric functions at each point for which they are defined follows from the continuity of the sine function.

18. List all of the results that are needed to show that $f(x) = (x + x^2)/(2 + \sin^3(6x))$ is continuous on $\mathbb{R}$.

19. Explain why the function $g(x) = \sin(1/x)$ is continuous for all nonzero real numbers.

20. Determine all of the points at which the function $h(x) = x/(1 + 2 \sin x)$ is not continuous.

21. Determine all of the points at which the function $f(x) = \cot(1/x)$ is not continuous.

22. Let f be the function defined by $f(x) = \lfloor x \rfloor + \lfloor -x \rfloor$. Prove that f has a limit at every integer n but that the value of the limit is not $f(n)$.

23. The functions f and g defined by

$$f(x) = \frac{x^3 - 3x^2 + 2}{x^2 - 1} \quad \text{and} \quad g(x) = \frac{x^2 + 4x - 5}{x^3 - 2x^2 + x}$$

are each undefined at two points. Determine whether or not the discontinuities at those points are removable.

24. Show that the function s defined by $s(x) = 1/x^2$ has neither a removable discontinuity nor a jump discontinuity at 0.

25. Give an example of a function $f: \mathbb{R} \to \mathbb{R}$ with the given property.

a) The function f has a removable discontinuity at 4.

b) The function f has a jump discontinuity at π.

c) The function f has a discontinuity at -3 that is neither a removable discontinuity nor a jump discontinuity.

d) The function f is bounded and has a discontinuity at 5 that is neither a removable discontinuity nor a jump discontinuity.

26. Let I be an interval, let c be an endpoint of I, and let f be a function defined on I except possibly at the point c. Define the concept of a removable discontinuity and a jump discontinuity for this situation. Give examples to illustrate your definitions.

27. Find and classify all of the discontinuities of f, where $f(x) = \lfloor x \rfloor + \lfloor -2x \rfloor$.

28. Give an example of a bounded function that is defined on $\mathbb{R}$, continuous at each point except the points 1 and -1, and has neither a removable discontinuity nor a jump discontinuity at 1 and -1.

29. Define functions f and g on $[-1, 1]$ by

$$f(x) = \begin{cases} x\cos(1/x), & \text{if } x \neq 0; \\ 0, & \text{if } x = 0; \end{cases} \quad \text{and} \quad g(x) = \begin{cases} \cos(1/x), & \text{if } x \neq 0; \\ 0, & \text{if } x = 0. \end{cases}$$

Prove that f is continuous at 0 and that g is not continuous at 0. Explain why these functions are continuous at every other point in $[-1, 1]$.

30. Give an example of a function $f\colon \mathbb{R} \to \mathbb{R}$ with the given property.

a) The function f is continuous everywhere except at the points 1, 2, and 3.

b) The function f is continuous everywhere except for the set of positive integers.

c) The function f is continuous everywhere except for the set $\{1/n : n \in \mathbb{Z}^+\}$.

d) The function f is continuous at 1 and 2 only.

31. Let $\{r_n\}$ be a listing of all the rational numbers and let $\{v_n\}$ be a sequence of nonzero numbers that converges to 0. Define a function f by $f(x) = 0$ if x is irrational and $f(r_n) = v_n$ for all n. Show that f is continuous everywhere except for the set of rational numbers.

3.3 Intermediate and Extreme Values

Let f be a continuous function defined on an interval $[a, b]$. Using the idea that the graph of a continuous function has no breaks or gaps, it is clear that the graph of f has a highest point and a lowest point. It is also clear that the graph intersects every horizontal line between the highest point and the lowest point. These two results concerning continuous functions are known as the Extreme Value Theorem and the Intermediate Value Theorem, respectively. They are usually stated without proof in calculus books since the proofs involve the Completeness Axiom. In this section, we will prove these two important theorems and discuss some of their implications.

As the name indicates, the Intermediate Value Theorem states that the range of a continuous function includes all of its intermediate values. To be explicit, if u and v are in the range of a continuous function f and $u < v$, then each real number y that satisfies $u < y < v$ is also in the range of f. In particular, if a continuous function is positive at one point and negative at another point, then the function equals zero somewhere in between those two points. The proof of the Intermediate Value Theorem given here uses the following fact: if S is a bounded set, then there exists a sequence $\{x_n\}$ in S that converges to $\sup S$. A proof of this fact will be left as an exercise.

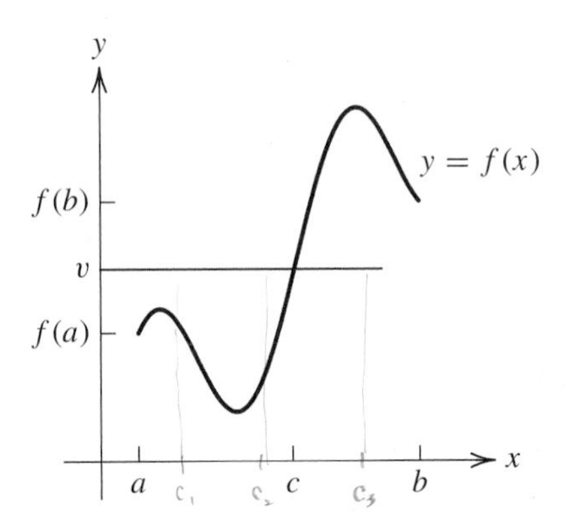

Figure 3.6 A geometric interpretation of the Intermediate Value Theorem

THEOREM 3.16 Intermediate Value Theorem Suppose that $f: [a, b] \to \mathbb{R}$ is continuous on $[a, b]$. If v is a number between $f(a)$ and $f(b)$, then there is a point $c \in (a, b)$ such that $f(c) = v$.

Proof. We will assume that $f(a) < v < f(b)$; a proof for the other case is similar. Let $S = \{x \in [a, b] : f(x) < v\}$. Since the set S is nonempty (it contains a) and bounded above (b is an upper bound), the Completeness Axiom asserts that S has a supremum. The point $c = \sup S$ satisfies $f(c) = v$. To verify this, let $\{c_n\}$ be a sequence in S such that $c_n \to c$. Since f is continuous at c, the sequence $\{f(c_n)\}$ converges to $f(c)$, and since $f(c_n) < v$ for all n, it follows that $f(c) \le v$. Now let $d_n = c + (b - c)/n$ for each positive integer n and note that $\{d_n\}$ is a sequence in $[a, b] \setminus S$ that converges to c. Since $f(d_n) \ge v$ for all n and $\{f(d_n)\}$ converges to $f(c)$, the inequality $f(c) \ge v$ must hold. We conclude that $f(c) = v$. ■

The Intermediate Value Theorem has a simple geometric interpretation. The graph of $y = v$ is a horizontal line that has a y-intercept that lies between $f(a)$ and $f(b)$. As indicated in Figure 3.6, the graph $y = f(x)$ of a continuous function that starts at the point $(a, f(a))$ and ends at the point $(b, f(b))$ must cross this horizontal line at some point c between a and b. The informal definition of a continuous function as a function whose graph can be sketched without raising the pencil is helpful here, but the Completeness Axiom is required to guarantee that the curve actually intersects the line at some point. The use of the Completeness Axiom is evident in the proof of the Intermediate Value Theorem, but the proof offers no practical method for finding the point c. Another approach to the proof of this theorem, which has the advantage of providing an algorithm for finding an approximation to the point c, is discussed in the exercises.

As a simple illustration of this theorem, consider the function f defined by $f(x) = x^3 - 4x - 3$. This function is continuous on the interval $[2, 3]$ with $f(2) = -3$ and $f(3) = 12$. By the Intermediate Value Theorem, there exists a point $c \in (2, 3)$ such that $f(c) = 0$. There also exists a point $d \in (2, 3)$ such that $f(d) = 5$. In general, for each real number v in the interval $(-3, 12)$, the equation $f(x) = v$ has a solution x in the interval $(2, 3)$.

As a second example, we will prove that the equation $4\sin x = x$ has a positive solution. Define a function g by $g(x) = 4\sin x - x$ for each real number x and note that g is continuous on $\mathbb{R}$. A positive solution to the equation $4\sin x = x$ is a positive number c such that $g(c) = 0$. Evaluating g at various points reveals that $g(\pi/2) > 0$ and $g(\pi) < 0$. By the Intermediate Value Theorem, there exists a number $c \in (\pi/2, \pi)$ such that $g(c) = 0$. Hence, the equation $4\sin x = x$ has a positive solution.

As indicated by the second example, many of the applications of the Intermediate Value Theorem use 0 as the intermediate value. The reason for this is a practical one because, in this case, it is only necessary to find a value where the function is positive and another value where the function is negative. The actual values of the function are not that important; only their signs are.

In addition to including all intermediate values, the range of a continuous function also includes its extreme values, at least when its domain is restricted to a closed and bounded interval. For this reason, this result is known as the Extreme Value Theorem. It states that a continuous function defined on a closed and bounded interval has a maximum output and a minimum output. Hence, the Extreme Value Theorem actually makes two assertions about a continuous function f defined on an interval $[a, b]$; namely, the range of f is a bounded set and the range of f includes both its infimum and its supremum. In other words, there is some output of the function f that is greater than or equal to all other outputs of the function, and there is some output of the function f that is less than or equal to all other outputs of the function (see Definition 1.33). As with the Intermediate Value Theorem, the proof of the Extreme Value Theorem depends on the Completeness Axiom; it appears in the form of the Bolzano-Weierstrass Theorem.

THEOREM 3.17 Extreme Value Theorem If $f : [a, b] \to \mathbb{R}$ is continuous on $[a, b]$, then there exist points c and d in $[a, b]$ such that $f(c) \le f(x) \le f(d)$ for all $x \in [a, b]$.

Proof. We will first prove that the continuous function f is bounded on $[a, b]$ by proving the contrapositive: if f is unbounded on $[a, b]$, then f is not continuous on $[a, b]$. Suppose that f is unbounded on $[a, b]$. Then for each positive integer n, there exists $x_n \in [a, b]$ such that $|f(x_n)| > n$. By the Bolzano-Weierstrass Theorem, the sequence $\{x_n\}$ contains a subsequence $\{x_{p_n}\}$ that converges to a point $z \in [a, b]$. Since $|f(x_{p_n})| > p_n$ for all n, the sequence $\{f(x_{p_n})\}$ is unbounded and cannot converge. By Theorem 3.10, the function f is not continuous at z.

Since f is bounded on the interval $[a, b]$, the set $f([a, b])$ is bounded. Let $\alpha = \inf(f([a, b]))$ and let $\beta = \sup(f([a, b]))$, and note that $\alpha \le f(x) \le \beta$ for all $x \in [a, b]$. We will prove that there exists a point $d \in [a, b]$ such that $f(d) = \beta$; a proof that there exists a point $c \in [a, b]$ such that $f(c) = \alpha$ is similar. For each positive integer n, there exists $d_n \in [a, b]$ such that

$$\beta - \frac{1}{n} < f(d_n) \le \beta.$$

By the Bolzano-Weierstrass Theorem, the sequence $\{d_n\}$ contains a subsequence $\{d_{q_n}\}$ that converges to a point $d \in [a, b]$. Since f is continuous at the point d, $f(d) = \lim_{n\to\infty} f(d_{q_n}) = \beta$. This completes the proof. ∎

The Extreme Value Theorem has three hypotheses; the function must be continuous, the interval must be closed, and the interval must be bounded. If any one of these hypotheses is not satisfied, then the conclusion of the Extreme Value Theorem may not be valid. The following three examples illustrate this point. In each case, two of the three hypotheses are satisfied.

1. The function f defined by $f(x) = 1/x$ for $x \neq 0$ and $f(0) = 0$ does not have a maximum value on the interval $[0, 1]$.
2. The function g defined by $g(x) = x^2$ does not have a maximum value on the interval $[-1, 2)$.
3. The function h defined by $h(x) = x^3$ does not have a maximum value on the interval $[0, \infty)$.

The next theorem states that continuous functions map intervals onto intervals. More precisely, if a continuous and nonconstant function is defined on some interval, then its range is an interval. If the domain happens to be a closed and bounded interval, then the Extreme Value Theorem shows that the range is also a closed and bounded interval. The details of the proofs of these results will be left as exercises. (The reader may wish to consult the definition of an interval found in Section 1.2).

THEOREM 3.18 Let I be any interval and let $f: I \to \mathbb{R}$.

a) If f is continuous and nonconstant on I, then $f(I)$ is an interval.

b) If f is continuous and nonconstant on I and if I is a closed and bounded interval, then $f(I)$ is a closed and bounded interval. ∎

A cautionary note concerning the statement of part (a) of this theorem should be made: the intervals I and $f(I)$ are not necessarily of the same type. For example, if $g(x) = \sin x$ and I is the open interval $(-\pi, \pi)$, then $g(I)$ is the closed interval $[-1, 1]$. As another example, if $h(x) = 1/x$ and $I = (0, 1)$, then $h(I) = (1, \infty)$. In this case, the interval I is bounded, but the interval $h(I)$ is unbounded. Part (b) gives one condition that guarantees that the intervals comprising the domain and range are of the same type.

The converse of part (a) of Theorem 3.18 is false; the range of a function on an interval can be an interval even if the function is not continuous on the interval. One simple example of this phenomenon is the function ϕ defined by $\phi(x) = x - \lfloor x \rfloor$ on the interval $[0, 2]$. However, if f is monotone on I and its range is an interval, then f is continuous on I. This represents a partial converse to part (a) of Theorem 3.18 and is the content of the next two results. Note that if f is increasing on an open interval I and $c \in I$, then

$$\sup\{f(x) : x \in I \text{ and } x < c\} \leq f(c) \leq \inf\{f(x) : x \in I \text{ and } x > c\}.$$

This fact will be used in the proofs of these results.

LEMMA 3.19 Let f be an increasing function defined on an open interval I and let $c \in I$. If $\sup\{f(x) : x \in I \text{ and } x < c\} = \inf\{f(x) : x \in I \text{ and } x > c\}$, then f is continuous at c.

Proof. Suppose that the given infimum and supremum are equal and note that, by the remark preceding the lemma, the common value must be $f(c)$. Let $\epsilon > 0$. Since $f(c) + \epsilon$ is not a lower bound of the set $\{f(x) : x \in I \text{ and } x > c\}$, there exists a point $v \in I$ such that $v > c$ and $f(v) < f(c) + \epsilon$. Since $f(c) - \epsilon$ is not an upper bound of the set $\{f(x) : x \in I \text{ and } x < c\}$, there exists a point $u \in I$ such that $u < c$ and $f(u) > f(c) - \epsilon$. If $u < x < v$, then the fact that f is increasing yields

$$f(c) - \epsilon < f(u) \leq f(x) \leq f(v) < f(c) + \epsilon.$$

Consequently, if $|x - c| < \min\{c - u, v - c\}$, then $|f(x) - f(c)| < \epsilon$, so the function f is continuous at c. This completes the proof. ∎

THEOREM 3.20 Let I be an interval and let $f: I \to \mathbb{R}$ be a monotone function. If $f(I)$ is an interval, then f is continuous on I.

Proof. Without loss of generality, we may assume that the interval I is open and that f is increasing on I (see the exercises). The proof will be a proof by contraposition. Suppose that f is not continuous on I and let $c \in I$ be a point at which f is not continuous. If

$$\alpha = \sup\{f(x) : x \in I \text{ and } x < c\} \quad \text{and} \quad \beta = \inf\{f(x) : x \in I \text{ and } x > c\},$$

then $\alpha < \beta$ by the previous lemma. Let u and v be points in I such that $u < c < v$ and let $y \in (\alpha, \beta) \setminus \{f(c)\}$. Then y is between $f(u)$ and $f(v)$, but y is not in the range of f. This shows that $f(I)$ is not an interval. ∎

Suppose that f is a continuous, one-to-one function defined on an interval I. Then f has an inverse f_{inv} (inverse functions are discussed in Appendix B) and it is natural to ask if f_{inv} is a continuous function. Since the graph of f_{inv} is simply the reflection of the graph of f across the line $y = x$, the graph of f_{inv} will have no breaks if the graph of f has no breaks. This observation does not constitute a proof that f_{inv} is continuous, but it does indicate that the inverse of a continuous function should be continuous.

THEOREM 3.21 Let I be an interval, let f be a continuous, strictly monotone function defined on I, and let the range of f be the interval J. Then the function f_{inv} is a continuous, strictly monotone function defined on J.

Proof. It is sufficient to consider the case in which the function f is strictly increasing on I (see the exercises). Suppose that u and v are two points in J such that $u < v$. Since f is strictly increasing on I, we must have $f_{\text{inv}}(u) < f_{\text{inv}}(v)$; this proves that f_{inv} is strictly increasing on J. By the properties of inverse functions, the range of f_{inv} is the interval I. By Theorem 3.20, the function f_{inv} is continuous on J. This completes the proof. ∎

The fact that the inverse of a continuous function is continuous makes it possible to increase the list of functions that we have proved to be continuous (see Theorem

3.14). Some familiar functions that fit into this category are included in the next theorem; the proof will be left as an exercise.

THEOREM 3.22

a) Let r be a rational number. The function f defined by $f(x) = x^r$ is continuous at every real number for which it is defined.

b) The six inverse trigonometric functions are continuous at every real number for which they are defined. ■

Using the algebraic properties of continuous functions and the short list of specific functions known to be continuous, it is possible to prove that complicated functions are continuous with very little effort. For example, the function g defined by

$$g(x) = \frac{5x \sin(4x)}{\sqrt[3]{x^4 + 3x^2 + 1}}$$

is continuous on $\mathbb{R}$. As should be immediately apparent, it would be extremely difficult to prove that this function is continuous directly from the definition.

The remainder of this section considers two concepts that are related to intermediate and extreme values: the intermediate value property and the notion of a locally bounded function. We will define these concepts, give some examples to illustrate them, and prove two theorems that involve these concepts. The exercises at the end of this section contain further properties and observations.

DEFINITION 3.23 A function f defined on an interval I has the **intermediate value property** on I if it satisfies the following condition: if a and b are distinct points in I and v is any number between $f(a)$ and $f(b)$, then there exists a point c between a and b such that $f(c) = v$.

A function with the intermediate value property is sometimes referred to as a **Darboux function** in honor of G. Darboux (1842–1917), but we will not use this term. To illustrate the definition and emphasize the importance of considering every subinterval, consider the function $f\colon [0, 2] \to \mathbb{R}$ defined by

$$f(x) = \begin{cases} x, & \text{if } 0 \le x < 1; \\ 4 - 2x, & \text{if } 1 \le x \le 2. \end{cases}$$

Although the range of f is an interval and every horizontal line between $y = 0$ and $y = 2$ intersects the graph of f, this function does not have the intermediate value property on $[0, 2]$. For example, with $a = 0.5$, $b = 1$, and $v = 1.5$, there is no number c between a and b such that $f(c) = v$ (see Figure 3.7).

If f is continuous on an interval I, then (by the Intermediate Value Theorem) f has the intermediate value property on I. The converse of this statement is false. Let g be the function defined by $g(x) = \sin(1/x)$ for $x \neq 0$ and $g(0) = 1$. This function is not continuous at 0, but it does have the intermediate value property on $[0, 1]$. To verify the latter statement, only intervals of the form $[0, b]$ must be checked since g is continuous on $(0, 1]$. Suppose that $b \in (0, 1]$ for which $\sin(1/b) \neq 1$ and let v be any number between $\sin(1/b)$ and 1. Since $v \in (-1, 1)$,

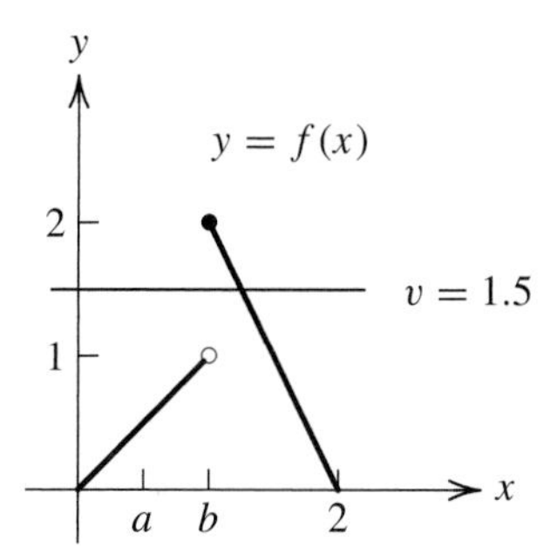

Figure 3.7 A function that does not have the intermediate value property on $[0, 2]$

there exists a positive number d such that $\sin d = v$. Choose a positive integer n such that $(d + 2n\pi)^{-1} < b$ and let $c = (d + 2n\pi)^{-1}$. Then $c \in (0, b)$ and $g(c) = v$. Hence, the function g has the intermediate value property on $[0, 1]$ even though it is not continuous on $[0, 1]$. In fact, there exist functions with the intermediate value property that are not continuous at any point; an example of such a function can be found in Gordon [8].

The following theorem gives one condition that is sufficient to guarantee that a function with the intermediate value property on an interval I is continuous on that interval. For the record, note that the function $\sin(1/x)$ assumes each value in its range a countably infinite number of times in the interval $(0, 1)$.

THEOREM 3.24 Suppose that f has the intermediate value property on an interval I. If the set $\{x \in I : f(x) = v\}$ is finite for each real number v, then f is continuous on I.

Proof. Let $c \in I$ and suppose that c is not an endpoint of I; the case in which c is an endpoint of I is a little easier. Let $\epsilon > 0$. Since the set

$$\{x \in I : f(x) = f(c) - \epsilon \text{ or } f(x) = f(c) + \epsilon\}$$

is finite, there exists $\delta > 0$ such that $(c - \delta, c + \delta) \subseteq I$ and $f(x)$ does not equal either $f(c) - \epsilon$ or $f(c) + \epsilon$ for all $x \in (c - \delta, c + \delta)$. Suppose that $|x - c| < \delta$. To show that $|f(x) - f(c)| < \epsilon$, we will consider one of four possible cases. Assume that $x > c$ and $f(x) < f(c) - \epsilon$. By the intermediate value property of f, there exists a point $z \in (c, x)$ such that $f(z) = f(c) - \epsilon$, a contradiction to the choice of δ. The other three cases also lead to contradictions and it follows that f is continuous at c. This completes the proof. ■

Unlike continuity, the intermediate value property is not preserved by all of the usual algebraic operations. For instance, the sum of two functions with the intermediate value property does not necessarily have the intermediate value property. The exercises request proofs for the results that are valid and provide counterexamples for those that are not. Interestingly enough, the derivative of a continuous function, which need not be continuous, has the intermediate value property. This fact will be proved in the next chapter.

The Extreme Value Theorem states that a continuous function defined on a closed and bounded interval is bounded. Rather than focusing on the function as a whole, it is possible to consider the behavior of the function near a particular point. This leads to the concept of a locally bounded function.

DEFINITION 3.25 Let f be a function defined on an interval I and let $c \in I$. The function f is **locally bounded** at c if there exist positive numbers M and δ such that $|f(x)| \leq M$ for all $x \in I$ that satisfy $|x - c| < \delta$.

The function f defined by $f(x) = \sin(1/x)$ for $x \neq 0$ and $f(0) = 0$ is locally bounded at 0, but this is no surprise since the function itself is bounded. For a more interesting example, the function g defined by $g(x) = 1/x$ is locally bounded at 0.002; let $\delta = 0.001$ and $M = 1000$. In fact, this function is locally bounded at each point of $(0, 1)$ (see the exercises). The main idea is that the graph does not go off the scale in some small open interval that contains the point. However, it does not follow that g is bounded on $(0, 1)$; each number $c \in (0, 1)$ generates a local bound M_c, but no single value of M can serve as a bound for g on $(0, 1)$. This example is worth pondering a bit; it illustrates that local properties of functions do not always extend to global properties.

As the next theorem indicates, a function that is continuous at a point c is locally bounded at that point. Notice the similarities between the following proof and the proof that a convergent sequence is bounded.

THEOREM 3.26 Let f be a function defined on an interval I and let $c \in I$. If f is continuous at c, then f is locally bounded at c.

Proof. Corresponding to $\epsilon = 1$, there exists $\delta > 0$ such that $|f(x) - f(c)| < 1$ for all $x \in I$ that satisfy $|x - c| < \delta$. For these same values of x,

$$|f(x)| \leq |f(x) - f(c)| + |f(c)| < 1 + |f(c)|.$$

This shows that f is locally bounded at c. ■

Some further examples of functions that are not locally bounded as well as some other properties of locally bounded functions can be found in the exercises.

Exercises

1. Let S be a bounded set. Prove that there is a sequence $\{x_n\}$ in S that converges to $\sup S$.
2. As mentioned in the text, the proof of the Intermediate Value Theorem does not give a practical method for finding the point c. This exercise discusses a technique known as the **bisection method** for approximating the number c. Let f be a continuous function defined on an interval $[a, b]$ and let v be any number between $f(a)$ and $f(b)$.
   - **a)** Explain why it is sufficient to consider the case in which $f(a)$ and $f(b)$ have opposite signs and $v = 0$.
   - **b)** Let m be the midpoint of $[a, b]$. Unless $f(m) = 0$ (which is unlikely), show that either $f(a)f(m) < 0$ or $f(m)f(b) < 0$. Thus, the length of the interval in which the point c lies has been shortened by a factor of 2.

c) By continuing the process started in part (b), show that there exists a nested sequence $\{[a_n, b_n]\}$ of closed intervals such that $f(a_n)f(b_n) < 0$ for all n and

$$b_{n+1} - a_{n+1} = \frac{b_n - a_n}{2}$$

for all n. This is where the term "bisection" comes into play; each interval is bisected and one half of it is used to form the next interval. For notational convenience, it helps to let $[a, b] = [a_0, b_0]$. (By the way, if $f(a_n)f(b_n) = 0$ for some n then we are extremely lucky and the problem is solved.)

d) Let c be the unique point in all of the intervals $[a_n, b_n]$ (see the Nested Intervals Theorem). Prove that $f(c) = 0$.

e) Find a formula for $b_n - a_n$ in terms of a, b, and n, then use it to find an overestimate for the quantity $\left|(a_n + b_n)/2 - c\right|$. This represents an estimate for the accuracy of the approximation $(a_n + b_n)/2$ to the number c.

3. Consider the function $f(x) = x^3 + 2x - 1$ on the interval $[0, 1]$ and let $v = 0$. Use the bisection method and the notation of the previous exercise to find $[a_4, b_4]$.

4. Consider the function f defined by $f(x) = (x^2 - 2)(x^2 - 3)(x^2 - 5)$. Which root of f is "found" by the bisection method if the first interval is $[0, 3]$? How about $[1, 3]$? Find an interval containing all three positive roots so that the bisection method locates the root $\sqrt{3}$.

5. Prove that the polynomial $x^6 + x^4 - 5x^2 + 1$ has at least four real roots.

6. Prove that the equation $x^2 = \sin x$ has a positive solution.

7. Let P be a polynomial of odd degree.

a) Prove that P has at least one real root.

b) Prove that the range of P is all real numbers.

8. Prove that there exists a real number $x > 2\pi$ such that $\tan x = x$.

9. Suppose that $f\colon [a, b] \to [a, b]$ is continuous. (Note that the range of f is a subset of $[a, b]$.) Prove that there exists at least one point $x \in [a, b]$ such that $f(x) = x$. A point with this property is known as a **fixed point** of f.

10. Find the maximum and minimum values of the function $f(x) = 2|2x - 1| - |3x - 1|$ on the interval $[0, 1]$.

11. Does the function $f(x) = x^2 - 2x + 3$ have either a maximum value or a minimum value on the interval $(0, 3)$?

12. Consider the function g defined by $g(x) = (\sin x)/\sqrt{x}$. Prove that g has a maximum value and a minimum value on $(0, \pi]$.

13. Give an example of a function $f\colon (0, 1) \to \mathbb{R}$ such that f is not continuous on $(0, 1)$, but f satisfies the conclusion of the Extreme Value Theorem.

14. The proof of the Extreme Value Theorem has two parts: the first part shows that f is bounded on $[a, b]$ and the second part shows that the supremum and infimum of the range are actually in the range. Here is an alternate way to prove the second part once the first part has been proved. Suppose that $\alpha = \inf(f([a, b]))$ is not in the range of f. Define a function $g\colon [a, b] \to \mathbb{R}$ by $g(x) = 1/(f(x) - \alpha)$. Prove that g is unbounded on $[a, b]$ and obtain a contradiction.

15. Prove Theorem 3.18.

16. Suppose that f is a nonconstant continuous function defined on an interval I.

a) Prove that $f(I)$ contains both rational and irrational numbers.

b) Prove that $f(I)$ is an uncountable set.

17. Let $I = (0, 1)$ and let Q be a polynomial of degree 2.

a) Find a specific polynomial Q so that $Q(I)$ is an open interval.

b) Find a specific polynomial Q so that $Q(I)$ is a half-open interval.

c) Is there a polynomial Q so that $Q(I)$ is a closed interval?

18. Give an example of a continuous increasing function $f: (0, 1) \to \mathbb{R}$ such that $f((0, 1))$ is a closed interval.

19. Let f be a continuous function defined on an interval I and let $J \subseteq I$ be an open interval. As we have seen, the set $f(J)$ is an interval but not necessarily an open interval. Is there a further condition that can be placed on f that will guarantee that $f(J)$ is an open interval whenever J is an open interval? If so, find such a condition.

20. Give an example of a continuous function $f: (0, 1) \to \mathbb{R}$ whose range is all real numbers.

21. State and prove the analogue of Lemma 3.19 for decreasing functions.

22. Let f be an increasing function defined on an open interval I and let $c \in I$. Suppose that f is continuous at c. Prove that the quantities

$$\sup\{f(x) : x \in I \text{ and } x < c\} \quad \text{and} \quad \inf\{f(x) : x \in I \text{ and } x > c\}$$

both equal $f(c)$. Note that this is the converse of Lemma 3.19.

23. This exercise concerns the "without loss of generality" statement made at the beginning of the proof of Theorem 3.20. Suppose that $f: [a, b] \to \mathbb{R}$ is increasing on $[a, b]$.

a) Prove that there exists an increasing function $g: \mathbb{R} \to \mathbb{R}$ such that $g(x) = f(x)$ for all $x \in [a, b]$.

b) Suppose that f is strictly increasing on $[a, b]$. Prove that there exists a strictly increasing function $g: \mathbb{R} \to \mathbb{R}$ such that $g(x) = f(x)$ for all $x \in [a, b]$.

c) Show how to extend these results for the cases in which the domain of f is some type of half-open interval. (Note that it may be possible to extend the function in only one direction!)

d) Explain the "without loss of generality" statement made at the beginning of the proof of Theorem 3.20.

24. Explain why it is sufficient to just consider the case in which the function is strictly increasing for the proof of Theorem 3.21.

25. Prove part (a) of Theorem 3.22. Consider first the case in which $r = 1/q$ for some positive integer q.

26. Prove part (b) of Theorem 3.22.

27. Restrict the domain of the function $f(x) = 3x^2 - 6x - 1$ to the largest interval I on which f is strictly decreasing, then find the inverse of f on I.

28. List all of the results that are necessary to prove that the function h defined by $h(x) = \sqrt[5]{x^3 + x^2 \sin(3x)}/(1 + x^2)$ is continuous on $\mathbb{R}$.

29. Suppose that f has the intermediate value property on an interval I and let $J \subseteq I$ be an interval on which f is not constant. Prove that $f(J)$ is an interval.

30. Prove that the function f defined by $f(x) = x - \lfloor x \rfloor$ does not have the intermediate value property on $[0, 3]$.

31. Let f be defined by $f(x) = \sin(1/x)$ for $x \neq 0$ and $f(0) = d$. For what values of d does f have the intermediate value property on $[0, 1]$?

32. Let f be a function defined on $[a, b]$ and let $c \in (a, b)$. Suppose that f has the intermediate value property on the intervals $[a, c]$ and $[c, b]$. Prove that f has the intermediate value property on $[a, b]$.

33. Let f be defined by $f(x) = \sin(1/x)$ for $x \neq 0$ and $f(0) = 0$. Prove that f has the intermediate value property on $[-1, 1]$.

34. Suppose that f has the intermediate value property on an interval I and let k be a constant. Prove that kf has the intermediate value property on I.

35. Suppose that f has the intermediate value property on an interval J, that g has the intermediate value property on an interval I, and that $g(I) \subseteq J$. Prove that $f \circ g$ has the intermediate value property on I.

36. Define functions f and g on $\mathbb{R}$ by

$$f(x) = \begin{cases} \sin(1/x), & \text{if } x \neq 0; \\ 1, & \text{if } x = 0; \end{cases} \quad \text{and} \quad g(x) = \begin{cases} \sin(1/x), & \text{if } x \neq 0; \\ -1, & \text{if } x = 0. \end{cases}$$

Show that f and g have the intermediate value property on $[0, 1]$, but that $f - g$ and fg do not have the intermediate value property on $[0, 1]$.

37. A continuous function maps closed and bounded intervals into closed and bounded intervals (see part (b) of Theorem 3.18). Use the function f defined by $f(0) = 0$ and $f(x) = (1 - x)\sin(1/x)$ for $x \neq 0$ to show that this result is not valid for functions with the intermediate value property.

38. List the other three cases that occur in the proof of Theorem 3.24 and give a proof for one of them.

39. Give an example of a continuous function $f: [0, 1] \to \mathbb{R}$ for which the set $\{x \in [0, 1] : f(x) = v\}$ is infinite for at least one value of v.

40. Let f be defined on an interval I and suppose that f is one-to-one on I.

a) Give an example to show that f may not be monotone on I.

b) Give an example to show that f may not be monotone on any subinterval of I.

c) Suppose that f is continuous on I. Prove that f is monotone on I.

d) Suppose that f has the intermediate value property on I. Prove that f is monotone on I.

41. Consider the function f defined by $f(x) = 0$ if x is irrational and $f(x) = (-1)^p q$ if $x = p/q$, where p and q are integers with no common factors and $q > 0$. Prove that f is not locally bounded at any point of $(0, 1)$.

42. For each positive integer k, let

$$f_k(x) = \begin{cases} 1/(1 - kx), & \text{if } 0 \leq x < 1/k; \\ 0, & \text{if } x \geq 1/k; \end{cases}$$

and let f be the function defined by $f(x) = \sum_{k=1}^{\infty} f_k(x)$ for each $x \in [0, 1]$. Note that for each value of x, there are only a finite number of terms in this sum that are nonzero. Find, with proof, the set of points at which f is not locally bounded.

43. Let f be a function defined on an open interval I and let $c \in I$. Suppose that f has one-sided limits at c. Prove that f is locally bounded at c.

44. Let f be a function defined on an interval $[a, b]$ and suppose that f is locally bounded at each point of $[a, b]$. Prove that f is bounded on $[a, b]$ in each of the following ways.

a) Use the Bolzano-Weierstrass Theorem in a way similar to its use in the proof of the Extreme Value Theorem.

b) Let $S = \{x \in [a, b] : f \text{ is bounded on } [a, x]\}$. Prove that S is nonempty, that $\sup S \in S$, and that $\sup S = b$.

c) Suppose that f is not bounded on $[a, b]$. Using the bisection technique, show that there exists a nested sequence $\{[a_n, b_n]\}$ of intervals such that f is not bounded on $[a_n, b_n]$ for each n. Use the Nested Intervals Theorem to obtain a contradiction.

45. This exercise extends slightly the concept of a locally bounded function and looks at some of its implications.

a) Show that it is possible to extend Definition 3.25 to the case in which f is defined on I except at the point c. Write out the definition in this case.

b) Let I be an open interval, let $c \in I$, and let f be a function that is defined on I except at the point c. Suppose that f has a limit at c. Prove that f is locally bounded at c.

c) Let I be an open interval, let $c \in I$, and let f and g be functions that are defined on I except at the point c. Suppose that f and g have limits of S and T, respectively, at c. Use the inequality

$$|f(x)g(x) - ST| \leq |f(x)|\,|g(x) - T| + |T|\,|f(x) - S|$$

and part (b) to prove that fg has a limit at c.

3.4 UNIFORM CONTINUITY

Suppose that a function f is continuous on an interval I. This means that f is continuous at each point of I. By definition, for each $c \in I$ and $\epsilon > 0$, there exists $\delta(\epsilon, c) > 0$ such that $|f(x) - f(c)| < \epsilon$ for all $x \in I$ that satisfy $|x - c| < \delta(\epsilon, c)$. As indicated by the additional functional notation, the number δ depends, in general, on both ϵ and c. For a fixed ϵ, the required δ may vary from point to point in the interval. If the graph of the function f is steep near c, then δ will need to be smaller than it would need to be at a point where the graph is more level. The following two examples illustrate this idea.

1. Consider the function $f(x) = x^2$ and let $c \in \mathbb{R}$. Let $\epsilon > 0$ and choose $\delta = \min\{1, \epsilon/(2|c| + 1)\}$. If $|x - c| < \delta$, then

$$|x + c| \leq |x - c| + |2c| < 1 + 2|c|$$

and it follows that

$$\left|x^2 - c^2\right| = |x + c|\,|x - c| \leq (2|c| + 1)|x - c| < (2|c| + 1)\delta \leq \epsilon.$$

In this case, the value of δ is a function of ϵ and c. For a given $\epsilon > 0$, as the magnitude of c increases, the graph of f becomes steeper and the value of $\delta(\epsilon, c)$ decreases (see Figure 3.8).

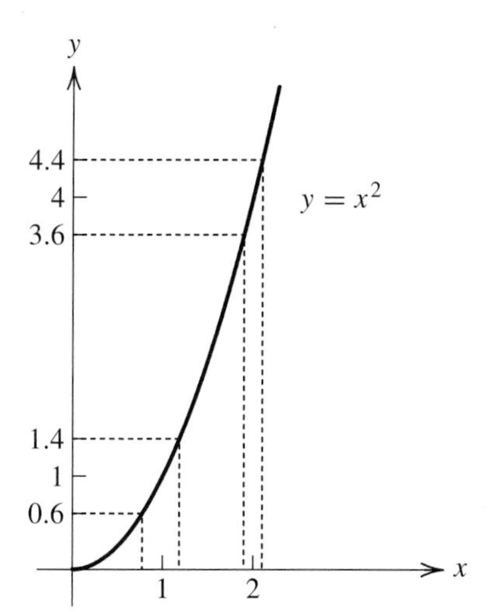

Figure 3.8 For a given $\epsilon > 0$, the value of δ decreases as c increases

2. Consider the function $g(x) = 4x - 7$ and let $c \in \mathbb{R}$. Let $\epsilon > 0$ and choose $\delta = \epsilon/4$. If $|x - c| < \delta$, then

$$|(4x - 7) - (4c - 7)| = 4|x - c| < 4\delta = \epsilon.$$

In this case, the value of δ depends only on ϵ. Since the graph of g has the same slope at every point, for a given $\epsilon > 0$, the same δ works at every point—the choice of δ is independent of c.

The situation in which δ depends only on ϵ and not on c is important enough to give a definition. Since one δ can be chosen that works for every number in the domain of the function, the continuity is said to be uniform.

DEFINITION 3.27 Let I be an interval. A function $f: I \to \mathbb{R}$ is **uniformly continuous** on I if for each $\epsilon > 0$ there exists $\delta > 0$ such that $|f(y) - f(x)| < \epsilon$ for all $x, y \in I$ that satisfy $|y - x| < \delta$.

It should be clear that a function is continuous on I whenever it is uniformly continuous on I. A function f is continuous at a point c if inputs near c generate outputs near $f(c)$ and continuous on I if f is continuous at each point $c \in I$. However, the meaning of the phrase "inputs near c" may vary from point to point in the interval. A function f is uniformly continuous on I if close inputs generate close outputs—the measure of "closeness" being the same throughout the interval.

As seen in Example (2), the function $g(x) = 4x - 7$ is uniformly continuous on $\mathbb{R}$. The function $\sin x$ is uniformly continuous on $\mathbb{R}$ as well since

$$|\sin y - \sin x| \leq |y - x|$$

for all $x, y \in \mathbb{R}$ (see Exercise 15 in Section 3.2). As another example, we will prove that the function $h(x) = 1/x$ is uniformly continuous on the interval $(0.1, 2)$. Let

$\epsilon > 0$ and choose $\delta = \epsilon/100$. If $|y - x| < \delta$ and $x, y \in (0.1, 2)$, then

$$|h(y) - h(x)| = \left|\frac{1}{y} - \frac{1}{x}\right| = \frac{|y - x|}{|xy|} < 100|y - x| < 100\delta = \epsilon.$$

The first inequality follows by substituting 0.1 for x and y; this is a positive number that is less than both x and y. Since the choice of δ is independent of x in the interval $(0.1, 2)$, the function h is uniformly continuous on this interval.

To prove that a function is not uniformly continuous on an interval requires some care. The first step is to negate the definition of uniform continuity. Let f be a continuous function on an interval I. Then f is not uniformly continuous on I if there exists a positive number ϵ such that for each $\delta > 0$ there exist points $x, y \in I$ such that $|y - x| < \delta$ and $|f(y) - f(x)| \geq \epsilon$. Example (1) indicates that x^2 may not be uniformly continuous on $\mathbb{R}$ but does not prove that this is the case. This is an important point. The example only shows that our method of choosing δ depends on both ϵ and c. However, there may be another way to choose δ that avoids dependence on c. To show that x^2 is not uniformly continuous on $\mathbb{R}$ requires a proof that for some $\epsilon > 0$ there is no appropriate choice for δ. Some experimentation yields the following proof. For each $\delta > 0$, let $x = 1/\delta$ and $y = \delta + 1/\delta$. Then $|y - x| \leq \delta$, but

$$|y^2 - x^2| = 2 + \delta^2 > 2.$$

This shows that x^2 is not uniformly continuous on $\mathbb{R}$. (Note that, as with the definition of a limit, it makes no real difference in this case if the inequalities are strict or not.)

As we have just seen, there are continuous functions that are not uniformly continuous on certain intervals. There are several conditions that guarantee that a continuous function is uniformly continuous; some of these will be scattered in the exercises. However, the most important result along these lines is contained in the following theorem, which states that a continuous function is uniformly continuous on a closed and bounded interval.

THEOREM 3.28 If $f\colon [a, b] \to \mathbb{R}$ is continuous on $[a, b]$, then f is uniformly continuous on $[a, b]$.

Proof. This will be a proof by contradiction. Suppose that f is continuous on $[a, b]$ but not uniformly continuous on $[a, b]$. Then there exists $\epsilon > 0$ such that for each $\delta > 0$ there exist points $x, y \in [a, b]$ such that $|y - x| < \delta$ and $|f(y) - f(x)| \geq \epsilon$. In particular, for each positive integer n there exist points x_n, y_n in $[a, b]$ such that $|y_n - x_n| < 1/n$ and $|f(y_n) - f(x_n)| \geq \epsilon$. Since $\{x_n\}$ is a sequence in $[a, b]$, by the Bolzano-Weierstrass Theorem it contains a subsequence $\{x_{p_n}\}$ that converges to a point $z \in [a, b]$. As

$$y_{p_n} = x_{p_n} + (y_{p_n} - x_{p_n})$$

for all n and $\{y_{p_n} - x_{p_n}\}$ converges to 0, the sequence $\{y_{p_n}\}$ converges to z as well. Since f is continuous at z, the sequence $\{f(y_{p_n}) - f(x_{p_n})\}$ converges to $f(z) - f(z) = 0$. This contradicts the fact that $|f(y_{p_n}) - f(x_{p_n})| \geq \epsilon$ for all n. Hence, the function f is uniformly continuous on $[a, b]$. ■

In the previous chapter, it was shown that every Cauchy sequence of real numbers converges. Consequently, it is possible to prove that a sequence converges by proving that it is a Cauchy sequence; the advantage of this lies in the fact that it is not necessary to know the limit to show convergence. The next theorem is the analogue of this result for the limit of a function. The proof of this theorem uses the fact that a Cauchy sequence converges and is an example of a common argument encountered in real analysis.

THEOREM 3.29 Cauchy Criterion for Limits Let I be an interval that either contains the point c or has c as one of its endpoints and suppose that f is a function that is defined on I except possibly at the point c. Then the function f has a limit at c if and only if for each $\epsilon > 0$ there exists $\delta > 0$ such that $|f(y) - f(x)| < \epsilon$ for all $x, y \in \{t \in I : 0 < |t - c| < \delta\}$. (Note that the limit at c is a one-sided limit if c is an endpoint of I.)

Proof. Suppose first that for each $\epsilon > 0$ there exists a positive number δ such that $|f(y) - f(x)| < \epsilon$ for all x and y in the set $\{t \in I : 0 < |t - c| < \delta\}$. It follows that for each positive integer n, there exists $\delta_n > 0$ such that $|f(y) - f(x)| < 1/n$ for all $x, y \in \{t \in I : 0 < |t - c| < \delta_n\}$. Without loss of generality, we may assume that $\{\delta_n\}$ is a decreasing sequence that converges to 0. For each n, choose $x_n \in I$ such that $0 < |x_n - c| < \delta_n$. The next step is to prove that the sequence $\{f(x_n)\}$ is a Cauchy sequence.

Let $\epsilon > 0$ and choose a positive integer N such that $1/N < \epsilon$. Suppose that $m, n \geq N$. Then

$$0 < |x_n - c| < \delta_n \leq \delta_N \quad \text{and} \quad 0 < |x_m - c| < \delta_m \leq \delta_N.$$

In other words, the points x_n and x_m belong to the set $\{t \in I : 0 < |t - c| < \delta_N\}$. It follows that $|f(x_n) - f(x_m)| < 1/N < \epsilon$. Hence, the sequence $\{f(x_n)\}$ is a Cauchy sequence. We will prove that $\lim_{x \to c} f(x) = L$, where L is the limit of the sequence $\{f(x_n)\}$.

Once again, let $\epsilon > 0$. Since $\{f(x_n)\}$ converges to L, there exists a positive integer K such that $1/K < \epsilon/2$ and $|f(x_n) - L| < \epsilon/2$ for all $n \geq K$. Let $\delta = \delta_K$. If $x \in I$ and $0 < |x - c| < \delta$, then $x, x_K \in \{t \in I : 0 < |x - c| < \delta_K\}$ and

$$|f(x) - L| \leq |f(x) - f(x_K)| + |f(x_K) - L| < 1/K + \epsilon/2 < \epsilon.$$

This proves that $\lim_{x \to c} f(x) = L$.

The proof of the converse will be left as an exercise. ∎

Let f be a function defined on (a, b) and suppose that f is continuous on (a, b). It is a simple matter to give the function f values at a and b so that f is defined on a closed interval $[a, b]$. A more important question is whether or not it is possible to define f at a and b in such a way that f becomes continuous on $[a, b]$. For example, it is not possible to define the function $f(x) = 1/x$ at 0 and have the resulting function be continuous on the interval $[0, 1]$, but it is possible to define the function $g(x) = \sqrt{x}\sin(1/x)$ at 0 so that it is continuous on $[0, 1]$. The deciding factor in such situations is whether or not the appropriate limit exists; for the preceding functions $\lim_{x\to 0^+} f(x)$ does not exist while $\lim_{x\to 0^+} g(x)$ does exist. It turns out that the appropriate limit always exists for uniformly continuous functions. The proof of this fact illustrates the use of the Cauchy criterion for the limit of a function.

THEOREM 3.30 Let f be a continuous function defined on an open interval (a, b). Then f is uniformly continuous on (a, b) if and only if f has one-sided limits at a and b.

Proof. Suppose that f is uniformly continuous on (a, b). Given $\epsilon > 0$, there exists $\delta > 0$ such that $|f(y) - f(x)| < \epsilon$ for all $x, y \in (a, b)$ that satisfy $|y - x| < \delta$. In particular, $|f(y) - f(x)| < \epsilon$ for all x, y in the set $\{t \in (a, b) : 0 < |t - a| < \delta\}$. By Theorem 3.29, the function f has a right-hand limit at a. In the same way, the function f has a left-hand limit at b. The proof of the converse will be left as an exercise. ■

Exercises

1. Consider the function f defined by $f(x) = 1/x$ and let $c > 0$. Given $\epsilon > 0$, find a positive number $\delta(\epsilon, c)$ such that $|x - c| < \delta(\epsilon, c)$ implies $|f(x) - f(c)| < \epsilon$.
2. Use the definition of uniform continuity to prove each statement.
   a) The function $f(x) = 23x + 47$ is uniformly continuous on $\mathbb{R}$.
   b) The function $g(x) = x^2 + 2x - 5$ is uniformly continuous on $[0, 3]$.
   c) The function $h(x) = 4/x^2$ is uniformly continuous on $[1, 5)$.
   d) The function $F(x) = \sqrt{x}$ is uniformly continuous on $[0, \infty)$.
   e) The function $G(x) = x^3$ is not uniformly continuous on $[0, \infty)$.
   f) The function $H(x) = x \sin x$ is not uniformly continuous on $\mathbb{R}$.
3. Give an example of a function $f: \mathbb{R} \to \mathbb{R}$ that is continuous and bounded on $\mathbb{R}$, but not uniformly continuous on $\mathbb{R}$. Be certain to include a proof that the function is not uniformly continuous.
4. Let I be an interval, let $f: I \to \mathbb{R}$, and suppose that there exists a positive constant M such that $|f(y) - f(x)| \leq M|y - x|$ for all $x, y \in I$. Prove that f is uniformly continuous on I.
5. Let I be an interval and let $f: I \to \mathbb{R}$ be uniformly continuous on I. Suppose that $\{x_n\}$ is a Cauchy sequence in I. Prove that $\{f(x_n)\}$ is a Cauchy sequence.
6. Suppose that f and g are uniformly continuous on an interval I.
   a) Prove that $f + g$ is uniformly continuous on I.
   b) Give an example to show that fg may not be uniformly continuous on I.
   c) Find additional hypotheses on f and g that will guarantee that fg is uniformly continuous on I. Include a proof of your conjecture.
   d) Under what conditions is $1/f$ uniformly continuous on I?
7. Suppose that g is uniformly continuous on an interval I and that f is uniformly continuous on an interval J that contains $g(I)$. Prove that $f \circ g$ is uniformly continuous on I.
8. Suppose that f is uniformly continuous on (a, b). Prove that f is bounded on (a, b).

9. Prove that $f(x) = 1/x$ is uniformly continuous on each closed subinterval of $(0, 1)$ but is not uniformly continuous on $(0, 1)$.

10. A function $f: \mathbb{R} \to \mathbb{R}$ is **periodic** if there exists a positive number p such that $f(x + p) = f(x)$ for all real numbers x. Suppose that $f: \mathbb{R} \to \mathbb{R}$ is a continuous periodic function. Prove that f is uniformly continuous on $\mathbb{R}$.

11. Prove the converse of Theorem 3.29.

12. State and prove an analogue of Theorem 3.29 for a limit of the form $\lim_{x\to\infty} f(x)$.

13. Prove the converse of Theorem 3.30.

14. Suppose that f is uniformly continuous on $[a, b]$. Given a positive integer n, let $x_i = a + i(b - a)/n$ for $0 \le i \le n$, and define functions s_n and ℓ_n on $[a, b]$ by

$$s_n(x) = \begin{cases} f(a), & \text{if } x = a; \\ f(x_i), & \text{if } x_{i-1} < x \le x_i; \end{cases}$$

and

$$\ell_n(x) = \begin{cases} f(a), & \text{if } x = a; \\ \dfrac{f(x_i) - f(x_{i-1})}{x_i - x_{i-1}}(x - x_i) + f(x_i), & \text{if } x_{i-1} < x \le x_i. \end{cases}$$

Begin by drawing a graph to indicate how the graphs of these functions are related to the graph of f. Then prove that for each $\epsilon > 0$, there exists a positive integer p such that $|s_p(x) - f(x)| < \epsilon$ and $|\ell_p(x) - f(x)| < \epsilon$ for all $x \in [a, b]$.

15. A function $f: [a, b] \to [a, b]$ is said to be a **contraction** on $[a, b]$ if there exists a constant $\alpha \in (0, 1)$ such that $|f(y) - f(x)| \le \alpha|y - x|$ for all $x, y \in [a, b]$. Note that the outputs $f(x), f(y)$ are closer together than the inputs x, y; thus the name contraction. (For the record, the notation $f: [a, b] \to [a, b]$ simply means that the range of f is contained in $[a, b]$; it need not be all of $[a, b]$.)

a) Let f be a contraction on $[a, b]$. Prove that f is uniformly continuous on $[a, b]$.

b) Exercise 9 in the previous section shows that a contraction on $[a, b]$ must have a fixed point on $[a, b]$. Prove that the fixed point of a contraction is unique.

c) Suppose that f is a contraction on $[a, b]$. Let x_0 be any point in $[a, b]$ and define a sequence $\{x_n\}$ recursively by $x_n = f(x_{n-1})$ for all $n \ge 1$. Prove that $\{f(x_n)\}$ converges to the fixed point of f.

d) Prove that $f(x) = \sqrt{3x + 2}$ is a contraction on $[1, 5]$ and find its fixed point.

e) Prove that $g(x) = \sqrt[3]{2x + 1}$ is a contraction on $[0, 2]$.

f) Find an interval on which $h(x) = x^3 + x^2/6$ is a contraction.

g) Use the trigonometric identity

$$\cos(a - b) - \cos(a + b) = 2 \sin a \sin b$$

and some properties of the sine function to prove that $\cos x$ is a contraction on $[0, 1]$. What is the fixed point of this contraction?

3.5 Monotone Functions

A monotone function is a function that is either increasing or decreasing. Because the output values of monotone functions exhibit regular behavior, monotone functions possess a number of nice properties that a typical function does not have. For convenience, the properties of monotone functions that are related to the topics in this book are listed here even though not all of the ideas have been discussed to this point. Let f be a monotone function on an interval $[a, b]$.

1. The function f has one-sided limits at each point of $[a, b]$.
2. The set of discontinuities of f in $[a, b]$ is countable.
3. The function f is differentiable almost everywhere on $[a, b]$.
4. The function f is Riemann integrable on $[a, b]$.

The first two properties of a monotone function will be proved in this section and the last one will be proved in Chapter 5. The third property states that monotone functions are differentiable at most points; the concept of "almost everywhere" will be made precise in the supplementary exercises of Chapter 5. This result does not seem all that surprising until one becomes aware of the existence of continuous nowhere differentiable functions (see Theorem 7.31). After realizing that such functions exist, any result that guarantees that a function is differentiable at many points is a positive result. The proof that monotone functions are differentiable almost everywhere involves concepts from measure theory and can be found in Gordon [8] or in almost any other graduate level analysis text.

As mentioned in Section 3.1, a function f has one-sided limits at each point of $[a, b]$ if f has both a left-hand limit and a right-hand limit at each point of (a, b), a right-hand limit at a, and a left-hand limit at b. Recall that $f(x-)$ represents the left-hand limit of f at x and that $f(x+)$ represents the right-hand limit of f at x. The next theorem shows that increasing functions have one-sided limits at each point of $[a, b]$ and gives a "formula" for the values of the limits; the analogous statement for decreasing functions will be left as an exercise.

THEOREM 3.31 If $f: [a, b] \to \mathbb{R}$ is an increasing function, then f has one-sided limits at each point of $[a, b]$. These limits are given by

$$f(x-) = \sup\{f(t) : t \in [a, x)\} \quad \text{and} \quad f(x+) = \inf\{f(t) : t \in (x, b]\}$$

for all x in the intervals $(a, b]$ and $[a, b)$, respectively. In addition, $f(a) \leq f(a+)$, $f(b) \geq f(b-)$, and $f(x-) \leq f(x) \leq f(x+)$ for all $x \in (a, b)$.

Proof. We will show that f has a right-hand limit at each point $x \in [a, b)$ and that

$$f(x) \leq f(x+) = \inf\{f(t) : t \in (x, b]\};$$

a proof that f has a left-hand limit at each point of $(a, b]$ will be left as an exercise. Let $x \in [a, b)$. Since f is increasing on $[a, b]$, the set $A = \{f(t) : x < t \leq b\}$ is bounded below by $f(x)$. Let $\alpha = \inf A$ and let $\epsilon > 0$. Since $\alpha + \epsilon$ is not a lower bound of the set A, there exists a point $t_1 \in (x, b]$ such that $f(t_1) < \alpha + \epsilon$. If $t \in (x, t_1)$, then

$$\alpha \leq f(t) \leq f(t_1) < \alpha + \epsilon$$

This shows that f has a right-hand limit at x and that $f(x+) = \alpha$. Since $f(x)$ is a lower bound of the set A, it follows that $f(x) \leq \alpha = f(x+)$. This completes the proof. ∎

In the previous theorem, it is not necessary that the domain of f be a closed and bounded interval, but that is the form of the theorem that will be most useful in this text. For the sake of completeness, a more general statement is recorded below; the minor observations required to prove the result will be left as an exercise.

THEOREM 3.32 Let I be an interval. If $f: I \to \mathbb{R}$ is a monotone function on I, then f has one-sided limits at each point of I. ∎

Since monotone functions have one-sided limits at each point in their domain, the discontinuities of a monotone function are jump discontinuities. Now suppose that f is increasing on $[a, b]$. If f is not continuous at some point $c \in (a, b)$, then the positive number $f(c+) - f(c-)$ gives the magnitude of the jump at c; the graph of the function f at the point c moves up from $f(c-)$ to $f(c+)$. Since the function f begins at $f(a)$ and increases to $f(b)$, the total increase in the function is $f(b) - f(a)$. It follows that the sum of all of the jumps of f must be less than or equal to $f(b) - f(a)$. Hence, the function f cannot have too many jumps of a given magnitude. As shown by the proof of the next theorem, this fact can be used to prove that the set of discontinuities of a monotone function is countable.

THEOREM 3.33 If $f: [a, b] \to \mathbb{R}$ is monotone, then the set of discontinuities of f in $[a, b]$ is countable.

Proof. Suppose that $f: [a, b] \to \mathbb{R}$ is an increasing function. According to Theorem 3.31, the function f has one-sided limits at each point of $[a, b]$. Let D be the set of all points $x \in (a, b)$ such that f is not continuous at x. For each positive integer n, let

$$D_n = \{x \in D : f(x+) - f(x-) \geq 1/n\}.$$

A proof of the fact that $D = \bigcup_{n=1}^{\infty} D_n$ will be left as an exercise. Since a countably infinite union of countable sets is still a countable set, it is sufficient to prove that each D_n is finite.

Fix n and choose a positive integer p such that $p/n > f(b) - f(a)$. Suppose that D_n contains more than p elements. Let $x_1, \ldots, x_p$ be points in D_n listed in increasing order and choose points $y_0, y_1, \ldots, y_p$ such that

$$a = y_0 < x_1 < y_1 < x_2 < y_2 < \cdots < x_p < y_p = b.$$

Now use the fact that f is increasing to compute

$$p/n \leq \sum_{i=1}^{p} \bigl(f(x_i+) - f(x_i-)\bigr) \leq \sum_{i=1}^{p} \bigl(f(y_i) - f(y_{i-1})\bigr) = f(b) - f(a).$$

This is a contradiction, so the set D_n contains at most p elements, that is, the set D_n is finite. This completes the proof. ∎

The final topic for this section is the concept of a function of bounded variation. Functions of bounded variation play a prominent role in the theory of integration, but such functions are also related to monotone functions in a simple way. We will pursue the connection between functions of bounded variation and monotone functions in this section and leave a discussion of applications of this topic to integration for the supplementary exercises of Chapter 5. The notion of a partition of an interval is needed in order to define the variation of a function. A **partition** P of an interval $[c, d]$ is a finite set of points $\{x_i : 0 \le i \le n\}$ such that

$$c = x_0 < x_1 < x_2 < \cdots < x_{n-1} < x_n = d.$$

A partition of an interval simply splits the interval into subintervals.

DEFINITION 3.34 Let $f: [a, b] \to \mathbb{R}$ be a function and let $[c, d]$ be any closed subinterval of $[a, b]$. The **variation** of f on $[c, d]$ is defined by

$$V(f, [c, d]) = \sup\left\{\sum_{i=1}^{n} |f(x_i) - f(x_{i-1})|\right\},$$

where the supremum is over all partitions $\{x_i : 0 \le i \le n\}$ of $[c, d]$. Note that the integer n is not fixed; the supremum is over all possible partitions of $[c, d]$. For the record, the variation of f on $[c, d]$ is ∞ if the corresponding set of numbers is not bounded above. The function f is of **bounded variation** on $[c, d]$ if $V(f, [c, d])$ is finite.

In a manner of speaking, the variation of a function measures how much its graph moves up and down. We first show that the variation of a monotone function is easy to compute. Suppose that f is increasing on the interval $[c, d]$ and let $\{x_i : 0 \le i \le n\}$ be any partition of $[c, d]$. Then

$$\sum_{i=1}^{n} \left|f(x_i) - f(x_{i-1})\right| = \sum_{i=1}^{n} \big(f(x_i) - f(x_{i-1})\big) = f(d) - f(c).$$

Since every partition of the interval $[c, d]$ generates the same sum, it follows that $V(f, [c, d]) = f(d) - f(c)$. For a slightly more complicated example, consider the function g defined by $g(x) = \sin x$ on the interval $[0, 2\pi]$. Using the partition $\{0, \pi/2, \pi, 3\pi/2, 2\pi\}$ of $[0, 2\pi]$, we obtain (see Figure 3.9)

$$\begin{aligned} V(g, [0, 2\pi]) \ge \big|\sin(\pi/2) - \sin 0\big| + \big|\sin(\pi) - \sin(\pi/2)\big| + \\ \big|\sin(3\pi/2) - \sin(\pi)\big| + \big|\sin(2\pi) - \sin(3\pi/2)\big| = 4. \end{aligned}$$

It turns out that $V(g, [0, 2\pi]) = 4$, but it is not easy to verify this without knowing some further properties of functions of bounded variation. A number of these basic properties are recorded in the next theorem.

THEOREM 3.35 Let f and g be functions defined on an interval $[a, b]$, let k be a constant, and suppose that f and g are of bounded variation on $[a, b]$. Then

- **a)** the function f is bounded on $[a, b]$;
- **b)** the function f is of bounded variation on every closed subinterval of $[a, b]$;
- **c)** the function kf is of bounded variation on $[a, b]$;
- **d)** the functions $f + g$ and $f - g$ are of bounded variation on $[a, b]$;
- **e)** the function fg is of bounded variation on $[a, b]$;

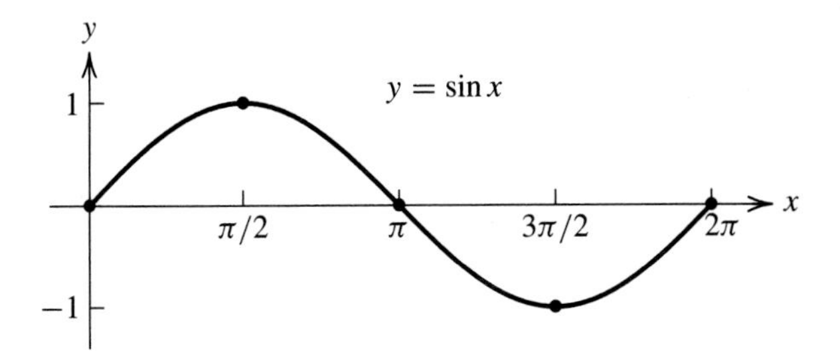

Figure 3.9 Computing the variation of the sine function

f) if $1/g$ is bounded on $[a, b]$, then the function f/g is of bounded variation on $[a, b]$.

Proof. To prove part (a), let $x \in [a, b]$. Since $\{a, x, b\}$ is a partition of $[a, b]$,

$$|f(x) - f(a)| + |f(b) - f(x)| \le V(f, [a, b]).$$

It follows that

$$|f(x)| \le |f(x) - f(a)| + |f(a)| \le V(f, [a, b]) + |f(a)|.$$

Since this inequality is valid for all $x \in [a, b]$, the function f is bounded on $[a, b]$.

To prove part (e), let M_f and M_g be bounds for the functions f and g, respectively, and let $\{x_i : 0 \le i \le n\}$ be any partition of $[a, b]$. Then

$$\begin{aligned}
&\sum_{i=1}^{n} \left| f(x_i)g(x_i) - f(x_{i-1})g(x_{i-1}) \right| \\
&\quad \le \sum_{i=1}^{n} \Big(\left| f(x_i)g(x_i) - f(x_i)g(x_{i-1}) \right| + \left| f(x_i)g(x_{i-1}) - f(x_{i-1})g(x_{i-i}) \right| \Big) \\
&\quad \le \sum_{i=1}^{n} M_f \left| g(x_i) - g(x_{i-1}) \right| + \sum_{i=1}^{n} M_g \left| f(x_i) - f(x_{i-1}) \right| \\
&\quad \le M_f\, V(g, [a, b]) + M_g\, V(f, [a, b]).
\end{aligned}$$

It follows that fg is of bounded variation on $[a, b]$. The proofs of the remaining parts of the theorem will be left as exercises. ∎

Let $c \in (a, b)$. Part (b) of the previous theorem shows that if f is of bounded variation on $[a, b]$, then f is of bounded variation on each of the intervals $[a, c]$ and $[c, b]$. The converse is true as well; if f is of bounded variation on the intervals $[a, c]$ and $[c, b]$, then f is of bounded variation on $[a, b]$. Furthermore, there is a simple relationship between the variations of f on these three intervals. This is the content of the next theorem.

THEOREM 3.36 Let $f: [a, b] \to \mathbb{R}$ and let $c \in (a, b)$. If f is of bounded variation on $[a, c]$ and $[c, b]$, then f is of bounded variation on $[a, b]$ and

$$V(f, [a, b]) = V(f, [a, c]) + V(f, [c, b]).$$

Proof. Let $\{x_i : 0 \le i \le n\}$ be any partition of $[a, b]$. Without loss of generality, we may assume that $c = x_q$ for some integer q; inserting the point c only makes the relevant sum larger (see the exercises). Since $\{x_i : 0 \le i \le q\}$ and $\{x_i : q \le i \le n\}$ are partitions of $[a, c]$ and $[c, b]$, respectively,

$$\begin{aligned}\sum_{i=1}^{n} |f(x_i) - f(x_{i-1})| &= \sum_{i=1}^{q} |f(x_i) - f(x_{i-1})| + \sum_{i=q+1}^{n} |f(x_i) - f(x_{i-1})| \\ &\le V(f, [a, c]) + V(f, [c, b]).\end{aligned}$$

Taking the supremum of the numbers $\sum_{i=1}^{n} |f(x_i) - f(x_{i-1})|$ over all partitions of the interval $[a, b]$ yields the inequality $V(f, [a, b]) \le V(f, [a, c]) + V(f, [c, b])$, which verifies that f is of bounded variation on $[a, b]$. To complete the proof, it is sufficient to establish the reverse inequality. Let $\epsilon > 0$ and choose partitions $\{u_i : 0 \le i \le p\}$ and $\{v_i : 0 \le i \le q\}$ of $[a, c]$ and $[c, b]$, respectively, such that

$$\sum_{i=1}^{p} |f(u_i) - f(u_{i-1})| > V(f, [a, c]) - \epsilon/2$$

and

$$\sum_{i=1}^{q} |f(v_i) - f(v_{i-1})| > V(f, [c, b]) - \epsilon/2.$$

Putting these two partitions together gives a partition of $[a, b]$. Thus,

$$\begin{aligned}V(f, [a, b]) &\ge \sum_{i=1}^{p} |f(u_i) - f(u_{i-1})| + \sum_{i=1}^{q} |f(v_i) - f(v_{i-1})| \\ &> V(f, [a, c]) + V(f, [c, b]) - \epsilon.\end{aligned}$$

Since $\epsilon > 0$ was arbitrary, $V(f, [a, b]) \ge V(f, [a, c]) + V(f, [c, b])$. This completes the proof. ■

Theorem 3.36 and the simple formula for the variation of a monotone function make it possible to find the variation of many functions without recourse to the definition of variation. If the interval on which the variation is to be found can be decomposed into a finite number of subintervals on each of which the function is monotone, then the variations on each of the subintervals can be added to give the total variation. For example,

$$\begin{aligned}&V(\sin x, [0, 2\pi]) \\ &\quad = V(\sin x, [0, \pi/2]) + V(\sin x, [\pi/2, 3\pi/2]) + V(\sin x, [3\pi/2, 2\pi]) \\ &\quad = 1 + 2 + 1 = 4.\end{aligned}$$

For more complicated functions, it can be more difficult to determine the variation, or even to decide whether or not the function is of bounded variation. In some cases, the derivative of a function can be used to determine if a function is of bounded variation; this will be considered in the next chapter. However, it is always possible to resort to the definition of variation to determine if a function is of bounded variation.

As an example of a more complicated function, let $f(x) = \sqrt{x}\cos(\pi/x)$ for $x \neq 0$ and $f(0) = 0$. This function is continuous on $[0, 1]$, but it is not of bounded variation on $[0, 1]$. To verify the latter statement, note that for each positive integer k, $\cos(\pi/x) = 0$ when $x = 2/(2k+1)$ and $|\cos(\pi/x)| = 1$ when $x = 1/k$. For each positive integer n, the intervals $[2/(2k+1), 1/k]$ for $1 \leq k \leq n$ are disjoint, and the function f is monotone on each of them. Therefore,

$$V(f, [0, 1]) \geq \sum_{k=1}^{n} V(f, [2/(2k+1), 1/k]) = \sum_{k=1}^{n} \frac{1}{\sqrt{k}} \geq \frac{n}{\sqrt{n}} = \sqrt{n}.$$

Since this inequality is valid for each positive integer n, the function f is not of bounded variation on $[0, 1]$.

A monotone function is of bounded variation on $[a, b]$ and, by part (d) of Theorem 3.35, the difference of two monotone functions is a function of bounded variation. Surprisingly enough, the converse is also true; a function of bounded variation can be written as the difference of two monotone functions. It will become apparent later in the text that this is a very useful result.

THEOREM 3.37 If $f: [a, b] \to \mathbb{R}$ is of bounded variation on $[a, b]$, then there exist increasing functions f_1 and f_2 such that $f = f_1 - f_2$.

Proof. Define a function f_1 on $[a, b]$ by $f_1(x) = V(f, [a, x])$ for $x \in (a, b]$ and $f_1(a) = 0$, and define a function f_2 by $f_2(x) = f_1(x) - f(x)$ for all $x \in [a, b]$. Then it is clear that $f = f_1 - f_2$ and that f_1 is increasing (see the exercises). Suppose that $a \leq x < y \leq b$ and use Theorem 3.36 to compute

$$f_1(y) - f_1(x) = V(f, [x, y]) \geq |f(y) - f(x)| \geq f(y) - f(x).$$

It follows that

$$f_2(y) = f_1(y) - f(y) \geq f_1(x) - f(x) = f_2(x),$$

which shows that f_2 is increasing as well. This completes the proof. ∎

For an arbitrary function f of bounded variation, it is not easy to visualize the functions f_1 and f_2 defined in the proof of Theorem 3.37. In most cases, the actual functions are not important; just the fact that the function is a difference of monotone functions is enough. An actual decomposition of a function of bounded variation is considered in Exercise 24.

Exercises

1. Finish the proof of Theorem 3.31 by considering left-hand limits.
2. State and prove the analogue of Theorem 3.31 for decreasing functions.
3. Use Theorem 3.31 to prove Lemma 3.19.
4. Prove Theorem 3.32.

5. This exercise concerns the proof of Theorem 3.33.

a) Explain why it is sufficient to consider only the discontinuities of the function f in the open interval (a, b).

b) Prove that f is continuous at $x \in (a, b)$ if and only if $f(x-) = f(x) = f(x+)$.

c) Prove that $D = \bigcup_{n=1}^{\infty} D_n$.

d) Show how to obtain the result for decreasing functions.

6. Here is an alternate proof of Theorem 3.33. Let f be an increasing function defined on the interval (a, b).

a) Suppose that $a < c < d < b$. Prove that $f(c+) \leq f(d-)$.

b) Let D be the set of all points in (a, b) at which f is not continuous. For each $c \in D$, choose a rational number r_c from the interval $(f(c-), f(c+))$. Show that $r_c \neq r_d$ if c and d are distinct points in D.

c) Use part (b) to find a one-to-one correspondence between D and a subset of the rational numbers and conclude that D is a countable set.

7. Let $f\colon \mathbb{R} \to \mathbb{R}$ be monotone and let D be the set of all points in $\mathbb{R}$ at which f is not continuous. Prove that D is a countable set.

8. Let $f\colon [a, b] \to \mathbb{R}$ be a monotone function. Prove that every subinterval of $[a, b]$ contains a point of continuity of f.

9. Give an example of an increasing function $f\colon \mathbb{R} \to \mathbb{R}$ that has the set D as its set of discontinuities.

a) $D = \{1, 2, 5\}$.

b) $D = \mathbb{Z}^+$.

c) $D = \mathbb{Z}$.

d) $D = \mathbb{Z}$ and the function f is bounded.

e) $D = \{1/n : n \in \mathbb{Z}^+\}$.

10. Let $f\colon [a, b] \to \mathbb{R}$, let $\{x_i : 0 \leq i \leq n\}$ be a partition of $[a, b]$, and let $\{y_i : 0 \leq i \leq p\}$ be a partition of $[a, b]$ that satisfies $\{x_i : 0 \leq i \leq n\} \subseteq \{y_i : 0 \leq i \leq p\}$. Prove that

$$\sum_{i=1}^{n} |f(x_i) - f(x_{i-1})| \leq \sum_{i=1}^{p} |f(y_i) - f(y_{i-1})|.$$

In other words, adding more points to a partition only increases the sum used to compute the variation of f.

11. Prove part (b) of Theorem 3.35.

12. Prove part (c) of Theorem 3.35.

13. Prove part (d) of Theorem 3.35.

14. Prove part (f) of Theorem 3.35.

15. Suppose that $f\colon [a, b] \to \mathbb{R}$ is of bounded variation on every closed subinterval of (a, b). Does it follow that f is of bounded variation on $[a, b]$?

16. Let $f\colon [a, b] \to \mathbb{R}$ and suppose that there is a partition $\{x_i : 0 \leq i \leq n\}$ of $[a, b]$ such that f is monotone on each of the intervals $[x_{i-1}, x_i]$. Find an expression for $V(f, [a, b])$.

17. Find $V(x - x^2, [-1, 2])$ and $V(\cos(4\pi x), [0, 1])$.

18. Let f be the function defined by $f(x) = 2|x| - |x - 2|$. Compute $V(f, [-1, 4])$.

19. Let g be the function defined by $g(x) = x - \lfloor x \rfloor$. Compute $V(g, [-2, 3])$.

20. Prove that the function f defined by $f(x) = 0$ if x is irrational and $f(x) = 1$ if x is rational is not of bounded variation on any interval $[a, b]$.

21. Let f be the function defined by $f(x) = \sqrt[3]{x}\sin(\pi/x)$ for $x \neq 0$ and $f(0) = 0$. Prove that f is not of bounded variation on $[0, 1]$.

22. Prove that the function f_1 defined in the proof of Theorem 3.37 is increasing.

23. Let $f\colon [a, b] \to \mathbb{R}$ be a function of bounded variation on $[a, b]$. Prove that f is the difference of two strictly increasing functions.

24. Let f be the function defined by $f(x) = x^2$. Define a function V by $V(-2) = 0$ and $V(x) = V(f, [-2, x])$ for $-2 < x \leq 5$. Find explicit formulas for V and $V - f$ and show that both of these functions are increasing on the interval $[-2, 5]$.

25. Let $f\colon [a, b] \to \mathbb{R}$ be a function of bounded variation on $[a, b]$ and define a function V on $[a, b]$ by $V(a) = 0$ and $V(x) = V(f, [a, x])$ for all $x \in (a, b]$.

a) Suppose that $a \leq x < y \leq b$. Prove that $V(y) - V(x) = V(f, [x, y])$.

b) Prove that V is increasing on $[a, b]$.

c) Suppose that V is continuous at $c \in [a, b]$. Prove that f is continuous at c.

d) Suppose that f is continuous at $c \in [a, b]$. Prove that V is continuous at c.

e) Suppose that f is continuous on $[a, b]$. Prove that f is the difference of two increasing continuous functions.

26. Let f be a function of bounded variation defined on $[a, b]$.

a) Prove that f has one-sided limits at each point of $[a, b]$.

b) Prove that the set of discontinuities of f is countable.

27. Define functions f and g by

$$f(x) = \begin{cases} 0, & \text{if } x \leq 0; \\ 1, & \text{if } x > 0; \end{cases} \quad \text{and} \quad g(x) = \begin{cases} (-1)^n/2^n, & \text{if } x = 1/n \text{ for } n \in \mathbb{Z}^+; \\ 0, & \text{otherwise.} \end{cases}$$

Prove that f is of bounded variation on $[-1, 1]$ and that g is of bounded variation on $[0, 1]$, but that $f \circ g$ is not of bounded variation on $[0, 1]$.

Remark. The rest of the exercises in this section are in no particular order.

28. Let D be a finite set. Write down an expression for an increasing function $f\colon \mathbb{R} \to \mathbb{R}$ that has the set D as its set of discontinuities.

29. This exercise assumes some knowledge of infinite series. Let $D = \{d_k : k \in \mathbb{Z}^+\}$ be a countably infinite set of distinct real numbers. For each real number x, let P_x be the set $\{k \in \mathbb{Z}^+ : d_k < x\}$. Define a function $f\colon \mathbb{R} \to \mathbb{R}$ by $f(x) = 0$ if $P_x = \emptyset$ and $f(x) = \sum\limits_{k \in P_x} 2^{-k}$ if $P_x \neq \emptyset$.

a) Prove that f is increasing on $\mathbb{R}$ and has the set D as its set of discontinuities.

b) Suppose that $D = \mathbb{Q}$. Prove that f is strictly increasing on $\mathbb{R}$.

30. Suppose that f and g are defined on an interval I and that each of these functions has a countable number of discontinuities. Prove that the functions $f + g$ and fg also have a countable number of discontinuities.

31. Give an example of a function $f\colon [0, 1] \to \mathbb{R}$ that has an uncountable number of discontinuities.

32. Suppose that f is defined on an open interval (a, b) and that for each $x \in (a, b)$, there exists a positive number r_x such that f is increasing on $(x - r_x, x + r_x)$. Prove that f is increasing on (a, b).

3.6 Supplementary Exercises

1. Let $f: [0, \infty) \to \mathbb{R}$ be an increasing function. Prove that either $\lim_{x \to \infty} f(x)$ exists or $\lim_{x \to \infty} f(x) = \infty$.

2. Let $f(x) = 1/x^2$ and let c be any nonzero real number. Use the definition of limit to prove that $\lim_{x \to c} f(x) = 1/c^2$. Show that the function f does not have a limit at 0.

3. Let P be a polynomial of even degree n and suppose that $a_n a_0 < 0$, where a_n is the coefficient of x^n and a_0 is the constant term. Prove that P has at least two roots.

4. Let I be an open interval that contains the point c and suppose that f is a function that is defined on I except possibly at the point c. Consider the following biconditional statement. The function f has a limit at c if and only if $\lim_{h \to 0^+} |f(c+h) - f(c-h)| = 0$. Determine whether or not either portion of this biconditional statement is valid. Give proofs and/or find counterexamples to support your answer.

5. Let $f: \mathbb{R} \to \mathbb{R}$ be a function that satisfies $f(x + y) = f(x)f(y)$ for all $x, y \in \mathbb{R}$. Suppose that f has a limit at 0. Prove that f has a limit at every real number.

6. Consider the function $f: [0, 1] \to \mathbb{R}$ defined by

$$f(x) = \begin{cases} \sqrt[n]{2}, & \text{if } x = 10^{-n} \text{ for } n \in \mathbb{Z}^+; \\ 0, & \text{otherwise.} \end{cases}$$

Prove that $\lim_{x \to 0^+} f(x)$ does not exist.

7. Let I be an open interval that contains the point c, let f and g be functions that are defined on I except possibly at c, and suppose that g is a bounded function.

a) Suppose that $\lim_{x \to c} f(x) = 0$. Prove that $\lim_{x \to c} (fg)(x) = 0$.

b) Suppose that f is a bounded function and that neither $\lim_{x \to c} f(x)$ nor $\lim_{x \to c} g(x)$ exist. Give an example to show that $\lim_{x \to c} (fg)(x)$ may still exist.

c) Suppose that $\lim_{x \to c} f(x) = \infty$. Prove that $\lim_{x \to c} (f + g)(x) = \infty$.

8. Let $f: [a, b) \to \mathbb{R}$ be a function and suppose that $\lim_{x \to b^-} f(x) = \infty$. Prove that there exists an increasing sequence $\{x_n\}$ in (a, b) such that $f(x_n) > n$ for each n.

9. Let $f: \mathbb{R} \to \mathbb{R}$ be a continuous periodic function. Prove that f has a fixed point. See Exercise 10 in Section 3.4 for the definition of a periodic function.

10. Let $f: [a, b] \to \mathbb{R}$ be continuous on $[a, b]$. Use Theorem 3.28 to give a direct proof that f is bounded on $[a, b]$.

11. Determine whether or not the sequence $\{\csc n\}$ is bounded.

12. Let I be an open interval that contains the point c and suppose that f is a function that is defined on I except possibly at the point c. Suppose that the sequence $\{f(x_n)\}$ converges for each sequence $\{x_n\}$ in $I \setminus \{c\}$ that converges to c. Prove that all the sequences $\{f(x_n)\}$ converge to the same value.

13. Let $f\colon (a,\infty) \to \mathbb{R}$ and suppose that $\lim_{x\to\infty} f(x) = L$. Prove that

$$\lim_{x\to\infty} \big(f(x+1) - f(x)\big) = 0 \quad \text{and} \quad \lim_{x\to\infty} \big(f(x^2) - f(x)\big) = 0.$$

14. Let f be the function defined by $f(x) = 2/x$ for $x \neq 0$ and $f(0) = 1$. Show that f does not satisfy the intermediate value property on the interval $[0, 2]$.

15. Explain why the functions f and g defined by the formulas

$$f(x) = \frac{x^2 - 2x - 3}{x^2 - 3x} \quad \text{and} \quad g(x) = \frac{x+1}{x}$$

are not the same, but that $\lim_{x\to 3} f(x) = \lim_{x\to 3} g(x)$.

16. Let $f\colon \mathbb{R} \to \mathbb{R}$ be continuous on $\mathbb{R}$ and define $g\colon \mathbb{R} \to \mathbb{R}$ by

$$g(x) = \begin{cases} 4, & \text{if } f(x) > 4; \\ f(x), & \text{if } |f(x)| \le 4; \\ -4, & \text{if } f(x) < -4. \end{cases}$$

Prove that g is continuous on $\mathbb{R}$ and that g is bounded by 4.

17. Let $f\colon \mathbb{R} \to \mathbb{R}$ be a function that satisfies $f(x+y) = f(x) + f(y)$ for all $x, y \in \mathbb{R}$.

a) Suppose that f is continuous at some point c. Prove that f is continuous on $\mathbb{R}$.

b) Suppose that f is continuous on $\mathbb{R}$ and that $f(1) = k$. Prove that $f(x) = kx$ for all $x \in \mathbb{R}$.

18. Find constants a and b so that $\displaystyle\lim_{x\to 2} \frac{ax^2 + bx - 8}{x^2 - x - 2} = 1$.

19. Suppose that $f\colon [a,b] \to \mathbb{R}$ is continuous on $[a,b]$ and define $g\colon [a,b] \to \mathbb{R}$ by

$$g(x) = \sup\{f(t) : a \le t \le x\}.$$

Prove that g is continuous and increasing on $[a,b]$.

20. Let f and g be functions defined on $[a,b]$ and let $c \in [a,b]$. Suppose that f is not continuous at c and that g is continuous at c. Prove that $f + g$ is not continuous at c.

21. Let $f\colon [a,b] \to \mathbb{R}$ be a continuous function with the following property: for each $x \in [a,b]$ there exists $y \in [a,b]$ such that $2|f(y)| \le |f(x)|$. Prove that there is a point c in $[a,b]$ such that $f(c) = 0$.

22. Let $f\colon [0,\infty) \to \mathbb{R}$ be a continuous function and suppose that $\lim_{x\to\infty} f(x)$ exists. Prove that f is uniformly continuous on $\mathbb{R}$.

23. Suppose that $f\colon I \to \mathbb{R}$ is a continuous function and that $|f|$ is a constant function. Prove that f is a constant function.

24. Prove that the function $x^{1/3}$ is continuous on $\mathbb{R}$ by using the identity

$$x^{1/3} - c^{1/3} = \frac{x - c}{x^{2/3} + x^{1/3}c^{1/3} + c^{2/3}}.$$

Use a similar identity to prove that the function $x^{1/4}$ is continuous on $[0,\infty)$.

25. Let $f\colon [a,b] \to \mathbb{R}$ be a continuous function and suppose that the range of f is a countable set. What can you conclude about the function f?

26. Consider the function $f\colon \mathbb{R} \to \mathbb{R}$ defined by

$$f(x) = \begin{cases} x^2 - 2, & \text{if } x \in \mathbb{Q}; \\ 2 - 3x, & \text{if } x \notin \mathbb{Q}. \end{cases}$$

Determine the points of continuity of f.

27. Suppose that P is a polynomial with even degree n and that the coefficient of x^n is positive. Prove that P has a minimum value on $\mathbb{R}$.

28. Prove that the function $f(x) = x^5 - 4x + 2$ has three real roots.

29. Let $f: [0, 2] \to \mathbb{R}$ be continuous on $[0, 2]$ and suppose that $f(0) = f(2)$. Prove that there exists $c \in [0, 1]$ such that $f(c) = f(c + 1)$.

30. Find a function $f: \mathbb{R} \to \mathbb{R}$ that is continuous only at $x = 2$.

31. Let $f: [0, \infty) \to \mathbb{R}$ be nonnegative and continuous with $\lim_{x\to\infty} f(x) = 0$. Prove that f has a maximum value on $[0, \infty)$.

32. For each positive integer n, let f_n be the function defined by

$$f_n(x) = x^{2n} + x^{2n-1} + x^{2n-2} + \cdots + x + 1$$

and let m_n be the minimum value of this function on the interval $[-1, 0]$. Prove that the sequence $\{m_n\}$ converges and find the limit of the sequence.

33. Let a and b be fixed nonzero real numbers and consider the polynomial P defined by $P(x) = 2b^2x^3 + abx^2 - (a^2 + 3b^2 + 1)x - 3ab$. Prove that P has three real roots and that one of these roots must lie in the interval $(-1, 1)$.

Remark. The limit concept is not easy and many calculus students are very confused by it. For example, it is not unusual to find the following equations/statements on homework assignments:

1. $\lim_{x\to 3} \dfrac{x^2 - 9}{x - 3} = x + 3 = 3 + 3$ so $\lim_{x\to 3} = 6$.
2. $\lim_{x\to\infty} \dfrac{x}{x + 1} \approx 1$.
3. $\lim_{x\to 0} x \sin(1/x) = \lim_{x\to 0} x \cdot \lim_{x\to 0} \sin(1/x) = 0$.
4. $\lim_{x\to c} f(x) = L$ means that $f(x)$ gets close to L but never equals L.
5. The graph of a function cannot cross one of its asymptotes.

The following exercises consider these statements.

34. Although the first example seems to be primarily a notational problem, I have found that students who continue to write such equations have deeper problems. For instance, does $\lim_{x\to 3}$ mean something by itself? Is $\lim_{x\to 3} f(x)$ a number or a function? How would you explain these points to a calculus student?

35. Explain why a student might write statement (2) and explain what is wrong with it.

36. Determine the error in (3) and provide a correct proof.

37. Give several examples (qualitatively different examples, not just a simple change in numbers or functions) that show that statement (4) is false.

38. Give several examples (qualitatively different examples, not just a simple change in numbers or functions) that show that statement (5) is false.

Remark. In Section 3.5, it was shown that a continuous function on an interval may not be of bounded variation on that interval. In some cases (the theory of Lebesgue integration is one instance), it becomes important to consider a stronger form of continuity which guarantees that the function is also of bounded variation. This type of continuity, which is known as absolute continuity, is defined below. (For the record, two intervals are said to be **non-overlapping** if their intersection consists of at most one point.) The remaining exercises in this section explore this type of continuity; some further properties will be considered in the supplementary exercises of the next chapter.

Let f be defined on an interval I. The function f is **absolutely continuous** on I if for each $\epsilon > 0$ there exists $\delta > 0$ such that $\sum_{i=1}^{n} |f(d_i) - f(c_i)| < \epsilon$ whenever $\{[c_i, d_i] : 1 \le i \le n\}$ is a finite collection of non-overlapping intervals in I that satisfy $\sum_{i=1}^{n} (d_i - c_i) < \delta$. (The letter n represents an arbitrary positive integer; the collection of intervals can contain any finite number of intervals.)

39. Prove that an absolutely continuous function is uniformly continuous.

40. Let $f: I \to \mathbb{R}$ be a function that satisfies $|f(y) - f(x)| \le M|y - x|$ for all $x, y \in I$. Prove that f is absolutely continuous on I.

41. Suppose that f is absolutely continuous on I. Prove that $|f|$ is absolutely continuous on I.

42. Prove that every function of the form $f(x) = ax + b$, where a and b are constants, is absolutely continuous on $\mathbb{R}$.

43. Prove that the given function is absolutely continuous on the interval $[0, 1]$.
(a) $f(x) = x^2$ (b) $g(x) = \sqrt{x}$ (c) $h(x) = \sin x$

44. Suppose that f and g are absolutely continuous on I. Prove that $f + g$ is absolutely continuous on I.

45. Suppose that f and g are absolutely continuous on $[a, b]$. Prove that fg is absolutely continuous on $[a, b]$.

46. Suppose that $f: [a, b] \to \mathbb{R}$ is absolutely continuous on $[a, b]$. Prove that f is of bounded variation on $[a, b]$.

47. Prove that an absolutely continuous function defined on $[a, b]$ is the difference of two continuous increasing functions.

48. Give an example of a function that is uniformly continuous on $[0, 1]$ but not absolutely continuous on $[0, 1]$. (For the record, there exist continuous increasing functions that are not absolutely continuous; the Cantor ternary function is one example. A study of such functions requires some knowledge of measure theory and can be found in graduate texts in real analysis.)

4

Differentiation

Suppose that a particle is moving along a straight line, let O be a fixed point on the line, and let $f(t)$ represent the distance of the particle from O at any time $t \geq 0$. A useful quantity to know is the velocity of the particle at some time $t_1 > 0$. The average velocity of an object over a time interval is defined as the change in distance divided by the change in time. However, at the instant t_1 there is no time change and hence no change in distance; the definition yields the meaningless ratio $0/0$ as the velocity of the particle at time t_1. The solution to this dilemma is to introduce the limit process. If t is a time value other than t_1, then the quantity

$$\frac{f(t) - f(t_1)}{t - t_1}$$

represents the average velocity of the particle over the time period from t_1 to t. This quantity is known as a **difference quotient** since it is the quotient of two differences. By letting $t \to t_1$ in the difference quotient, we can determine the velocity of the particle at the instant t_1, provided that the limit exists. When the limit

$$\lim_{t \to t_1} \frac{f(t) - f(t_1)}{t - t_1}$$

exists, it is usually referred to as the instantaneous velocity of the particle at time t_1. There are many other physical situations (such as acceleration, decay rate, and growth rate) that involve a limit of difference quotients.

There is also a classic problem in geometry associated with difference quotients of this type. Consider the graph of a "smooth" continuous function (the graph has no sharp corners or turns) and pick a point on the graph. If you look at the graph on a short interval containing the point, the graph resembles a straight line. As

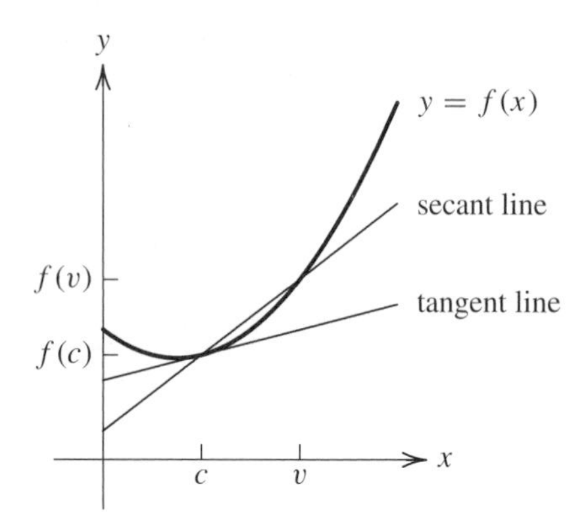

Figure 4.1 A secant line approximates the tangent line

the interval shrinks (and the graph is magnified), the effect becomes more dramatic. This sort of behavior is easy to see on a computer graph that has a zoom feature. The line resembling the graph is called the **tangent line** to the curve at the given point and its slope can be interpreted as the slope of the curve at the point. To compute the slope of the curve $y = f(x)$ at the point $(c, f(c))$, take any other point $(v, f(v))$ on the graph and determine the slope of the **secant line**—the line connecting the two points on the graph. The limit of the slopes of the secant lines as $v \to c$ is the slope of the tangent line at c (see Figure 4.1). Since the slope of the secant line is given by the difference quotient $(f(v) - f(c)/(v - c)$, the geometric problem of determining the slope of the tangent line reduces to the algebraic problem of evaluating the limit

$$\lim_{v \to c} \frac{f(v) - f(c)}{v - c}.$$

This is the same mathematical problem discussed in the first paragraph.

The mathematical concept involved in this discussion is the derivative. In this chapter, we will define the derivative of a function and explore some of its properties. Although the derivative concept has a number of practical applications (some of which should be familiar to the reader), the focus in this book will be on the mathematical aspects of the derivative. However, a few simple applications of the derivative will appear in the exercises.

4.1 The Derivative of a Function

In this section, we will discuss the definition of the derivative and record many of the familiar formulas for finding derivatives. Most of this material should be familiar to the reader. Consequently, a number of the proofs (all of which appear in calculus books since they do not involve any of the deeper ideas in analysis) will be left as exercises.

Let f be defined on an interval I that contains the point c. As discussed in the introduction to this chapter, the computation of limits of the form

$$\lim_{v\to c}\frac{f(v)-f(c)}{v-c}$$

appears frequently in certain types of applications of mathematics. Consequently, limits of this form are given a special name: the derivative of the function f at c. Before presenting this familiar definition, we note that the limits

$$\lim_{v\to c}\frac{f(v)-f(c)}{v-c} \quad\text{and}\quad \lim_{h\to 0}\frac{f(c+h)-f(c)}{h}$$

are equivalent. This means that if either one of these limits exists, then the other limit exists and both limits have the same value. (The reader should pause long enough to be convinced that these two limits are indeed equivalent.) Depending on the situation or the function, one of these limits may be easier to evaluate than the other, so it is a good idea to keep both versions in mind.

DEFINITION 4.1 Let I be an interval, let $f\colon I\to\mathbb{R}$, and let $c\in I$. The function f is **differentiable** at c provided that the limit

$$\lim_{v\to c}\frac{f(v)-f(c)}{v-c}$$

exists. (If c is an endpoint of I, then an appropriate one-sided limit is assumed.) If f is differentiable at c, we say that f has a derivative at c. The **derivative** of f at c is the value of the displayed limit and will be denoted by $f'(c)$. The function f is differentiable on an interval J if f is differentiable at each point of J.

We will illustrate the computations involved in finding a derivative with two simple examples. Suppose that $f(x)=x^2$ and $c=4$. Then

$$f'(4)=\lim_{h\to 0}\frac{(4+h)^2-4^2}{h}=\lim_{h\to 0}(8+h)=8.$$

Consequently (see the chapter introduction), the slope of the graph of $y=x^2$ at the point $(4,16)$ is 8. Now let $g(x)=1/x$ and let c be any nonzero real number. Since

$$\lim_{v\to c}\frac{1/v-1/c}{v-c}=\lim_{v\to c}\frac{1}{v-c}\left(\frac{c-v}{vc}\right)=\lim_{v\to c}\frac{-1}{vc}=-\frac{1}{c^2},$$

the function g is differentiable at c and $g'(c)=-1/c^2$.

Since the derivative of a function at a point is defined in terms of a limit, limit theorems can be used to compute derivatives. For instance, Theorem 3.2 yields the following result for derivatives; this result sometimes makes it easier to compute a derivative or to show that one does not exist.

THEOREM 4.2 Let I be an interval, let $f\colon I\to\mathbb{R}$, and let $c\in I$. The function f is differentiable at c with derivative $f'(c)=L$ if and only if for each sequence $\{x_n\}$ in $I\setminus\{c\}$ that converges to c, the sequence $\left\{\dfrac{f(x_n)-f(c)}{x_n-c}\right\}$ converges to L. ■

Definition 4.1 shows how to find the derivative of a function f at a particular point, but it is often useful to think of the derivative f' as a function. As will become clear shortly, in many instances we are more interested in the general behavior of the function f' than in its particular values. The domain of the function f' is the set of all points x in the domain of f for which $f'(x)$ exists. In other words, the derivative of a function f is another function f' defined by

$$f'(x) = \lim_{v \to x} \frac{f(v) - f(x)}{v - x}$$

for all values of x in the domain of f for which the limit exists. The function f' is called the derivative of f because its values are derived from the function f.

As discussed in the introduction to this chapter, the number $f'(c)$ represents the slope of the curve $y = f(x)$ at the point $(c, f(c))$; zooming in on the graph at the point $(c, f(c))$ reveals that the curve resembles a straight line and $f'(c)$ is the slope of this line (the tangent line). From a geometrical perspective, the following two statements make perfect sense.

1. A curve cannot have a slope at a point where it has a break.
2. A curve with no breaks may have a sharp corner (such as the peak of a curve shaped like $\wedge$) and thus not have a slope at a particular point.

In other words, curves with breaks or corners do not resemble straight lines no matter how much the graph is magnified. Statements (1) and (2) translate mathematically into the statements "a differentiable function is continuous" and "a continuous function may not be differentiable", respectively. The absolute value function at 0 provides an example of a continuous function that is not differentiable at a point (see the exercises). Although "obvious" geometrically, a proof that differentiability implies continuity must use the mathematical definitions of the derivative and of continuity. A proof of this fact will be left as an exercise.

THEOREM 4.3 Let I be an interval, let $f: I \to \mathbb{R}$, and let $c \in I$. If f is differentiable at c, then f is continuous at c. Consequently, if f is differentiable on an interval J, then f is continuous on J. ■

Except for some very simple functions, the computation of derivatives using the definition is a tedious and difficult process. Once the derivative of a function has been computed, there is no need to repeat the process every time that it is needed; the formula for the derivative can be recorded for future use. You probably remember most of the derivative formulas (and perhaps some of their proofs) from calculus. These formulas are of two types: formulas for the derivatives of specific functions and formulas for the derivatives of combinations of functions. We first consider algebraic combinations of functions, leaving the proofs of the following familiar results as exercises.

THEOREM 4.4 If f and g are differentiable functions defined on an interval I and k is any constant, then the functions kf, $f + g$, and $f - g$ are differentiable on I and

a) $(kf)'(x) = kf'(x)$ for all $x \in I$;

b) $(f+g)'(x) = f'(x) + g'(x)$ for all $x \in I$;

c) $(f-g)'(x) = f'(x) - g'(x)$ for all $x \in I$. ■

Since $(f+g)' = f' + g'$, it is tempting to write $(fg)' = f'g'$. Unfortunately, some calculus students do just this, over and over! The correct formula for $(fg)'$ is a bit more complicated. Here is one explanation for it. Let ℓ and w be two varying quantities (the length and width of a rectangle perhaps). If ℓ increases by a small amount $\Delta\ell$ and w increases by a small amount Δw, then the change in the product ℓw is

$$(\ell + \Delta\ell)(w + \Delta w) - \ell w = \ell\Delta w + w\Delta\ell + \Delta\ell\Delta w \approx \ell\Delta w + w\Delta\ell,$$

where the approximation can be made since the product $\Delta\ell\Delta w$ is "very small" compared to the other terms. The formula "the first times the change (derivative) of the second plus the second times the change (derivative) of the first" thus appears. This result is known as the product rule. The key step in its proof (see below) is the manipulation of the difference quotient for fg so that the difference quotients for f and g appear. A proof of the quotient rule, a rule for differentiating the quotient of two differentiable functions, will be left as an exercise.

THEOREM 4.5 Product Rule If f and g are differentiable functions defined on an interval I, then fg is differentiable on I and

$$(fg)'(x) = f(x)g'(x) + g(x)f'(x)$$

for all $x \in I$.

Proof. Let x be any point in the interval I. Some simple algebra yields

$$\begin{aligned}\frac{f(v)g(v) - f(x)g(x)}{v - x} &= \frac{f(v)g(v) - f(v)g(x) + f(v)g(x) - f(x)g(x)}{v - x} \\ &= f(v)\,\frac{g(v) - g(x)}{v - x} + g(x)\,\frac{f(v) - f(x)}{v - x}\end{aligned}$$

for all $v \in I \setminus \{x\}$. Each of the limits

$$\lim_{v \to x} \frac{f(v) - f(x)}{v - x} \quad \text{and} \quad \lim_{v \to x} \frac{g(v) - g(x)}{v - x}$$

exist since the functions f and g are differentiable at x, and $\lim_{v \to x} f(v) = f(x)$ since f is continuous at x (see Theorem 4.3). Using the algebraic properties of limits, we obtain

$$\begin{aligned}(fg)'(x) &= \lim_{v \to x} \frac{f(v)g(v) - f(x)g(x)}{v - x} \\ &= \lim_{v \to x} f(v) \lim_{v \to x} \frac{g(v) - g(x)}{v - x} + g(x) \lim_{v \to x} \frac{f(v) - f(x)}{v - x} \\ &= f(x)g'(x) + g(x)f'(x).\end{aligned}$$

This completes the proof. ■

THEOREM 4.6 Quotient Rule If f and g are differentiable functions defined on an interval I and g is nonzero on I, then f/g is differentiable on I and

$$\left(\frac{f}{g}\right)'(x) = \frac{g(x)f'(x) - f(x)g'(x)}{(g(x))^2}$$

for all $x \in I$. ■

The previous three theorems provide derivative formulas for combinations of functions that involve the operations of addition, subtraction, multiplication, and division. Another way to combine functions is composition, and it seems reasonable to postulate that the derivative of $f \circ g$ is related to the derivatives of the functions f and g in some way. This is indeed the case, and the relationship is a fairly simple one. The derivative formula for the composition of two functions is known as the Chain Rule; it is an extremely important and useful derivative formula.

Suppose that f and g are defined on $\mathbb{R}$, that g is differentiable at c, and that f is differentiable at $g(c)$. We would like to prove that $f \circ g$ is differentiable at c and find a formula for $(f \circ g)'(c)$. The definition of the derivative yields

$$\begin{aligned}\lim_{x\to c} \frac{f(g(x)) - f(g(c))}{x - c} &= \lim_{x\to c} \frac{f(g(x)) - f(g(c))}{g(x) - g(c)} \cdot \frac{g(x) - g(c)}{x - c} \\ &= f'(g(c))g'(c).\end{aligned}$$

The last equality follows from the definition of the derivative of f at $g(c)$ and of g at c. (Note that $\lim_{x\to c} g(x) = g(c)$ by Theorem 4.3.) This gives a proof that $f \circ g$ is differentiable at c and a formula for $(f \circ g)'(c)$ in two easy steps. However, the insertion of the quantity $g(x) - g(c)$ into the denominator of a fraction is not valid if $g(x) = g(c)$. In rare cases, it may happen that $g(x) = g(c)$ for values of x arbitrarily close to c. A proof of the Chain Rule that allows for this possibility is not as intuitive as the preceding proof, but it still only involves basic properties of the derivative.

THEOREM 4.7 Chain Rule Let I be an interval, let $g\colon I \to \mathbb{R}$, and let f be a function defined on an interval J that contains $g(I)$. If g is differentiable on I and f is differentiable on J, then the function $f \circ g$ is differentiable on I and $(f \circ g)'(x) = f'(g(x))g'(x)$ for all $x \in I$.

Proof. Fix $c \in I$ and define a function $F\colon J \to \mathbb{R}$ by

$$F(x) = \begin{cases} \dfrac{f(x) - f(g(c))}{x - g(c)}, & \text{if } x \neq g(c); \\ f'(g(c)), & \text{if } x = g(c). \end{cases}$$

Since f is differentiable at $g(c)$, the function F is continuous at $g(c)$. Since the equality

$$F(x)(x - g(c)) = f(x) - f(g(c))$$

is valid for all $x \in J$, it follows that

$$F(g(x))(g(x) - g(c)) = f(g(x)) - f(g(c))$$

7. If $f(x) = \csc x$, then $f'(x) = -\csc x \cot x$.

8. If $f(x) = \arcsin x$, then $f'(x) = \dfrac{1}{\sqrt{1-x^2}}$.

9. If $f(x) = \arctan x$, then $f'(x) = \dfrac{1}{1+x^2}$.

10. If $f(x) = \operatorname{arcsec} x$, then $f'(x) = \dfrac{1}{|x|\sqrt{x^2-1}}$. ■

If a function f is differentiable on an interval I, then its derivative f' is another function defined on I. Consequently, it makes sense to consider the derivative of f'. This function, if it exists, is known as the second derivative of f and is denoted by f''. The third derivative of f is the derivative of f'' and is denoted by f'''. This process can be continued indefinitely as long as the derivatives are defined on some interval. Collectively, all these derivatives are known as the **higher derivatives** of f. Since the "prime" notation becomes awkward, the nth derivative of f is usually denoted by $f^{(n)}$.

Exercises

1. Use the definition of derivative to find $f'(1)$.
a) $f(x) = x^4 - 2x^2$ **b)** $f(x) = \sqrt{x}$ **c)** $f(x) = x/\sqrt{x^2+1}$

2. Prove that the absolute value function is not differentiable at 0. Note, however, that this function is continuous at 0.

3. Prove that the function $f\colon [0,1] \to \mathbb{R}$ defined by $f(x) = |x|$ is differentiable on $[0, 1]$. (Read the definition carefully.)

4. Prove that the function $f(x) = \begin{cases} x^2, & \text{if } x \le 1; \\ 2x-1, & \text{if } x > 1; \end{cases}$ is differentiable at 1.

5. Prove that the function f defined by $f(x) = x^r \cos(1/x)$ for $x \neq 0$ and $f(0) = 0$ is differentiable at 0 if $r = 2$ and not differentiable at 0 if $r = 1$.

6. Use the definition of derivative to find $f'(x)$.
a) $f(x) = x^3 - 5x^2$ **b)** $f(x) = 1/x^2$ **c)** $f(x) = \sqrt{x}$

7. Suppose that f is defined on an open interval I and that f is differentiable at a point $c \in I$. Prove that the sequence $\{n(f(c+1/n) - f(c))\}$ converges to $f'(c)$.

8. Prove Theorem 4.3.

9. Prove Theorem 4.4.

10. How would you compute the derivative of the sum of n functions, where n is an integer greater than 2? Formulate your answer as a theorem and prove it.

11. Use the discussion involving the quantities ℓ and w, which can be found prior to the statement of the product rule, to give a different proof of the product rule. Start by letting $\ell = f(c)$ and $\Delta\ell = f(x) - f(c)$.

12. Suppose that f is differentiable on I. Use the product rule and mathematical induction to show that f^n (the function f raised to the nth power) is differentiable on I for every positive integer n and find a formula for $(f^n)'$.

13. Prove the quotient rule in the following way.

a) Use the definition of the derivative to find a formula for the derivative of $1/g$.

b) Use part (a) and the product rule to prove the quotient rule.

14. Show that the "quick" proof of the Chain Rule is valid if $g'(c) \neq 0$ and fill in all of the details of this proof.

15. State the Chain Rule for the case in which the functions are only differentiable at a single point.

16. Suppose that f and g are differentiable functions and that $f(2) = 4$, $g(2) = 2$, $f'(2) = 5$, and $g'(2) = -3$. Find $(fg)'(2)$, $(f/g)'(2)$, and $(f \circ g)'(2)$.

17. Suppose that f is a strictly monotone differentiable function defined on an interval I. Assume that f_{inv} is differentiable and apply the Chain Rule to the function $f \circ f_{\text{inv}}$ to derive the formula for $(f_{\text{inv}})'$.

18. Assume that the function f defined by $f(x) = 5x + \sin(\pi x)$ is strictly increasing on $\mathbb{R}$ and find $(f_{\text{inv}})'(10)$.

19. Show that the function f defined by $f(x) = x^3 + 3x + 1$ is strictly increasing on $\mathbb{R}$, then find an equation for the line tangent to the graph of $y = f_{\text{inv}}(x)$ when $x = 5$.

20. Prove part (1) of Theorem 4.9 with the following steps.

a) If r is a positive integer, then use the definition of derivative and either the binomial theorem or a factoring formula.

b) If r is a negative integer, then use either the definition or the quotient rule.

c) If r is the reciprocal of a positive integer, then use Theorem 4.8.

d) For the general case, write $r = p/q$ and use the Chain Rule.

21. Find, with proof, the values of each of the following limits. Part (a) requires some geometry and the definitions of the trigonometric functions. The limits in parts (b) and (c) can be determined using the answer to part (a).

a) $\displaystyle\lim_{x\to 0} \frac{\sin x}{x}$ **b)** $\displaystyle\lim_{x\to 0} \frac{\cos x - 1}{x}$ **c)** $\displaystyle\lim_{x\to 0} \frac{\sin x}{\tan 3x}$

22. Use some of the limits in the previous exercise to prove part (2) of Theorem 4.9.

23. Use the fact that $\cos x = \sin\big((\pi/2) - x\big)$ to prove part (3) of Theorem 4.9.

24. Use the quotient rule to prove parts (4), (5), (6), and (7) of Theorem 4.9.

25. Use Theorem 4.8 to prove part (8) of Theorem 4.9.

26. Use the Chain Rule to prove parts (9) and (10) of Theorem 4.9.

Remark. The rest of the exercises in this section are in no particular order.

27. Define a function $f\colon [0, 2] \to \mathbb{R}$ by $f(x) = \inf\{|x - 1/n| : n \in \mathbb{Z}^+\}$. Determine the set of points at which f is differentiable.

28. Suppose that f has a third derivative on $\mathbb{R}$ and that $f(a) = a$ and $f'(a) = -1$ for some real number a. Let $g(x) = f(f(x))$. Compute $g'(a)$ and $g''(a)$. (You should get numerical answers.) Find a simple formula for $g'''(a)$.

29. Let q be a positive integer and let $r = 1/q$. Use the factorization

$$a^q - b^q = (a - b)\big(a^{q-1} + a^{q-2}b + a^{q-3}b^2 + \cdots + ab^{q-2} + b^{q-1}\big),$$

with $a = v^{1/q}$ and $b = x^{1/q}$, to evaluate the limit

$$\lim_{v\to x} \frac{v^{1/q} - x^{1/q}}{v - x}$$

and conclude that the derivative of x^r is rx^{r-1}.

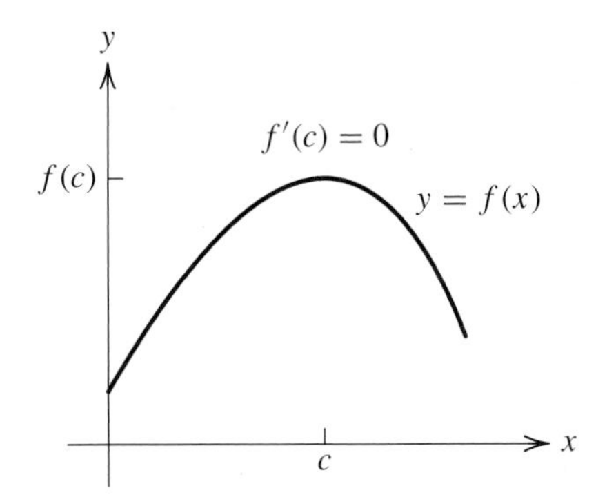

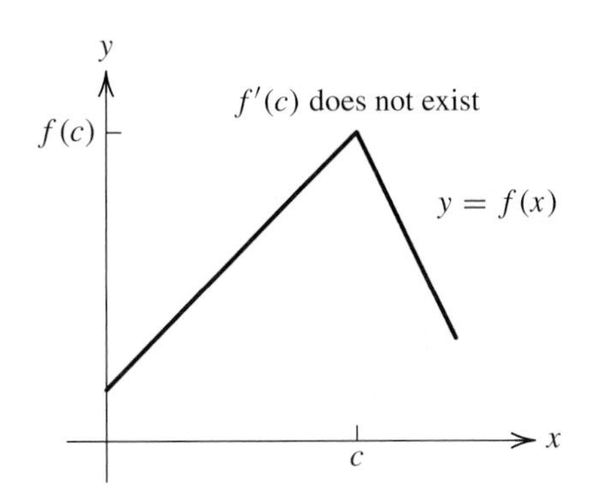

Figure 4.2 Two possibilities for the derivative when a relative maximum value occurs

30. Find the derivative of the function $f(x) = \sqrt[3]{6x^4 + x^2\cos(4x)}$ and note all of the formulas that you use. Can you imagine using the definition to compute this derivative?

31. Evaluate $\lim_{h \to 0} \dfrac{(3a + 2h)^{16/5} - (3a)^{16/5}}{h}$, where a is a constant.

4.2 THE MEAN VALUE THEOREM

Almost all of the basic theorems in differential calculus have relatively simple proofs, and these proofs can be found in most current calculus textbooks. In this section, we will consider a number of these results and some of their applications. The most important of these results, from a theoretical perspective, is the Mean Value Theorem.

The extreme values and relative extreme values of a function (these values were defined in Section 1.5) are important in many applications and, as the reader may recall from calculus, can often be found using the derivative of the function. Suppose that f has a relative extreme value at a point $c \in (a, b)$. It follows from the definition of a relative extreme value that the difference quotients of f at c have opposite signs on either side of c. Hence, the derivative of f at c must be 0 if it exists (see Figure 4.2). This is the content of the next theorem; the details of the proof will be left as an exercise.

THEOREM 4.10 Let I be an interval, let $f: I \to \mathbb{R}$, and suppose that $c \in I$ is not an endpoint of I. If f has a relative extreme value at c, then either f is not differentiable at c or $f'(c) = 0$. ■

Theorem 4.10 has a simple geometric interpretation. If $f: [a, b] \to \mathbb{R}$ has a relative extreme value at a point $c \in (a, b)$ and the graph of $y = f(x)$ has a tangent line at c, then the tangent line at c is horizontal. For a physical interpretation, consider the common experience of tossing a ball into the air. When the ball reaches its maximum height, it has zero velocity for an instant. If $f(t)$ represents the height of the ball at time t, then $f'(t)$ represents the velocity of the ball at time t. The common sense statement that the ball reaches its maximum height at the instant its velocity is zero translates mathematically into the statement that the function f has a maximum value at a time c for which $f'(c) = 0$.

The Extreme Value Theorem (Theorem 3.17) states that a continuous function has a maximum value and a minimum value on a closed and bounded interval, but the proof of the theorem does not provide any method for finding the points where such values occur. It is possible that the extreme values of a continuous function f on an interval $[a, b]$ occur at the endpoints of the interval. Theorem 4.10 shows that if the extreme values occur at a point in the interior of the interval, then f' is either zero or does not exist at those points. A point x in the domain of f for which either $f'(x) = 0$ or $f'(x)$ does not exist is called a **critical point** of f. Hence, the extreme values of a continuous function f on an interval $[a, b]$ occur either at the endpoints a and b or at the critical points of f that belong to the interval $[a, b]$. This makes the problem of finding the extreme values of many functions very easy.

As an example, consider the function $f(x) = (10x - x^2)^{2/3}$ on the interval $[-1, 8]$. The simplified derivative of this function is

$$f'(x) = \frac{4(5 - x)}{3(10x - x^2)^{1/3}}.$$

Now $f'(x)$ does not exist when $x = 0$ or $x = 10$ and $f'(x) = 0$ when $x = 5$. Since 10 is not in the interval $[-1, 8]$, the only relevant critical points are 0 and 5. Since

$$f(-1) = \sqrt[3]{121}, \quad f(0) = 0, \quad f(5) = \sqrt[3]{625}, \text{ and } f(8) = \sqrt[3]{256},$$

we see that the maximum value of f on $[-1, 8]$ is $\sqrt[3]{625}$ and the minimum value of f on $[-1, 8]$ is 0.

The properties of the derivative of a function provide quite a bit of information about the function itself. The main result that links the properties of a function to its derivative is known as the Mean Value Theorem. The Mean Value Theorem is a rather simple theorem, but it has far-reaching consequences, and it well deserves the title "Fundamental Theorem of Differential Calculus". The proof of the Mean Value Theorem is elementary and is included in almost every calculus textbook. It uses a common proof strategy: prove a special case of the theorem and show that the general case reduces to the special case. The special case of the Mean Value Theorem is known as Rolle's Theorem, named after Michel Rolle (1652–1719).

THEOREM 4.11 Rolle's Theorem Let $f\colon [a, b] \to \mathbb{R}$ be continuous on $[a, b]$ and differentiable on (a, b). If $f(a) = f(b)$, then there exists a point $c \in (a, b)$ such that $f'(c) = 0$.

Proof. By the Extreme Value Theorem, the function f has a maximum value and a minimum value on the interval $[a, b]$. Unless $f(x) = f(a)$ for all $x \in [a, b]$ (in which case the conclusion is obvious), at least one of these extreme values occurs at a point $c \in (a, b)$. It then follows from Theorem 4.10 that $f'(c) = 0$. ■

In applications involving Rolle's Theorem, it is often the case that both $f(a)$ and $f(b)$ are 0. This version of Rolle's Theorem has a strong geometric flavor since it states that a differentiable function with two x-intercepts has a horizontal tangent line somewhere between the x-intercepts. Although it is occasionally mentioned in

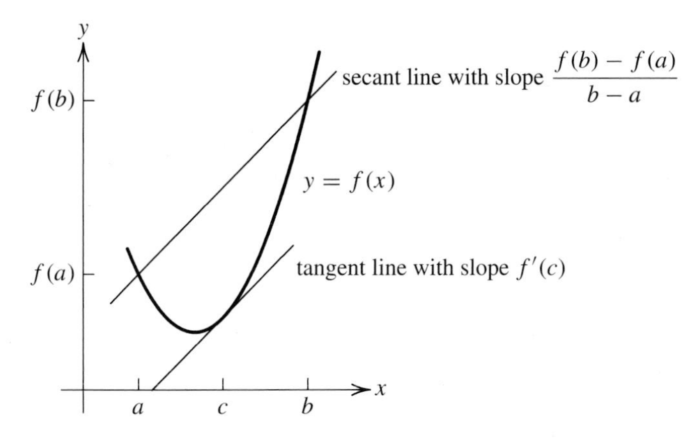

Figure 4.3 A geometric interpretation of the Mean Value Theorem

problems, the main purpose of Rolle's Theorem is as a stepping stone to the Mean Value Theorem.

THEOREM 4.12 Mean Value Theorem If $f\colon [a, b] \to \mathbb{R}$ is continuous on $[a, b]$ and differentiable on (a, b), then there exists a point $c \in (a, b)$ such that $f'(c) = (f(b) - f(a))/(b - a)$.

Proof. Define a function $g\colon [a, b] \to \mathbb{R}$ by

$$g(x) = f(x) - \left(\frac{f(b) - f(a)}{b - a}(x - a) + f(a)\right).$$

It is easy to verify that g satisfies all of the hypotheses of Rolle's Theorem. Consequently, there exists a point $c \in (a, b)$ such that $g'(c) = 0$. Plugging c into the formula for g', it follows that $f'(c) = (f(b) - f(a))/(b - a)$. ∎

The Mean Value Theorem guarantees the existence of a point c with a certain property. Like most existence theorems in real analysis, it has its roots in the Completeness Axiom; the reader is encouraged to search the proof of the Mean Value Theorem to spot the appearance of the Completeness Axiom. It is worthwhile to point out that the conclusion of the Mean Value Theorem can be phrased in other ways. The last equation can be written as

$$f(b) - f(a) = f'(c)(b - a) \quad \text{or} \quad f(b) = f(a) + f'(c)(b - a).$$

Of course, this is just simple algebra, but each version provides a slightly different way of looking at the conclusion.

Figure 4.3 provides a geometric interpretation for the Mean Value Theorem. The number $(f(b) - f(a))/(b - a)$ represents the slope of the secant line connecting the points $(a, f(a))$ and $(b, f(b))$ on the graph of f, and the number $f'(c)$ represents the slope of the tangent line to the graph of f at $(c, f(c))$. The conclusion of the Mean Value Theorem states that there exists a point in the interval where the tangent line is parallel to the secant line. The slope of the secant line can be interpreted as the average rate of change of the function f on the interval $[a, b]$. It is from this perspective that the adjective "mean" appears in the Mean Value Theorem; there exists a point c in the interval at which the instantaneous rate of change of f at c equals the average rate of change of f on the interval $[a, b]$.

The Mean Value Theorem can be used to establish useful and/or interesting inequalities. As an example, we will prove that $|\tan b - \tan a| \geq |b - a|$ for all a, b in the open interval $(-\pi/2, \pi/2)$. Let a and b be any two such points and suppose that $a < b$. The function $f(x) = \tan x$ is continuous on the interval $[a, b]$ and differentiable on the interval (a, b). By the Mean Value Theorem, there exists a point $c \in (a, b)$ such that

$$|\tan b - \tan a| = |(\sec^2 c)\,(b - a)|.$$

Since $\sec^2 c \geq 1$ for all $c \in (-\pi/2, \pi/2)$, it follows that $|\tan b - \tan a| \geq |b - a|$.

The Mean Value Theorem is the basis for a number of important results that are used frequently in differentiation theory; the next theorem illustrates one such application. It is known as a monotonicity theorem since it gives conditions that guarantee a function is monotone. Its conclusion is probably familiar to you. The theorem is stated only for the case of an increasing function; the analogous statement for decreasing functions is left to the reader, as is the proof of the corollary that follows the theorem.

THEOREM 4.13 Monotonicity Theorem Let $f\colon [a, b] \to \mathbb{R}$ be continuous on $[a, b]$ and differentiable on (a, b).

a) If $f' > 0$ on (a, b), then f is strictly increasing on $[a, b]$.

b) If $f' \geq 0$ on (a, b), then f is increasing on $[a, b]$.

c) If $f' = 0$ on (a, b), then f is constant on $[a, b]$.

Proof. We will prove part (a); the proofs of parts (b) and (c) are similar. Suppose that $a \leq x < y \leq b$. Applying the Mean Value Theorem to the function f on the interval $[x, y]$, there exists a point $c \in (x, y)$ such that

$$f(y) - f(x) = f'(c)(y - x).$$

Since $f'(c)$ and $y - x$ are positive, it follows that $f(y) > f(x)$. Hence, the function f is strictly increasing on $[a, b]$. ■

COROLLARY 4.14 Let f and g be two continuous functions that are defined on $[a, b]$ and differentiable on (a, b). If $f' = g'$ on (a, b), then there exists a constant k such that $f(x) = g(x) + k$ for all $x \in [a, b]$. ■

As a simple example of this familiar theorem from calculus, let f be the function defined by $f(x) = 2x + \sin x$. Since $f'(x) = 2 + \cos x \geq 1$ for all x, the function f is strictly increasing on $\mathbb{R}$. As for the corollary, suppose we need to find a function g such that $g'(x) = 3x^2$ for all x. It is clear that $g(x) = x^3$ is one such function and that $g(x) = x^3 + 4$ is another. Is there another, possibly very complicated and/or unusual function, whose derivative is also $3x^2$? The answer is no; Corollary 4.14 guarantees that every function whose derivative is $3x^2$ for all values of x is of the form $x^3 + k$ for some constant k.

Theorem 4.13 appears to be an almost trivial result, especially if one is encouraged to visualize a graph as resembling its tangent lines: if all of the tangent lines have positive slopes, then the function must be increasing, and if all of the tangent lines are horizontal, then the function must be constant. However, after considering all of the results that lie behind it, it becomes clear that these results are not trivial. In fact, Theorem 4.13 (c), which probably seems to be the clearest of all, requires the Completeness Axiom at some point in its proof. It is certainly possible to prove these monotonicity results without using the Mean Value Theorem (see Gordon [9] for one way to do this), but there is no truly simple proof. In recent years, various mathematicians have debated the role of the Mean Value Theorem in first year calculus and its use in the proof of the Monotonicity Theorem; for a taste of this debate, see the articles by Tucker [27] and Swann [25]. By the way, Exercise 26 in this section shows that a function can have a positive derivative at a point without being an increasing function in an open interval containing that point.

At the beginning of this section, we stated that the derivative of a function with a relative extreme value at a point in the interior of an interval either is zero or does not exist (see Theorem 4.10). The converse of this result is false; a function may not have a relative extreme value at one of its critical points. A simple example of this situation is the function f defined by $f(x) = x^3$, which has a critical point at 0 but does not have a relative extreme value at 0. Therefore, after finding the critical points of a function f, it is necessary to determine if these points actually correspond to relative extreme values. The First Derivative Test provides one method for determining whether or not a relative extreme value occurs at a critical point, and the Second Derivative Test provides a method for the particular case in which $f'(c) = 0$. The proofs of these tests involve Theorem 4.13 and will be left as exercises. Since the reader has probably encountered these tests before and since they are not difficult to apply, no examples will be given.

THEOREM 4.15 First Derivative Test Suppose that f is continuous on an open interval (a, b) and differentiable on (a, b) except possibly at the point $c \in (a, b)$ and assume that c is a critical point of f.

a) If f' is positive on (a, c) and negative on (c, b), then f has a relative maximum value at c.

b) If f' is negative on (a, c) and positive on (c, b), then f has a relative minimum value at c. ■

THEOREM 4.16 Second Derivative Test Suppose that f is twice differentiable on an open interval (a, b) and that $f'(c) = 0$ for some point $c \in (a, b)$.

a) If $f''(c) < 0$, then f has a relative maximum value at c.

b) If $f''(c) > 0$, then f has a relative minimum value at c. ■

We conclude this section with a brief look at L'Hôpital's Rule. This rule, named after Guillaume de L'Hôpital (1661–1704), provides a simple method for evaluating some limits. Although the rule is easy to apply, its proof, at least for the general case, is rather complicated. In addition, there are many different variations of L'Hôpital's

Rule, which makes it rather difficult to state every possible version. We will be content to look at one particular case and leave an exploration of other cases to the exercises. Although L'Hôpital's Rule is sometimes useful, it is not very important as a theoretical tool in real analysis. For this reason, it will be covered lightly here; the interested reader can find a general proof of this result in Rudin [22].

L'Hôpital's Rule states that under appropriate hypotheses,

$$\lim_{x\to c}\frac{f(x)}{g(x)}=\lim_{x\to c}\frac{f'(x)}{g'(x)}.$$

The limit may be a two-sided limit, a one-sided limit, or a limit as x goes to ∞ or $-\infty$. The appropriate hypotheses include the fact that either

$$\lim_{x\to c} f(x)=0=\lim_{x\to c} g(x) \quad \text{or} \quad \lim_{x\to c} f(x)=\pm\infty=\lim_{x\to c} g(x).$$

As the following equations reveal, there is very little mystery to L'Hôpital's Rule for the $0/0$ case:

$$\lim_{x\to c}\frac{f(x)}{g(x)}=\lim_{x\to c}\frac{\dfrac{f(x)-f(c)}{x-c}}{\dfrac{g(x)-g(c)}{x-c}}=\frac{f'(c)}{g'(c)}=\lim_{x\to c}\frac{f'(x)}{g'(x)}.$$

The determination of the hypotheses required for these equations to be valid will be left as an exercise. The version of L'Hôpital's Rule that we will prove is slightly more general than this and requires the following extension of the Mean Value Theorem. The proof of this generalization appears to be very easy; the hard part of the proof is buried in the discovery of the function h.

THEOREM 4.17 Cauchy Mean Value Theorem If f and g are continuous on $[a, b]$ and differentiable on (a, b), then there exists a point $c \in (a, b)$ such that

$$(f(b)-f(a))g'(c)=(g(b)-g(a))f'(c).$$

Proof. To prove this result, apply Rolle's Theorem to the function $h\colon [a, b] \to \mathbb{R}$ defined by

$$h(x)=\big(f(b)-f(a)\big)g(x)-\big(g(b)-g(a)\big)f(x). \qquad \blacksquare$$

THEOREM 4.18 L'Hôpital's Rule Let f and g be continuous on $[a, b]$ and differentiable on (a, b) except possibly at the point $c \in (a, b)$, and suppose that $g' \neq 0$ on $(a, b) \setminus \{c\}$. If $\lim_{x\to c} f(x) = 0 = \lim_{x\to c} g(x)$ and $\lim_{x\to c} f'(x)/g'(x)$ exists, then $\lim_{x\to c} f(x)/g(x)$ exists and $\lim_{x\to c}\dfrac{f(x)}{g(x)}=\lim_{x\to c}\dfrac{f'(x)}{g'(x)}$.

Proof. Since f and g are continuous at c, we see that $f(c) = 0 = g(c)$. By the Cauchy Mean Value Theorem, for each $x \in (a, b) \setminus \{c\}$ there exists a point z_x

between c and x such that

$$\frac{f'(z_x)}{g'(z_x)} = \frac{f(x) - f(c)}{g(x) - g(c)}.$$

(Note that $g'(z_x) \neq 0$ by hypothesis and that $g(x) - g(c) \neq 0$ by the Mean Value Theorem.) Since $z_x \to c$ as $x \to c$,

$$\lim_{x \to c} \frac{f(x)}{g(x)} = \lim_{x \to c} \frac{f(x) - f(c)}{g(x) - g(c)} = \lim_{x \to c} \frac{f'(z_x)}{g'(z_x)} = \lim_{x \to c} \frac{f'(x)}{g'(x)}.$$

This completes the proof. ■

Exercises

1. Prove Theorem 4.10. Begin by assuming that f has a relative extreme value at c and that f is differentiable at c.
2. Give an example of a function $f: [0, 1] \to \mathbb{R}$ such that f is continuous on $[0, 1]$, the extreme values of f occur at points in the interval $(0, 1)$, and the derivative of f is never 0. If possible, find a formula for $f(x)$, but at least give a graphical example.
3. Find all of the critical points of the given function.
   a) $f(x) = x^{2/3}(x - 2)^2$ **b)** $g(x) = \lfloor x \rfloor$ **c)** $h(x) = (x + 2)^2(2x - 3)^5$
4. Find the maximum and minimum values of $f(x) = \sin^2 x + \cos x$ on $[0, \pi]$.
5. Let a be a positive constant and define a function f by $f(x) = a^2x^2 - x^4$. Find the maximum and minimum values of f on the interval $[0, 2a]$.
6. Does the function f defined by $f(x) = x^3 - 12x^2$ have a maximum or a minimum value on the interval $(0, 10)$? If so, find it; if not, explain why.
7. Let a be a positive constant. Among all the possible choices for two nonnegative numbers whose sum is a, find the largest product of one number with the cube of the other number.
8. Suppose that $a < b$ and define a function f by $f(x) = (x - a)^2(x - b)^4$. Find the point c guaranteed by Rolle's Theorem on the interval $[a, b]$.
9. Consider the function f defined by $f(x) = 10\sqrt{x} - x$ on the interval $[0, 100]$. Examine both the hypotheses and the conclusion of Rolle's Theorem.
10. Prove that the equation $x^7 + x^5 + x^3 + 1 = 0$ has exactly one real solution. You should use Rolle's Theorem at some point in the proof.
11. Suppose that P is a polynomial of degree n and that P has n distinct real roots. Prove that $P^{(k)}$ has $n - k$ distinct real roots for $1 \leq k \leq n - 1$.
12. Referring to Figure 4.3, give a geometric interpretation of the function g that appears in the proof of the Mean Value Theorem.
13. Consider the function f defined by $f(x) = x^{2/3}$ on the interval $[-1, 27]$. Examine both the hypotheses and the conclusion of the Mean Value Theorem.
14. Let Q be an arbitrary polynomial of degree two and let $[a, b]$ be any interval. Find the point c guaranteed by the Mean Value Theorem. Do you notice anything interesting about this point?
15. Let f be the function defined by $f(x) = 1/x$ and let $[a, b]$ be any interval with $a > 0$. Find the point c guaranteed by the Mean Value Theorem. Do you notice anything interesting about this point?

16. A trip in a car covers 80 miles and takes one hour. Explain why the car was traveling at exactly 80 miles per hour at least one instant during the trip.

17. Use the Mean Value Theorem to prove that $\sqrt{1+x} < 1+x/2$ for all $x > 0$. Generalize this result to the function $(1+x)^r$, where $r < 1$ is a positive rational number.

18. Suppose that $f: \mathbb{R} \to \mathbb{R}$ is differentiable on $\mathbb{R}$ and that f' is bounded on $\mathbb{R}$. Prove that f is uniformly continuous on $\mathbb{R}$.

19. Suppose that f is continuous on $[a, b]$ and differentiable on (a, b). For each x in $[a, b]$, let $d(x)$ be the shortest distance from the point $(x, f(x))$ to the line joining the points $(a, f(a))$ and $(b, f(b))$. Show that the critical points of d are precisely the points guaranteed by the Mean Value Theorem.

20. Suppose that $f: [a, b] \to \mathbb{R}$ is an increasing function on $[a, b]$ and that f is differentiable on $[a, b]$. Prove that $f'(x) \geq 0$ for each $x \in [a, b]$.

21. Let f and g be nonzero differentiable functions defined on an interval I and suppose that

$$\frac{f'(x)}{f(x)} = \frac{g'(x)}{g(x)}$$

for all $x \in I$. Find a relationship between the functions f and g. (Do not use properties of logarithms.)

22. Finish the proof of Theorem 4.13.

23. State and prove the analogue of Theorem 4.13 for decreasing functions.

24. Prove Corollary 4.14.

25. Let a be a positive number and define a function f by $f(x) = x/(x^2+a^2)$. Determine the intervals on which f is increasing and those on which f is decreasing.

26. Consider the function $f: [-1, 1] \to \mathbb{R}$ defined by

$$f(x) = \begin{cases} x/2 + x^2 \sin(1/x), & \text{if } x \neq 0; \\ 0, & \text{if } x = 0. \end{cases}$$

Prove that $f'(0)$ is positive, but that f is not increasing on any open interval that contains 0.

27. Prove that the function f defined by $f(x) = x + \sin x$ is strictly increasing on $\mathbb{R}$.

28. Let n be any positive integer. Prove that $(1 - x^2)^n \geq 1 - nx^2$ for all $x \in [0, 1]$.

29. Prove the First Derivative Test.

30. Prove the Second Derivative Test.

31. Suppose that $f''(c) = 0$ in the Second Derivative Test. Can any conclusion be drawn?

32. Use the First Derivative Test to determine the nature of the critical points of the function $f: \mathbb{R} \to \mathbb{R}$ defined by $f(x) = x^5 - 5a^2x^3$, where a is a positive constant.

33. Although my calculus students seem to like the Second Derivative Test, it is actually a weaker test than the First Derivative Test. See if you can list some of the advantages of the First Derivative Test over the Second Derivative Test.

34. Let a and b be constants with $a > 0$, let f be defined by $f(x) = ax^3 + bx$, and let

$$\kappa(x) = \frac{f''(x)}{\left(1 + (f'(x))^2\right)^{3/2}}.$$

(The function κ represents the curvature of f.) Find the critical points of κ and use the First Derivative Test to classify them. What are the absolute extreme values of κ? Would you want to use the Second Derivative Test here?

35. Determine the hypotheses that are required to make the simple proof of L'Hôpital's Rule valid and fill in the details of this proof.

36. Fill in the details of the proof of the Cauchy Mean Value Theorem.

37. Evaluate $\lim_{x \to a} \dfrac{\sqrt{2a^3x - x^4} - a\sqrt[3]{a^2x}}{a - \sqrt[4]{ax^3}}$, which is one of L'Hôpital's original examples.

38. Evaluate each of the following limits.

a) $\lim_{x \to 0} \dfrac{2x + \arctan x}{x + \arctan 3x}$ **b)** $\lim_{x \to 0} \dfrac{x^2 \sin x}{\sin x - x \cos x}$ **c)** $\lim_{x \to -1} \dfrac{x^7 - x^5 + x^3 - x}{x^8 + x^6 + x^4 - 3x^2}$

39. Find values for the constants a and b so that $\lim_{x \to 0} (x^{-3} \sin(3x) + ax^{-2} + b) = 0$.

4.3 Further Topics on Differentiation

This section contains several unrelated properties and applications of the derivative: the intermediate value property of the derivative, the relationship between the graph of a function and the sign of its second derivative, and a discussion of convex functions. The intermediate value property of derivatives will be used later in this book, most prominently in the treatment of numerical integration; the other two topics will not be mentioned again, except in occasional exercises.

If f is differentiable on an interval I, then f is continuous on I. However, the function f' may not be continuous on I. As an example, consider the function f defined by

$$f(x) = \begin{cases} x^2 \sin(1/x), & \text{if } x \neq 0; \\ 0, & \text{if } x = 0. \end{cases}$$

This function is differentiable for all values of x and, as the reader should verify,

$$f'(x) = \begin{cases} 2x \sin(1/x) - \cos(1/x), & \text{if } x \neq 0; \\ 0, & \text{if } x = 0. \end{cases}$$

Since $\cos(1/x)$ does not have a limit at 0, the function f' is not continuous at 0. It is possible to construct much more complicated examples—there are functions whose derivatives exist at every point but have an uncountable number of discontinuities. An example of such a function can be found in Gordon [8], but some amount of measure theory (an extension of the concept of length of an interval to measure the size of arbitrary sets of real numbers) is necessary to understand the construction of this function. Given that a derivative may have many points of discontinuity, it is quite remarkable that derivatives have the intermediate value property.

THEOREM 4.19 If f is differentiable on an interval I, then f' has the intermediate value property on I.

Proof. Suppose that $[a, b] \subseteq I$ and assume, without loss of generality, that $f'(a) < f'(b)$. Let v be any number between $f'(a)$ and $f'(b)$. The function $g\colon [a, b] \to \mathbb{R}$ defined by $g(x) = f(x) - vx$ is differentiable (and hence continuous)

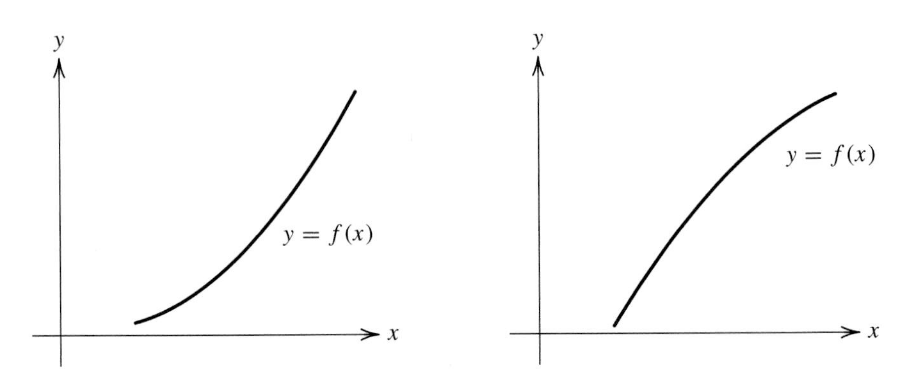

Figure 4.4 Two possible "shapes" for the graph of an increasing function

on $[a, b]$ and $g'(a) < 0 < g'(b)$. Since

$$\lim_{x \to a^+} \frac{g(x) - g(a)}{x - a} < 0 \quad \text{and} \quad \lim_{x \to b^-} \frac{g(x) - g(b)}{x - b} > 0,$$

there exist points $s \in (a, b)$ and $t \in (s, b)$ such that $g(s) < g(a)$ and $g(t) < g(b)$. (The proof of this statement will be left as an exercise.) By the Extreme Value Theorem, the function g has a minimum value on $[a, b]$. Since $g(s) < g(a)$ and $g(t) < g(b)$, the minimum value of g does not occur at either a or b. Hence, the function g has its minimum value at a point $c \in (a, b)$. By Theorem 4.10, we find that $g'(c) = 0$, and it follows that $f'(c) = v$. This completes the proof. ■

The most common application of differentiation discussed in first semester calculus is as a tool to reveal the nature of the graph of a function. Suppose that f is a differentiable function. The sign of f' on various intervals indicates whether the function is increasing or decreasing on those intervals, and the zeroes of f' are points at which the function may have a local extremum. As indicated in Figure 4.4, the graph of an increasing differentiable function can have two different "shapes". The distinction between the two graphs lies in the behavior of f'. The function f' is increasing (the slopes of the tangent lines are increasing) in the graph on the left, while f' is decreasing (the slopes of the tangent lines are decreasing) in the graph on the right. There is a similar difference in the graphs of decreasing differentiable functions; the reader should sketch graphs of decreasing functions f for which f' is increasing and some for which f' is decreasing. In order to distinguish between these two types of graphs, we make the following definition.

DEFINITION 4.20 Let f be a differentiable function defined on an interval I.

a) The function f is **concave up** on I if f' is increasing on I.

b) The function f is **concave down** on I if f' is decreasing on I.

Since $-f$ is concave up when f is concave down, each of the properties of concave up functions can be translated into a property of concave down functions. Hence, we will only consider functions that are concave up. Since a differentiable

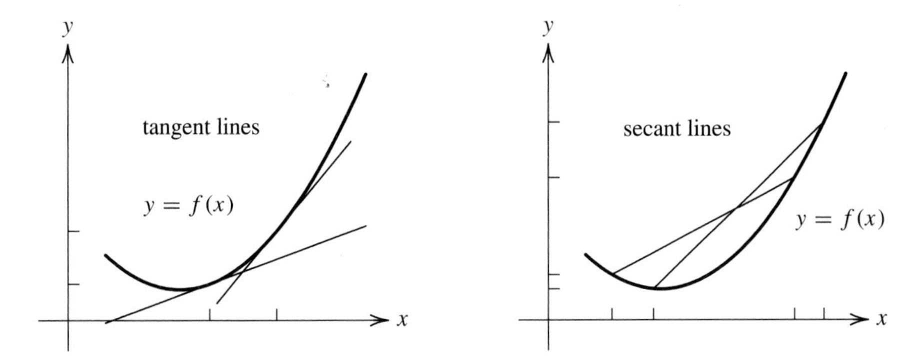

Figure 4.5 Tangent lines and secant lines for a concave up function

function is increasing if and only if its derivative is nonnegative, the following result is an immediate consequence of the definition.

THEOREM 4.21 Suppose that f is twice differentiable on an interval I. Then f is concave up on I if and only if f'' is nonnegative on I. ■

By focusing on the graph of a function, we can use geometry to discover some further properties of concave up functions. Let f be concave up on an interval I and consider the graph of $y = f(x)$ on I. Looking at the tangent lines to f on I, it is evident that the graph of the function lies above the tangent lines. In addition, looking at the secant lines for f on I, it is evident that the graph of the function lies below the secant lines. Figure 4.5 illustrates these last two comments. These results are valid for an arbitrary concave up function and will be proved after defining more precisely what it means for a graph to lie above its tangent lines or below its secant lines on an interval.

DEFINITION 4.22 Let f be a function defined on an interval I.

a) Suppose that f is differentiable on I. The graph of f **lies above its tangent lines** on I if for each $c \in I$, the inequality $f(x) \geq T_c(x)$ is valid for all $x \in I$, where T_c represents the function whose graph is the line tangent to the graph of $y = f(x)$ when $x = c$.

b) The graph of f **lies below its secant lines** on I if for each pair of distinct points $a, b \in I$ with $a < b$, the inequality $f(x) \leq S_{ab}(x)$ is valid for all $x \in [a, b]$, where S_{ab} represents the function whose graph is the line joining the points $(a, f(a))$ and $(b, f(b))$.

THEOREM 4.23 Let f be a differentiable function defined on an interval I. Then f is concave up on I if and only if the graph of f lies above its tangent lines on I.

Proof. Suppose first that f is concave up on I and let $c \in I$. The function T_c that represents the tangent line to $y = f(x)$ when $x = c$ is defined by

$$T_c(x) = f'(c)(x - c) + f(c).$$

We must show that $f(x) \geq T_c(x)$ for all $x \in I$. Let $x \in I$ and assume that $x > c$; the case in which $x < c$ is similar and will be left as an exercise. By the Mean Value Theorem, there exists $z_x \in (c, x)$ such that

$$f(x) - f(c) = f'(z_x)(x - c).$$

Since f' is increasing on I and $z_x > c$, we find that $f'(z_x) \geq f'(c)$. Using the fact that $x - c > 0$,

$$f(x) = f'(z_x)(x - c) + f(c) \geq f'(c)(x - c) + f(c) = T_c(x),$$

as desired.

Now suppose that the graph of f lies above its tangent lines on I and let a and b be two points in I with $a < b$. We must show that $f'(a) \leq f'(b)$. The functions T_a and T_b defined by

$$T_a(x) = f'(a)(x - a) + f(a) \quad \text{and} \quad T_b(x) = f'(b)(x - b) + f(b)$$

represent the tangent lines to $y = f(x)$ when $x = a$ and $x = b$, respectively. By hypothesis, $f(a) \geq T_b(a)$ and $f(b) \geq T_a(b)$. Substituting a and b into the formulas for T_b and T_a, respectively, and rearranging the resulting inequalities yields

$$f'(a) \leq \frac{f(b) - f(a)}{b - a} \leq f'(b).$$

This completes the proof. ■

THEOREM 4.24 Let f be a differentiable function defined on an interval I. Then f is concave up on I if and only if the graph of f lies below its secant lines on I.

Proof. Suppose first that f is concave up on I and let $[a, b] \subseteq I$. The function S_{ab} that represents the secant line joining the points $(a, f(a))$ and $(b, f(b))$ is defined by

$$S_{ab}(x) = \frac{f(b) - f(a)}{b - a}(x - a) + f(a).$$

Define a function D on $[a, b]$ by

$$D(x) = \begin{cases} \dfrac{f(x) - f(a)}{x - a}, & \text{if } a < x \leq b; \\ f'(a), & \text{if } x = a. \end{cases}$$

Since f is continuous on I and differentiable at a, the function D is continuous on $[a, b]$. In addition, it can be shown that $D'(x) \geq 0$ for all $x \in (a, b)$ (this will be left as an exercise). By Theorem 4.13, the function D is increasing on $[a, b]$. In particular, for each $x \in (a, b)$, we have $D(x) \leq D(b)$, which is equivalent to

$$f(x) \leq \frac{f(b) - f(a)}{b - a}(x - a) + f(a) = S_{ab}(x).$$

Therefore, the graph of f lies below its secant lines on I.

Now suppose that the graph of f lies below its secant lines on I and let $a, b \in I$ with $a < b$. We must show that $f'(a) \leq f'(b)$. By the Mean Value Theorem, there exists a point $c \in (a, b)$ such that $f(b) - f(a) = f'(c)(b - a)$. Consequently, the function S_{ab} that represents the secant line joining the points $(a, f(a))$ and $(b, f(b))$ can be expressed as either

$$S_{ab}(x) = f'(c)(x-a) + f(a) \quad \text{or} \quad S_{ab}(x) = f'(c)(x-b) + f(b).$$

Since the graph of f lies below its secant lines,

$$f(x) \leq S_{ab}(x) = f'(c)(x-a) + f(a)$$

for all $x \in [a, b]$. Hence,

$$f'(a) = \lim_{x \to a^+} \frac{f(x) - f(a)}{x - a} \leq f'(c).$$

The similar proof that $f'(b) \geq f'(c)$ will be left as an exercise. It follows that $f'(b) \geq f'(a)$. This completes the proof. ■

The definition of a function which lies below its secant lines does not require the function to be differentiable. It is clear geometrically, and can be shown algebraically, that the absolute value function lies below its secant lines on the interval $(-1, 1)$ even though it is not differentiable on $(-1, 1)$ (see the exercises). We are thus led to a more general condition for a function to be concave up. To avoid a confusion of terms and to be consistent with standard terminology in real analysis, such functions will be called convex functions.

DEFINITION 4.25 Let f be a function defined on an interval I. The function f is **convex** on I if for each interval $[c, d] \subseteq I$, the inequality

$$f\big((1-t)c + td\big) \leq (1-t)f(c) + tf(d)$$

is valid for $0 \leq t \leq 1$.

The definition of a convex function looks different than the definition of a function which lies below its secant lines, but the two definitions are equivalent; a proof of this fact will be left as an exercise. It is a good idea to keep both forms of the definition in mind as sometimes one form is easier to use than the other.

THEOREM 4.26 Let f be a function defined on an interval I. Then f is convex on I if and only if the graph of f lies below its secant lines on I. ■

After sketching some graphs of convex functions (we leave this to the reader), it appears that the graph of a convex function cannot oscillate back and forth and that the graph cannot have any interior breaks in it. This leads to the conjecture that convex functions are continuous. To prove this property of convex functions, as well as most other properties of convex functions, we need the following lemma.

LEMMA 4.27 Let f be a convex function defined on an interval I. If a, b, and c are points in I such that $a < b < c$, then

$$\frac{f(b) - f(a)}{b - a} \leq \frac{f(c) - f(a)}{c - a} \leq \frac{f(c) - f(b)}{c - b}.$$

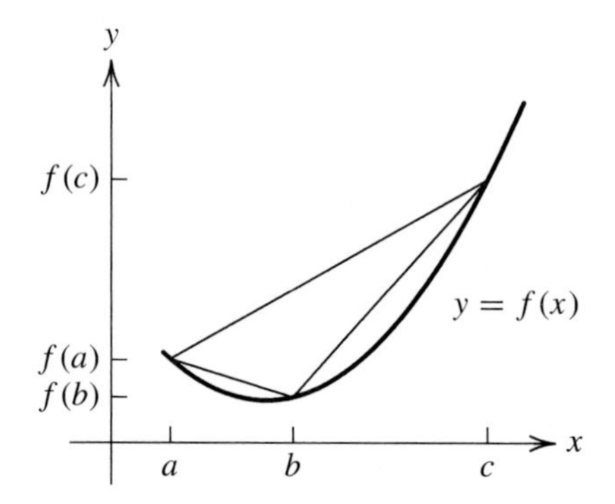

Figure 4.6 Comparing the slopes of secant lines for a convex function

Proof. The conclusion simply compares the slopes of various secant lines for the graph of $y = f(x)$; see Figure 4.6. Let S_{ac} be the function that represents the secant line joining the points $(a, f(a))$ and $(c, f(c))$. Then S_{ac} has the dual representations

$$\frac{f(c) - f(a)}{c - a}(x - a) + f(a) = S_{ac}(x) = \frac{f(c) - f(a)}{c - a}(x - c) + f(c)$$

for $a \leq x \leq c$. Since f is a convex function, $f(b) \leq S_{ac}(b)$. Using this fact and the two expressions for $S_{ac}(b)$ yields the desired inequality. ■

THEOREM 4.28 If f is a convex function defined on an open interval (a, b), then f is continuous on (a, b).

Proof. Let $[c, d]$ be any closed subinterval of (a, b) and choose points c_1 and d_1 in (a, b) so that $a < c_1 < c < d < d_1 < b$. If $x, y \in [c, d]$ with $x < y$, then Lemma 4.27 implies that

$$\frac{f(y) - f(x)}{y - x} \leq \frac{f(d) - f(y)}{d - y} \leq \frac{f(d_1) - f(d)}{d_1 - d} \quad \text{and}$$
$$\frac{f(y) - f(x)}{y - x} \geq \frac{f(x) - f(c)}{x - c} \geq \frac{f(c) - f(c_1)}{c - c_1}.$$

This shows that the set

$$\left\{ \left| \frac{f(y) - f(x)}{y - x} \right| : c \leq x < y \leq d \right\}$$

is bounded. Hence, there exists $M > 0$ such that $|f(y) - f(x)| \leq M|y - x|$ for all $x, y \in [c, d]$. It follows that f is uniformly continuous on $[c, d]$. Since $[c, d]$ was an arbitrary closed interval in (a, b), the function f is continuous on (a, b) (see the exercises). ■

Although a convex function defined on an open interval I must be continuous on I, it is not true that a convex function defined on a closed interval J must be continuous on J; an example is given in the exercises. Now that it is known that convex functions are continuous on open intervals, it makes sense to ask whether or not convex functions are differentiable. The following definition, which is consistent with the notion of one-sided limits, is required in order to discuss the differentiability properties of convex functions.

DEFINITION 4.29 Let f be defined on an interval I and let $c \in I$. The **left and right derivatives** of f at c are denoted and defined by

$$f^-(c) = \lim_{x \to c^-} \frac{f(x) - f(c)}{x - c} \quad \text{and} \quad f^+(c) = \lim_{x \to c^+} \frac{f(x) - f(c)}{x - c},$$

respectively, provided that the appropriate limits exist. (If c is an endpoint of I, only one of these expressions is defined.) To say that f has **one-sided derivatives** at c means that both $f^-(c)$ and $f^+(c)$ exist.

It should be clear that a function f is differentiable at c if and only if f has one-sided derivatives at c and $f^-(c) = f^+(c)$. As an example for which the two one-sided derivatives are not equal, consider the function f defined by $f(x) = |x|$. For this function, $f^-(0) = -1$ and $f^+(0) = 1$. Further properties of one-sided derivatives are considered in the exercises. Two interesting facts concerning convex functions on open intervals are that they have one-sided derivatives at each point of the interval and that the derivative actually exists at all but a countable number of points. The remaining results in this section prove these facts.

THEOREM 4.30 If f is a convex function defined on an open interval (a, b), then f has one-sided derivatives at each point of (a, b), $f^-(x) \le f^+(x)$ for all $x \in (a, b)$, and the functions f^- and f^+ are increasing on (a, b).

Proof. Fix $c \in (a, b)$. Define a function ϕ on $(a, b) \setminus \{c\}$ by

$$\phi(x) = \frac{f(x) - f(c)}{x - c},$$

let $A = \{\phi(x) : a < x < c\}$, and let $B = \{\phi(x) : c < x < b\}$. By Lemma 4.27, the function ϕ is increasing on each of the intervals (a, c) and (c, b), each element of A is a lower bound for the set B, and each element of B is an upper bound for the set A. Using these facts and Theorem 3.31,

$$f^-(c) = \phi(c-) = \sup A \le \inf B = \phi(c+) = f^+(c).$$

This shows that f has one-sided derivatives at c and that $f^-(c) \le f^+(c)$. Since c was an arbitrary point in (a, b), the function f has one-sided derivatives at each point of (a, b) and $f^-(x) \le f^+(x)$ for all $x \in (a, b)$.

Now suppose that $x_1, x_2 \in (a, b)$ with $x_1 < x_2$. Let v be any point between x_1 and x_2. Using Lemma 4.27 and the results from the first part of the proof, we find that

$$f^-(x_1) \le f^+(x_1) \le \frac{f(v) - f(x_1)}{v - x_1} \le \frac{f(v) - f(x_2)}{v - x_2} \le f^-(x_2) \le f^+(x_2).$$

This shows that both of the functions f^- and f^+ are increasing on (a, b). This completes the proof. ■

THEOREM 4.31 If f is a convex function defined on an open interval (a, b), then the set of points in (a, b) at which f is not differentiable is countable.

Proof. Since the function f^+ is increasing on (a, b) by the previous theorem, Theorem 3.33 shows that f^+ has a countable number of discontinuities. To complete the proof, it is sufficient to prove that f is differentiable at each point of continuity of f^+. Let c be a point of continuity of f^+ and let $\epsilon > 0$. Since f^+ is continuous at c, there exists $\delta > 0$ such that $|f^+(x) - f^+(c)| < \epsilon$ for all $x \in (c - \delta, c + \delta)$. Suppose that $c - \delta < x < c$. The last inequality in the proof of Theorem 4.30 shows that $f^+(x) \leq f^-(c)$. It follows that

$$f^+(c) - \epsilon < f^+(x) \leq f^-(c) \leq f^+(c).$$

Since $\epsilon > 0$ was arbitrary, we find that $f^-(c) = f^+(c)$. This shows that f is differentiable at c. ∎

Since a countable set (such as the set of rational numbers) can be quite large, the fact that a convex function is differentiable except at a countable number of points may not seem all that impressive. However, as Theorem 7.31 will demonstrate, there exist continuous functions that are not differentiable at any point. Thus, convex functions are much better behaved than the typical continuous function.

Exercises

1. Prove the existence of the points s and t used in the proof of Theorem 4.19.
2. Consider the function g defined by

$$g(x) = \begin{cases} x^2 \sin(\pi/x) + (x - 1)^2 \sin(\pi/(x - 1)), & \text{if } x \neq 0, 1; \\ 0, & \text{if } x = 0, 1. \end{cases}$$

Prove that g is differentiable at each real number and that g' is not continuous at either 0 or 1.

3. Give an example of a function that is differentiable on $\mathbb{R}$ and whose derivative has five points of discontinuity.
4. Suppose that f is differentiable on an interval I and that f' is nonzero on I. Prove that f is strictly monotone on I.
5. Let $f : [a, b] \to \mathbb{R}$ be differentiable on $[a, b]$ and suppose that f' is monotone on $[a, b]$. Prove that f' is continuous on $[a, b]$.
6. Explain why there is no function f such that $f'(x) = \lfloor x \rfloor$ for all $x > 0$.
7. Let $f : [0, 2] \to \mathbb{R}$ be continuous on $[0, 2]$ and differentiable on $(0, 2)$. Suppose that $f(0) = 0$, $f(1) = 0$, and $f(2) = 3$. Prove that there exist points $a, b, c \in (0, 2)$ such that $f'(a) = 0$, $f'(b) = 3$, and $f'(c) = 1$.
8. Suppose that f is differentiable on I and let $x_1, x_2, x_3 \in I$. Prove that there exists a point $c \in I$ such that $f'(c) = \big(f'(x_1) + 2f'(x_2) + 3f'(x_3)\big)/6$.
9. Let a be a positive constant. Find the intervals on which the function f defined by $f(x) = (x^2 + a^2)^{-1}$ is concave up and those on which it is concave down.
10. Give an example of a polynomial that is concave up on $\mathbb{R}$ and whose second derivative has at least two distinct roots.

11. Fill in the details of the first part of the proof of Theorem 4.23 by considering the case in which $x < c$.

12. Prove that the function D defined in the proof of Theorem 4.24 satisfies $D'(x) \geq 0$ for all $x \in (a, b)$.

13. Prove that $f'(b) \geq f'(c)$ as indicated in the second part of the proof of Theorem 4.24.

14. Let f be a function defined on an interval $[c, d]$.

a) Suppose that $0 \leq t \leq 1$. Prove that $c \leq (1-t)c + td \leq d$.

b) Let S_{cd} be the function that represents the line through the points $(c, f(c))$ and $(d, f(d))$ and suppose that $0 \leq t \leq 1$. Prove that

$$S_{cd}\big((1-t)c + td\big) = (1-t)f(c) + tf(d).$$

c) Prove Theorem 4.26.

15. Use the definition of a convex function to prove that $f(x) = x^2$ is convex.

16. Use the definition of a convex function to prove that $f(x) = |x|$ is convex.

17. Let f be a convex function on an interval I and let k be a positive constant. Prove that kf is convex on I.

18. Prove that the sum of two convex functions is a convex function.

19. Fill in the details of the proof of Lemma 4.27.

20. Give an example of a function which is convex and unbounded on $(0, 1)$.

21. Show that the function f defined by

$$f(x) = \begin{cases} 2, & \text{if } x = -1; \\ x^2, & \text{if } -1 < x < 2; \\ 5, & \text{if } x = 2; \end{cases}$$

is convex on $[-1, 2]$ even though it is not continuous on $[-1, 2]$.

22. Fill in the details behind the last sentence of the proof of Theorem 4.28.

23. Suppose that f is a convex function defined on $\mathbb{R}$. Prove that f is continuous on $\mathbb{R}$.

24. Suppose that f is a bounded convex function defined on (a, b). Prove that f has one-sided limits at a and b.

25. Give an example of a function f defined on $\mathbb{R}$ such that $f^-(2) = 3$ and $f^+(2) = 8$.

26. Let f be defined on an open interval I and let $c \in I$. Suppose that f has one-sided derivatives at c. Prove that f is continuous at c.

27. Let f be defined on an open interval I and let $c \in I$. Suppose that f has a relative extreme value at c and has one-sided derivatives at c. What can be said about $f^-(c)$ and $f^+(c)$?

28. Are the product rule and quotient rule valid for one-sided derivatives?

29. Find the one-sided derivatives of the function f defined by $f(x) = |x^4 - x|$ at the points 0 and 1.

30. Find the one-sided derivatives of the function f defined by $f(x) = |\sin(\pi x^2)|$ at each of its x-intercepts.

31. Let n be a positive integer.

a) Prove that the function f defined by $f(x) = \sum_{i=0}^{n} |x - i|$ is convex on $\mathbb{R}$.

b) Find the points at which f is not differentiable.

c) Prove that the function g defined by $g(x) = \sum_{i=0}^{n} |nx - i|$ is convex on $[0, 2]$.

d) Find the points at which g is not differentiable.

32. Suppose that f is a convex function defined on $\mathbb{R}$. Prove that the set of points at which f is not differentiable is countable.

33. Suppose that f is continuous and convex on $[a, b]$. Show that the conclusion of Theorem 4.30 can be extended to the closed interval $[a, b]$.

34. Let f be a continuous function defined on $[a, b]$, let $c \in (a, b)$, and suppose that f is convex on each of the intervals $[a, c]$ and $[c, b]$.

a) Give an example to show that f may not be convex on $[a, b]$.

b) Suppose that f is linear on each of the intervals $[a, c]$ and $[c, b]$ and that the slope of f on $[a, c]$ is less than the slope of f on $[c, b]$. Prove that f is convex on $[a, b]$.

c) Suppose that $f^-(c) \le f^+(c)$. Prove that f is convex on $[a, b]$.

35. For each positive integer n, let $x_n = (2^n - 1)/2^n$ and $y_n = n/(n + 1)$. Let f be the piecewise linear function that joins the points (x_n, y_n). Prove that f is convex on the interval $[0.5, 1)$ and that the set of points at which f is not differentiable is countably infinite. Part (b) or part (c) of the previous exercise may be helpful.

4.4 Supplementary Exercises

1. Suppose that a, b, c, and d are real numbers that satisfy $12a + 6b + 4c + 3d = 0$. Prove that the equation $a + bx + cx^2 + dx^3 = 0$ has a real solution.

2. Suppose that f is defined on $(0, b)$ and that f is differentiable at $c \in (0, b)$. (Do not assume that f is differentiable at any other points.) Evaluate each of the following limits; the answers will be in terms of c, $f'(c)$, etc.

a) $\displaystyle\lim_{x\to c} \frac{f(x) - f(c)}{x^2 - c^2}$ **b)** $\displaystyle\lim_{x\to c} \frac{xf(c) - cf(x)}{x - c}$ **c)** $\displaystyle\lim_{x\to \sqrt{c}} \frac{f(x^2) - f(c)}{x - \sqrt{c}}$

3. Consider the function $f\colon \mathbb{R} \to \mathbb{R}$ defined by $f(x) = x^3 - 2x^2 - x + 5$. Prove that f is monotone on an open interval that contains 1 and conclude that f_{inv} exists on this interval. Then prove that f_{inv} is differentiable at 3 and find an equation for the tangent line to the graph of $y = f_{\text{inv}}(x)$ at the point $(3, 1)$.

4. Let f and g be differentiable functions defined on $\mathbb{R}$. Suppose that $f' = g$ and $g' = -f$ on $\mathbb{R}$. Prove that $f^2 + g^2$ is a constant function.

5. Let $f\colon \mathbb{R} \to \mathbb{R}$ be twice differentiable on $\mathbb{R}$ and suppose that $f'' > 0$ on $\mathbb{R}$. Prove that for each real number L, the set $\{x \in \mathbb{R} : f(x) = L\}$ contains at most two points.

6. Give an example of a differentiable function $f\colon \mathbb{R} \to \mathbb{R}$ such that f is strictly increasing on $\mathbb{R}$ but f' is not monotone.

7. Consider the function f defined by $f(x) = x^2/2 + 15 - 12/x^2$. Find the minimum slope of the graph of $y = f(x)$ on the interval $[2, 10]$.

8. Find real numbers a and b so that the function f defined by

$$f(x) = \begin{cases} 3x^2, & \text{if } x \le 1; \\ ax + b, & \text{if } x > 1; \end{cases}$$

is differentiable at 1.

9. Let r be a rational number greater than 1 and let $f\colon \mathbb{R} \to \mathbb{R}$ be a function that satisfies $|f(y) - f(x)| \le (y - x)^r$ for all $x, y \in \mathbb{R}$. Prove that f is constant on $\mathbb{R}$.

10. Show that the function f defined by $f(x) = \sqrt{|x|}$ is uniformly continuous on $\mathbb{R}$ even though its derivative is unbounded.

11. Let $f(x) = x^2 \sin(1/x^2)$ for $x \neq 0$ and let $f(0) = 0$. Show that f is differentiable on $[0, 1]$, but that its derivative is unbounded on $[0, 1]$.

12. Suppose that $f: [a, b] \to \mathbb{R}$ is differentiable on $[a, b]$ and that f' is continuous on $[a, b]$. Prove that the set $\left\{ \dfrac{f(y) - f(x)}{y - x} : x, y \in [a, b], \ x \neq y \right\}$ is bounded.

13. Let $f: [a, b] \to \mathbb{R}$ be continuous on $[a, b]$ and differentiable on (a, b). Suppose that $|f'(x)| < 1$ for all $x \in (a, b)$. Prove that f has at most one fixed point.

14. Let $f: [a, b] \to \mathbb{R}$ be a differentiable function and let (c, d) be some point not on the graph of $y = f(x)$. Define a function D on $[a, b]$ by letting $D(x)$ be the distance between the points $(x, f(x))$ and (c, d).

a) Explain why the function D must have a minimum value on $[a, b]$.

b) Suppose the minimum value of D occurs at $t \in (a, b)$. Prove that the tangent line to f at t is perpendicular to the line joining the points $(t, f(t))$ and (c, d).

c) Find the minimum distance from the curve $y = x^2$ to the point $(14, 1)$.

15. Let $f: [a, b] \to \mathbb{R}$ be differentiable on $[a, b]$. Referring to the definition of uniform continuity, define the concept of **uniform differentiability** on $[a, b]$. Prove that f' is continuous on $[a, b]$ if and only if f is uniformly differentiable on $[a, b]$.

16. Let I be an interval, let $g: I \to \mathbb{R}$, let $c \in I$, let f be a function defined on an interval J that contains the set $g(I)$, and let $h = f \circ g$. Suppose that g is continuous at c, that h is differentiable at c, that f is differentiable at $g(c)$, and that $f'(g(c)) \neq 0$. Prove that g is differentiable at c and $g'(c) = h'(c)/f'(g(c))$.

17. Let p be a positive integer and define a function f on $\mathbb{R}$ by

$$f(x) = \min\{|x - k/p!| : 0 \leq k \leq p!\}.$$

Determine the set of points at which f is differentiable.

18. Let $f: [a, b] \to \mathbb{R}$ be continuous on $[a, b]$ and let $c \in (a, b)$. Suppose that f is differentiable on $(a, b) \setminus \{c\}$ and that $\lim_{x \to c} f'(x) = L$. Prove that f is differentiable at c and that $f'(c) = L$.

19. Let $f: [a, b] \to \mathbb{R}$ be continuous on $[a, b]$ and let $c \in (a, b)$. Suppose that f is differentiable on $(a, b) \setminus \{c\}$ and that $f' = 0$ on $(a, b) \setminus \{c\}$. Prove that f is constant on $[a, b]$. Try to generalize this result by allowing the number of points at which f is not differentiable to increase.

20. Suppose that f is differentiable on the interval $(0, \infty)$ and that $\lim_{x \to \infty} f'(x) = 0$. Let $g(x) = f(x + 2) - f(x)$. Prove that $\lim_{x \to \infty} g(x) = 0$. Give an example to show that f may be unbounded.

21. Let $f: [a, b] \to \mathbb{R}$ be differentiable at each point of $[a, b]$. Prove that there exists a sequence $\{x_n\}$ in (a, b) such that $\{x_n\}$ converges to a and $\{f'(x_n)\}$ converges to $f'(a)$. Note that f' may not be continuous.

22. Suppose that f is continuous on $[a, b]$ and differentiable on (a, b), and assume that the set $\{x \in (a, b) : f'(x) < 0\}$ is countable. Prove that f is increasing on I.

23. Let a and b be positive constants. Find the intervals on which the function f defined by $f(x) = ax^2 - (b/x)$ is concave up and those on which it is concave down.

24. Let f be a function defined on an interval (a, b) and let $c \in (a, b)$.

a) Suppose that f is differentiable at c. Prove that

$$f'(c) = \lim_{h \to 0^+} \frac{f(c+h) - f(c-h)}{2h}.$$

Give a geometric interpretation of this difference quotient.

b) Show that the limit in (a) may exist even if f is not differentiable at c.

c) Suppose that f is twice differentiable at c. Prove that

$$f''(c) = \lim_{h \to 0^+} \frac{f(c+h) - 2f(c) + f(c-h)}{h^2}.$$

Note that this formula for f'' does not involve f'.

25. Establish the following inequalities, valid for all $x \in [0, 1]$.

a) $0 \leq \sin x \leq x$

b) $1 - \frac{1}{2}x^2 \leq \cos x \leq 1$

c) $x - \frac{1}{6}x^3 \leq \sin x \leq x$

d) $1 - \frac{1}{2}x^2 \leq \cos x \leq 1 - \frac{1}{2}x^2 + \frac{1}{24}x^4$

e) Assuming that the pattern continues, what are the next two sets of inequalities?

26. Suppose that f is a function defined on an interval $[a, b]$, that $f^{(n)}$ exists on $[a, b]$ for some positive integer n, and that there is a constant M such that $f^{(n)}(x) \leq M$ for all $x \in [a, b]$. Establish the following string of inequalities, valid for all $x \in [a, b]$.

$$f^{(n-1)}(x) \leq M(x-a) + f^{(n-1)}(a);$$

$$f^{(n-2)}(x) \leq \frac{M}{2}(x-a)^2 + f^{(n-1)}(a)(x-a) + f^{(n-2)}(a);$$

$$\vdots$$

$$f(x) \leq \frac{M}{n!}(x-a)^n + \sum_{k=0}^{n-1} \frac{f^{(k)}(a)}{k!}(x-a)^k.$$

Suppose there is a constant m such that $m \leq f^{(n)}(x)$ for all $x \in [a, b]$. Establish a similar string of inequalities in this case.

27. Here is another proof of the Mean Value Theorem. Apply Rolle's Theorem to the function F defined by $F(x) = f(x) + A(b - x)$ for all $x \in [a, b]$, where the constant A is chosen so that $F(a) = f(b)$. Fill in the details of this proof.

28. Let $f\colon [a, b] \to \mathbb{R}$ and suppose that f' is continuous on $[a, b]$ and differentiable on (a, b). Prove that there exist points $c, d \in (a, b)$ such that

$$f(b) = f(a) + f'(a)(b-a) + \frac{f''(c)}{2}(b-a)^2,$$

$$f(a) = f(b) + f'(b)(a-b) + \frac{f''(d)}{2}(a-b)^2.$$

For the first equality, start with a function F defined by

$$F(x) = f(x) + f'(x)(b-x) + A(b-x)^2$$

for all $x \in [a, b]$, where A is a constant chosen so that $F(a) = f(b)$.

Remark. Several applications of the derivative appear in the next few exercises, but (except for the use of parameters rather than numbers) these problems could just as easily make an appearance in first semester calculus. However, you should not approach these problems the same way a typical calculus student would. First, make certain that you understand the why behind every single step—especially those steps that require results from this chapter. Second, write up the solution using complete sentences. Many students complain that they cannot understand their calculus books. Write your solution so that a calculus student would not make this complaint about your solution.

29. Let a and b be positive constants. Find an equation for the line that passes through the point (a, b) and cuts off the least area in the first quadrant. Find the area of this smallest triangle as well as the intercepts of the line.

30. Let a and b be positive constants and define a function f by $f(x) = b - ax^2$. For each value of x between 0 and the positive x-intercept of f, the tangent line to the graph cuts off a triangle in the first quadrant. Find the intercepts of the tangent line that cuts off the triangle of least area and the area of this triangle.

31. Find the volume of the largest cylinder that can be inscribed in a sphere of radius r. What is the ratio of height to radius for this cylinder? Repeat this problem for a cone.

32. Let a, b, and d be positive numbers. Suppose that two vertical poles of heights a and b are separated by a distance d (all units are the same). Find the length of the shortest wire that can run from the top of one pole to a point on the ground between the poles and up to the top of the other pole. Do you notice anything interesting about the two angles formed by the optimal wire and the ground?

33. Let a and d be positive numbers. Suppose that two light sources are separated by a distance d and that one source is a times as bright as the other. Find the point between the light sources at which there is the least amount of light. Use the fact that the intensity of the light at a point is proportional to the reciprocal of the square of the distance from the light source.

34. Find the volume of the largest cone with surface area S. Consider both the case in which the cone has a base and the case in which it does not.

Remark. Suppose that f is differentiable on an interval (a, b), that the equation $f(x) = 0$ has one root r in (a, b), and that $f'(r) \neq 0$. The existence of r may be guaranteed by the Intermediate Value Theorem or some other result, but the actual value of r is not known. **Newton's method** for finding approximations to r proceeds as follows. Begin by making a "good" initial approximation x_0 to r. Since the tangent line to the curve $y = f(x)$ when $x = x_0$ resembles the graph of the function for values of x near x_0, the x-intercept x_1 of the tangent line should be closer to r than the initial guess. Try sketching several graphs to convince yourself that this is a reasonable approach for obtaining a better approximation to the root. Similarly, the x-intercept x_2 of the tangent line at $x = x_1$ should be closer to r than x_1. This process can be continued to generate a sequence of (hopefully) better approximations to r. The next few exercises consider Newton's method for approximating roots.

35. Show that the sequence generated by Newton's method is defined recursively by

$$x_{n+1} = x_n - \frac{f(x_n)}{f'(x_n)}.$$

36. Suppose that f' is continuous on (a, b) and that all of the terms of the sequence $\{x_n\}$ generated by Newton's method are in the interval (a, b). Prove that if $\{x_n\}$ converges, its limit must be r. However, there is no guarantee that the sequence converges (see Exercise 39).

37. One method for approximating the square root of a positive integer n that has been around for centuries is to start with an initial guess s and let the average of the numbers s and n/s be the next approximation. Explain why $\sqrt{n}$ is always between s and n/s, then show that Newton's method applied to the equation $x^2 - n = 0$ yields this same formula.

38. Use Newton's Method and computations without a calculator to obtain rational approximations (written as the ratio of two integers) for $\sqrt{2}$, $\sqrt{3}$, and $\sqrt{5}$. Let x_0 be the integer that is closest to the root and find x_1, x_2, and x_3 in each case.

39. Suppose that f is twice differentiable on $[a, b]$, that $f(a)$ and $f(b)$ have opposite signs, and that there exist positive numbers α and β such that $|f'(x)| \ge \alpha$ and $|f''(x)| \le \beta$ for all $x \in [a, b]$. Let r be the unique root of f in (a, b) (see part (a) below) and choose $\delta > 0$ so that $[r - \delta, r + \delta] \subseteq [a, b]$ and $\delta < 2\alpha/\beta$. Let x_0 be any number in $[r - \delta, r + \delta]$ and let $\{x_n\}$ be the sequence generated by Newton's method. The following steps outline a proof that the sequence $\{x_n\}$ is contained in the interval $[r - \delta, r + \delta]$, converges to r, and satisfies $|x_{n+1} - r| \le \beta |x_n - r|^2/2\alpha$ for all n.

a) Explain why f has a unique root in (a, b).

b) Use Exercise 28 to show that there exists a point $c_n \in (a, b)$ such that

$$x_{n+1} - r = \frac{f''(c_n)}{2f'(x_n)}(x_n - r)^2.$$

c) Finish the proof by taking advantage of the definition of δ.

d) Referring to part (b), note that $|x_{n+1} - r| \le C|x_n - r|^2$ for some constant C. Suppose that x_0 is within $1/10$ of r. How accurate is x_4 guaranteed to be? This gives some indication of how quickly Newton's method can find an accurate approximation to a root.

Remark. Although it may come as a surprise, there are other types of derivatives. Examples include symmetric derivatives, strong derivatives, approximate derivatives, qualitative derivatives, and path derivatives. The last few get rather technical (the interested reader can consult Bruckner [4]), but the first two are fairly simple and are defined in the following exercises. These other derivatives are used primarily for theoretical purposes, but they lead to some interesting analysis.

40. Let f be a function defined on an interval (a, b) and let $c \in (a, b)$. The **symmetric derivative** of f at c is defined by

$$f_s'(c) = \lim_{h \to 0^+} \frac{f(c+h) - f(c-h)}{2h},$$

provided that the limit exists. Prove that $f_s'(c)$ exists whenever $f'(c)$ exists, but that it is possible for $f_s'(c)$ to exist even when $f'(c)$ does not exist.

41. Let f be a function defined on an interval (a, b), let $c \in (a, b)$, and suppose that f has a symmetric derivative at c. Does it follow that f is continuous at c?

42. Let $f\colon [a, b] \to \mathbb{R}$, let $c \in (a, b)$, and suppose that f is differentiable at c. Prove that for each $\epsilon > 0$ there is a $\delta > 0$ such that

$$\left| \frac{f(y) - f(x)}{y - x} - f'(c) \right| < \epsilon$$

for all points $x \neq y$ that satisfy $c - \delta < x \leq c \leq y < c + \delta$. This result is sometimes called the **Straddle Lemma** since the point c is straddled by the points x and y.

43. Let f be a function defined on an interval (a, b) and let $c \in (a, b)$. The number L is the **strong derivative** of f at c if for each $\epsilon > 0$ there exists $\delta > 0$ such that

$$\left|\frac{f(y) - f(x)}{y - x} - L\right| < \epsilon$$

for all points x, y that satisfy $c - \delta < x < y < c + \delta$. Note that (in contrast to the previous exercise) it is not necessary for c to belong to the interval $[x, y]$. We will use $f'_{st}(c)$ to represent the strong derivative of f at c. Prove that f is differentiable at c if it has a strong derivative at c and that $f'(c) = f'_{st}(c)$.

44. Show that the function $f : [-1, 1] \to \mathbb{R}$ defined by

$$f(x) = \begin{cases} x^2 \sin(\pi/x), & \text{if } x \neq 0; \\ 0, & \text{if } x = 0; \end{cases}$$

is differentiable at 0, but that $f'_{st}(0)$ does not exist.

45. Prove that a function $f : [a, b] \to \mathbb{R}$ is strongly differentiable at each point of $[a, b]$ if and only if f' is continuous on $[a, b]$.

Remark. The last few exercises involve the concept of absolute continuity, which was defined in the supplementary exercises of the previous chapter. The first exercise provides one of the simplest ways to prove that a function is absolutely continuous.

46. Suppose that f is continuous on $[a, b]$, that f is differentiable on (a, b), and that f' is bounded on (a, b). Prove that f is absolutely continuous on $[a, b]$.

47. Show that the function $\sqrt{x}$ is absolutely continuous on $[0, 1]$ even though its derivative is not bounded on $(0, 1)$.

48. Consider the function f defined by $f(x) = x^2 \,|\sin(1/x)|$ for $x \neq 0$ and $f(0) = 0$. Prove that f is absolutely continuous on $[0, 1]$.

49. Use the functions in the last two exercises to show that the composition of two absolutely continuous functions may not be absolutely continuous.

50. Consider the function f defined by $f(x) = x^2 \sin(1/x^2)$ for $x \neq 0$ and $f(0) = 0$. Determine whether or not f is absolutely continuous on $[0, 1]$.

5

Integration

The concept of the integral has its origins in the geometric problem of finding the area of a region with curved boundaries. This sort of problem arose in connection with trade and property rights, but the less practical problem of attempting to find the exact area of a circle also captured the attention of various individuals over the centuries. Theoretical solutions to the area problem extend back at least as far as ancient Greece. In general, the area problem is more difficult and opens the door to more applications than the tangent problem. The discovery by Isaac Newton (1642–1727) and Gottfried Leibniz (1646–1716) that the area problem can be interpreted as an inverse tangent problem is typically considered to be the beginning of calculus. As a result of this connection, tangents and inverse tangents came to the forefront and, in a rush to solve a number of applied problems, a concern for a rigorous definition of area was set aside for many years. This is illustrated by the fact that although calculus was discovered in the late seventeenth century, a rigorous definition of the integral did not appear until the middle of the nineteenth century. This definition and its consequences are the focus of the current chapter.

The simplest version of the geometric problem of finding the area of a region with curved boundaries can be stated as follows: suppose that a function f is continuous and nonnegative on an interval $[a, b]$ and let R be the region under the graph $y = f(x)$ and above the x-axis on the interval $[a, b]$. In order to approximate the area A of the region R (assuming the area exists), the region R is approximated with rectangles since the area of a rectangle is easy to determine. For some positive integer n, choose points x_i for $0 \leq i \leq n$ and points t_i for $1 \leq i \leq n$ so that

$$a = x_0 < x_1 < x_2 < \cdots < x_{n-1} < x_n = b$$

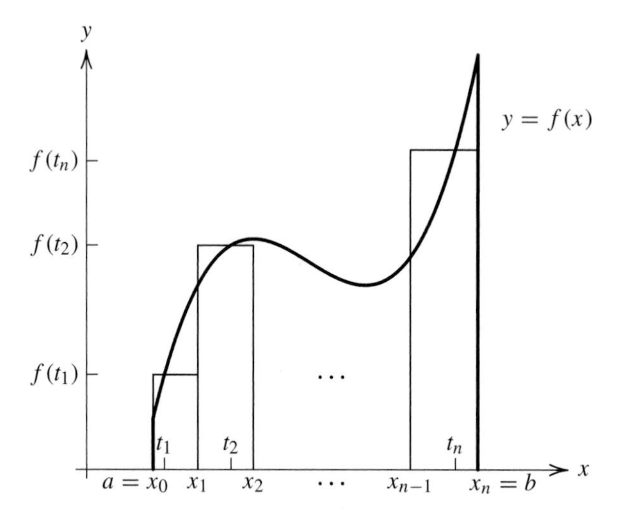

Figure 5.1 An approximation to the area under the curve $y = f(x)$

and $x_{i-1} \le t_i \le x_i$ for $1 \le i \le n$. The area of the rectangle with height $f(t_i)$ and width $x_i - x_{i-1}$ is an approximation to the area of R on the interval $[x_{i-1}, x_i]$ (see Figure 5.1). The sum of the areas of all the rectangles is an approximation to A:

$$A \approx \sum_{i=1}^{n} f(t_i)(x_i - x_{i-1}).$$

Intuitively, as n increases and the rectangles get thinner, the approximation to A improves. It should therefore be possible to approximate A to any degree of accuracy by choosing n large enough. In other words, under certain restrictions on the widths of the rectangles, any approximations generated by this method should be close to the value of A, assuming of course that the area A exists.

A somewhat different approach to this problem involves circumscribed and inscribed rectangles. With the x_i's defined as in the previous paragraph, let M_i and m_i be the maximum and minimum values, respectively, of the function f on the interval $[x_{i-1}, x_i]$. In this case (see Figure 5.2),

$$\sum_{i=1}^{n} m_i(x_i - x_{i-1}) \le A \le \sum_{i=1}^{n} M_i(x_i - x_{i-1}).$$

As n increases and the rectangles get thinner, the inscribed and circumscribed sums should "squeeze" the area A; that is, the sum

$$\sum_{i=1}^{n} (M_i - m_i)(x_i - x_{i-1})$$

should get closer to 0. This approach actually gives a very useful criterion for the existence of an area A: if the displayed sum can be made arbitrarily close to 0 by choosing thin enough rectangles, then, as we will see, a number A representing the area of R exists by the Completeness Axiom.

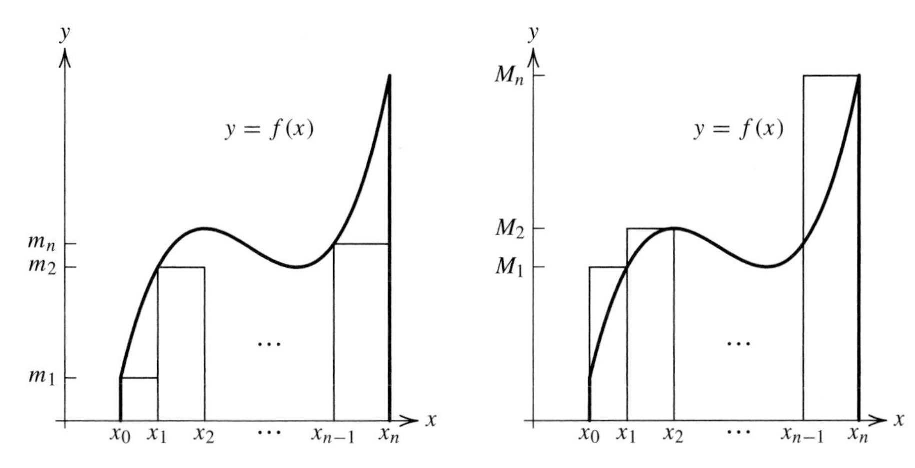

Figure 5.2 Area approximations with inscribed and circumscribed rectangles

We thus have two ways of looking at the area problem: one involving arbitrary rectangles and one involving inscribed and circumscribed rectangles. The first approach is useful for computations and also in applications because the products $f(t_i)(x_i - x_{i-1})$ can often be assigned a physical interpretation such as volume, mass, or work. The second approach provides a simple criterion for the existence of area (or the integral in general) and is extremely useful in the theory of integration. It should come as no surprise that these two approaches to the area problem are equivalent, but a proof of this fact is a bit tedious. This proof will appear in Section 5.2 after the integral is defined formally in Section 5.1. Further properties of the integral, including the Fundamental Theorem of Calculus and numerical techniques for approximating the value of an integral, will be discussed in the later sections.

5.1 The Riemann Integral

There are a number of different integration processes, but the two most common ones are the Riemann integral and the Lebesgue integral. The integral that appears in traditional calculus courses is the Riemann integral, named after Bernhard Riemann (1826–1866). Since this integral has some limitations, primarily from a theoretical perspective, Henri Lebesgue (1875–1942) was motivated to define an integral that extended the Riemann integral and overcame its deficiencies. (This means that every Riemann integrable function is Lebesgue integrable and the integrals have the same value, but there are Lebesgue integrable functions that are not Riemann integrable.) Because of its abstract nature, the Lebesgue integral is generally the subject of a graduate course in real analysis. The focus in this text is the Riemann integral. In this section, we will present the necessary terminology and definitions for this integral.

Compared with other concepts defined thus far in this book, the definition of the Riemann integral is more involved and complicated. To begin a discussion of this topic, it is necessary to develop additional terminology and notation; the next three definitions do just that.

Figure 5.3 A tagged partition of $[a, b] = [x_0, x_n]$

DEFINITION 5.1 A **partition** P of an interval $[a, b]$ is a finite set of points $\{x_i : 0 \le i \le n\}$ such that

$$a = x_0 < x_1 < x_2 < \cdots < x_{n-1} < x_n = b.$$

The **norm** of a partition P, denoted by $\| P \|$, is the largest of the numbers $x_i - x_{i-1}$, that is, $\| P \| = \max\{x_i - x_{i-1} : 1 \le i \le n\}$. If P_1 and P_2 are partitions of $[a, b]$ and $P_1 \subseteq P_2$, then P_2 is a **refinement** of P_1.

It is always assumed that the points of a partition of an interval are in increasing order with the first point corresponding to the left endpoint of the interval and the last point corresponding to the right endpoint of the interval. A partition of an interval splits the interval into subintervals and the norm of the partition is the length of the largest of these subintervals. A refinement of a partition is just the partition with some more points added; some or all of the subintervals are further subdivided in the refinement.

DEFINITION 5.2 A **tagged partition** ${}^tP$ of an interval $[a, b]$ consists of a partition $P = \{x_i : 0 \le i \le n\}$ of $[a, b]$ along with a set $\{t_i : 1 \le i \le n\}$ of points, known as **tags**, that satisfy $x_{i-1} \le t_i \le x_i$ for $1 \le i \le n$. We will express a tagged partition ${}^tP$ of $[a, b]$ by ${}^tP = \{(t_i, [x_{i-1}, x_i]) : 1 \le i \le n\}$, say that ${}^tP$ is a tagged partition formed from P, and refer to P as the partition associated with ${}^tP$. The norm $\|{}^tP\|$ of a tagged partition ${}^tP$ is the norm of the partition P associated with ${}^tP$.

Tagged partitions of an interval are not a difficult concept: the interval is simply split into subintervals with one point chosen (or tagged) from each subinterval (see Figure 5.3). As an example, divide the interval [0, 1] into five subintervals of equal length and choose the midpoint of each subinterval as the tag. This yields the tagged partition

$${}^tP = \{(.1, [0, .2]), (.3, [.2, .4]), (.5, [.4, .6]), (.7, [.6, .8]), (.9, [.8, 1])\}.$$

We can use tagged partitions to form the type of sum discussed in the introduction to this chapter.

DEFINITION 5.3 Let $f\colon [a, b] \to \mathbb{R}$ and let ${}^tP = \{(t_i, [x_{i-1}, x_i]) : 1 \le i \le n\}$ be a tagged partition of $[a, b]$. The **Riemann sum** $S(f, {}^tP)$ of f associated with ${}^tP$ is defined by

$$S(f, {}^tP) = \sum_{i=1}^{n} f(t_i)(x_i - x_{i-1}).$$

Let f be the function defined by $f(x) = x^2$ and let ${}^tP$ be the tagged partition of $[0, 1]$ defined prior to Definition 5.3. Then

$$S(f, {}^tP) = (.1)^2(.2) + (.3)^2(.2) + (.5)^2(.2) + (.7)^2(.2) + (.9)^2(.2) = 0.33.$$

Although the computations here involve nothing more than simple arithmetic, the notation associated with tagged partitions and Riemann sums can be a little overwhelming at first. With a little practice and experience, these concepts and their notation will become second nature.

In order for a function to be Riemann integrable, its Riemann sums must approach a limit as the norms of the tagged partitions go to 0. This corresponds to making the rectangles discussed in the introduction very thin, thereby obtaining a better approximation to the area under the curve. This concept is made precise in the following definition.

DEFINITION 5.4 A function $f: [a, b] \to \mathbb{R}$ is **Riemann integrable** on $[a, b]$ if there exists a number L with the following property: for each $\epsilon > 0$ there exists $\delta > 0$ such that $|S(f, {}^tP) - L| < \epsilon$ for all tagged partitions ${}^tP$ of $[a, b]$ that satisfy $\|{}^tP\| < \delta$. The number L is called the **Riemann integral** of f on $[a, b]$ and is denoted by the symbol $\int_a^b f(x)\, dx$, or simply $\int_a^b f$. The adjective "Riemann" will occasionally be dropped when discussing the integral or integrable functions.

This is a complicated limit process. For each $\delta > 0$, there are many (uncountably many in fact) tagged partitions ${}^tP$ of $[a, b]$ that satisfy $\|{}^tP\| < \delta$. The subintervals can be chosen in any manner as long as the length of each subinterval is less than δ. Once these subintervals have been chosen, the tag from each subinterval may also be chosen at random. As a result of all this variability, it is tedious and/or difficult to prove that a function is Riemann integrable on an interval using the definition unless the function has a very simple form.

As an illustration of the definition, consider the function f defined by $f(0) = 4$ and $f(x) = 0$ for all $x \neq 0$. We will prove that the function f is Riemann integrable on $[0, 1]$ and that $\int_0^1 f = 0$. (The value 0 comes from the area interpretation of the integral; sketch the graph.) Let $\epsilon > 0$ and choose $\delta = \epsilon/4$. Suppose that ${}^tP = \{(t_i, [x_{i-1}, x_i]) : 1 \le i \le n\}$ is a tagged partition of $[0, 1]$ with $\|{}^tP\| < \delta$. Since $t_i > 0$ for $2 \le i \le n$, it follows that $f(t_i) = 0$ for these values of i. It is also clear that $f(t_1)$ is either 0 or 4. Therefore,

$$\left|S(f, {}^tP)\right| = f(t_1)(x_1 - x_0) \le 4(x_1 - x_0) < 4\delta = \epsilon.$$

This shows that f is Riemann integrable on $[0, 1]$ and the value of the integral is 0.

In general, working with the definition of the Riemann integral requires a lot of attention to details. Using the definition of the Riemann integral helps to develop an understanding of this concept, but it is not a practical method for computing the value of an integral. Fortunately, there are ways of finding the value of an integral that bypass the definition; several such methods will be developed later in this chapter.

Suppose that f is Riemann integrable on $[a, b]$. It should be clear that the number $\int_a^b f$ is unique and that the value of the Riemann integral depends only on the function f and the interval $[a, b]$; the name of the variable is therefore irrelevant. In other words,

$$\int_a^b f(x)\,dx = \int_a^b f(t)\,dt = \int_a^b f(v)\,dv,$$

and so on. Variables such as these are known as **dummy variables** since they do not play a role in the value of the integral. Consequently, there is no real need to include the variable in the notation of the integral; the symbol $\int_a^b f$ contains all of the relevant information—namely, the function f and the interval $[a, b]$. The reason that the variable is often included follows from the symbolic statement

$$\int_a^b f(x)\,dx = \lim_{\|{}'P\|\to 0} \sum_{i=1}^{n} f(t_i)(x_i - x_{i-1}).$$

The symbol $\int$ is an elongated S, the first letter of the Latin word for sum, and the product $f(x)\,dx$ is reminiscent of the product $f(t_i)(x_i - x_{i-1}) = f(t_i)\Delta_i x$, where $\Delta_i x$ is the change in x in the ith subinterval. (This notation for the integral is due to Leibniz.) In applications, the numbers $f(t_i)$ and $\Delta_i x$ have physical meanings and, if you are willing to accept differentials, the symbols $f(x)$ and dx each have physical meanings as well. Since the focus in this chapter is the theory of the integral rather than its applications, the notation $\int_a^b f$ will be used most of the time.

The following theorem lists some of the algebraic properties of the Riemann integral. These properties follow easily from the definition of the integral, so the proofs will be left as exercises.

THEOREM 5.5 Suppose that f and g are Riemann integrable functions defined on $[a, b]$ and that k is a constant. Then

- **a)** kf is Riemann integrable on $[a, b]$ and $\int_a^b kf = k\int_a^b f$;
- **b)** $f + g$ is Riemann integrable on $[a, b]$ and $\int_a^b (f + g) = \int_a^b f + \int_a^b g$;
- **c)** $\int_a^b f \le \int_a^b g$ if $f(x) \le g(x)$ for all $x \in [a, b]$;
- **d)** $\left|\int_a^b f\right| \le M(b - a)$ if $|f(x)| \le M$ for all $x \in [a, b]$. ■

Two approaches to the area problem were mentioned in the introduction to this chapter; the one involving arbitrary rectangles has been used as the basis for the definition of the integral. In order to consider the second approach, which involves inscribed and circumscribed rectangles, we need to introduce one more concept.

DEFINITION 5.6 Let f be a bounded function defined on an interval $[a, b]$. The **oscillation** of f on $[a, b]$ is defined by

$$\omega(f, [a, b]) = \sup\{f(x) : x \in [a, b]\} - \inf\{f(x) : x \in [a, b]\}.$$

Note that the oscillation of f is defined on any subinterval $[c, d]$ of $[a, b]$ and that $\omega(f, [c, d]) \le \omega(f, [a, b])$.

The oscillation of a continuous function on $[a, b]$ is simply its maximum value minus its minimum value. Since an arbitrary bounded function may not have a maximum or minimum value, it is necessary to use the supremum and infimum of its range on the appropriate interval. For example, consider the function f defined by $f(x) = (1 - x)\sin(1/x)$ for $x \neq 0$ and $f(0) = 0$. On the interval $[0, 1]$, the range of this function is $(-1, 1)$, so $\omega(f, [0, 1]) = 2$. The importance of the oscillation of a function in integration theory and its relationship to inscribed and circumscribed rectangles will become apparent in the next section.

Exercises

1. For $0 \leq i \leq 5$, let $x_i = (i^2 + i)/30$ and define $P = \{x_i : 0 \leq i \leq 5\}$.

   a) Show that P is a partition of $[0, 1]$ and find $\|P\|$.

   b) Find a refinement P_0 of P such that $\|P_0\| = 1/15$.

2. Find a partition P of the interval $\left[1, \sqrt{10}\right]$ such that the points in P are equally spaced and $\|P\| < 1/10$. How would you proceed if $1/10$ is replaced with an arbitrary positive number δ?

3. Let P_1 and P_2 be two partitions of $[a, b]$. Find a partition P_0 of $[a, b]$ such that P_0 is a refinement of both P_1 and P_2.

4. Find a tagged partition ${}^tP$ of $[0, 1]$ so that $\|{}^tP\| = 0.4$ and all of the tags of ${}^tP$ are irrational.

5. Let f be the function defined by $f(x) = x^2 + x$ and divide the interval $[-1, 2]$ into six subintervals of equal length. For each of the following choices of tags, write out the corresponding tagged partition ${}^tP$ of $[-1, 2]$ and find $S(f, {}^tP)$.

   a) The left endpoint of each subinterval is the tag.

   b) The right endpoint of each subinterval is the tag.

   c) The midpoint of each subinterval is the tag.

6. Prove that the number L in the definition of the Riemann integral is unique.

7. Let k be a constant and suppose that $f(x) = k$ for all $x \in [a, b]$. Prove that f is Riemann integrable on $[a, b]$ and that $\int_a^b f = k(b - a)$.

8. Prove that the function f defined by $f(x) = 0$ if $x \neq 1$ and $f(1) = 8$ is Riemann integrable on $[0, 2]$.

9. Prove that the function g defined by $g(x) = 1$ if $x < 1/2$ and $g(x) = 2$ if $x \geq 1/2$ is Riemann integrable on $[0, 1]$.

10. Prove that the function $h(x) = x$ is Riemann integrable on $[0, 1]$.

11. Prove parts (a) and (b) of Theorem 5.5.

12. Suppose that f and g are Riemann integrable on $[a, b]$. Prove that $f - g$ is Riemann integrable on $[a, b]$.

13. Prove parts (c) and (d) of Theorem 5.5.

14. A calculus student writes $\int_\pi^{2\pi} \sin^2(x^3)\, dx \approx 3.38$. Use results in this section to explain why this cannot be correct.

15. Let f and g be bounded functions defined on $[a, b]$.

   a) Suppose that f is equal to zero on $[a, b]$ except for one point. Prove that f is Riemann integrable on $[a, b]$ and that $\int_a^b f = 0$.

b) Suppose that f is equal to zero on $[a, b]$ except for a finite set of points. Prove that f is Riemann integrable on $[a, b]$ and that $\int_a^b f = 0$.

c) Suppose that f and g are equal on $[a, b]$ except for a finite number of points and that f is Riemann integrable on $[a, b]$. Prove that g is Riemann integrable on $[a, b]$ and that the integrals of f and g are equal.

16. Let a be a positive number and consider the function f defined by $f(x) = x^3 - a^2x$. Find $\omega(f, [0, 2a])$.

17. Let f be a monotone function defined on $[a, b]$. Find a simple representation for $\omega(f, [a, b])$.

18. For the function f defined by $f(x) = \sin(1/x)$ if $x \neq 0$ and $f(0) = 2$, find $\omega(f, [0, 1])$ and $\omega(f, [0.1, 0.25])$.

19. Let f be bounded on $[a, b]$.

a) Prove that $\omega(|f|, [a, b]) \leq \omega(f, [a, b])$.

b) Suppose that f is bounded by M. Prove that $\omega(f^2, [a, b]) \leq 2M\omega(f, [a, b])$.

c) Suppose that f is bounded below by m and that m is a positive number. Prove that $\omega(1/f, [a, b]) \leq \omega(f, [a, b])/m^2$.

20. Let f be a bounded function defined on $[a, b]$. Prove that

$$\omega(f, [a, b]) = \sup\{|f(t) - f(s)| : s, t \in [a, b]\}.$$

21. Suppose that f and g are bounded on $[a, b]$. Prove that

$$\omega(f + g, [a, b]) \leq \omega(f, [a, b]) + \omega(g, [a, b]).$$

Give an example to show that the inequality may be strict and one to show that equality may occur.

5.2 Conditions for Riemann Integrability

In this section, we will show that two different approaches to integration are in fact equivalent. The essence of each approach can be described geometrically, as in the introduction to this chapter, but the actual formulations of these approaches are completely independent of geometry. The equivalence of the two versions provides another way to view the Riemann integrability of a function, but the proof of the equivalence requires careful attention to details. The effort involved in proving the equivalence is justified by the fact that having two equivalent ways of viewing the integral will be extremely useful. For many results in integration theory, one of the two approaches offers a proof that is much simpler than a proof using the other approach.

A number of preliminary results, some of which are interesting in their own right, are required before the desired condition (see Theorem 5.10) can be established. The first theorem states that a function that is Riemann integrable on $[a, b]$ must be bounded on $[a, b]$. The proof of this fact is not difficult, but it does require a firm understanding of tagged partitions.

THEOREM 5.7 If $f: [a, b] \to \mathbb{R}$ is Riemann integrable on $[a, b]$, then f is bounded on $[a, b]$.

Proof. Since f is Riemann integrable on $[a,b]$, there exists a positive number δ such that $|S(f, {}^tP) - \int_a^b f| < 0.5$ for all tagged partitions ${}^tP$ of $[a,b]$ that satisfy $\|{}^tP\| < \delta$. By the Archimedean property of the real numbers (just a reminder), there exists a positive integer q such that $\beta = (b-a)/q < \delta$. Let $x_j = a + j\beta$ for $0 \le j \le q$ and let ${}^tP_0 = \{(x_j, [x_{j-1}, x_j]) : 1 \le j \le q\}$. Note that ${}^tP_0$ is a tagged partition of $[a,b]$ with $\|{}^tP_0\| = \beta < \delta$. We will prove that the number M defined by

$$M = \frac{1}{\beta} + \max\{|f(x_j)| : 0 \le j \le q\}$$

is a bound for f on $[a,b]$.

Let $x \in [a,b]$. Choose an integer j such that $x_{j-1} \le x \le x_j$ and define

$${}^tP = \left({}^tP_0 \setminus \big\{(x_j, [x_{j-1}, x_j])\big\}\right) \cup \big\{(x, [x_{j-1}, x_j])\big\}.$$

Since ${}^tP$ is simply ${}^tP_0$ with one tag changed, ${}^tP$ is a tagged partition of $[a,b]$ with $\|{}^tP\| = \|{}^tP_0\| < \delta$. By the definition of ${}^tP$,

$$S(f, {}^tP) - S(f, {}^tP_0) = f(x)(x_j - x_{j-1}) - f(x_j)(x_j - x_{j-1}) = \big(f(x) - f(x_j)\big)\beta.$$

Note also that

$$\left|S(f, {}^tP) - S(f, {}^tP_0)\right| \le \left|S(f, {}^tP) - \int_a^b f\right| + \left|\int_a^b f - S(f, {}^tP_0)\right| < 1.$$

Therefore,

$$\begin{aligned} |f(x)| &= \left|\frac{1}{\beta}\big(S(f, {}^tP) - S(f, {}^tP_0)\big) + f(x_j)\right| \\ &\le \frac{1}{\beta}\left|S(f, {}^tP) - S(f, {}^tP_0)\right| + |f(x_j)| \\ &< \frac{1}{\beta} + |f(x_j)| \le M. \end{aligned}$$

Since $|f(x)| \le M$ for all $x \in [a,b]$, the number M is a bound for f on $[a,b]$. ■

The converse of this theorem is false. The function f defined by

$$f(x) = \begin{cases} 0, & \text{if } x \notin \mathbb{Q}; \\ 1, & \text{if } x \in \mathbb{Q}; \end{cases}$$

is an example of a bounded function that is not Riemann integrable on any interval. A proof of this statement will be left as an exercise.

The difficulty with using the definition of the Riemann integral to prove that a function is Riemann integrable is the fact that the value of the integral must be known in advance. However, at this stage of the game, we have very few ways of determining this value. As with sequences, we need a method for determining that a limit exists without knowing the value of the limit. There are several methods for establishing the Riemann integrability of a function without knowing the value of its integral. The following theorem presents one such method; it is a direct consequence of the definition of the integral. We will refer to it as the Cauchy criterion for Riemann integrability or simply as the Cauchy criterion.

THEOREM 5.8 Cauchy Criterion for Riemann Integrability A bounded function f is Riemann integrable on $[a, b]$ if and only if for each $\epsilon > 0$ there exists $\delta > 0$ such that $|S(f, {}^tP_1) - S(f, {}^tP_2)| < \epsilon$ for all tagged partitions ${}^tP_1$ and ${}^tP_2$ of $[a, b]$ with norms less than δ.

Proof. Suppose that for each $\epsilon > 0$ there exists a positive number δ such that $|S(f, {}^tP_1) - S(f, {}^tP_2)| < \epsilon$ for all tagged partitions ${}^tP_1$ and ${}^tP_2$ of $[a, b]$ with norms less than δ. For each positive integer n, choose a positive number δ_n such that $|S(f, {}^tP_1) - S(f, {}^tP_2)| < 1/n$ for all tagged partitions ${}^tP_1$ and ${}^tP_2$ of $[a, b]$ with norms less than δ_n. We may assume that the sequence $\{\delta_n\}$ is decreasing. For each n, let ${}^tP_n$ be a tagged partition of $[a, b]$ that satisfies $\|{}^tP_n\| < \delta_n$. If m and n are both greater than some positive integer K, then the tagged partitions ${}^tP_n$ and ${}^tP_m$ have norms less than δ_K (this is where the fact that $\{\delta_n\}$ is decreasing is used) and it follows that

$$\left|S(f, {}^tP_n) - S(f, {}^tP_m)\right| < 1/K.$$

This shows that $\{S(f, {}^tP_n)\}$ is a Cauchy sequence. Let L be the limit of this sequence. We will prove that $\int_a^b f = L$.

Let $\epsilon > 0$. Since $\{S(f, {}^tP_n)\}$ converges to L, there exists a positive integer N such that $1/N < \epsilon/2$ and $|S(f, {}^tP_n) - L| < \epsilon/2$ for all $n \geq N$. Let $\delta = \delta_N$ and suppose that ${}^tP$ is a tagged partition of $[a, b]$ that satisfies $\|{}^tP\| < \delta$. We then have

$$|S(f, {}^tP) - L| \leq |S(f, {}^tP) - S(f, {}^tP_N)| + |S(f, {}^tP_N) - L| < 1/N + \epsilon/2 < \epsilon.$$

Hence, the function f is Riemann integrable on $[a, b]$ and $\int_a^b f = L$.

The proof of the converse is routine and will be left as an exercise. ∎

Although the Cauchy criterion eliminates the need for a value of the integral in order to determine whether or not a function is Riemann integrable, it is not all that helpful for proving that specific functions are Riemann integrable. The difficulty lies in the fact that the quantity $S(f, {}^tP_1) - S(f, {}^tP_2)$ is a big mess when ${}^tP_1$ and ${}^tP_2$ have very few, if any, intervals in common. The Cauchy criterion is more of a theoretical tool than a practical one, and it is easiest to use when it is not necessary to actually compute any of the Riemann sums. In fact, the Cauchy criterion will play a prominent role in the proof of the main result of this section, which appears after the following lemma. (The oscillation of a function, which was defined at the end of Section 5.1, will be used in the next few results.)

LEMMA 5.9 Let f be a bounded function defined on an interval $[a, b]$. Suppose that $P_1 = \{z_i : 0 \leq i \leq n\}$ is a partition of $[a, b]$ and that P_2 is a refinement of P_1. If ${}^tP_1$ and ${}^tP_2$ are tagged partitions of $[a, b]$ formed from the partitions P_1 and P_2, respectively, then

$$|S(f, {}^tP_2) - S(f, {}^tP_1)| \leq \sum_{i=1}^{n} \omega(f, [z_{i-1}, z_i])(z_i - z_{i-1}).$$

Proof. Since P_2 is a refinement of P_1, the set $P_2 \cap [z_{i-1}, z_i]$ is a partition of $[z_{i-1}, z_i]$. Consequently, since we can work with each interval $[z_{i-1}, z_i]$ separately, it is sufficient to consider the case in which $P_1 = \{a, b\}$; this simplifies the notation in the proof considerably. Let ${}^tP_1 = (v, [a, b])$, let $P_2 = \{x_i : 0 \le i \le p\}$, and let

$$ {}^tP_2 = \{(t_i, [x_{i-1}, x_i]) : 1 \le i \le p\}. $$

It follows that

$$ \begin{aligned} |S(f, {}^tP_2) - S(f, {}^tP_1)| &= \left| \sum_{i=1}^{p} f(t_i)(x_i - x_{i-1}) - f(v)(b - a) \right| \\ &= \left| \sum_{i=1}^{p} \big(f(t_i) - f(v)\big)(x_i - x_{i-1}) \right| \\ &\le \sum_{i=1}^{p} \big| f(t_i) - f(v) \big| (x_i - x_{i-1}) \\ &\le \sum_{i=1}^{p} \omega(f, [a, b])(x_i - x_{i-1}) = \omega(f, [a, b])(b - a). \end{aligned} $$

This completes the proof. ■

The next theorem is a very important result in the theory of the Riemann integral since it gives a condition for the existence of the integral that is equivalent to the definition of the integral. It shows that the two geometric approaches to the area problem discussed in the introduction to the chapter are indeed equivalent. However, this equivalence is not immediately apparent from the statement of the theorem; sorting out the geometrical meaning of this theorem will be left as an exercise. It should be pointed out that the proof of this result is rather tedious, so read slowly and carefully.

THEOREM 5.10 Let f be a bounded function defined on $[a, b]$. Then f is Riemann integrable on $[a, b]$ if and only if for each $\epsilon > 0$ there exists a partition $P = \{x_i : 0 \le i \le n\}$ of $[a, b]$ such that

$$ \sum_{i=1}^{n} \omega(f, [x_{i-1}, x_i])(x_i - x_{i-1}) < \epsilon. $$

Proof. Suppose first that for each $\epsilon > 0$ there exists a partition P of $[a, b]$ satisfying the inequality displayed in the statement of the theorem. Let $M > 1$ be a bound for f on $[a, b]$ and let $0 < \epsilon < 1$. By hypothesis, there exists a partition $P_\epsilon = \{z_k : 0 \le k \le N\}$ of $[a, b]$ such that $\|P_\epsilon\| < 1$ (see the exercises) and

$$ \sum_{k=1}^{N} \omega(f, [z_{k-1}, z_k])(z_k - z_{k-1}) < \epsilon. $$

Let ${}^tP_\epsilon$ be any tagged partition of $[a, b]$ that is formed from the partition P_ϵ, let $d = \min\{z_k - z_{k-1} : 1 \le k \le N\}$, and let $\delta = d\epsilon/(2MN)$. Now suppose that ${}^tP = \{(t_i, [x_{i-1}, x_i]) : 1 \le i \le p\}$ is any tagged partition of $[a, b]$ that satisfies $\|{}^tP\| < \delta$. Since $\delta < d$, the set $(x_{i-1}, x_i) \cap P_\epsilon$ contains at most one point for any

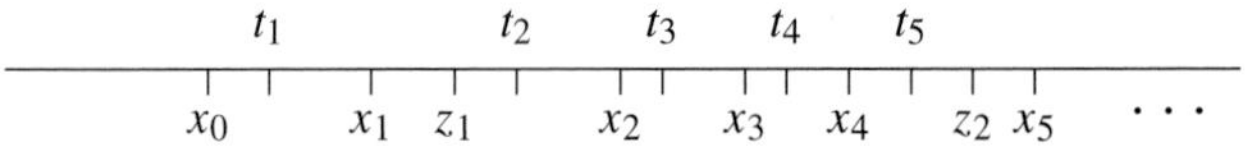

$$\begin{array}{l} \qquad t_1 \qquad\qquad t_2 \quad t_3 \quad t_4 \quad t_5 \\ x_0 \quad x_1 \; z_1 \quad x_2 \quad x_3 \quad x_4 \quad z_2\; x_5 \quad \ldots \end{array}$$

$${}^tP = \Big\{(t_1, [x_0, x_1]), (t_2, [x_1, x_2]), (t_3, [x_2, x_3]), \ldots\Big\}$$

$${}^tP_0 = \Big\{(t_1, [x_0, x_1]), (z_1, [x_1, z_1]), (z_1, [z_1, x_2]), (t_3, [x_2, x_3]), \ldots\Big\}$$

Figure 5.4 The formation of ${}^tP_0$

value of i. Let $A = \{i : (x_{i-1}, x_i) \cap P_\epsilon \neq \emptyset\}$ and let $(x_{i-1}, x_i) \cap P_\epsilon = \{z_{k_i}\}$ for each $i \in A$. Since P_ϵ contains $N - 1$ points in (a, b), the set A contains at most $N - 1$ elements. We will form a new tagged partition ${}^tP_0$ of $[a, b]$ so that most of the tagged intervals in ${}^tP_0$ are the same as those in ${}^tP$ and the partition associated with ${}^tP_0$ is a refinement of P_ϵ. Form ${}^tP_0$ using the following criteria:

if $i \notin A$, then $(t_i, [x_{i-1}, x_i])$ belongs to ${}^tP_0$;
if $i \in A$, then $(z_{k_i}, [x_{i-1}, z_{k_i}])$ and $(z_{k_i}, [z_{k_i}, x_i])$ belong to ${}^tP_0$.

(Figure 5.4 may help clarify how ${}^tP_0$ is formed.) Using the previous lemma and the fact that the set A has fewer than N elements,

$$\begin{aligned} &|S(f, {}^tP) - S(f, {}^tP_\epsilon)| \\ &\qquad \le |S(f, {}^tP) - S(f, {}^tP_0)| + |S(f, {}^tP_0) - S(f, {}^tP_\epsilon)| \\ &\qquad = \Big|\sum_{i \in A}\big(f(t_i) - f(z_{k_i})\big)(x_i - x_{i-1})\Big| + |S(f, {}^tP_0) - S(f, {}^tP_\epsilon)| \\ &\qquad \le \sum_{i \in A}\big|f(t_i) - f(z_{k_i})\big|(x_i - x_{i-1}) + \sum_{i=1}^{N} \omega(f, [z_{i-1}, z_i])(z_i - z_{i-1}) \\ &\qquad < 2MN\delta + \epsilon = d\,\epsilon + \epsilon < 2\epsilon. \end{aligned}$$

If ${}^tP_1$ and ${}^tP_2$ are any two tagged partitions of $[a, b]$ with norms less than δ, then

$$\begin{aligned} |S(f, {}^tP_1) - S(f, {}^tP_2)| &\le |S(f, {}^tP_1) - S(f, {}^tP_\epsilon)| + |S(f, {}^tP_\epsilon) - S(f, {}^tP_2)| \\ &< 2\epsilon + 2\epsilon = 4\epsilon. \end{aligned}$$

By the Cauchy criterion, the function f is Riemann integrable on $[a, b]$.

The proof of the converse will be left as an exercise. ■

The reason that it is easier to prove that a function is Riemann integrable using Theorem 5.10 rather than the definition of the integral lies in the fact that Theorem 5.10 only requires something "good" to happen for a single partition, as opposed to every tagged partition with small norm. This greatly reduces the amount of work required to prove that a function is or is not Riemann integrable. This condition should at least be considered for any problem that asks whether or not a function is Riemann integrable. Theorem 5.10 is not much help in actually finding the value of an integral; this problem will be addressed in the next section.

To illustrate the use of Theorem 5.10, we will prove that continuous functions are Riemann integrable on closed and bounded intervals. The key idea is that continuous functions are uniformly continuous on such intervals (see Theorem 3.28).

THEOREM 5.11 If $f: [a, b] \to \mathbb{R}$ is continuous on $[a, b]$, then f is Riemann integrable on $[a, b]$.

Proof. Let $\epsilon > 0$. Since f is uniformly continuous on $[a, b]$, there exists $\delta > 0$ such that $|f(y) - f(x)| < \epsilon/(b - a)$ for all $x, y \in [a, b]$ that satisfy $|y - x| < \delta$. Choose a positive integer n such that $\beta = (b - a)/n < \delta$ and define $x_i = a + i\beta$ for $0 \le i \le n$. Let $P = \{x_i : 0 \le i \le n\}$ and note that $\omega(f, [x_{i-1}, x_i]) < \epsilon/(b - a)$ for $1 \le i \le n$ (see the exercises). Then

$$\sum_{i=1}^{n} \omega(f, [x_{i-1}, x_i])(x_i - x_{i-1}) < \sum_{i=1}^{n} \frac{\epsilon}{b - a}(x_i - x_{i-1}) = \epsilon,$$

so f is Riemann integrable on $[a, b]$ by Theorem 5.10. ■

Since the problem of finding the area of a region under the graph of a continuous nonnegative function is one motivation for the definition of the Riemann integral, it is not surprising that continuous functions are Riemann integrable. The proof of this fact may seem too difficult (the difficulty lies in the proof of Theorem 5.10) for a seemingly obvious result. It is important to remember that continuous functions come in all shapes and sizes; the proof must use only the properties of continuous functions, not some sort of appeal to geometric areas. In fact, the area of the region under an arbitrary continuous nonnegative function is defined by the Riemann integral (the integral defines area as opposed to merely computing a quantity already known to exist). Note that the existence of the area of a region under the graph of a continuous nonnegative function involves the Completeness Axiom (see the proof of Theorem 5.8).

The set of all continuous functions on an interval $[a, b]$ is thus one large collection of Riemann integrable functions. Another large collection of Riemann integrable functions is the set of all monotone functions on an interval $[a, b]$; the proof of this fact will be left as an exercise.

THEOREM 5.12 If $f: [a, b] \to \mathbb{R}$ is monotone on $[a, b]$, then f is Riemann integrable on $[a, b]$. ■

It follows easily from the definition of the integral that the sum and difference of Riemann integrable functions are Riemann integrable (see Theorem 5.5). Theorem 5.10 is useful in proving that the product and quotient (with some restrictions) of Riemann integrable functions are Riemann integrable. The proof of these results, which involve the properties of the oscillation of a function found in Exercise 19 of the previous section, will be left as exercises.

THEOREM 5.13 Suppose that f and g are Riemann integrable on $[a, b]$.

a) The function $|f|$ is Riemann integrable on $[a, b]$ and $\left|\int_a^b f\right| \le \int_a^b |f|$.

b) The function fg is Riemann integrable on $[a, b]$.

c) Let k be a positive constant. If $g(x) \geq k$ for all $x \in [a, b]$, then f/g is Riemann integrable on $[a, b]$. ■

Exercises

1. Use the definition of the integral to prove that the function f defined by $f(x) = 1/x$ for $x \neq 0$ and $f(0) = 0$ is not Riemann integrable on $[0, 1]$.
2. Prove that the bounded function f defined by $f(x) = 0$ if x is irrational and $f(x) = 1$ if x is rational is not Riemann integrable on any interval $[a, b]$.
3. Verify the assertion that the sequence $\{\delta_n\}$ in the proof of the Cauchy criterion may be assumed to be decreasing.
4. Prove the converse of the Cauchy criterion.
5. Let $f: [a, b] \to \mathbb{R}$ be bounded, let $P_1 = \{x_i : 1 \leq i \leq p\}$ and $P_2 = \{z_i : 1 \leq i \leq q\}$ be partitions of $[a, b]$, and suppose that P_2 is a refinement of P_1. Prove that

$$\sum_{i=1}^{q} \omega(f, [z_{i-1}, z_i])(z_i - z_{i-1}) \leq \sum_{i=1}^{p} \omega(f, [x_{i-1}, x_i])(x_i - x_{i-1}).$$

   Note that this result is used in the proof of Theorem 5.10.
6. Suppose that f is continuous and nonnegative on $[a, b]$. Give an area interpretation for the sum that appears in the statement of Theorem 5.10. Then explain how Theorem 5.10 shows that the two geometric approaches to the area under a continuous nonnegative function are equivalent.
7. Prove the converse of Theorem 5.10.
8. Adopting the notation from the proof of Theorem 5.11, explain why the inequality $\omega(f, [x_{i-1}, x_i]) < \epsilon/(b - a)$ is strict. (Explain first why it may not be strict.)
9. Prove that the function $g: [0, 1] \to \mathbb{R}$ defined by $g(x) = 1 + \sqrt{x}\sin(1/x^8)$ for $x \neq 0$ and $g(0) = 1$ is continuous and hence Riemann integrable on $[0, 1]$. Note that the existence of an area under this continuous curve is not obvious from the graph.
10. Prove Theorem 5.12.
11. Let $f: [a, b] \to \mathbb{R}$ and let $g: [c, d] \to [a, b]$.
    a) Suppose that f and g are continuous. Prove that $f \circ g$ is integrable on $[c, d]$.
    b) Suppose that f and g are monotone. Prove that $f \circ g$ is integrable on $[c, d]$.
12. Give an example of a function $f: [a, b] \to \mathbb{R}$ such that $|f|$ is Riemann integrable on $[a, b]$, but f is not Riemann integrable on $[a, b]$.
13. Prove part (a) of Theorem 5.13.
14. Complete the following steps to prove part (b) of Theorem 5.13.
    a) Use properties of oscillation to prove that f^2 is Riemann integrable on $[a, b]$.
    b) Express the function fg in terms of the squares of the functions $f + g$ and $f - g$.
    c) Use the results from parts (a) and (b) to prove that fg is Riemann integrable on $[a, b]$.
15. Prove part (c) of Theorem 5.13.

Remark. The rest of the exercises in this section are in no particular order.

16. Let $f: [a, b] \to \mathbb{R}$ be a function of bounded variation. Prove that f is Riemann integrable on $[a, b]$.

17. Suppose that $f\colon [a, b) \to \mathbb{R}$ is a bounded function and that f is Riemann integrable on $[a, c]$ for all $a < c < b$. Let $\{b_n\}$ be a sequence in (a, b) that converges to b. Prove that $\left\{\int_a^{b_n} f\right\}$ is a Cauchy sequence.

18. For each positive integer n, let

$$\gamma_n = 1 + \frac{1}{2} + \cdots + \frac{1}{n} - \int_1^n \frac{1}{x}\, dx.$$

Prove that the sequence $\{\gamma_n\}$ converges.

19. Prove that the function $f\colon [0, 1] \to \mathbb{R}$ defined by

$$f(x) = \begin{cases} 1, & \text{if } x = 1/n,\ n \in \mathbb{Z}^+; \\ 0, & \text{otherwise}; \end{cases}$$

is Riemann integrable on $[0, 1]$.

20. Let $\{r_n\}$ be a listing of the rational numbers in the interval $[a, b]$ and let $\{v_n\}$ be a sequence of nonzero real numbers that converges to 0. Define $f\colon [a, b] \to \mathbb{R}$ by

$$f(x) = \begin{cases} v_n, & \text{if } x = r_n; \\ 0, & \text{if } x \notin Q. \end{cases}$$

Prove that f is Riemann integrable on $[a, b]$. Note that f has a discontinuity in every interval and that f is not monotone on any interval.

5.3 The Fundamental Theorem of Calculus

As is evident from the previous section, Theorem 5.10 provides a very useful criterion for showing that a function is Riemann integrable. However, neither this result nor the definition of the integral gives any indication how to find the value of an integral. In this section, we will discuss the Fundamental Theorem of Calculus and see how it provides a simple way to evaluate many integrals.

One of the difficulties with the definition of the Riemann integral is the amount of variability in the formation of Riemann sums; for a given $\delta > 0$, there are simply too many tagged partitions of an interval $[a, b]$ whose norms are less than δ. The first theorem in this section shows that it is possible to find the value of an integral by looking at only one sequence of tagged partitions. The proof of this theorem will be left as an exercise.

THEOREM 5.14 Suppose that f is Riemann integrable on $[a, b]$. If $\{{}^tP_n\}$ is any sequence of tagged partitions of $[a, b]$ for which $\{\|{}^tP_n\|\}$ converges to zero, then the sequence $\{S(f, {}^tP_n)\}$ converges to $\int_a^b f$. ∎

For example, the theorem shows that if f is Riemann integrable on $[a, b]$, then

$$\int_a^b f = \lim_{n\to\infty} \sum_{i=1}^{n} f\big(a + i(b-a)/n\big)\frac{b-a}{n}.$$

In other words, using tagged partitions for which every subinterval has the same length and the right endpoint of each subinterval is used as the tag is sufficient to find the value of an integral. As indicated by the exercises for this section, this limit can be evaluated fairly easily for some functions. It should be pointed out that the existence of this limit does not guarantee that a function is integrable (see the exercises).

Before presenting the Fundamental Theorem of Calculus, it is necessary to discuss the relationship between integration and subintervals. This is the content of the next theorem, which looks at integration from a slightly different perspective by focusing on the interval rather than the function. It states that if a function is Riemann integrable on an interval $[a, b]$, then it is Riemann integrable on every subinterval of $[a, b]$, and that if a function is Riemann integrable on the intervals $[a, c]$ and $[c, b]$, then it is Riemann integrable on $[a, b]$. Neither of these results is surprising considering the geometric motivation for the integral, but they are hard to prove directly from the definition of the integral. Once again, Theorem 5.10 comes to the rescue.

THEOREM 5.15 Let $f : [a, b] \to \mathbb{R}$ and let $c \in (a, b)$.

a) If f is Riemann integrable on $[a, b]$, then f is Riemann integrable on each subinterval of $[a, b]$.

b) If f is Riemann integrable on each of the intervals $[a, c]$ and $[c, b]$, then f is Riemann integrable on $[a, b]$ and $\int_a^b f = \int_a^c f + \int_c^b f$.

Proof. The proof of part (a) will be left as an exercise. Suppose that f is Riemann integrable on $[a, c]$ and $[c, b]$ and let $\epsilon > 0$. By Theorem 5.10, there exist partitions $P_a = \{v_i : 0 \le i \le p\}$ of $[a, c]$ and $P_b = \{z_i : 0 \le i \le q\}$ of $[c, b]$ such that

$$\sum_{i=1}^{p} \omega(f, [v_{i-1}, v_i])(v_i - v_{i-1}) < \epsilon/2$$

and

$$\sum_{i=1}^{q} \omega(f, [z_{i-1}, z_i])(z_i - z_{i-1}) < \epsilon/2.$$

Let $P = P_a \cup P_b$, write $P = \{x_i : 0 \le i \le n\}$, and note that P is a partition of $[a, b]$ with $x_p = c$. Then

$$\begin{aligned}&\sum_{i=1}^{n} \omega(f, [x_{i-1}, x_i])(x_i - x_{i-1})\\ &\qquad = \sum_{i=1}^{p} \omega(f, [v_{i-1}, v_i])(v_i - v_{i-1}) + \sum_{i=1}^{q} \omega(f, [z_{i-1}, z_i])(z_i - z_{i-1}) < \epsilon.\end{aligned}$$

By Theorem 5.10, the function f is Riemann integrable on $[a, b]$.

Now let $\{{}^tP_n\}$ be any sequence of tagged partitions of $[a, b]$ such that $\{\|{}^tP_n\|\}$ converges to 0 and c is an element of the partition associated with each ${}^tP_n$. Then ${}^tP_n = {}^tP_n^a \cup {}^tP_n^b$, where ${}^tP_n^a$ is a tagged partition of $[a, c]$ and ${}^tP_n^b$ is a tagged partition of $[c, b]$. Since the sequences $\{\|{}^tP_n^a\|\}$ and $\{\|{}^tP_n^b\|\}$ both converge to 0, Theorem

5.14 yields

$$\int_a^b f = \lim_{n\to\infty} S(f, {}^t P_n) = \lim_{n\to\infty} \left(S(f, {}^t P_n^a) + S(f, {}^t P_n^b) \right)$$
$$= \lim_{n\to\infty} S(f, {}^t P_n^a) + \lim_{n\to\infty} S(f, {}^t P_n^b) = \int_a^c f + \int_c^b f.$$

This completes the proof. ■

The Riemann integral is defined for a function f on an interval $[a, b]$, where it is assumed that $a < b$. For such functions, it is a standard convention to define

$$\int_a^a f = 0 \quad \text{and} \quad \int_b^a f = -\int_a^b f.$$

This convention yields the following corollary; its proof will be left as an exercise.

COROLLARY 5.16 If f is Riemann integrable on an interval that contains the points a, b, and c, then

$$\int_a^b f = \int_a^c f + \int_c^b f.$$ ■

The Fundamental Theorem of Calculus provides a relationship between the important concepts of integration and differentiation. As it is usually stated, this theorem has two parts. One part deals with the properties of the integral when it is treated as a function. Suppose that f is Riemann integrable on $[a, b]$. By Theorem 5.15, the function f is Riemann integrable on every subinterval of $[a, b]$. In particular, a function F can be defined on $[a, b]$ by $F(x) = \int_a^x f$. The values of the function F may be difficult to compute, but it represents a perfectly legitimate function. It is the properties of this function that are addressed in part (a) of the Fundamental Theorem of Calculus.

The other part of the Fundamental Theorem of Calculus provides a way to calculate Riemann integrals in some cases. It is perhaps the most familiar result in calculus, and its discovery in the late seventeenth century by Newton and Leibniz is usually taken as the beginning of calculus. A function G is an **antiderivative** of a function f on an interval I if $G'(x) = f(x)$ for all $x \in I$. Part (b) of the Fundamental Theorem of Calculus states that the Riemann integral of f can be evaluated by simply plugging the endpoints of the appropriate interval into the antiderivative G. This result is the most common method for finding the value of a Riemann integral.

THEOREM 5.17 Fundamental Theorem of Calculus Suppose that f is Riemann integrable on $[a, b]$.

a) If a function F is defined by $F(x) = \int_a^x f$ for each $x \in [a, b]$, then F is continuous on $[a, b]$ and differentiable at each point $x \in [a, b]$ for which f is continuous. At these points, $F'(x) = f(x)$.

b) If G is an antiderivative of f on $[a, b]$, then $\int_a^b f = G(b) - G(a)$.

Proof. A proof of part (b) will be left as an exercise. To prove part (a), we first show that F is uniformly continuous on $[a, b]$. By Theorem 5.7, the function f is bounded on $[a, b]$. Let M be a bound for f on $[a, b]$, let $\epsilon > 0$, and choose $\delta = \epsilon/M$. Suppose that $x, y \in [a, b]$ and that $|y - x| < \delta$. By Corollary 5.16 and part (d) of Theorem 5.5,

$$|F(y) - F(x)| = \left|\int_a^y f - \int_a^x f\right| = \left|\int_x^y f\right| \leq M|y - x| < M\delta = \epsilon.$$

Hence, the function F is uniformly continuous on $[a, b]$.

Now suppose that f is continuous at $x \in (a, b)$ (it is easy to modify the proof if x is either a or b) and let $\epsilon > 0$. Since f is continuous at x, there exists a positive number δ such that $|f(t) - f(x)| < \epsilon$ for all t that satisfy $|t - x| < \delta$. Suppose that $0 < |z - x| < \delta$. Then

$$\left|\frac{f(t) - f(x)}{z - x}\right| < \frac{\epsilon}{|z - x|}$$

for all t between z and x, and

$$f(x) = \frac{1}{z - x}\int_x^z f(x)\,dt$$

since $f(x)$ represents a constant when t is the variable of integration. Using Corollary 5.16 and part (d) of Theorem 5.5, we find that

$$\begin{aligned}
\left|\frac{F(z) - F(x)}{z - x} - f(x)\right| &= \left|\frac{1}{z - x}\int_x^z f(t)\,dt - \frac{1}{z - x}\int_x^z f(x)\,dt\right| \\
&= \left|\int_x^z \frac{f(t) - f(x)}{z - x}\,dt\right| \\
&\leq \frac{\epsilon}{|z - x|}|z - x| = \epsilon.
\end{aligned}$$

It follows that $F'(x) = f(x)$. This completes the proof. ■

The second part of the Fundamental Theorem of Calculus provides a method for determining the value of an integral. Suppose that f is Riemann integrable on $[a, b]$ and that an antiderivative G of f can be found. Then

$$\int_a^b f = G(b) - G(a) = G(x)\Big|_a^b.$$

The last symbol is the standard abbreviation that is used in most calculus books. For example,

$$\int_1^3 \left(4x - \frac{3}{x^2}\right) dx = \left(2x^2 + \frac{3}{x}\right)\Big|_1^3 = (18 + 1) - (2 + 3) = 14.$$

However, the process of finding an antiderivative is not always this easy. There are a number of techniques available to help find an antiderivative, but they do not always work. In fact, there are Riemann integrable functions that do not have an antiderivative at all. This means that there exists a Riemann integrable function f

for which there is no function F such that $F' = f$; an example of such a function is requested in the exercises. In order to evaluate the integral of an integrable function when either the function has no antiderivative or the values of its antiderivative are not known, we may need to resort to numerical techniques. For example, if f is Riemann integrable on $[a, b]$, then

$$\int_a^b f = \lim_{n\to\infty} \sum_{i=1}^{n} f(a + i(b-a)/n)\, \frac{b-a}{n}$$

(see the remarks following Theorem 5.14). For large values of n, this sum provides an estimate for the value of the integral. There are other (better and faster) methods to evaluate integrals numerically; some of these will be considered in Section 5.5.

While discussing the second part of the Fundamental Theorem of Calculus, it should also be pointed out that not all derivatives are Riemann integrable; an example of such a function is provided in the exercises. The existence of such functions illustrates one of the deficiencies of the Riemann integral.

The Fundamental Theorem of Calculus is fundamental because it links the two major concepts of calculus: differentiation and integration. The two parts of this theorem can be stated as

$$\left(\int_a^x f\right)' = f(x) \quad \text{and} \quad \int_a^x F' = F(x) - F(a).$$

The first equation is valid for all $x \in [a, b]$, provided that f is continuous on $[a, b]$. It shows that integrating a continuous function and then differentiating the result recovers the original function. The second equation is valid for all $x \in [a, b]$ provided that F' is Riemann integrable on $[a, b]$. It shows that differentiation followed by integration recovers the original function up to an additive constant. Hence, in the ways just described, differentiation and integration are inverse operations. Considering the complicated nature of the definition of the Riemann integral, it is amazing that integration can often be reduced to antidifferentiation.

We conclude this section with two familiar results from calculus which follow easily from the Fundamental Theorem of Calculus. The proofs of these well-known techniques of integration will be left as exercises.

THEOREM 5.18 Integration by Parts Suppose that f and g are differentiable functions defined on an interval $[a, b]$. If f' and g' are Riemann integrable on $[a, b]$, then $f'g$ and fg' are Riemann integrable on $[a, b]$ and

$$\int_a^b f'g = f(b)g(b) - f(a)g(a) - \int_a^b g'f. \qquad \blacksquare$$

THEOREM 5.19 Integration by Substitution Let $g: [a, b] \to [c, d]$ be differentiable on $[a, b]$ and let $f: [c, d] \to \mathbb{R}$ be continuous on $[c, d]$. If g' is Riemann integrable on $[a, b]$, then $(f \circ g)g'$ is Riemann integrable on $[a, b]$ and

$$\int_a^b (f \circ g)g' = \int_{g(a)}^{g(b)} f. \qquad \blacksquare$$

Exercises

1. Prove Theorem 5.14.

2. Let f be defined by $f(x) = 1$ if x is rational and $f(x) = 0$ if x is irrational. For each of the following conditions, give an example of a sequence $\{{}^tP_n\}$ of tagged partitions of $[0, 1]$ for which $\{\|{}^tP_n\|\}$ converges to zero and $\{S(f, {}^tP_n)\}$ satisfies the condition.

a) $\{S(f, {}^tP_n)\}$ converges to 0.

b) $\{S(f, {}^tP_n)\}$ converges to 1.

c) $\{S(f, {}^tP_n)\}$ converges to $1/2$.

d) $\{S(f, {}^tP_n)\}$ does not converge.

3. Use the function in the previous exercise to illustrate that the limit mentioned following Theorem 5.14 cannot be used to prove that a function is Riemann integrable.

4. Suppose that f is bounded on $[a, b]$. Prove that f is Riemann integrable on $[a, b]$ if and only if the sequence $\{S(f, {}^tP_n)\}$ converges for every sequence $\{{}^tP_n\}$ of tagged partitions of $[a, b]$ for which $\{\|{}^tP_n\|\}$ converges to zero.

5. For a fixed but arbitrary value of n, use mathematical induction to prove that

$$\sum_{i=1}^{n} i^r = \frac{n^{r+1}}{r+1} + \frac{n^r}{2} + P_r(n)$$

for each positive integer r, where P_r is a polynomial of degree less than r. (See Exercise 22 in Section 1.2 for a suggestion as to how to proceed.)

6. Use the previous exercise and the limit formula for the integral following Theorem 5.14 to evaluate $\int_0^b x^r\,dx$, where r is a positive integer.

7. Let a be a positive constant. Assume that a^x is defined for all real numbers x and that its derivative is $(\ln a)a^x$. Use the limit formula for the integral recorded after Theorem 5.14 to evaluate $\int_0^b a^x\,dx$, where b is a positive number.

8. Prove part (a) of Theorem 5.15.

9. Prove Corollary 5.16.

10. Suppose that f is continuous on $[a, b]$ and that $\int_a^b f = 0$.

a) Prove that there exists a point $c \in [a, b]$ such that $f(c) = 0$.

b) Suppose that f is nonnegative on $[a, b]$. Prove that $f(x) = 0$ for all x in $[a, b]$.

11. Suppose that f and g are continuous on $[a, b]$ and that $\int_a^b f = \int_a^b g$. Prove that there exists a point $c \in [a, b]$ such that $f(c) = g(c)$.

12. Suppose that f is differentiable on $[a, b]$. Prove that for each $\delta > 0$ there exists a tagged partition ${}^tP$ of $[a, b]$ such that $\|{}^tP\| < \delta$ and $S(f', {}^tP) = f(b) - f(a)$. Note that this result is valid even if f' is not Riemann integrable on $[a, b]$.

13. Use the previous exercise to prove part (b) of the Fundamental Theorem of Calculus.

14. Assuming that f is continuous on $[a, b]$, use part (a) of the Fundamental Theorem of Calculus to prove part (b).

15. Find the derivative of the function F defined by $F(x) = \displaystyle\int_0^{x^2} t^2 \sin t^2\,dt$.

16. Evaluate $\displaystyle\int_{1/9}^{1} \frac{\sin(\pi\sqrt{t}/2)}{\sqrt{t}}\,dt$.

17. Suppose that f is continuous on $[a, b]$ and define a function F by $F(x) = \int_x^b f$ for all $x \in [a, b]$. Show that F is differentiable on $[a, b]$ and find $F'(x)$.

18. Suppose that f is continuous on $[a, b]$ and let $c \in (a, b)$. Define a function F on $[a, b]$ by $F(x) = \int_c^x f$. Prove that $F'(x) = f(x)$ for all $x \in [a, b]$.

19. Suppose that f is a continuous function on $\mathbb{R}$ and define a function F on $[0, \infty)$ by $F(x) = \int_{-x}^x f$. Find a simple expression for $F'(x)$.

20. Suppose that f is a continuous function on $\mathbb{R}$, that ϕ and ψ are differentiable functions on $\mathbb{R}$, and define a function F on $\mathbb{R}$ by $F(x) = \int_{\phi(x)}^{\psi(x)} f$. Find a simple expression for $F'(x)$.

21. Find a function $f: \mathbb{R} \to \mathbb{R}$ such that $f(1) = 0$ and $f'(x) = x + \sin x^2$ for all x.

22. Suppose that f is Riemann integrable on $[a, b]$ and define a function F on $[a, b]$ by $F(x) = \int_a^x f$. Show that F may be differentiable even at points for which f is not continuous. After finding one example, see if you can find an example for which the set of points of differentiability of F differs as much as possible from the set of points of continuity of f.

23. Suppose that $f: [a, b] \to \mathbb{R}$ is continuous. Prove that f has an antiderivative on $[a, b]$. Does a continuous function $f: \mathbb{R} \to \mathbb{R}$ have an antiderivative on $\mathbb{R}$?

24. Give an example of a Riemann integrable function that does not have an antiderivative.

25. Use the function $f(x) = x^2 \sin(1/x^2)$ for $x \neq 0$ and $f(0) = 0$ to show that there are derivatives that are not Riemann integrable. There are even functions whose derivatives are bounded but not Riemann integrable, but these examples are more difficult to construct; see Gordon [8].

26. Prove Theorem 5.18.

27. Use Theorem 5.18 to evaluate $\displaystyle\int_0^{\pi/3} x \sin x \, dx$ and $\displaystyle\int_0^1 \arctan x \, dx$.

28. Prove Theorem 5.19.

29. Use Theorem 5.19 to evaluate $\displaystyle\int_1^3 \frac{6}{\sqrt{x}(1+x)} \, dx$.

30. Use Theorem 5.19 to prove that

$$\int_0^b \frac{x^3}{\sqrt{1-x^2}} \, dx = \int_0^{\arcsin b} \sin^3 x \, dx$$

for any real number $b \in (0, 1)$.

5.4 Further Properties of the Integral

The first three sections of this chapter contain the definition of the integral, a useful necessary and sufficient condition for the existence of the integral, and a method for finding the exact value of many integrals. The reader should be familiar with some of the applications of the integral that are related to geometry and physics (such as computing volume, arc length, center of mass, and work), since these are typically discussed in a calculus course. A few exercises concerning applications can be found at the end of the chapter. However, there are a number of other properties of the Riemann integral; this section presents a variety of these results.

The determination of which functions are Riemann integrable is an important aspect of the theory of the Riemann integral. As we have seen, continuous functions and monotone functions are Riemann integrable, but there exist Riemann integrable functions that are neither continuous nor monotone. (For an extreme example, see Exercise 20 in Section 5.2.) The next few results involve the collection of functions that are known to be integrable.

THEOREM 5.20 Let f be a bounded function defined on $[a, b]$. If f is Riemann integrable on each closed subinterval of (a, b), then f is Riemann integrable on $[a, b]$ and

$$\int_a^b f = \lim_{r \to 0^+} \int_{a+r}^{b-r} f.$$

Proof. Let M be a bound for the function f. Let $\epsilon > 0$ and let r be any positive number that satisfies $r < \min\{\epsilon/M, (b-a)/2\}$. By hypothesis, the function f is Riemann integrable on the interval $[a+r, b-r]$. According to Theorem 5.10, there exists a partition $P_r = \{z_i : 0 \le i \le n\}$ of $[a+r, b-r]$ such that

$$\sum_{i=1}^{n} \omega(f, [z_{i-1}, z_i])(z_i - z_{i-1}) < \epsilon.$$

The set P defined by $P = \{a\} \cup P_r \cup \{b\} = \{x_i : 0 \le i \le n+2\}$ is a partition of $[a, b]$ and

$$\begin{aligned} &\sum_{i=1}^{n+2} \omega(f, [x_{i-1}, x_i])(x_i - x_{i-1}) \\ &\qquad = \omega(f, [a, a+r])\, r + \sum_{i=1}^{n} \omega(f, [z_{i-1}, z_i])(z_i - z_{i-1}) + \omega(f, [b-r, b])\, r \\ &\qquad < 2Mr + \epsilon + 2Mr < 5\epsilon. \end{aligned}$$

Hence, the function f is Riemann integrable on $[a, b]$ by Theorem 5.10. To establish the limit, simply note that

$$\left| \int_a^b f - \int_{a+r}^{b-r} f \right| \le \left| \int_a^{a+r} f \right| + \left| \int_{b-r}^{b} f \right| \le Mr + Mr = 2Mr$$

for all values of r that satisfy $0 < r < (b-a)/2$. This completes the proof. ■

THEOREM 5.21 Let f be a bounded function defined on $[a, b]$.

a) If f has a finite number of discontinuities in $[a, b]$, then f is Riemann integrable on $[a, b]$.

b) If f has a countably infinite number of discontinuities in $[a, b]$ and these points form the terms of a convergent sequence, then f is Riemann integrable on $[a, b]$.

Proof. We will prove part (b) and leave the proof of part (a) as an exercise. Suppose that $\{d_n\}$ is a listing of the discontinuities of f in $[a, b]$ and that $\{d_n\}$

converges to d. Assume that $d \in (a, b)$; the cases in which $d = a$ or $d = b$ are easier. Let r be a positive number such that $a < d - r < d + r < b$. Since $\{d_n\}$ converges to d, there exists a positive integer N such that $d - r < d_n < d + r$ for all $n \geq N$. This means that f has only a finite number of discontinuities in the intervals $[a, d - r]$ and $[d + r, b]$. By part (a), the function f is Riemann integrable on both $[a, d - r]$ and $[d + r, b]$. Since $r > 0$ is arbitrary, Theorem 5.20 shows that f is Riemann integrable on $[a, d]$ and $[d, b]$. Finally, by Theorem 5.15, the function f is Riemann integrable on $[a, b]$. ■

The final word on the class of Riemann integrable functions is the following: a function f is Riemann integrable on $[a, b]$ if and only if f is bounded on $[a, b]$ and the set of discontinuities of f in $[a, b]$ has measure zero. The term "measure zero", which is defined in the supplementary exercises, essentially means that the set of discontinuities is not too large. This result indicates that continuity plays an important role in the existence of the Riemann integral. One method for proving this condition for the existence of the Riemann integral is outlined in the supplementary exercises of this chapter. For the record, note that a function may have a discontinuity at each rational number and still be Riemann integrable; see Exercise 20 in Section 5.2. In fact, this type of function makes an appearance in the next paragraph.

The composition of two continuous functions is continuous and, by the Chain Rule, the composition of two differentiable functions is differentiable. However, the composition of two Riemann integrable functions may not be Riemann integrable. Let $\{r_n\}$ be a listing of the rational numbers in an interval $[a, b]$ and let $\{v_n\}$ be a sequence in $(0, 1)$ that converges to 0. Consider the functions f and g defined by

$$f(x) = \begin{cases} 0, & \text{if } x = 0; \\ 1, & \text{if } x \neq 0; \end{cases} \quad \text{and} \quad g(x) = \begin{cases} v_n, & \text{if } x = r_n; \\ 0, & \text{otherwise.} \end{cases}$$

The function f is Riemann integrable on $[0, 1]$ and the function g is Riemann integrable on $[a, b]$, but, as the reader may verify, the function $f \circ g$ is not Riemann integrable on $[a, b]$. The next theorem provides a condition that guarantees that the composition of two Riemann integrable functions is Riemann integrable.

THEOREM 5.22 Suppose that $g\colon [a, b] \to [c, d]$ is Riemann integrable on $[a, b]$. If $f\colon [c, d] \to \mathbb{R}$ is continuous on $[c, d]$, then $f \circ g$ is Riemann integrable on $[a, b]$.

Proof. Let M be a bound for the continuous function f on $[c, d]$, let $\epsilon > 0$, and let $\epsilon_1 = \epsilon/(b - a + 2M)$. Since f is uniformly continuous on $[c, d]$, there exists a positive number $\eta < \epsilon_1$ such that $|f(y) - f(x)| < \epsilon_1$ for all $x, y \in [c, d]$ that satisfy $|y - x| \leq \eta$. Since g is Riemann integrable on $[a, b]$, Theorem 5.10 guarantees the existence of a partition $P = \{x_i : 0 \leq i \leq n\}$ of $[a, b]$ such that

$$\sum_{i=1}^{n} \omega(g, [x_{i-1}, x_i])(x_i - x_{i-1}) < \eta^2.$$

Let $S_1 = \{i : \omega(g, [x_{i-1}, x_i]) \leq \eta\}$ and $S_2 = \{i : \omega(g, [x_{i-1}, x_i]) > \eta\}$ and note that $\omega(f \circ g, [x_{i-1}, x_i]) \leq \epsilon_1$ for $i \in S_1$ and $\omega(f \circ g, [x_{i-1}, x_i]) \leq 2M$ for $i \in S_2$.

By the choice of P and the definition of the set S_2,

$$\begin{aligned}\sum_{i\in S_2}(x_i - x_{i-1}) &= \frac{1}{\eta}\sum_{i\in S_2}\eta\,(x_i - x_{i-1})\\ &< \frac{1}{\eta}\sum_{i\in S_2}\omega(g,[x_{i-1},x_i])(x_i - x_{i-1})\\ &\le \frac{1}{\eta}\sum_{i=1}^{n}\omega(g,[x_{i-1},x_i])(x_i - x_{i-1}) < \frac{1}{\eta}\cdot\eta^2 = \eta.\end{aligned}$$

It follows that

$$\begin{aligned}\sum_{i=1}^{n}\omega(f\circ g,[x_{i-1},x_i])(x_i - x_{i-1}) &= \sum_{i\in S_1}\omega(f\circ g,[x_{i-1},x_i])(x_i - x_{i-1})\\ &\quad + \sum_{i\in S_2}\omega(f\circ g,[x_{i-1},x_i])(x_i - x_{i-1})\\ &\le \sum_{i\in S_1}\epsilon_1\,(x_i - x_{i-1}) + \sum_{i\in S_2}2M(x_i - x_{i-1})\\ &< \epsilon_1\,(b-a) + 2M\eta\\ &< \epsilon_1\,(b-a+2M) = \epsilon.\end{aligned}$$

By Theorem 5.10, the function $f \circ g$ is Riemann integrable on $[a, b]$. ■

We next consider the average value of a function. It is easy to find the average of n numbers: simply add them up and divide by n. How can this idea be extended to find the average value of a function on an interval? There are, in general, an infinite number of output values of a function, so adding them up and dividing by the total number is not an option. One possible approach is to take a sample of the output values of the function and use their average as an estimate for the average value of the function. It should be clear that this estimate will improve as more output values are taken, assuming that these values are representative of all of the output values of the function. Hence, it appears that a limit of some sort can be used to define the average value of a function. Suppose that f is Riemann integrable on $[a, b]$, let n be a positive integer, and let $a_i = a + i(b-a)/n$ for $0 \le i \le n$. Then an estimate (see the exercises for a comment on this choice for the estimate) for the average value of f on $[a, b]$ is

$$\frac{f(a_1) + f(a_2) + \cdots + f(a_n)}{n} = \frac{1}{b-a}\sum_{i=1}^{n} f(a_i)\,\frac{b-a}{n}.$$

Letting n tend to infinity and using the limit formula for the value of an integral following Theorem 5.14, the **average value** of a Riemann integrable function f on an interval $[a, b]$ is defined by

$$\frac{1}{b-a}\int_a^b f.$$

A geometric interpretation of the average value of a function is considered in the exercises.

The Mean Value Theorem for integrals asserts that a continuous function equals its average value at some point in the interval. As this theorem is a simple consequence of the more general theorem that follows it, the proof will be omitted.

THEOREM 5.23 Mean Value Theorem for Integrals If f is continuous on $[a, b]$, then there exists a point $c \in [a, b]$ such that $f(c)(b - a) = \int_a^b f$. ■

THEOREM 5.24 Generalized Mean Value Theorem for Integrals If f is continuous on $[a, b]$ and g is a nonnegative Riemann integrable function on $[a, b]$, then there exists a point $c \in [a, b]$ such that $f(c) \int_a^b g = \int_a^b fg$.

Proof. By the Extreme Value Theorem, there exist points $u, v \in [a, b]$ such that $f(u) \leq f(x) \leq f(v)$ for all $x \in [a, b]$. Since g is nonnegative on $[a, b]$, the inequality $f(u)g(x) \leq f(x)g(x) \leq f(v)g(x)$ is valid for all $x \in [a, b]$. By Theorem 5.5,

$$f(u) \int_a^b g \leq \int_a^b fg \leq f(v) \int_a^b g.$$

If $\int_a^b g = 0$, then $\int_a^b fg = 0$ and the result is trivial. If $\int_a^b g \neq 0$, then $\int_a^b fg / \int_a^b g$ is a number between $f(u)$ and $f(v)$, and the result follows from the Intermediate Value Theorem. ■

The last topic for this section concerns step functions and some of their properties. Step functions are an especially simple class of functions and are sometimes used as the basis for the definition of the Riemann integral (see Gordon [8]).

DEFINITION 5.25 A function $\phi\colon [a, b] \to \mathbb{R}$ is a **step function** if there exists a partition $\{x_i : 0 \leq i \leq n\}$ of $[a, b]$ such that ϕ is constant on each of the intervals (x_{i-1}, x_i). The values of ϕ at the partition points can be arbitrary.

The range of a step function is a finite set and the inverse image of each element of the range is either a single point, an interval, or a finite union of single points and intervals. The graph of a step function (especially one that is monotone) looks like a set of steps: the function "steps" from one constant value to the next as the inputs increase along the domain of the function. It is easy to construct examples of step functions either with a formula or a graph; Figure 5.5 gives the graph of one such function. The greatest integer function $\lfloor x \rfloor$ (on any closed and bounded interval) is an example of a step function that has appeared earlier in this book. Although the range of a step function is a finite set, keep in mind that the range can still be rather large—a step function may assume millions of different values.

Some properties of step functions are recorded in the next two theorems; the proofs of these results will be left as exercises.

THEOREM 5.26 Let $\phi\colon [a, b] \to \mathbb{R}$ be a step function. Then

a) the function ϕ has one-sided limits at every point of $[a, b]$;

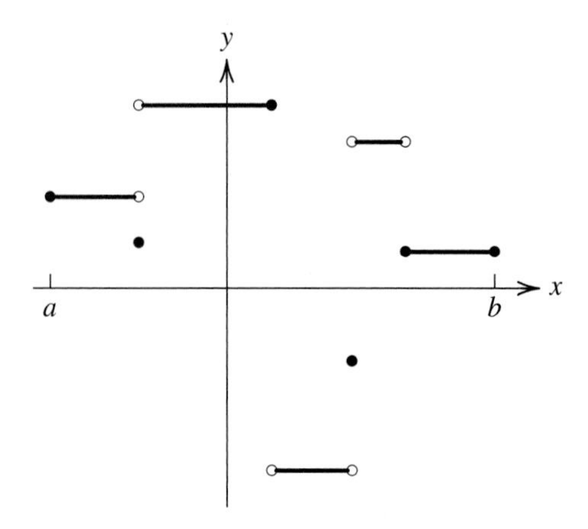

Figure 5.5 The graph of a step function on the interval $[a, b]$

b) the function ϕ has at most a finite number of discontinuities in $[a, b]$;

c) the function ϕ assumes its maximum and minimum values on $[a, b]$. ■

THEOREM 5.27 Let $\phi: [a, b] \to \mathbb{R}$ be a step function, let $P = \{x_i : 0 \leq i \leq n\}$ be a partition of $[a, b]$ such that ϕ is constant on each of the intervals (x_{i-1}, x_i), and let $\phi(x) = c_i$ for all $x \in (x_{i-1}, x_i)$. Then ϕ is Riemann integrable on $[a, b]$ and

$$\int_a^b \phi = \sum_{i=1}^{n} c_i (x_i - x_{i-1}).$$ ■

Given a step function ϕ on $[a, b]$, there are many possible choices for a partition $P = \{x_i : 0 \leq i \leq n\}$ of $[a, b]$ such that ϕ is constant on each of the intervals (x_{i-1}, x_i). As a very simple example, consider the function ψ defined by $\psi(x) = 1$ if $x < 2$ and $\psi(x) = 5$ if $x \geq 2$. Then

$$\psi(x) = \begin{cases} 1, & \text{if } 0 \leq x < 2; \\ 5, & \text{if } 2 \leq x \leq 3; \end{cases} \quad \text{and} \quad \psi(x) = \begin{cases} 1, & \text{if } 0 \leq x < 1; \\ 1, & \text{if } 1 \leq x < 2; \\ 5, & \text{if } 2 \leq x \leq 3; \end{cases}$$

are two different representations of ψ on $[0, 3]$. There are actually situations for which it is helpful to have more than one representation for a step function (see the exercises), but these multiple representations of step functions will not play a major role in this text. Note that the Riemann integral of a step function is independent of the particular representation of the step function since the Riemann integral of a function is unique.

The next theorem links Riemann integrable functions with step functions. It shows that every Riemann integrable function is trapped between two step functions whose integrals are arbitrarily close together.

THEOREM 5.28 A function $f: [a, b] \to \mathbb{R}$ is Riemann integrable on $[a, b]$ if and only if for each $\epsilon > 0$ there exist step functions ϕ and ψ defined on $[a, b]$ such that $\phi(x) \le f(x) \le \psi(x)$ for all $x \in [a, b]$ and $\int_a^b (\psi - \phi) < \epsilon$.

Proof. Suppose first that f is Riemann integrable on $[a, b]$ and let $\epsilon > 0$. By Theorem 5.10, there exists a partition $P = \{x_i : 0 \le i \le n\}$ of $[a, b]$ such that

$$\sum_{i=1}^{n} \omega(f, [x_{i-1}, x_i])(x_i - x_{i-1}) < \epsilon.$$

Define step functions ϕ and ψ on $[a, b]$ by $\phi(a) = f(a) = \psi(a)$ and, for $1 \le i \le n$,

$$\phi(x) = \inf\{f(x) : x_{i-1} \le x \le x_i\} \quad \text{and} \quad \psi(x) = \sup\{f(x) : x_{i-1} \le x \le x_i\}$$

for $x \in (x_{i-1}, x_i]$. Then $\phi(x) \le f(x) \le \psi(x)$ for all $x \in [a, b]$ and

$$\int_a^b (\psi - \phi) = \sum_{i=1}^{n} \omega(f, [x_{i-1}, x_i])(x_i - x_{i-1}) < \epsilon,$$

as desired.

We will use the Cauchy criterion to prove the converse of the theorem. Let $\epsilon > 0$. By hypothesis, there exist step functions ϕ and ψ defined on $[a, b]$ such that $\phi(x) \le f(x) \le \psi(x)$ for all $x \in [a, b]$ and $\int_a^b (\psi - \phi) < \epsilon$. Since ϕ and ψ are Riemann integrable on $[a, b]$, there exists $\delta > 0$ such that

$$\left| S(\phi, {}^tP) - \int_a^b \phi \right| < \epsilon \quad \text{and} \quad \left| S(\psi, {}^tP) - \int_a^b \psi \right| < \epsilon$$

for every tagged partition ${}^tP$ of $[a, b]$ that satisfies $\|{}^tP\| < \delta$. For any such tagged partition ${}^tP$,

$$\int_a^b \phi - \epsilon < S(\phi, {}^tP) \le S(f, {}^tP) \le S(\psi, {}^tP) < \int_a^b \psi + \epsilon.$$

If ${}^tP_1$ and ${}^tP_2$ are two tagged partitions of $[a, b]$ with norms less than δ, then both of the Riemann sums $S(f, {}^tP_1)$ and $S(f, {}^tP_2)$ lie in the interval

$$\left[\int_a^b \phi - \epsilon, \int_a^b \psi + \epsilon \right],$$

whose length is

$$\left(\int_a^b \psi + \epsilon \right) - \left(\int_a^b \phi - \epsilon \right) = \int_a^b (\psi - \phi) + 2\epsilon < 3\epsilon.$$

It follows that $|S(f, {}^tP_1) - S(f, {}^tP_2)| < 3\epsilon$. Hence, the function f is Riemann integrable on $[a, b]$ by the Cauchy criterion. This completes the proof. ∎

Exercises

1. Suppose that f is a function defined on $[a, b]$ and that f is Riemann integrable on each closed subinterval of (a, b). Give an example to show that f may not be Riemann integrable on $[a, b]$.
2. Let f be a bounded function defined on $[a, b]$ and suppose that f is Riemann integrable on each closed subinterval of $[a, b)$. Prove that f is Riemann integrable on $[a, b]$ and that $\int_a^b f = \lim_{x \to b^-} \int_a^x f$.
3. Prove part (a) of Theorem 5.21.
4. Let $\{x_n\}$ be a sequence of distinct numbers in $[a, b]$ and let $f: [a, b] \to \mathbb{R}$ be a bounded function such that $f(x) = 0$ unless $x = x_n$ for some n. The values for $f(x_n)$ can be chosen arbitrarily as long as f is bounded.
   a) Suppose that the sequence $\{x_n\}$ converges. Prove that f is Riemann integrable on $[a, b]$ and that $\int_a^b f = 0$.
   b) Suppose that the sequence $\{x_n\}$ does not converge. Give an example for which f is not Riemann integrable on $[a, b]$ and an example for which f is Riemann integrable on $[a, b]$.
5. Let f be a bounded function defined on $[a, b]$ and suppose that the set D of all the discontinuities of f in $[a, b]$ is countably infinite. Let $D = \{d_n : n \in \mathbb{Z}^+\}$ and suppose that the set $\{x : x \text{ is a limit of a subsequence of } \{d_n\}\}$ is finite. Prove that f is Riemann integrable on $[a, b]$.
6. Let $g: [a, b] \to [c, d]$ and $f: [c, d] \to \mathbb{R}$ be functions such that g is a polynomial and f is monotone. Prove that $f \circ g$ is Riemann integrable on $[a, b]$.
7. Find Riemann integrable functions $f: [0, 1] \to \mathbb{R}$ and $g: [0, 1] \to [0, 1]$ such that neither f nor g is continuous on $[0, 1]$, but $f \circ g$ is Riemann integrable on $[0, 1]$.
8. Suppose that f is Riemann integrable on $[a, b]$. Use Theorem 5.22 to prove that f^2 and $|f|$ are Riemann integrable on $[a, b]$. What further conditions must be placed on f in order to use Theorem 5.22 to prove that $1/f$ is Riemann integrable on $[a, b]$?
9. In the derivation of the average value of a function, it probably makes more sense to use
$$\frac{f(a_0) + f(a_1) + f(a_2) + \cdots + f(a_n)}{n + 1}$$
as an estimate for the average value of f on $[a, b]$. Show that, although a little more work is involved, the limit of this quantity gives the same formula for the average value of a function.
10. Suppose that f is Riemann integrable on $[a, b]$ and let V be the average value of f on $[a, b]$.
    a) Prove that $\displaystyle\int_a^b V = \int_a^b f.$
    b) Suppose that f is nonnegative. Use the area interpretation of the Riemann integral to give a geometric interpretation of V.
11. Find the average value of the function $f(x) = x \cos x$ on the interval $[0, \pi/2]$.
12. Find a positive number b so that the average value of the function f defined by $f(x) = x^2 + x + 1$ on the interval $[0, b]$ is 10.

13. Let f be a continuous and positive function defined on $[a, b]$. Derive an integral expression that represents the harmonic mean of the function on $[a, b]$. (See Exercise 30 in Section 1.2 for the definition of the harmonic mean of a set of numbers.)

14. Show that the Mean Value Theorem for integrals is a special case of the generalized Mean Value Theorem for integrals.

15. Give a geometric interpretation of the Mean Value Theorem for integrals in the case in which the function is nonnegative.

16. Use the Fundamental Theorem of Calculus to prove the Mean Value Theorem for integrals.

17. Use the Mean Value Theorem for integrals to prove part (a) of the Fundamental Theorem of Calculus for the case in which f is continuous. Be certain to avoid circular reasoning.

18. Give an example of a function $f : [a, b] \to \mathbb{R}$ such that f has a finite range, but f is not a step function.

19. Give an example of a step function $\phi : [a, b] \to \mathbb{R}$ such that the inverse image of an element of its range is neither a single point nor an interval.

20. Give an example of a function $f : [0, 1] \to \mathbb{R}$ such that the range of f is countably infinite and the inverse image of each element of the range of f is either a single point or an interval.

21. Prove Theorem 5.26.

22. Prove Theorem 5.27.

23. Find the average value of the function $f(x) = \lfloor x^2 \rfloor$ on the interval $[0, 2.4]$.

24. This exercise shows how various representations of a step function can be useful. Let ϕ and ψ be two step functions defined on an interval $[a, b]$.

    a) Show that there exists a partition $P = \{x_i : 0 \le i \le n\}$ of $[a, b]$ such that both ϕ and ψ are constant on each of the intervals (x_{i-1}, x_i).

    b) Use part (a) to prove that $\phi + \psi$ and $\phi\psi$ are step functions on $[a, b]$.

25. Suppose that $g : [a, b] \to [c, d]$ is Riemann integrable on $[a, b]$ and $f : [c, d] \to \mathbb{R}$ is Riemann integrable on $[c, d]$. Prove that $f \circ g$ is Riemann integrable on $[a, b]$ if either f or g is a step function.

26. Adopting the notation of the proof of the second part of Theorem 5.28, determine whether or not it is possible to conclude that the equation

$$\sum_{i=1}^{n} \omega(f, [x_{i-1}, x_i])(x_i - x_{i-1}) \le \int_a^b \psi - \int_a^b \phi$$

    is valid.

27. Let f be a continuous function on $[a, b]$ and let $\epsilon > 0$. Using only properties of continuous functions and no integration theory, prove that there exists a step function ϕ on $[a, b]$ such that $\phi(x) \le f(x) \le \phi(x) + \epsilon$ for all $x \in [a, b]$.

5.5 Numerical Integration

Suppose that f is Riemann integrable on $[a, b]$ and that a numerical value for $\int_a^b f$ is needed. If f has an antiderivative F and the values $F(a)$ and $F(b)$ are known, then $\int_a^b f = F(b) - F(a)$ by the Fundamental Theorem of Calculus. For example,

$$\int_2^5 \frac{1}{1+x^2}\,dx = \arctan x \Big|_2^5 = \arctan 5 - \arctan 2 \approx 0.266252.$$

A moment's reflection concerning this calculation might create a little discomfort. How is it that we know the values arctan 5 and arctan 2? Most likely, these values are found by pushing buttons on some machine, but what numerical procedures does the machine use to find the answer? Most people would be unable to evaluate the above integral without some kind of assistance from tables and/or a calculator. In addition to this sort of difficulty, some continuous functions—the function e^{-x^2} is one example—do not have antiderivatives that can be expressed in terms of algebraic, trigonometric, exponential, or logarithmic functions. (It is difficult to prove that a function does not have an elementary antiderivative; one reference is Ritt [21].) This means that the Fundamental Theorem of Calculus offers no assistance in finding the value of an integral of this type. In this section, we will discuss ways of approximating the value of $\int_a^b f$ when the familiar method of plugging endpoints into an antiderivative breaks down.

As we have seen, the definition of the Riemann integral is not a very efficient way to prove that a function is Riemann integrable. However, once it is known that a function f is Riemann integrable on some interval $[a, b]$, a modification of the definition makes it possible to evaluate the integral as a simple limit:

$$\int_a^b f = \lim_{n\to\infty} S(f, {}^tP_n),$$

where $\{{}^tP_n\}$ is any sequence of tagged partitions of $[a, b]$ whose norms converge to 0 (see Theorem 5.14). In actual practice, the tagged partitions are usually formed in some patterned way. The following method is the most common way to do this. Fix n and let $P = \{x_i : 0 \le i \le n\}$ be a partition of $[a, b]$ with $x_i = a + i(b-a)/n$. Then P divides the interval $[a, b]$ into n subintervals with equal lengths of $(b - a)/n$. A tagged partition of $[a, b]$ can be formed from P by choosing a tag from each interval; simple choices for the tag include the right endpoint, the left endpoint, or the midpoint of each interval. These choices for the tags generate the nth **right endpoint approximation** R_n, the nth **left endpoint approximation** L_n, and the nth **midpoint approximation** M_n to $\int_a^b f$. It is easy to verify that these approximations are given by the formulas

$$R_n = \frac{b-a}{n}\sum_{i=1}^n f(x_i),\quad L_n = \frac{b-a}{n}\sum_{i=1}^n f(x_{i-1}),\quad \text{and } M_n = \frac{b-a}{n}\sum_{i=1}^n f(m_i),$$

where $m_i = (x_{i-1} + x_i)/2$. By Theorem 5.14, each of the sequences $\{R_n\}$, $\{L_n\}$, and $\{M_n\}$ converges to $\int_a^b f$. It follows that the numbers R_n, L_n, and M_n should be close to $\int_a^b f$ for large values of n. These formulas thus give practical (and easily programmable) ways to approximate the value of an integral.

Mathematicians are always searching for faster and more efficient ways to do things, and they have found better methods for approximating integrals than the

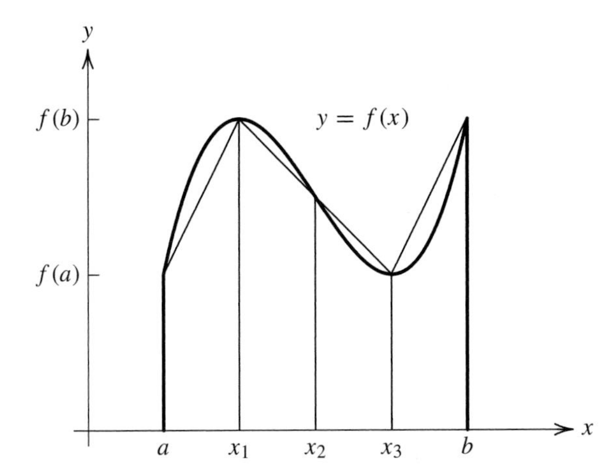

Figure 5.6 An approximation to the area under a curve using trapezoids

special Riemann sums discussed in the previous paragraph. The simplest of these methods, the trapezoid rule and Simpson's rule, will be discussed in the rest of the section. We will also consider the accuracy with which these two rules approximate an integral.

Since the definition of the integral was motivated by the problem of finding the area under a curve, the idea of area can sometimes provide helpful information. Suppose that f is a continuous and nonnegative function defined on $[a, b]$. Then $\int_a^b f$ represents the area under the curve $y = f(x)$ and above the x-axis on the interval $[a, b]$. Very little creative insight is required to recognize that this area appears to be better approximated with trapezoids (see Figure 5.6) rather than with the rectangles that are used to motivate the definition. To obtain a formula to approximate an integral using this idea, fix n and let $P = \{x_i : 0 \leq i \leq n\}$ be a partition of $[a, b]$ with $x_i = a + i(b - a)/n$. Then the nth **trapezoidal approximation** T_n to $\int_a^b f$ is defined by

$$\begin{aligned} T_n &= \frac{1}{2}(x_1 - x_0)\big(f(x_0) + f(x_1)\big) + \frac{1}{2}(x_2 - x_1)\big(f(x_1) + f(x_2)\big) + \cdots \\ &\quad + \frac{1}{2}(x_n - x_{n-1})\big(f(x_{n-1}) + f(x_n)\big) \\ &= \frac{b-a}{2n}\big(f(x_0) + 2f(x_1) + 2f(x_2) + \cdots + 2f(x_{n-1}) + f(x_n)\big). \end{aligned}$$

The fact that $\{T_n\}$ converges to $\int_a^b f$ when f is Riemann integrable on $[a, b]$ will be left as an exercise. The formula for T_n is often known as the **trapezoid rule**.

The trapezoid rule appears to be a good method for approximating $\int_a^b f$ when its value cannot be found by other means. But if $\int_a^b f$ is an unknown number, how do we know when the terms T_n are close to $\int_a^b f$? Questions concerning the accuracy of numerical estimations, including the theory behind the analysis of numerical algorithms, are addressed in a branch of mathematics called numerical analysis. Since most numerical algorithms require the use of a computer, numerical analysis must consider the methods of computation (evaluation techniques, round-off errors, etc.) as well as the mathematical theory of approximation. Only questions concerning the mathematical accuracy of numerical integration will be addressed here.

The following information about the trapezoid rule is not hard to determine and provides some idea about the errors involved in using it to approximate an integral.

1. The trapezoid rule gives exact answers for linear functions.
2. The accuracy of the trapezoid rule generally improves as n increases.
3. The trapezoid rule gives overestimates when the curve is concave up and underestimates when the curve is concave down.
4. The error when using the trapezoid rule depends roughly on how much the graph of the function curves.

While these observations are useful, they do not quantify the error involved when the trapezoid rule is used to approximate an integral. However, the next set of results does give a measure of the accuracy of the trapezoid rule for approximating the value of an integral for certain functions. In addition, the proofs of these results give some indication of how the theory of the derivative and the integral is used in analysis.

LEMMA 5.29 If g is twice differentiable on an interval $[-r, r]$ for some positive constant r, then there exists a point z in the interval $(-r, r)$ such that

$$\int_{-r}^{r} g - r\big(g(r) + g(-r)\big) = -\frac{(2r)^3}{12}\, g''(z).$$

Proof. Let k be the constant that satisfies the equation

$$\int_{-r}^{r} g - r\big(g(r) + g(-r)\big) = k\,(2r)^3.$$

(In other words, the number k is the left side of the displayed equation divided by $8r^3$.) We must show that $k = -g''(z)/12$ for some point $z \in (-r, r)$. Define a function $G: [0, r] \to \mathbb{R}$ by

$$G(x) = \int_{-x}^{x} g - x\big(g(x) + g(-x)\big) - k(2x)^3.$$

Since $G(0) = 0 = G(r)$, Rolle's Theorem guarantees the existence of a point $c \in (0, r)$ such that $G'(c) = 0$. Using the Fundamental Theorem of Calculus to find G', we obtain

$$\begin{aligned} G'(x) &= \big(g(x) + g(-x)\big) - \big(g(x) + g(-x)\big) - x\big(g'(x) - g'(-x)\big) - 24kx^2 \\ &= -x\big(g'(x) - g'(-x) + 24kx\big). \end{aligned}$$

Since $G'(c) = 0$ and $c \neq 0$, it follows that $g'(c) - g'(-c) + 24kc = 0$. Applying the Mean Value Theorem to the function g' on the interval $[-c, c]$ yields

$$k = -\frac{1}{12} \cdot \frac{g'(c) - g'(-c)}{2c} = -\frac{1}{12} g''(z),$$

where $z \in (-c, c) \subseteq (-r, r)$. This completes the proof. ■

LEMMA 5.30 If f is twice differentiable on an interval $[c, d]$, then there exists a point v in the interval (c, d) such that

$$\int_c^d f - \frac{d-c}{2}\big(f(d) + f(c)\big) = -\frac{(d-c)^3}{12} f''(v).$$

Proof. Let $m = (d + c)/2$, let $r = (d - c)/2$, and define a function g on $[-r, r]$ by $g(x) = f(x + m)$. Note that

$$\int_c^d f - \frac{d-c}{2}\big(f(d) + f(c)\big) = \int_{-r}^r g - r\big(g(r) + g(-r)\big).$$

By the previous lemma, there exists a point $z \in (-r, r)$ such that

$$\int_c^d f - \frac{d-c}{2}\big(f(d) + f(c)\big) = -\frac{(2r)^3}{12} g''(z) = -\frac{(d-c)^3}{12} f''(v),$$

where $v = z + m$ is a point in the interval (c, d). ■

THEOREM 5.31 If f is twice differentiable on an interval $[a, b]$ and n is a positive integer, then there exists a point v in the interval (a, b) such that

$$\int_a^b f - T_n = -\frac{(b-a)^3}{12n^2} f''(v),$$

where T_n is the nth trapezoidal estimate to the integral.

Proof. Fix a positive integer n and let $x_i = a + i(b - a)/n$ for $0 \leq i \leq n$. Using the previous lemma, we obtain

$$\begin{aligned}\int_a^b f - T_n &= \sum_{i=1}^n \left(\int_{x_{i-1}}^{x_i} f - \frac{x_i - x_{i-1}}{2}\big(f(x_i) + f(x_{i-1})\big)\right) \\ &= \sum_{i=1}^n -\frac{(x_i - x_{i-1})^3}{12} f''(v_i) \\ &= -\frac{(b-a)^3}{12n^3} \sum_{i=1}^n f''(v_i),\end{aligned}$$

where $v_i \in (x_{i-1}, x_i)$ for each i. By Theorem 4.19, the function f'' has the intermediate value property on the interval $[v_1, v_n]$. Since the average

$$\frac{f''(v_1) + f''(v_2) + \cdots + f''(v_n)}{n}$$

is between $\min\{f''(v_i) : 1 \le i \le n\}$ and $\max\{f''(v_i) : 1 \le i \le n\}$, there exists a point $v \in (v_1, v_n) \subseteq (a, b)$ such that

$$\int_a^b f - T_n = -\frac{(b-a)^3}{12n^2} \cdot \frac{1}{n} \sum_{i=1}^{n} f''(v_i) = -\frac{(b-a)^3}{12n^2} f''(v).$$

This completes the proof. ■

COROLLARY 5.32 If f is twice differentiable on an interval $[a, b]$, B_2 is a bound for f'' on $[a, b]$, and n is a positive integer, then

$$\left| \int_a^b f - T_n \right| \le \frac{B_2(b-a)^3}{12n^2},$$

where T_n is the nth trapezoidal estimate to the integral.

Proof. The corollary follows immediately from Theorem 5.31. ■

The trapezoid rule error estimate can be used to either determine the accuracy of a given estimate T_n or to determine a value of n for which T_n will give a desired degree of accuracy. The most difficult part of the process is finding a value for B_2, the maximum value of the absolute value of the second derivative. In most of the cases in which the trapezoid rule is used, the second derivative of the integrand is a complicated function, so finding its maximum value can be a difficult problem. Using a calculator to graph the second derivative is one way to estimate its maximum value; another method is to do some simple overestimation. We will leave an exploration of such questions to the exercises.

The trapezoidal approximation to $\int_a^b f$ is obtained by using straight lines to approximate the function on each subinterval. If curves rather than lines are used to approximate the function, it is reasonable to expect an even better approximation to the integral. Two simple curves that come to mind are circles and parabolas—each of these curves is uniquely determined by three non-collinear points. However, an approximation using circles leads to complicated formulas whereas an approximation using parabolas leads to a formula almost as simple as the trapezoid result. Let Q be an arbitrary polynomial of degree two and let $[a, b]$ be any interval. A simple but somewhat tedious computation (left as an exercise) shows that

$$\int_a^b Q = \frac{b-a}{6}\big(Q(a) + 4Q(c) + Q(b)\big),$$

where c is the midpoint of $[a, b]$. In other words, the integral of a quadratic function depends only on the values of the function at the endpoints and at the midpoint of the interval and on the length of the interval. The quantity $\big(Q(a) + 4Q(c) + Q(b)\big)/6$ is known as a **weighted average**; it represents the average of six numbers, but the value of the function at the midpoint is used four times.

Using this property of quadratic functions to approximate an integral results in a formula known as Simpson's rule, named after the English mathematician Thomas Simpson (1710–1761). To derive this formula, fix a positive integer n and let $P = \{x_i : 0 \le i \le n\}$ be a partition of $[a, b]$ with $x_i = a + i(b-a)/n$.

Consider first the points $(x_0, f(x_0))$, $(x_1, f(x_1))$, and $(x_2, f(x_2))$. These three points determine a unique parabola $y = Q_1(x)$ that approximates the graph of $y = f(x)$ on the interval $[x_0, x_2]$. Without actually finding a formula for $Q_1(x)$, we know that

$$\begin{aligned}\int_{x_0}^{x_2} f \approx \int_{x_0}^{x_2} Q_1 &= \frac{x_2 - x_0}{6}\big(f(x_0) + 4f(x_1) + f(x_2)\big)\\ &= \frac{b-a}{3n}\big(f(x_0) + 4f(x_1) + f(x_2)\big).\end{aligned}$$

Similarly, the points $(x_2, f(x_2))$, $(x_3, f(x_3))$, and $(x_4, f(x_4))$ determine a unique parabola $y = Q_2(x)$ that approximates the graph of $y = f(x)$ on the interval $[x_2, x_4]$ and

$$\begin{aligned}\int_{x_2}^{x_4} f \approx \int_{x_2}^{x_4} Q_2 &= \frac{x_4 - x_2}{6}\big(f(x_2) + 4f(x_3) + f(x_4)\big)\\ &= \frac{b-a}{3n}\big(f(x_2) + 4f(x_3) + f(x_4)\big).\end{aligned}$$

If n is even (so that the parabolic approximations "match up right"), then this process can be repeated along the entire interval $[a, b]$. For each even positive integer n, it follows that S_n, the nth **Simpson's rule** approximation, is

$$\begin{aligned}S_n = \frac{b-a}{3n}\big(&f(x_0) + 4f(x_1) + 2f(x_2) + 4f(x_3) + 2f(x_4) + \cdots\\ &+ 2f(x_{n-2}) + 4f(x_{n-1}) + f(x_n)\big).\end{aligned}$$

It will be left as an exercise to show that $\{S_{2n}\}$ converges to $\int_a^b f$.

As with the trapezoid rule, an error estimate for the accuracy of Simpson's rule is needed. After spending some time working with Simpson's rule (as some of the exercises ask you to do), we can make the following observations:

1. Simpson's rule gives exact answers for polynomials of degree 3 or less.
2. The accuracy of Simpson's rule generally improves as n increases.
3. Simpson's rule appears to be much more accurate than the trapezoid rule.

The following set of results provides a measure of the accuracy of Simpson's rule for approximating the value of an integral for certain functions. Most of the proofs, which are similar to those for the trapezoid rule, will be left as exercises.

LEMMA 5.33 If g has a fourth derivative on an interval $[-r, r]$ for some positive constant r, then there exists a point z in the interval $(-r, r)$ such that

$$\int_{-r}^{r} g - \frac{r}{3}\big(g(-r) + 4g(0) + g(r)\big) = -\frac{r^5}{90}\, g''''(z).$$

Proof. Let k be the constant that satisfies the equation

$$\int_{-r}^{r} g - \frac{r}{3}\big(g(-r) + 4g(0) + g(r)\big) = kr^5.$$

Define a function G on the interval $[0, r]$ by

$$G(x) = \int_{-x}^{x} g - \frac{x}{3}\big(g(-x) + 4g(0) + g(x)\big) - kx^5.$$

Then $G'''(x) = -\frac{x}{3}\big(g'''(x) - g'''(-x) + 180kx\big)$ and it can be shown that there is a point $c \in (0, r)$ for which $G'''(c) = 0$. (Proofs of these statements are left as exercises.) Since $c \neq 0$, it follows that

$$k = \frac{g'''(c) - g'''(-c)}{-180c} = -\frac{g''''(z)}{90},$$

where the existence of the point $z \in (-c, c) \subseteq (-r, r)$ is guaranteed by the Mean Value Theorem. This completes the proof. ■

LEMMA 5.34 If f has a fourth derivative on an interval $[c, d]$, then there exists a point v in the interval (c, d) such that

$$\int_{c}^{d} f - \frac{d-c}{6}\big(f(c) + 4f(m) + f(d)\big) = -\frac{(d-c)^5}{32 \cdot 90} f''''(v),$$

where $m = (c + d)/2$ is the midpoint of the interval $[c, d]$. ■

THEOREM 5.35 If f has a fourth derivative on an interval $[a, b]$ and n is an even positive integer, then there exists a point v in the interval (a, b) such that

$$\int_{a}^{b} f - S_n = -\frac{(b-a)^5}{180n^4} f''''(v),$$

where S_n is the nth Simpson's rule estimate to the integral. ■

COROLLARY 5.36 If f has a fourth derivative on an interval $[a, b]$, B_4 is a bound for f'''' on $[a, b]$, and n is an even positive integer, then

$$\left|\int_{a}^{b} f - S_n\right| \le \frac{B_4(b-a)^5}{180n^4},$$

where S_n is the nth Simpson's rule estimate to the integral. ■

Exercises

1. Let f be a nonnegative continuous function defined on an interval $[a, b]$. Sketch some graphs to indicate area interpretations for the approximations R_n, L_n, and M_n.
2. Prove that $T_n = (L_n + R_n)/2$ for each positive integer n, then use this fact to prove that $\{T_n\}$ converges to $\int_a^b f$.
3. Suppose that f is continuous on $[a, b]$ and let n be a positive integer. Show that T_n is actually a Riemann sum of f.
4. For the function $f(x) = 1/x$ on the interval $[1, 2]$, find L_4, R_4, M_4, and T_4. Express the answers as rational numbers, that is, as ratios of integers.

5. Does Theorem 5.31 remain valid if f is continuous on $[a, b]$ and twice differentiable on (a, b)?

6. Let f be the function defined by $f(x) = \sqrt{1 + x^3}$.

a) By making some overestimations, show that $|f''(x)| \le 3.75$ for all $x \in [0, 1]$.

b) Use a calculator or a computer to show that $|f''(x)| \le 1.5$ on $[0, 1]$.

c) Use either part (a) or part (b) to estimate $\left|\int_0^1 f - T_{10}\right|$.

d) Use either part (a) or part (b) to determine a value of n so that $\left|\int_0^1 f - T_n\right| < 10^{-6}$.

7. Discuss how Theorem 5.31 explains items (1) through (4) listed prior to Lemma 5.29.

8. Let Q be a quadratic polynomial and let $[a, b]$ be any interval. Show that

$$\int_a^b Q = \frac{b-a}{6}\big(Q(a) + 4Q(c) + Q(b)\big),$$

where c is the midpoint of $[a, b]$.

9. Show that Simpson's rule gives exact answers for cubic polynomials by first proving

$$\int_a^b x^3\,dx = \frac{b-a}{6}\left(a^3 + 4c^3 + b^3\right),$$

where c is the midpoint of $[a, b]$. Give an example to show that Simpson's rule may not give exact answers for quartic polynomials.

10. The previous exercise is rather interesting since there are an infinite number of cubic polynomials that go through three distinct points (with different x-coordinates). However, if the x-coordinates of those points are such that the middle one is the midpoint of the other two, then all of the functions must have the same integral over the given interval. To illustrate this, find a representation for all of the cubic polynomials that pass through the points $(0, 0)$, $(1, 1)$, and $(2, 8)$, then show (by integrating this representation) that they all have the same integral on the interval $[0, 2]$.

11. Prove that $S_{2n} = \frac{1}{3}T_n + \frac{2}{3}M_n$ for each positive integer n. Use this result to prove that the sequence $\{S_{2n}\}$ converges to $\int_a^b f$.

12. Fill in the details in the proof of Lemma 5.33.

13. Prove Lemma 5.34.

14. Prove Theorem 5.35.

15. What happens (roughly) to the error in Simpson's rule when n is doubled?

16. Let $f(x) = \sqrt{1 + x^3}$. It can be shown that $|f''''(x)| \le 7.1$ for $0 \le x \le 1$. Find a value of n so that $\left|\int_0^1 f - S_n\right| < 10^{-6}$. Compare this value of n with the one found for the trapezoid rule in Exercise 6.

17. For each of the given integrals, find

i) the maximum values of the absolute value of the second derivative and the fourth derivative of the integrand on the appropriate interval;

ii) the accuracy of T_{10} and S_{10} (based upon the error estimates);

iii) a value of n for which T_n has better than 10^{-6} accuracy and a value of n for which S_n has better than 10^{-6} accuracy (based upon the error estimates).

a) $\displaystyle\int_1^2 x^{-1}\,dx$ **b)** $\displaystyle\int_0^2 x^4\,dx$ **c)** $\displaystyle\int_0^1 e^{-2x}\,dx$

18. For each of the given integrals, compute $|A - T_n|$, $|A - M_n|$, and $|A - S_n|$, where A is the actual value of the integral, for $n = 2, 4, 8, 16$. Give all of the answers accurate to 8 decimal places. What quantitative conclusions can you draw about the errors and how these errors change as n increases?

a) $\displaystyle\int_0^1 x^4\,dx$ **b)** $\displaystyle\int_0^{\pi} \sin x\,dx$ **c)** $\displaystyle\int_1^2 x^{-2}\,dx$

19. Although rectangles do not appear to approximate the area under a curve all that well, the midpoint rule actually does a pretty good job of approximating the value of an integral. If you did all of the parts of the previous exercise, you might conjecture that the midpoint error estimate is half the trapezoid rule error estimate. Verify this conjecture by first proving the following:

If g is twice differentiable on an interval $[-r, r]$ for some positive constant r, then there exists a point z in the interval $(-r, r)$ such that

$$\int_{-r}^{r} g - 2rg(0) = \frac{(2r)^3}{24}\, g''(z).$$

Start by letting k be the constant that satisfies $\int_{-r}^{r} g - 2rg(0) = k\,(2r)^3$, then define a function $G\colon [0, r] \to \mathbb{R}$ by

$$G(x) = \int_{-x}^{x} g - 2xg(0) - k(2x)^3.$$

The proof is similar to the trapezoid rule proof, but there are some differences.

20. Suppose that f is twice differentiable on $[a, b]$ and that f'' is negative on $[a, b]$. Prove that $T_n \le \int_a^b f \le M_n$ for each positive integer n.

21. Simpson's rule or the trapezoid rule are useful for approximating the values of integrals of functions that arise from experimental data since, in these cases, the values of the functions are only known at certain points. As a simple illustration, consider the following table of values for a function f.

x	1	2	3	4	5
$f(x)$	2	3	5	7	2

Assume that f is a nonnegative continuous function defined on $[1, 5]$ and let R be the region under the curve $y = f(x)$ and above the x-axis on the interval $[1, 5]$. Use Simpson's rule to approximate each of the following:

a) the area of R;

b) the volume generated when R is revolved around the x-axis;

c) the volume generated when R is revolved around the y-axis.

22. Suppose that $f\colon [0, 1] \to \mathbb{R}$ is convex, positive, and decreasing on $[0, 1]$.

a) Let n be a positive integer and let i be any integer in the set $\{1, 2, \ldots, n\}$. Prove that

$$f\left(\frac{i}{n}\right) \le \frac{n-i}{n}\, f\left(\frac{i}{n+1}\right) + \frac{i}{n}\, f\left(\frac{i+1}{n+1}\right).$$

b) Prove that the sequence $\{R_n\}$ of right endpoint approximations to $\int_0^1 f$ is increasing. (This result appears in Stein [23].)

c) Extend this result to an arbitrary interval $[a, b]$.

5.6 SUPPLEMENTARY EXERCISES

1. Suppose that $f\colon [1,\infty) \to \mathbb{R}$ is positive and decreasing for all $x \geq 1$. For each positive integer n, let

$$x_n = f(1) + f(2) + \cdots + f(n) - \int_1^n f.$$

Prove that the sequence $\{x_n\}$ converges.

2. Let $f\colon [a,b] \to \mathbb{R}$ be a continuous function and suppose that $\int_a^b fg = 0$ for every Riemann integrable function $g\colon [a,b] \to \mathbb{R}$. Prove that $f(x) = 0$ for all $x \in [a,b]$.

3. Let $f\colon [a,b] \to \mathbb{R}$ be Riemann integrable on $[a,b]$ and suppose that F is a continuous function on $[a,b]$ such that $F'(x) = f(x)$ on $[a,b]$ except for a finite number of points. Prove that $\int_a^b f = F(b) - F(a)$.

4. By interpreting the sum as a Riemann sum, evaluate $\lim_{n\to\infty} \sum_{k=1}^{n} \frac{n}{k^2+n^2}$.

5. Let f be a nonnegative continuous function defined on $[a,b]$. For each positive integer n, let v_n be the nth root of $\int_a^b f^n$. Prove that the sequence $\{v_n\}$ converges to the maximum value of f on $[a,b]$.

6. Show that the function f defined by $f(x) = x$ if x is rational and $f(x) = 0$ if x is irrational is not Riemann integrable on $[0,1]$.

7. Prove that $\int_1^2 \frac{1}{x}\,dx = \lim_{n\to\infty} \sum_{i=1}^{n} \frac{1}{n+i}$.

8. Suppose that $f\colon [a,b] \to \mathbb{R}$ is monotone. Prove that there exists a point $c \in (a,b)$ such that

$$\int_a^b f = f(a)(c-a) + f(b)(b-c).$$

Give an example to show that the conclusion may be false if f is not monotone.

9. Suppose that f and g are Riemann integrable on $[a,b]$. Prove that $f \vee g$ and $f \wedge g$ are Riemann integrable on $[a,b]$. (See Exercise 29 in Section 1.5 for the definitions of these functions.)

10. Let f be a continuous function defined on $[a,b]$.

a) Prove that for each tagged partition ${}^tP$ of $[a,b]$ there exists a point $c \in [a,b]$ such that $S(f, {}^tP) = f(c)(b-a)$.

b) Assume that f is differentiable on $[a,b]$. Prove that for each tagged partition ${}^tP$ of $[a,b]$ there exists a point $c \in [a,b]$ such that $S(f', {}^tP) = f'(c)(b-a)$.

11. Let f be a continuous function defined on $[a,b]$. Find $\lim_{n\to\infty} \int_a^{a+1/n} nf$.

12. Suppose that a function f satisfies $f(x) = 3 + \int_1^x 2f^2$ for all x in some interval I. Find the function f and an appropriate interval I.

13. Suppose that f is Riemann integrable on $[a,b]$. Prove that there exists a continuous function g such that $\int_a^b |f-g| < \epsilon$.

Remark. Although the definition of the integral is motivated by the concept of area, the integral has a wide variety of applications in geometry, physics, and other fields. The reader should have seen some of these applications in a calculus course, so familiarity with some

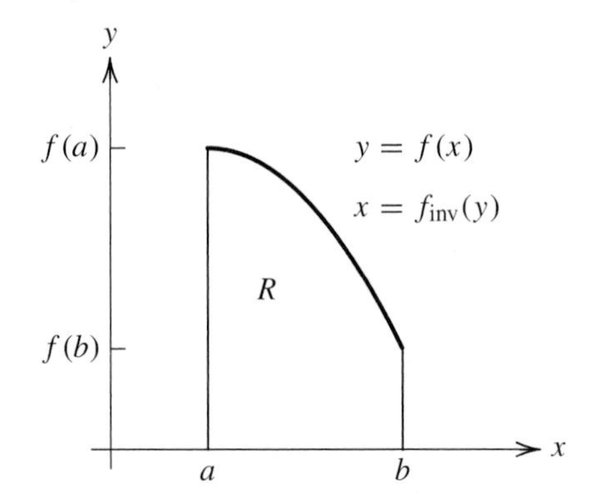

Figure 5.7 Comparing the disk and shell methods

applications is assumed in the next few exercises. The first two exercises provide a sampling of how integration appears in applications; the solution to the problem is approximated with a Riemann sum, and the corresponding integral gives the exact value of the desired quantity. The remaining exercises focus more on the theory behind some applied formulas than on the formulas themselves.

14. Suppose that a large cylindrical tank is lying on its side (in a position to roll if not held in place). The diameter of the tank is 12 feet and its height (or length in this case) is 30 feet. The tank contains diesel fuel and the fuel level is 4 feet above the ground. How many gallons of fuel are in the tank? (There are about 231 cubic inches in a gallon).

15. When a liquid flows through a cylindrical tube, friction at the wall of the tube tends to slow its motion. This results in liquid flowing faster near the center of the tube than at the wall. According to Poiseuille's Law, the liquid can be considered as flowing in cylindrical layers which have a constant velocity given by $v(r) = k(R^2 - r^2)$, where R is the radius of the tube, r is the distance from the center of the tube, and k is a constant that depends on the liquid and the tube. Show that the total volume of flow (that is, the volume per time flowing past a certain cross section) in the tube is $k\pi R^4/2$.

16. There are two different methods for finding the volume of a solid of revolution: the disk method and the shell method. A proof that the two methods actually compute the same number is generally omitted in calculus books. Suppose that a function f is nonnegative, decreasing, and has a continuous derivative on an interval $[a, b]$ with $a > 0$. Let R be the region under the graph of $y = f(x)$ and above the x-axis on the interval $[a, b]$ (see Figure 5.7). The region R sweeps out a volume when it is revolved around the x-axis. The disk method and the shell method for computing the volume of this solid give the formulas

$$V_{\text{disk}} = \int_a^b \pi\big(f(x)\big)^2\,dx \quad \text{and}$$

$$V_{\text{shell}} = \pi\big(f(b)\big)^2(b-a) + \int_{f(b)}^{f(a)} 2\pi y\big(f_{\text{inv}}(y) - a\big)\,dy.$$

Show that these two numbers are equal by starting with V_{shell}, splitting the integral into two parts, making a substitution, and using integration by parts. Thus, at least for this special case, the shell method and the disk method compute the same number.

17. Let f be a continuous function defined on $[a, b]$. The length of the curve $y = f(x)$ on the interval $[a, b]$ is defined by

$$L = \sup\Big\{\sum_{i=1}^{n} \sqrt{(x_i - x_{i-1})^2 + \big(f(x_i) - f(x_{i-1})\big)^2}\Big\},$$

where the supremum is taken over all partitions $P = \{x_i : 0 \le i \le n\}$ of $[a, b]$; the positive integer n is not fixed but may vary with P. If the set in question is unbounded, then the length of the curve is infinite.

a) Consider the function f defined by $f(x) = \sqrt{x}\cos(\pi/x)$ for $x \neq 0$ and $f(0) = 0$. Prove that the length of this curve on $[0, 1]$ is infinite.

b) Prove that the length of a curve is finite if and only if f is of bounded variation on $[a, b]$.

c) For the remainder of this exercise, assume that f has a continuous derivative on $[a, b]$. Let $[c, d]$ be any subinterval of $[a, b]$ and define

$$r = \sqrt{(d - c)^2 + \big(f(d) - f(c)\big)^2}, \quad \alpha = \frac{d - c}{r}, \quad \text{and} \quad \beta = \frac{f(d) - f(c)}{r}.$$

Explain the following steps.

$$\begin{aligned}\sqrt{(d - c)^2 + \big(f(d) - f(c)\big)^2} &= \int_c^d \big(\alpha + \beta \cdot f'(x)\big)\,dx \\ &\le \int_c^d \big|\alpha \cdot 1 + \beta \cdot f'(x)\big|\,dx \\ &\le \int_c^d \sqrt{\alpha^2 + \beta^2}\sqrt{1 + \big(f'(x)\big)^2}\,dx \\ &= \int_c^d \sqrt{1 + \big(f'(x)\big)^2}\,dx.\end{aligned}$$

d) Let $P = \{x_i : 0 \le i \le n\}$ be any partition of $[a, b]$. Use part (c) to show that

$$\sum_{i=1}^{n} \sqrt{(x_i - x_{i-1})^2 + \big(f(x_i) - f(x_{i-1})\big)^2} \le \int_a^b \sqrt{1 + \big(f'(x)\big)^2}\,dx.$$

e) Show that for every partition $P = \{x_i : 0 \le i \le n\}$ of $[a, b]$ there exists a tagged partition ${}^tP = \{(t_i, [x_{i-1}, x_i]) : 1 \le i \le n\}$ associated with P such that

$$\sum_{i=1}^{n} \sqrt{(x_i - x_{i-1})^2 + \big(f(x_i) - f(x_{i-1})\big)^2} = \sum_{i=1}^{n} \sqrt{1 + \big(f'(t_i)\big)^2}\,(x_i - x_{i-1}).$$

f) Use the previous results to prove that $L = \displaystyle\int_a^b \sqrt{1 + \big(f'(x)\big)^2}\,dx$.

18. Let f and g be continuous functions defined on $[a, b]$. For each positive integer n, let $P_n = \{x_i : 0 \le i \le n\}$ be the partition of $[a, b]$ with $x_i = a + i(b - a)/n$. For each $1 \le i \le n$, let s_i and t_i be points in $[x_{i-1}, x_i]$, and let

$$A_n = \sum_{i=1}^{n} f(s_i)g(t_i)\frac{b - a}{n}.$$

Prove that the sequence $\{A_n\}$ converges to $\int_a^b fg$.

19. Consult a standard calculus textbook and look up the derivation of the formula for the surface area of a solid of revolution. Write up a careful proof for this derivation and note the need of the result in the previous exercise.

Remark. I have chosen not to give rigorous definitions of the trigonometric functions and other transcendental functions in this book. However, the next few exercises present one way to define the functions $\ln x$, e^x, a^x and x^x. Although many calculus books define the natural logarithm function as in Exercise 20, most students in calculus fail to appreciate the need for such a definition. At the level of undergraduate analysis, the development of the function $\ln x$ as an integral should be both interesting and enlightening.

20. Define a function $L: (0, \infty) \to \mathbb{R}$ by $L(x) = \int_1^x (1/t)\, dt$. Verify the following properties of L.

a) L is continuous on $(0, \infty)$.

b) L is strictly increasing on $(0, \infty)$.

c) L is differentiable on $(0, \infty)$.

d) $L(ab) = L(a) + L(b)$ for all positive numbers a and b.

e) $L(a^b) = bL(a)$ for all positive numbers a and rational numbers b.

f) $L(a/b) = L(a) - L(b)$ for all positive numbers a and b.

g) $\lim_{x \to 0^+} L(x) = -\infty$ and $\lim_{x \to \infty} L(x) = \infty$.

h) There is a unique real number, denoted by e, such that $L(e) = 1$.

The function L is called the **natural logarithm** function and is often denoted by $\ln x$.

21. Since the function L defined in the previous exercise is strictly increasing on $(0, \infty)$, it has an inverse. Let $E(x) = L_{\text{inv}}(x)$ and note that the domain of E is all real numbers.

a) Using the properties of L from the previous exercise, find and prove a list of properties possessed by the function E.

b) Find a formula for $E'(x)$.

c) Show that $E(x) = e^x$ for all $x \in \mathbb{R}$, where e is the number defined in part (h) of the previous exercise. (The Maclaurin series for E will reveal that the number e is the same as the number considered in Exercise 35 of Section 2.2.)

22. Let r be a positive rational number and define a function f by $f(x) = x^{-r} \ln x$.

a) Prove that f is eventually decreasing and therefore has a limit as x tends to infinity.

b) Use properties of the natural logarithm and results listed in Theorem 2.14 to find the limit of the sequence $\{f(2^n)\}$.

c) Use the results in parts (a) and (b) to evaluate $\lim_{x \to \infty} x^{-r} \ln x$.

d) Use the result of part (c) to evaluate $\lim_{x \to 0^+} x^r \ln x$.

23. Prove that $\lim_{x \to \infty} x^r e^{-x} = 0$ for each rational number r.

24. To this point, we have avoided functions such as 2^x because of the difficulty that occurs with irrational exponents. However, note that $2^x = e^{x \ln 2}$ for all real numbers x. In general, let a be a positive number and define the expression a^x by $a^x = e^{x \ln a}$ for all real numbers x.

a) Show that the function a^x is continuous and differentiable on $\mathbb{R}$ and find the derivative of this function.

b) Explain how to define a function x^r, where the exponent r is irrational.

c) Show that the derivative of x^r is rx^{r-1} for any real number r.

d) Explain how to define the function x^x, then find its derivative.

25. Let f be a continuous and positive function defined on $[a, b]$. Derive an integral expression that represents the geometric mean of the function on $[a, b]$.

Remark. The definition of the Riemann integral requires a bounded function f defined on a closed and bounded interval $[a, b]$. An integral is said to be an **improper integral** if either the function f is unbounded on $[a, b]$ or the interval of integration is unbounded. For example, the integrals

$$\int_0^1 \frac{1}{x^2}\, dx \quad \text{and} \quad \int_2^\infty e^{-x}\, dx$$

are both improper: the first because $1/x^2$ is unbounded on $[0, 1]$ and the second because the interval of integration is unbounded. The limit concept makes it possible to assign a meaning to integrals of these types. Let a be a real number and suppose that f is Riemann integrable on $[a, b]$ for all $b > a$. Then

$$\int_a^\infty f = \lim_{b\to\infty} \int_a^b f,$$

provided the limit exists. When the limit exists, the integral is said to converge or be convergent; if not, the integral is said to diverge or be divergent. Now suppose that f is a function that is Riemann integrable on each of the intervals $[a, c]$ for all $c \in (a, b)$ and that f is unbounded at b. Then

$$\int_a^b f = \lim_{c\to b^-} \int_a^c f,$$

provided the limit exists. The next set of exercises deals with improper integrals.

26. Evaluate $\int_0^\infty 12xe^{-3x}\, dx$.

27. For what values of r does the integral $\int_1^\infty x^r\, dx$ converge?

28. Provide a definition for $\int_{-\infty}^b f$.

29. Let a be a real number, suppose that f and g are Riemann integrable on $[a, b]$ for all $b > a$, and suppose that $0 \le f(x) \le g(x)$ for all $x \ge a$.

a) Suppose that the integral $\int_a^\infty g$ converges. Prove that the integral $\int_a^\infty f$ converges.

b) Suppose that the integral $\int_a^\infty f$ diverges. Prove that the integral $\int_a^\infty g$ diverges.

30. Use the previous exercise to determine whether or not the following integrals converge.

a) $\displaystyle\int_0^\infty e^{-x}\sin^2 x\, dx$ **b)** $\displaystyle\int_1^\infty \frac{\sqrt{x^4+10}}{x^3}\, dx$ **c)** $\displaystyle\int_3^\infty \frac{12}{\sqrt{x^4+4x^3+5}}\, dx$

31. Consider an integral of the form $\int_{-\infty}^\infty f$.

a) What properties must f possess in order to make sense of this integral?

b) By definition, the integral $\int_{-\infty}^\infty f$ converges if both of the integrals $\int_{-\infty}^a f$ and $\int_a^\infty f$ converge, where a is any real number, and

$$\int_{-\infty}^\infty f = \int_{-\infty}^a f + \int_a^\infty f.$$

Show that this definition is independent of the choice of a.

c) Evaluate $\displaystyle\int_{-\infty}^{\infty} \frac{c^3}{c^2 + x^2}\, dx$, where c is any positive constant.

d) It is tempting to make the definition

$$\int_{-\infty}^{\infty} f = \lim_{a \to \infty} \int_{-a}^{a} f,$$

provided the limit exists. This value is known as the **Cauchy principal value** of the integral, but it may exist even when the integral does not converge according to the definition in part (b). Find a function that illustrates this fact.

e) Suppose that $\int_{-\infty}^{\infty} f$ converges. Prove that the Cauchy principal value of the integral exists and that the two values are the same.

32. Prove that $\int_0^{\infty} x^n e^{-x}\, dx = n!$ for each positive integer n. (As usual, the symbol $n!$ represents n factorial.) To be consistent, how should $0!$ be defined?

33. Evaluate $\int_0^1 \ln x\, dx$.

34. For what values of r does the integral $\int_0^1 x^r\, dx$ converge?

35. Provide a definition for the improper integral of a function that is unbounded at the left endpoint of a closed and bounded interval.

36. Consider the integral $\displaystyle\int_0^1 \frac{dx}{\sqrt[3]{4x - 1}}$. Give a general definition that covers integrals of this type, then apply it to this particular integral.

Remark. Another common way to define the Riemann integral is through the use of upper and lower sums. This method of defining the integral is sometimes referred to as the Darboux approach, and we will use this adjective to avoid confusion with the terminology associated with the Riemann integral. Let f be a bounded function defined on an interval $[a, b]$. Given a partition $P = \{x_i : 0 \le i \le n\}$ of $[a, b]$, let

$$m_i = \inf\{f(x) : x_{i-1} \le x \le x_i\} \quad \text{and} \quad M_i = \sup\{f(x) : x_{i-1} \le x \le x_i\}$$

for $1 \le i \le n$. Define the **lower and upper Darboux sums** of f for the partition P by

$$L(f, P) = \sum_{i=1}^{n} m_i(x_i - x_{i-1}) \quad \text{and} \quad U(f, P) = \sum_{i=1}^{n} M_i(x_i - x_{i-1}),$$

respectively. The **upper and lower Darboux integrals** of f on $[a, b]$ are defined by

$$(D)\overline{\int_a^b} f = \inf\{U(f, P) : P \text{ is a partition of } [a, b]\} \quad \text{and}$$

$$(D)\underline{\int_a^b} f = \sup\{L(f, P) : P \text{ is a partition of } [a, b]\},$$

respectively. If these two integrals are equal, then f is **Darboux integrable** on $[a, b]$ and the common value is denoted by $(D)\int_a^b f$. The prefix (D) is used here to distinguish this integral from the Riemann integral, but, as we will see, the two integrals are equivalent.

37. Suppose that $m \le f(x) \le M$ for all $x \in [a, b]$. Prove that

$$m(b - a) \le L(f, P) \le U(f, P) \le M(b - a)$$

for all partitions P of $[a, b]$.

38. Let f be the function defined by $f(x) = 3x^3 - x$, let $P_1 = \{0, 0.1, 0.3, 0.7, 1\}$ be a partition of $[0, 1]$, and let $P_2 = P_1 \cup \{0.2, 0.4\}$. Find $L(f, P_1)$, $U(f, P_1)$, $L(f, P_2)$, and $U(f, P_2)$.

39. Suppose that f is bounded on $[a, b]$.

a) Show that an upper Darboux sum of f may not be a Riemann sum of f.

b) Suppose that f is continuous on $[a, b]$. Prove that every upper Darboux sum is a Riemann sum.

40. Let f be a bounded function defined on $[a, b]$ and let P_1, P_2 be partitions of $[a, b]$.

(a) Suppose that P_1 is a refinement of P_2. Prove that $L(f, P_1) \geq L(f, P_2)$ and $U(f, P_1) \leq U(f, P_2)$.

(b) Suppose that P_1 and P_2 are arbitrary partitions. Prove that $L(f, P_1) \leq U(f, P_2)$.

41. Let f be a bounded function defined on $[a, b]$.

a) Prove that the set $\{U(f, P) : P \text{ is a partition of } [a, b]\}$ is bounded below and conclude that the infimum in the definition of the upper Darboux integral exists.

b) Prove that the supremum in the definition of the lower Darboux integral exists.

c) Prove that the upper Darboux integral of f on $[a, b]$ is always greater than or equal to the lower Darboux integral of f on $[a, b]$.

42. Consider the function f defined by $f(x) = 0$ if x is irrational and $f(x) = 1$ if x is rational. Find the upper and lower Darboux integrals of f on the interval $[1, 3]$.

43. Consider the function g defined by $g(x) = x$ if x is irrational and $g(x) = -x$ if x is rational. Find the upper and lower Darboux integrals of g on the interval $[0, 1]$.

44. Let f be bounded on $[a, b]$.

a) Prove that f is Darboux integrable on $[a, b]$ if and only if for each $\epsilon > 0$ there exists a partition P of $[a, b]$ such that $U(f, P) - L(f, P) < \epsilon$.

b) Let $P = \{x_i : 0 \leq i \leq n\}$. Prove that

$$\sum_{i=1}^{n} \omega(f, [x_{i-1}, x_i])(x_i - x_{i-1}) = U(f, P) - L(f, P).$$

c) Use parts (a) and (b) to prove that a function is Darboux integrable on $[a, b]$ if and only if it is Riemann integrable on $[a, b]$. Prove that the values of the integrals are the same.

Remark. As pointed out in this chapter, a function is Riemann integrable on $[a, b]$ if and only if it is bounded and continuous almost everywhere on $[a, b]$. To prove this result, it is first necessary to clarify the term "almost everywhere". For any interval I with endpoints a and b ($a < b$), define the **length** $\ell(I)$ of I by $\ell(I) = b - a$. A set E of real numbers has **measure zero** if for each $\epsilon > 0$ there exists a sequence (finite or infinite) $\{I_k\}$ of open intervals such that $E \subseteq \bigcup_{k=1}^{\infty} I_k$ and $\sum_{k=1}^{\infty} \ell(I_k) < \epsilon$. (We assume that the reader has some familiarity with infinite series.) A property is said to hold **almost everywhere** on $[a, b]$ if it holds at each point of $[a, b] \setminus E$, where E is a set of measure zero.

45. Prove that every finite set has measure zero.

46. Prove that every countably infinite set has measure zero. For the record, there are uncountable sets of measure zero; see the discussion of the Cantor set in Chapter 8.

47. Use the previous exercise to find a distinction between the sizes of the set of rational numbers and the set of irrational numbers in the interval $[0, 1]$. (A little hand-waving may be needed here, but the general idea should be clear.)

48. Prove that the union of two sets of measure zero has measure zero.

49. Prove that a countable union of sets of measure zero has measure zero.

50. Suppose that $f: [a, b] \to \mathbb{R}$ is monotone. Prove that f is continuous almost everywhere on $[a, b]$.

51. Let $f: [a, b] \to \mathbb{R}$ be a bounded function and let $c \in (a, b)$. The **oscillation** of the function f at the point c is defined by

$$\omega(f, c) = \lim_{r \to 0^+} \omega(f, [c - r, c + r]).$$

Explain why the limit in question is guaranteed to exist.

52. For the function f defined by $f(x) = \sin(1/x)$ for $x \neq 0$ and $f(0) = 0$, find $\omega(f, 0)$.

53. Let $f: [a, b] \to \mathbb{R}$ be a bounded function and let $c \in (a, b)$. Prove that the function f is continuous at c if and only if $\omega(f, c) = 0$.

54. Suppose that f is Riemann integrable on $[a, b]$, fix a positive integer n, and let $D_n = \{x \in (a, b) : \omega(f, x) \geq 1/n\}$. Prove that D_n has measure zero. You may find an idea used in the proof of Theorem 5.22 helpful.

55. Suppose that f is Riemann integrable on $[a, b]$. Use the results in the previous two exercises to prove that f is continuous almost everywhere on $[a, b]$.

56. Suppose that f is Riemann integrable on $[a, b]$ and define a function F by $F(x) = \int_a^x f$ for each $x \in [a, b]$. Prove that $F' = f$ almost everywhere on $[a, b]$.

57. Suppose that f is bounded and continuous almost everywhere on $[a, b]$. There are several methods for proving that f is Riemann integrable on $[a, b]$, but I favor the one found in Gordon [9]. Rather than outline that proof here, I will simply refer the reader to that article.

6

Infinite Series

A study of infinite series is usually included as part of a standard calculus course, but it is one of the more abstract topics and students often resort to following procedures with little or no clue as to what the calculations actually mean. The confusion surrounding infinite series is often the result of a lack of understanding of the basic concepts of sequences of real numbers. With a thorough knowledge of sequences and a higher level of mathematical sophistication, the topic of infinite series is much more accessible. In this chapter, we will consider the usual results concerning convergence tests for infinite series and focus on the proofs of these tests, but there will also be some discussion of results (primarily in the exercises) that may be new and surprising to the reader.

Many of the more interesting results on infinite series arose from questions involving power series and Fourier series; both types of series play an important role in pure and applied mathematics. A **power series** is an expression of the form

$$a_0 + a_1 x + a_2 x^2 + a_3 x^3 + a_4 x^4 + a_5 x^5 + \cdots,$$

where the a_k's are constants, and a **Fourier series** is an expression of the form

$$a_0 + a_1 \cos x + b_1 \sin x + a_2 \cos 2x + b_2 \sin 2x + a_3 \cos 3x + b_3 \sin 3x + \cdots,$$

where the a_k's and b_k's are constants. Some properties of power series will be discussed in the next chapter; the origins and applications of Fourier series will be left to another course. The important point for this discussion is that each expression reduces to an infinite series of real numbers when a real number is substituted for x. Before considering infinite series of functions, it is therefore a good idea to develop a firm understanding of infinite series of real numbers.

6.1 Convergence of Infinite Series

An **infinite series** of real numbers is an expression of the form

$$\sum_{k=1}^{\infty} a_k = a_1 + a_2 + a_3 + a_4 + a_5 + \cdots,$$

where each a_k is a real number. The numbers a_k are called the **terms** of the infinite series. An infinite series of real numbers thus represents the sum of an infinite number of real numbers. Since it is not possible to actually perform infinitely many additions, this expression must be more precisely defined in order to have mathematical significance. A natural approach to an infinite sum is to just start adding the terms and see what happens. This process yields the sequence

$$\begin{aligned}
s_1 &= a_1, \\
s_2 &= a_1 + a_2, \\
s_3 &= a_1 + a_2 + a_3, \\
s_4 &= a_1 + a_2 + a_3 + a_4, \\
&\vdots \\
s_n &= a_1 + a_2 + \cdots + a_n = \sum_{k=1}^{n} a_k,
\end{aligned}$$

which is known as the sequence of **partial sums** of the infinite series. An infinite series is thus reduced to a sequence, and the series is said to converge if its sequence of partial sums converges. This discussion is summarized in the following definition.

DEFINITION 6.1 An infinite series $\sum_{k=1}^{\infty} a_k$ **converges** if its corresponding sequence $\{s_n\}$ of partial sums converges. If S is the limit of the sequence $\{s_n\}$, then the series $\sum_{k=1}^{\infty} a_k$ converges to S; we write $\sum_{k=1}^{\infty} a_k = S$ and refer to S as the sum of the series. If the sequence $\{s_n\}$ does not converge, then the series is said to **diverge**.

As indicated in the definition, we will often write "series" rather than "infinite series". The equality $\sum_{k=1}^{\infty} a_k = S$ is an abuse of notation because the symbol $\sum_{k=1}^{\infty} a_k$ is used to represent both the series (which is actually a sequence) and the sum of the series (which is a real number). For sequences, it would be confusing to write $\{x_n\} = L$ when $\{x_n\}$ converges to L. However, the notation for infinite series has good motivation; it comes from the symbolic manipulation

$$S = \lim_{n\to\infty} s_n = \lim_{n\to\infty} \sum_{k=1}^{n} a_k = \sum_{k=1}^{\infty} a_k.$$

The reader needs to be aware of this notational dichotomy and to study the context to know whether the infinite series (a sequence) or the sum of the infinite series (a number) is meant.

Although it is common for the index of an infinite series to begin with $k = 1$, it is certainly not necessary. It is convenient in many cases to start a series with $k = 0$; almost every series in this book will begin with either $k = 0$ or $k = 1$. However, an infinite series can start at any given integer. The index n in the sequence $\{s_n\}$ of partial sums will refer to the upper limit of the sum, not the number of terms that are added together. For example, the first term of the sequence $\{s_n\}$ of partial sums for the series $\sum_{k=8}^{\infty} a_k$ would be s_8.

Since an infinite series is really just a sequence in disguise, all of the theorems about sequences can be used in a study of infinite series. However, a potential source of confusion is the fact that there are two sequences related to an infinite series. For a series $\sum_{k=1}^{\infty} a_k$, there is the sequence $\{a_k\}$ of the terms of the series and the sequence $\{s_n\}$ of the partial sums of the series. The sequence $\{a_k\}$ represents the numbers that are to be added together. The sequence $\{s_n\}$, where

$$s_n = a_1 + a_2 + \cdots + a_n,$$

is used to determine whether or not the series converges and to find the sum of the series when it does converge. It is important to distinguish between these two sequences related to a given infinite series as well as to take great care in the use of the words "series" and "sequence".

To illustrate these ideas with a simple example, consider the infinite series

$$\sum_{k=1}^{\infty} \frac{1}{2^k} = \frac{1}{2} + \frac{1}{4} + \frac{1}{8} + \frac{1}{16} + \frac{1}{32} + \cdots.$$

In this case, the sequence $\{s_n\}$ of partial sums is defined by

$$s_n = \sum_{k=1}^{n} \frac{1}{2^k} = \frac{2^n - 1}{2^n},$$

where the second equality comes from the formula for a finite geometric sum (see Theorem 1.10). Since $\{s_n\}$ converges to 1, the series converges to 1; hence

$$\sum_{k=1}^{\infty} \frac{1}{2^k} = \frac{1}{2} + \frac{1}{4} + \frac{1}{8} + \frac{1}{16} + \frac{1}{32} + \cdots = 1.$$

We have thus found the sum of an infinite series by finding a convenient formula for the sequence of partial sums and using this formula to determine the limit.

As a second example, let $\{a_k\} = \big\{1/k(k+1)\big\}$ and consider the series $\sum_{k=1}^{\infty} a_k$. Using the technique of partial fractions (that is, expressing a rational function as a sum of simpler fractions), we find that

$$s_n = \sum_{k=1}^{n} \frac{1}{k(k+1)} = \sum_{k=1}^{n} \left(\frac{1}{k} - \frac{1}{k+1}\right)$$
$$= \left(\frac{1}{1} - \frac{1}{2}\right) + \left(\frac{1}{2} - \frac{1}{3}\right) + \cdots + \left(\frac{1}{n} - \frac{1}{n+1}\right) = 1 - \frac{1}{n+1}.$$

A sum that collapses like this is known as a **telescoping sum**. Since the sequence $\{s_n\}$ converges to 1, the series $\sum_{k=1}^{\infty} a_k$ converges and $\sum_{k=1}^{\infty} a_k = 1$.

Since the convergence of an infinite series depends on the convergence of its sequence of partial sums and since sequences have already been studied in detail (see Chapter 2), it might seem as though there is little left to do. However, the two examples that have been given thus far have been misleading. For both examples, it was possible to find an expression for s_n that did not involve a sum, then use this expression to find the limit of the sequence. In most cases involving infinite series, it is not possible to express s_n in a form which makes the limit easy to find; it is this fact that makes the study of series necessary and quite interesting.

Much of the theory of infinite series is concerned with determining whether or not a series converges. In this regard, it should be clear that a series cannot converge if its terms do not tend to 0. This observation leads to the following theorem.

THEOREM 6.2 If the series $\sum_{k=1}^{\infty} a_k$ converges, then the sequence $\{a_k\}$ converges to 0.

Proof. By hypothesis, the sequence $\{s_n\}$ of partial sums for the series converges; let S be the limit of this sequence. Note that

$$a_k = (a_1 + a_2 + \cdots + a_k) - (a_1 + a_2 + \cdots + a_{k-1}) = s_k - s_{k-1}$$

for each $k \geq 2$. Since the sequence $\{s_n\}$ converges to S, the sequence $\{a_k\}$ converges to $S - S = 0$. ■

The converse of Theorem 6.2 is false; convergence of the terms to 0 is no guarantee that the series converges. To see this, consider the series $\sum_{k=1}^{\infty} 1/\sqrt[3]{k}$, whose terms clearly converge to 0. Replacing each term of the sum by the smallest term, we find that

$$s_n = \sum_{k=1}^{n} \frac{1}{\sqrt[3]{k}} = \frac{1}{\sqrt[3]{1}} + \frac{1}{\sqrt[3]{2}} + \cdots + \frac{1}{\sqrt[3]{n}}$$
$$\geq \frac{1}{\sqrt[3]{n}} + \frac{1}{\sqrt[3]{n}} + \cdots + \frac{1}{\sqrt[3]{n}} = \frac{n}{\sqrt[3]{n}} = n^{2/3}$$

for all n. Since the sequence $\{s_n\}$ is unbounded, the series diverges. Further examples of series of this type can be found in the exercises.

The contrapositive of Theorem 6.2, which is often called the **Divergence Test**, provides a simple method for proving that a series diverges: if the terms of a series do not converge to 0, then the series diverges. For example, the series $\sum_{k=1}^{\infty} \sqrt[k]{2}$ diverges since its terms converge to 1. In summary, if the terms of a series converge to 0, then the series may converge, but it is necessary to do some further checking to know for certain.

To decide whether or not an infinite series converges, it is necessary to determine whether or not its sequence of partial sums converges. Since a convenient formula for the sequence of partial sums is generally not available, we are left with the problem of deciding if a sequence converges when its limit is unknown. What results are available to prove that a sequence of real numbers converges when its limit is unknown? Two facts should come to mind: every Cauchy sequence converges and every bounded monotone sequence converges. These facts lead to the following two theorems; the proof of the second will be left as an exercise.

THEOREM 6.3 The series $\sum_{k=1}^{\infty} a_k$ converges if and only if for each $\epsilon > 0$ there exists a positive integer N such that $\left| \sum_{k=m+1}^{n} a_k \right| < \epsilon$ for all positive integers m and n that satisfy $n > m \geq N$.

Proof. For each positive integer n, let $s_n = \sum_{k=1}^{n} a_k$. If m and n are positive integers with $n > m$, then that

$$s_n - s_m = \sum_{k=1}^{n} a_k - \sum_{k=1}^{m} a_k = \sum_{k=m+1}^{n} a_k.$$

The theorem follows from the fact that $\{s_n\}$ converges if and only if it is a Cauchy sequence. ■

THEOREM 6.4 A series with nonnegative terms converges if and only if its sequence of partial sums is bounded. ■

One conceptual difficulty that some people have with series is that it does not seem to be possible to add together an infinite number of positive terms and get a finite number. By looking at the sequence of partial sums, it can be seen that this amounts to believing that an increasing sequence cannot converge. It is probably easier to first convince someone that an increasing sequence converges, then use this fact to show that an infinite sum of positive numbers can converge. Another convincing argument for the novice is the series

$$\sum_{k=1}^{\infty} \frac{1}{10^k} = 0.1 + 0.01 + 0.001 + 0.0001 + 0.00001 + \cdots.$$

The sum is easily seen to be $0.11111\ldots$, which many people recognize as $1/9$.

Since infinite series are sequences of partial sums, the usual algebraic properties of multiplication by a constant and addition, which are valid for sequences, are valid for series as well. This is the content of the next theorem, whose proof will be left as an exercise.

THEOREM 6.5 Let $\sum_{k=1}^{\infty} a_k$ and $\sum_{k=1}^{\infty} b_k$ be two convergent series and let c be any real number. Then

a) the series $\sum_{k=1}^{\infty} ca_k$ converges and $\sum_{k=1}^{\infty} ca_k = c \sum_{k=1}^{\infty} a_k$;

b) the series $\sum_{k=1}^{\infty} (a_k + b_k)$ converges and $\sum_{k=1}^{\infty} (a_k + b_k) = \sum_{k=1}^{\infty} a_k + \sum_{k=1}^{\infty} b_k$. ■

Exercises

1. Find a simple expression for the sequence of partial sums of the given series, then determine whether or not the series converges.

a) $\sum_{k=1}^{\infty} (-1)^{k+1}$ **b)** $\sum_{k=1}^{\infty} (3/4)^k$ **c)** $\sum_{k=1}^{\infty} \ln\big((k+1)/k\big)$

2. Use the technique of partial fractions to find a formula without sums for the sequence of partial sums of the given series, then find the sum of the series.

a) $\sum_{k=1}^{\infty} \frac{1}{k(k+2)}$ **b)** $\sum_{k=3}^{\infty} \frac{1}{k^2-4}$ **c)** $\sum_{k=1}^{\infty} \frac{1}{4k^2-1}$

3. Let p be a positive integer. Find the sum of the series $\sum_{k=1}^{\infty} \frac{1}{k(k+p)}$.

4. Prove that the series $\sum_{k=1}^{\infty} (a_k - a_{k+1})$ converges if and only if the sequence $\{a_k\}$ converges.

5. Suppose that $-1 < a < 1$. Prove that the series $\sum_{k=1}^{\infty} 1/(1+a^k)$ diverges.

6. Determine whether or not the series $\sum_{k=1}^{\infty} \sin k$ converges.

7. Let $\sum_{k=1}^{\infty} a_k$ be a series and suppose that there exist constants $c > 0$ and $r \in (0, 1)$ such that $|a_k| < cr^k$ for all k. Use Theorem 6.3 to prove that the series converges.

8. Prove Theorem 6.4.

9. Use Theorem 6.4 to prove that the series $\sum_{k=1}^{\infty} k^{-k}$ converges.

10. Prove Theorem 6.5.

11. State and prove a theorem concerning the difference of two convergent series.

12. Give an example to show that the sum of two divergent series may be convergent.

13. Prove that the sum of a divergent series and a convergent series diverges.

14. Suppose $\sum_{k=1}^{\infty} a_k$ diverges. Show that $\sum_{k=1}^{\infty} ca_k$ diverges, where c is a nonzero real number.

Remark. The rest of the exercises in this section are in no particular order.

15. Use results in this section to find the sum of the series $\sum_{k=1}^{\infty} \left(\frac{3}{2^k} - \frac{4}{k(k+1)}\right)$.

16. Let p be a positive real number and consider the series $\sum_{k=1}^{\infty} 1/k^p$.

a) Use Theorem 6.4 to prove that the series diverges when $p = 1/2$.

b) Use Theorem 6.3 to prove that the series diverges when $p = 1$.

c) Use the fact that $\dfrac{1}{k^2} \leq \dfrac{2}{k(k+1)}$ for all k to prove that the series converges when $p = 2$.

17. Find the sum of the series $\sum_{k=1}^{\infty} (-1)^{k+1} \dfrac{2k+1}{k^2+k}$.

18. Determine whether or not the series $\sum_{k=1}^{\infty} \dfrac{1}{\sqrt{k+1}+\sqrt{k}}$ converges.

19. Let $\sum_{k=1}^{\infty} a_k$ and $\sum_{k=1}^{\infty} b_k$ be two series and suppose that $a_k = b_k$ except for a finite number of positive integers k. Prove that the two series either both converge or both diverge. In other words, prove that changing a finite number of the terms of an infinite series does not affect its convergence or divergence.

20. Suppose that $\sum_{k=1}^{\infty} a_k$ is a convergent series.

a) Prove that the series $\sum_{k=n}^{\infty} a_k$ converges for each positive integer n.

b) Let $t_n = \sum_{k=n}^{\infty} a_k$. Prove that the sequence $\{t_n\}$ converges to 0.

21. Let $\sum_{k=1}^{\infty} a_k$ be a convergent series with nonnegative terms and let $\{a_{n_k}\}$ be a subsequence of $\{a_k\}$. Prove that the series $\sum_{k=1}^{\infty} a_{n_k}$ converges.

22. Let $\sum_{k=1}^{\infty} a_k$ be a series and let $\{a_{n_k}\}$ be the subsequence of $\{a_k\}$ that consists of all the nonzero terms of the sequence. Prove that the series $\sum_{k=1}^{\infty} a_k$ and $\sum_{k=1}^{\infty} a_{n_k}$ either both converge or both diverge and that their sums are the same when they converge.

23. Let $\sum_{k=1}^{\infty} a_k$ be a series and consider the series $\sum_{k=1}^{\infty} (a_{2k-1} + a_{2k})$.

a) Explain the statement: the second series is a regrouping of the first series.

b) Prove that the series $\sum_{k=1}^{\infty} (a_{2k-1} + a_{2k})$ converges if $\sum_{k=1}^{\infty} a_k$ converges.

c) Discuss the two series for the case in which $a_k = (-1)^k$.

d) Suppose that $\{a_k\}$ converges to 0. Prove that $\sum_{k=1}^{\infty} a_k$ converges if the series $\sum_{k=1}^{\infty} (a_{2k-1} + a_{2k})$ converges.

e) Repeat this analysis for the series $\sum_{k=1}^{\infty} (a_{3k-2} + a_{3k-1} + a_{3k})$.

6.2 THE COMPARISON TESTS

There are two important questions generated by an infinite series.

1. Does the series converge?
2. If a series converges, what is its sum?

Although we will make a few comments about the numerical value of the sum of a convergent series, our primary concern will be the first question. There are a number of ways to determine whether or not a series converges, but the focus in this section will be a simple comparison test that applies to series with nonnegative terms. In order to have some series available for the sake of comparison, two large collections of infinite series will also be presented: namely, geometric series and p-series.

The sequence of partial sums of a series with nonnegative terms is increasing and, by Theorem 6.4, the series converges if the sequence of partial sums is bounded. It is therefore reasonable to believe that a series of this type whose terms are smaller than the terms of a convergent series must itself converge. This is the main idea of the next theorem, whose proof will be left as an exercise.

THEOREM 6.6 Comparison Test Let $\sum_{k=1}^{\infty} a_k$ and $\sum_{k=1}^{\infty} b_k$ be two series with nonnegative terms and suppose that $a_k \leq b_k$ for all $k \geq K$ for some integer K.

a) If the series $\sum_{k=1}^{\infty} b_k$ converges, then the series $\sum_{k=1}^{\infty} a_k$ converges.

b) If the series $\sum_{k=1}^{\infty} a_k$ diverges, then the series $\sum_{k=1}^{\infty} b_k$ diverges. ■

It is important to read the statement of this theorem carefully in order to be certain what it does and does not say. To use the Comparison Test, we either need a divergent series whose terms are smaller than the given series or a convergent series whose terms are larger than the given series. This sounds simple enough, but the inequalities involved do require some care. As an example, consider the series $\sum_{k=1}^{\infty} k^2/(k^3+5k+2)$. The series $\sum_{k=1}^{\infty} 1/(3k)$ diverges (see Exercise 16 in the previous section) and

$$\frac{k^2}{k^3+5k+2} > \frac{k^2}{k^3+k^3+k^3} = \frac{1}{3k}$$

for all $k \geq 3$. Consequently, the series $\sum_{k=1}^{\infty} k^2/(k^3+5k+2)$ diverges by the Comparison Test.

A variation of the Comparison Test is a test known as the Limit Comparison Test. It is often easier to apply than the Comparison Test since there is no need to use inequalities. To state this test concisely, we will say that a sequence $\{r_k\}$ of positive numbers has a **limit in the extended sense** if either $r_k \to \infty$ or $r_k \to L$ for some real number L.

THEOREM 6.7 Limit Comparison Test Let $\sum_{k=1}^{\infty} a_k$ and $\sum_{k=1}^{\infty} b_k$ be two series with positive terms. Suppose that the sequence $\{a_k/b_k\}$ has a limit in the extended sense and let α be the limit of this sequence.

a) If $\sum_{k=1}^{\infty} b_k$ converges and $0 \le \alpha < \infty$, then $\sum_{k=1}^{\infty} a_k$ converges.

b) If $\sum_{k=1}^{\infty} b_k$ diverges and $0 < \alpha \le \infty$, then $\sum_{k=1}^{\infty} a_k$ diverges.

Proof. Suppose that the series $\sum_{k=1}^{\infty} b_k$ converges and that $0 \le \alpha < \infty$. Since the sequence $\{a_k/b_k\}$ converges, it is bounded. Let M be a positive number such that $a_k/b_k \le M$ for all k. Since $a_k \le Mb_k$ for all k and the series $\sum_{k=1}^{\infty} Mb_k$ converges, the series $\sum_{k=1}^{\infty} a_k$ converges by the Comparison Test.

Now suppose that $\sum_{k=1}^{\infty} b_k$ diverges and that $0 < \alpha \le \infty$. Since the limit of the sequence $\{a_k/b_k\}$ is positive, there exists a positive integer K and a positive number m such that $a_k/b_k > m$ for all $k \ge K$. Since $a_k \ge mb_k$ for all $k \ge K$, the conclusion follows from the Comparison Test. This completes the proof. ■

As with the Comparison Test, it is important to read the statement of this theorem carefully. To use the Limit Comparison Test, let the terms of the series in question be determined by the sequence $\{a_k\}$ and look for a simple sequence $\{b_k\}$ whose terms are similar to the terms of $\{a_k\}$ when k is large. The test essentially states that the two series either both converge or both diverge. To illustrate the Limit Comparison Test, consider once again the series $\sum_{k=1}^{\infty} k^2/(k^3+5k+2)$. For large values of k, the numbers $5k$ and 2 are small in comparison to k^3, so the terms of the series are similar to $k^2/k^3 = 1/k$ when k is large. The series $\sum_{k=1}^{\infty} 1/k$ diverges and

$$\lim_{k\to\infty}\left(\frac{k^2}{k^3+5k+2} \div \frac{1}{k}\right) = \lim_{k\to\infty}\left(\frac{k^3}{k^3+5k+2}\right) = 1.$$

Hence, the series $\sum_{k=1}^{\infty} k^2/(k^3+5k+2)$ diverges by the Limit Comparison Test.

In order for the comparison tests to be of much use, a large collection of convergent and divergent series must be available. Two such collections, the geometric series and the p-series, appear frequently in real analysis and other branches of mathematics and have terms with particularly simple patterns. A **geometric series** is a series that can be put in the form $\sum_{k=0}^{\infty} ar^k$, where $a \ne 0$ and r are constants.

Writing out this series yields

$$a + ar + ar^2 + ar^3 + ar^4 + ar^5 + \cdots;$$

each term of the series is r times the previous term. The series

$$3 - 2 + \frac{4}{3} - \frac{8}{9} + \frac{16}{27} - \cdots$$

is an example of a geometric series with $a = 3$ and $r = -2/3$. Geometric series occur in a variety of contexts and, in contrast to most other series, there is a simple formula for the sum of the series.

THEOREM 6.8 Geometric Series Suppose that $a \neq 0$. The geometric series $\sum_{k=0}^{\infty} ar^k$ converges if $|r| < 1$ and diverges if $|r| \geq 1$. In addition, when $|r| < 1$,

$$\sum_{k=0}^{\infty} ar^k = \frac{a}{1-r}.$$

Proof. If $|r| \geq 1$, then the terms of the series do not converge to 0. By Theorem 6.2, the series diverges. Suppose that $|r| < 1$. For each positive integer n, let $s_n = \sum_{k=0}^{n} ar^k$. Then

$$s_n = a + ar + ar^2 + \cdots + ar^{n-1} + ar^n;$$
$$rs_n = ar + ar^2 + ar^3 + \cdots + ar^n + ar^{n+1}.$$

Subtracting these two equations and solving for s_n yields

$$s_n = \frac{a(1 - r^{n+1})}{1 - r}.$$

Since $|r| < 1$, the sequence $\{r^n\}$ converges to 0. Consequently, the sequence $\{s_n\}$ converges to $a/(1-r)$. This completes the proof. ■

It is important to note that the formula for the sum of a geometric series as stated in the theorem is valid only when the index begins with 0. It is always possible to rewrite an infinite series so that the index begins with 0. For example,

$$\sum_{k=2}^{\infty} \frac{3 \cdot 4^k}{5^{k+1}} = \sum_{k=2}^{\infty} \frac{3}{5}\left(\frac{4}{5}\right)^k = \sum_{k=0}^{\infty} \frac{3}{5}\left(\frac{4}{5}\right)^{k+2} = \sum_{k=0}^{\infty} \frac{48}{125}\left(\frac{4}{5}\right)^k = \frac{48/125}{1-(4/5)} = \frac{48}{25}.$$

This process can get a little tedious, but the technique of rewriting a series with a different starting point is a useful one to know. However, to avoid such computations when possible, the reader should verify that the sum of a geometric series is always the first term of the series divided by $1 - r$.

A ***p*-series** is a series of the form $\sum_{k=1}^{\infty} 1/k^p$, where p is a fixed real number. For $p \leq 0$, the series clearly diverges since the terms do not converge to 0. Exercise 16 in Section 6.1 indicates that the p-series diverges for $p = 1$ and converges for

$p = 2$. By the Comparison Test, it follows that the p-series diverges for $0 < p \leq 1$ and converges for $p \geq 2$. As the next theorem indicates, the p-series also converges for values of p between 1 and 2.

THEOREM 6.9 The p-series $\sum_{k=1}^{\infty} 1/k^p$ converges if $p > 1$ and diverges if $p \leq 1$.

Proof. We have already noted that a p-series diverges if $p \leq 1$, so suppose that $p > 1$. Since the sequence $\{s_n\}$ of partial sums for the series $\sum_{k=1}^{\infty} 1/k^p$ is increasing, the series converges if $\{s_n\}$ is bounded above. Using some convenient overestimation, we find that

$$\begin{aligned} s_{2^n-1} &= 1 + \left(\frac{1}{2^p} + \frac{1}{3^p}\right) + \left(\frac{1}{4^p} + \frac{1}{5^p} + \frac{1}{6^p} + \frac{1}{7^p}\right) + \cdots \\ &\quad + \left(\frac{1}{\left(2^{n-1}\right)^p} + \frac{1}{\left(2^{n-1}+1\right)^p} + \cdots + \frac{1}{\left(2^{n-1}+2^{n-1}-1\right)^p}\right) \\ &\leq 1 + \frac{2}{2^p} + \frac{4}{4^p} + \cdots + \frac{2^{n-1}}{\left(2^{n-1}\right)^p} \\ &= \sum_{k=0}^{n-1} \frac{2^k}{\left(2^k\right)^p} \end{aligned}$$

for each positive integer n. Since $2^{1-p} < 1$, the formula for the sum of a geometric series yields

$$s_n \leq s_{2^n-1} \leq \sum_{k=0}^{n-1} \frac{2^k}{\left(2^k\right)^p} < \sum_{k=0}^{\infty} \left(2^{1-p}\right)^k = \frac{1}{1-2^{1-p}}$$

for all n. Since the sequence $\{s_n\}$ is bounded above, the p-series converges for $p > 1$. This completes the proof. ■

Unlike the geometric series, there is no simple method for finding the sum of a p-series. A number of interesting techniques, often involving power series, Fourier series, or complex variables, have determined the sum of a p-series for some values of p. (Some of these techniques can be found in Knopp [12].) For example, it has been shown that

$$\sum_{k=1}^{\infty} \frac{1}{k^2} = \frac{\pi^2}{6} \quad \text{and} \quad \sum_{k=1}^{\infty} \frac{1}{k^4} = \frac{\pi^4}{90}.$$

It may come as a surprise that the number π appears in these sums. One method for verifying the first sum can be found in Exercise 45 of Section 7.7.

Exercises

1. Prove the Comparison Test.

2. Let $\sum_{k=1}^{\infty} a_k$ be a series and suppose that $0 < ka_k < 1$ for all k. Does the Comparison Test offer any information on the convergence or divergence of this series?

3. Adopting the notation of the Limit Comparison Test, give an example for which

a) $\sum_{k=1}^{\infty} b_k$ converges, $\alpha = \infty$, and $\sum_{k=1}^{\infty} a_k$ diverges;

b) $\sum_{k=1}^{\infty} b_k$ diverges, $\alpha = 0$, and $\sum_{k=1}^{\infty} a_k$ converges.

4. Show that part (b) of the Limit Comparison Test is the contrapositive of part (a). (This requires some knowledge of logic).

5. After studying the proof of the Limit Comparison Test, it should be clear that the hypothesis that the sequence $\{a_k/b_k\}$ converges is not necessary.

a) Find conditions on the sequence $\{a_k/b_k\}$ so that part (a) remains valid.

b) Find conditions on the sequence $\{a_k/b_k\}$ so that part (b) remains valid.

6. Let P and Q be polynomials of degree p and q, respectively. Suppose that the coefficient of x^p in P is positive, the coefficient of x^q in Q is positive, and that $Q(k) \neq 0$ for all positive integers k. Prove that the series $\sum_{k=1}^{\infty} P(k)/Q(k)$ converges if and only if $p < q - 1$. (Note that some of the terms of this series may be negative.)

7. Rewrite the series $\sum_{k=3}^{\infty} \frac{3^{k+2}}{4^{k-1}}$ in the form $\sum_{k=0}^{\infty} ar^k$, then find the sum of the series.

8. Find the sum of the given series.

a) $\sum_{k=1}^{\infty} \frac{(-1)^{k+1}}{(1.2)^k}$ **b)** $\sum_{k=4}^{\infty} \frac{10 \cdot 2^{k+1}}{5^{k-3}}$ **c)** $\sum_{k=1}^{\infty} \left(\frac{3}{2^k} - \frac{2}{3^{k+1}} \right)$

9. Find $\lim_{r \to 1^-} \sum_{k=0}^{\infty} r^k$ and $\lim_{r \to -1^+} \sum_{k=0}^{\infty} r^k$.

10. Although the series $\sum_{k=0}^{\infty} (-1)^k$ does not converge, it is sometimes assigned the value $1/2$. Why does this make a certain amount of sense?

11. Prove that the series $\sum_{k=1}^{\infty} \left(\frac{k^2 + 10}{k^4} \right)$ converges and the series $\sum_{k=1}^{\infty} \left(\frac{2^k - k}{k2^k} \right)$ diverges. Do not use the Comparison Test.

12. For what values of p does the series $\sum_{k=1}^{\infty} k^{p-p^2}$ converge?

13. This exercise requires a careful reading of the proof of Theorem 6.9.

a) Suppose that $p > 1$. Prove that $\sum_{k=1}^{\infty} 1/k^p < \frac{2^p}{2^p - 2}$.

b) Prove that $\sum_{k=1}^{\infty} k^{-3/2} < 2 + \sqrt{2}$.

c) Use part (a) to find overestimates for the sums of the series $\sum_{k=1}^{\infty} k^{-2}$ and $\sum_{k=1}^{\infty} k^{-4}$. How do these values compare with the actual sums of these series?

d) Using ideas similar to those found in the proof of Theorem 6.9, find a simple formula for an underestimate of the sum of a p-series for $p > 1$.

e) Prove that $\lim_{p\to\infty} \sum_{k=1}^{\infty} \frac{1}{k^p} = 1$ and $\lim_{p\to 1^+} \sum_{k=1}^{\infty} \frac{1}{k^p} = \infty$.

14. Use one of the comparison tests (or both if you wish to compare them) to determine whether or not $\sum_{k=1}^{\infty} a_k$ converges, where a_k is given by

a) $\dfrac{1}{k^2 + 2k + 6}$ **b)** $\dfrac{4}{\sqrt{k^3 + 4}}$ **c)** $\dfrac{2}{k + \sqrt{k}}$

d) $\dfrac{1}{k!}$ **e)** $\dfrac{2^k + k^2}{3^k + k}$ **f)** $\dfrac{1}{\ln(k + 1)}$

Remark. The rest of the exercises in this section are in no particular order.

15. Let $a > 1$ and determine whether or not the series $\sum_{k=1}^{\infty} \left(\sqrt[k]{a} - 1\right)$ converges.

16. Does the series $\sum_{k=1}^{\infty} \left(\sqrt[k]{k} - 1\right)$ converge?

17. Find a value of r for which $\sum_{k=0}^{\infty} r^k = 7/8$ and a value of r for which $\sum_{k=1}^{\infty} r^k = 7/8$.

18. Is there a value of r for which $\sum_{k=0}^{\infty} r^k = 1/4$? Explain your answer.

19. Find the values of r for which the series $\sum_{k=0}^{\infty} 3^k r^{2k}$ converges.

20. Find the values of r for which the series $\sum_{k=0}^{\infty} r^k(1 + r^k)$ converges and find the sum of the series for these values of r. What is the minimum possible sum?

21. Use the formulas at the end of this section to find the sum of each series.

a) $1 + \dfrac{1}{3^2} + \dfrac{1}{5^2} + \dfrac{1}{7^2} + \dfrac{1}{9^2} + \dfrac{1}{11^2} + \dfrac{1}{13^2} + \cdots$

b) $1 + \dfrac{1}{2^4} + \dfrac{1}{4^4} + \dfrac{1}{5^4} + \dfrac{1}{7^4} + \dfrac{1}{8^4} + \dfrac{1}{10^4} + \cdots$

22. Suppose that $|r| < 1$.

a) Prove that $\displaystyle\sum_{i=k}^{n} r^i = \frac{r^k - r^{n+1}}{1 - r}$ for positive integers k and n with $k \le n$.

b) Prove that $\displaystyle\sum_{k=1}^{n} kr^k = \sum_{k=1}^{n}\sum_{i=k}^{n} r^i$.

c) Find a formula for the sum of the series $\displaystyle\sum_{k=1}^{\infty} kr^k$.

d) Prove that $\displaystyle\sum_{k=1}^{n} k^2 r^k = \sum_{k=1}^{n}(2k - 1)\sum_{i=k}^{n} r^i$.

e) Find a formula for the sum of the series $\sum_{k=1}^{\infty} k^2 r^k$.

f) Use similar reasoning to find a formula for the sum of the series $\sum_{k=1}^{\infty} k^3 r^k$.

23. Consider the geometric series $\sum_{k=0}^{\infty} ar^k$ with $a = 2$ and $r = 0$. What is the sum of this series? Do you see a notational problem in this case?

6.3 Absolute Convergence

In general, a series with positive or nonnegative terms is easier to analyze than a series with both positive and negative terms. This is a consequence of the fact that the sequence of partial sums is increasing for a series with nonnegative terms and increasing sequences either converge or are unbounded. Given any series $\sum_{k=1}^{\infty} a_k$, the series $\sum_{k=1}^{\infty} |a_k|$ has nonnegative terms. In some cases, we can use the second series to determine whether or not the original series converges. The following theorem is the basis for this statement.

THEOREM 6.10 Let $\sum_{k=1}^{\infty} a_k$ be a series of real numbers. If the series $\sum_{k=1}^{\infty} |a_k|$ converges, then the series $\sum_{k=1}^{\infty} a_k$ converges.

Proof. This theorem is a simple consequence of Theorem 6.3 and the inequality

$$\left| \sum_{k=m+1}^{n} a_k \right| \leq \sum_{k=m+1}^{n} |a_k|,$$

which is valid for all positive integers m and n with $n > m$. ■

The converse of this theorem is false; an example will be given later in this section. The following definition distinguishes between convergent series $\sum_{k=1}^{\infty} a_k$ for which $\sum_{k=1}^{\infty} |a_k|$ converges and those for which it does not.

DEFINITION 6.11 Let $\sum_{k=1}^{\infty} a_k$ be a convergent series.

a) The series $\sum_{k=1}^{\infty} a_k$ **converges absolutely** if $\sum_{k=1}^{\infty} |a_k|$ converges.

b) The series $\sum_{k=1}^{\infty} a_k$ **converges nonabsolutely** if $\sum_{k=1}^{\infty} |a_k|$ diverges.

Many books use the term "**conditionally convergent**" rather than the term "nonabsolutely convergent". The word "conditional" does not seem quite as descriptive of this situation as the word "nonabsolute", so we will adopt the term "nonabsolute convergence" for convergent series that do not converge absolutely.

For series with all but a finite number of terms having the same sign, there is no distinction between convergence and absolute convergence (see the exercises). If a series has an infinite number of both positive and negative terms, then it is usually a good idea to first check for absolute convergence. The reason for doing so is that there are more tests available for series with nonnegative terms and, if the series converges absolutely, then it certainly converges. The comparison tests, which were discussed in the previous section, are tests for series with nonnegative terms. The Root Test and the Ratio Test are two other important tests for the absolute convergence of an infinite series. To state and prove these tests in a general form, it is necessary to use the concepts of limit inferior and limit superior of a sequence, which were introduced in Chapter 2. The proof of the Ratio Test will be given in detail while the proof of the Root Test, which is similar and easier, will be left as an exercise.

THEOREM 6.12 Ratio Test Let $\sum_{k=1}^{\infty} a_k$ be a series with nonzero terms.

a) If $\limsup_{k\to\infty} |a_{k+1}/a_k| < 1$, then the series $\sum_{k=1}^{\infty} a_k$ converges absolutely.

b) If $\liminf_{k\to\infty} |a_{k+1}/a_k| > 1$, then the series $\sum_{k=1}^{\infty} a_k$ diverges.

Proof. Suppose that $\limsup_{k\to\infty} |a_{k+1}/a_k| < 1$ and let r be a real number that satisfies $\limsup_{k\to\infty} |a_{k+1}/a_k| < r < 1$. By part (b) of Theorem 2.21, there exists an integer p such that $|a_{k+1}/a_k| < r$ for all $k \geq p$. It follows that

$$|a_{p+1}| < r\,|a_p| = \left(\frac{|a_p|}{r^p}\right) r^{p+1},$$

$$|a_{p+2}| < r\,|a_{p+1}| < \left(\frac{|a_p|}{r^p}\right) r^{p+2},$$

$$|a_{p+3}| < r\,|a_{p+2}| < \left(\frac{|a_p|}{r^p}\right) r^{p+3},$$

and, in general, $|a_k| < \left(\frac{|a_p|}{r^p}\right) r^k$ for all $k > p$. Since the series $\sum_{k=1}^{\infty}\left(\frac{|a_p|}{r^p}\right) r^k$ is a convergent geometric series, the series $\sum_{k=1}^{\infty} a_k$ converges absolutely by the Comparison Test. This proves part (a).

Now suppose that $\liminf_{k\to\infty}\left|a_{k+1}/a_k\right| > 1$. By part (c) of Theorem 2.21, there exists a positive integer q such that $\left|a_{k+1}/a_k\right| > 1$ for all $k \geq q$. It follows that

$$\begin{aligned}
|a_{q+1}| &> |a_q|, \\
|a_{q+2}| &> |a_{q+1}| > |a_q|, \\
|a_{q+3}| &> |a_{q+2}| > |a_q|,
\end{aligned}$$

and so on. Since $|a_k| > |a_q| > 0$ for all $k > q$, the sequence $\{a_k\}$ does not converge to 0. By Theorem 6.2, the series $\sum_{k=1}^{\infty} a_k$ diverges. This completes the proof. ■

THEOREM 6.13 Root Test Let $\sum_{k=1}^{\infty} a_k$ be an arbitrary series.

a) If $\limsup_{k\to\infty} \sqrt[k]{|a_k|} < 1$, then the series $\sum_{k=1}^{\infty} a_k$ converges absolutely.

b) If $\limsup_{k\to\infty} \sqrt[k]{|a_k|} > 1$, then the series $\sum_{k=1}^{\infty} a_k$ diverges. ■

Since roots are usually more difficult to compute than ratios, the Ratio Test is often easier to apply than the Root Test. However, the Root Test is a stronger test than the Ratio Test. This means that it is easier for a series to fall between the cracks—that is, for the test to give no information—when the Ratio Test rather than the Root Test is used. The reason for this lies in the fact that both parts of the Root Test involve the limit superior whereas the Ratio Test involves both the limit superior and the limit inferior. The statement, "the Root Test is stronger than the Ratio Test", is made precise in the exercises, and several examples are given to illustrate this fact. Neither test provides a subtle test for divergence: if either test shows that the series diverges, it is because the terms of the series do not converge to 0. An improvement on the divergence part of the Ratio Test can also be found in the exercises.

One of the advantages of the Root and Ratio Tests over the comparison tests is that there is no need to mention another series. These two tests are quite useful for series in which the appropriate limits are not too difficult to calculate and have values other than 1. In many instances, the sequences $\{\sqrt[k]{|a_k|}\}$ or $\{|a_{k+1}/a_k|\}$ actually converge; calculus books usually state the Root and Ratio Tests under this assumption. For example, the series $\sum_{k=1}^{\infty} 5^k/k!$ converges by the Ratio Test since

$$\lim_{k\to\infty}\left|\frac{a_{k+1}}{a_k}\right| = \lim_{k\to\infty}\left(\frac{5^{k+1}}{(k+1)!}\cdot\frac{k!}{5^k}\right) = \lim_{k\to\infty}\left(\frac{5}{k+1}\right) = 0.$$

The series $\sum_{k=1}^{\infty}(-k)^9/2^k$ converges absolutely by the Root Test since

$$\lim_{k\to\infty}\sqrt[k]{|a_k|} = \lim_{k\to\infty}\sqrt[k]{\left|\frac{(-k)^9}{2^k}\right|} = \lim_{k\to\infty}\frac{(\sqrt[k]{k})^9}{2} = \frac{1}{2}.$$

The sequence $\{\sqrt[k]{k}\}$ (or slight variations on this sequence) often appears when using the Root Test. By Theorem 2.14, this sequence converges to 1.

In general, a series with an infinite number of both positive and negative terms that does not converge absolutely is difficult to analyze. However, in actual practice, many series of this type have terms that alternate signs. For such series, it may be possible to use the following test.

THEOREM 6.14 Alternating Series Test If $\{a_k\}$ is a decreasing sequence of positive numbers that converges to 0, then the series $\sum_{k=1}^{\infty}(-1)^{k+1}a_k$ converges. Furthermore, if S is the sum of the series and s_n is the nth partial sum of the series, then $|s_n - S| \leq a_{n+1}$ for each positive integer n.

Proof. Let $\{s_n\}$ be the sequence of partial sums for the series $\sum_{k=1}^{\infty}(-1)^{k+1}a_k$. We first show that the sequence $\{s_{2n}\}$ is increasing and bounded above. For each positive integer n,

$$s_{2n+2} = s_{2n} + \left(a_{2n+1} - a_{2n+2}\right) \geq s_{2n}$$

since $\{a_k\}$ is a decreasing sequence. This shows that $\{s_{2n}\}$ is an increasing sequence. Furthermore,

$$s_{2n} = a_1 + (a_3 - a_2) + (a_5 - a_4) + \cdots + (a_{2n-1} - a_{2n-2}) - a_{2n} \leq a_1$$

for all n, so the sequence $\{s_{2n}\}$ is bounded above. It follows that $\{s_{2n}\}$ converges. Since $s_{2n-1} = s_{2n} + a_{2n}$ and $\{a_{2n}\}$ converges to 0, the sequence $\{s_{2n-1}\}$ converges and has the same limit as $\{s_{2n}\}$. It follows that the sequence $\{s_n\}$ converges (see Exercise 6 in Section 2.3). Hence, the alternating series converges.

To verify the estimate in the second part of the theorem, let S be the sum of the series. Since $\{s_{2n}\}$ is an increasing sequence that converges to S and $\{s_{2n-1}\}$ is a decreasing sequence that converges to S (see the exercises), it follows that S is between s_n and s_{n+1} for each positive integer n. Thus,

$$|s_n - S| \leq |s_{n+1} - s_n| = a_{n+1}$$

for each positive integer n. This completes the proof. ■

It is easy to show that the series $\sum_{k=1}^{\infty}(-1)^{k+1}/k$ converges by the Alternating Series Test: simply note that the sequence $\{1/k\}$ is decreasing and converges to 0. Since this series does not converge absolutely, it is an example of a series that converges nonabsolutely.

The second part of the Alternating Series Test provides an estimate for how closely the partial sums approximate the sum of the series. When the sum of a series is unknown, an error estimate such as this is very useful. For example, the Alternating Series Test asserts that

$$\left|\sum_{k=1}^{n}\frac{(-1)^{k+1}}{k^3} - \sum_{k=1}^{\infty}\frac{(-1)^{k+1}}{k^3}\right| < \frac{1}{(n+1)^3}$$

for each positive integer n. To approximate the sum of this series with an accuracy of 10^{-4}, we would compute s_n, where n is chosen so that $(n+1)^{-3} < 10^{-4}$. The smallest value of n that satisfies this inequality is $n = 21$. Therefore, the number

$$s_{21} = \sum_{k=1}^{21} \frac{(-1)^{k+1}}{k^3} \approx 0.9015928$$

is within 10^{-4} of the sum of the series. In fact, upper and lower estimates for the sum of the series can be found since the sum of this series is between s_{21} and s_{22}:

$$0.9014989 \approx s_{22} < \sum_{k=1}^{\infty} \frac{(-1)^{k+1}}{k^3} < s_{21} \approx 0.9015928.$$

Exercises

1. Prove the following statement made in the text: if all but a finite number of the terms of a series have the same sign, then the series converges if and only if it converges absolutely.

2. Are all convergent geometric series absolutely convergent?

3. Use the inequality $0 \le a_k + |a_k| \le 2|a_k|$ and the Comparison Test to give a different proof of Theorem 6.10.

4. Suppose that $\sum_{k=1}^{\infty} a_k$ converges absolutely. Prove that the series $\sum_{k=1}^{\infty} a_k^2$ converges.

5. Let $\sum_{k=1}^{\infty} a_k$ be a convergent series.

a) Suppose that $\sum_{k=1}^{\infty} a_k$ converges absolutely and that the sequence $\{b_k\}$ is bounded. Prove that the series $\sum_{k=1}^{\infty} a_k b_k$ converges absolutely.

b) Suppose that $\sum_{k=1}^{\infty} a_k$ converges nonabsolutely. Prove that there exists a bounded sequence $\{b_k\}$ such that the series $\sum_{k=1}^{\infty} a_k b_k$ diverges.

6. Prove that the series $\sum_{k=1}^{\infty} \frac{\sin k}{k^2}$ converges.

7. In the proof of part (b) of the Ratio Test, it was shown that the sequence $\{a_k\}$ does not converge to 0. Show that in fact $|a_k| \to \infty$ under the hypothesis of part (b).

8. Let $\sum_{k=1}^{\infty} a_k$ be a series with nonzero terms and suppose that $|a_{k+1}/a_k| \ge 1$ for all $k \ge K$. Prove that the series diverges. Note that this hypothesis is weaker than the hypothesis of part (b) of the Ratio Test.

9. Apply the result of the previous exercise to the series $\sum_{k=1}^{\infty} \frac{4^k (k!)^2}{(2k)!}$.

10. Prove part (a) of the Root Test.

11. Prove part (b) of the Root Test. Show that $|a_k| \to \infty$ under the hypothesis of part (b).

12. Determine whether or not the following series converge.

a) $\sum_{k=1}^{\infty} k^5 5^{-k}$ **b)** $\sum_{k=1}^{\infty} \frac{10^k}{k!}$ **c)** $\sum_{k=1}^{\infty} \frac{(k!)^2}{(2k)!}$

d) $\sum_{k=1}^{\infty} \frac{(-3)^k\, k!}{k^k}$ **e)** $\sum_{k=1}^{\infty} (\sqrt[k]{k} - 1)^k$ **f)** $\sum_{k=1}^{\infty} \left(\frac{k}{2}\sin(1/k)\right)^k$

13. Show that the Root and Ratio Tests provide no information on the p-series.

14. Compute the limit inferior and the limit superior of a_{k+1}/a_k and $\sqrt[k]{a_k}$ for the sequences $a_k = \left(2 + (-1)^k\right)2^{-k}$ and $a_k = \left(3 + (-1)^k\right)^{-k}$. In each case, what do the Root and Ratio Tests tell you about the convergence of the series $\sum_{k=1}^{\infty} a_k$?

15. Let $\{a_k\}$ be a sequence of positive numbers. Prove that

$$\liminf_{k\to\infty} \frac{a_{k+1}}{a_k} \le \liminf_{k\to\infty} \sqrt[k]{a_k} \le \limsup_{k\to\infty} \sqrt[k]{a_k} \le \limsup_{k\to\infty} \frac{a_{k+1}}{a_k}.$$

Give an example of a sequence $\{a_k\}$ for which each of the quantities is finite and all of the inequalities are strict.

16. To say that the Root Test is a stronger test than the Ratio Test means that the Ratio Test implies convergence or divergence for any series for which the Root Test implies convergence or divergence, but that there are series for which the Ratio Test provides no information but the Root Test yields convergence or divergence. Use the previous exercise to justify this statement and find examples of series for which the Root Test "works" but the Ratio Test does not.

17. Find the sum of the series $\sum_{k=1}^{\infty} (-1)^{k+1}/2^k$. Then find the first 8 terms of the sequence of partial sums and note their relationship to each other and to the sum.

18. Show that under the hypotheses of the Alternating Series Test, the sequence $\{s_{2n-1}\}$ is decreasing and bounded below.

19. For what values of p does the series $\sum_{k=1}^{\infty} \frac{(-1)^{k+1}}{k^p}$ converge?

20. For what values of p does the series $\sum_{k=1}^{\infty} \frac{(-1)^{k+1} k^p}{k+6}$ converge?

21. Suppose that the series $\sum_{k=1}^{\infty} a_k$ converges. Give an example to show that the series $\sum_{k=1}^{\infty} a_k^2$ may not converge.

22. For the series $\sum_{k=1}^{\infty} (-1)^{k+1}/k^2$, find s_4 and s_5 as rational numbers with a common denominator and determine an interval in which the sum of the series lies. Which partial sum is closer to the actual sum and how good is its approximation?

23. Use Theorem 6.14 to find an integer n so that

$$\left| \sum_{k=1}^{n} \frac{(-1)^{k+1}}{k^4} - \sum_{k=1}^{\infty} \frac{(-1)^{k+1}}{k^4} \right| < 10^{-8}.$$

24. Give an example to show that the condition that $\{a_k\}$ is a decreasing sequence is essential in the Alternating Series Test.

25. The adjective "alternating" in the Alternating Series Test is intended to refer to the signs of the terms. By looking at the relationship between the partial sums and the sum of the series, give another interpretation of this adjective.

26. Let $\sum_{k=1}^{\infty} a_k$ be a nonabsolutely convergent series and for each positive integer k define

$$p_k = \begin{cases} a_k, & \text{if } a_k > 0; \\ 0, & \text{if } a_k \le 0; \end{cases} \quad \text{and} \quad q_k = \begin{cases} a_k, & \text{if } a_k < 0; \\ 0, & \text{if } a_k \ge 0. \end{cases}$$

Prove that both of the series $\sum_{k=1}^{\infty} p_k$ and $\sum_{k=1}^{\infty} q_k$ diverge.

27. Determine whether or not the given series converges.

a) $1 - \frac{1}{2} - \frac{1}{3} + \frac{1}{4} - \frac{1}{5} - \frac{1}{6} + \frac{1}{7} - \frac{1}{8} - \frac{1}{9} + \frac{1}{10} - \frac{1}{11} - \frac{1}{12} + \cdots$

b) $1 + \frac{1}{2} - \frac{1}{3} - \frac{1}{4} + \frac{1}{5} + \frac{1}{6} - \frac{1}{7} - \frac{1}{8} + \frac{1}{9} + \frac{1}{10} - \frac{1}{11} - \frac{1}{12} + \cdots$

c) $1 + \frac{1}{2} + \frac{1}{3} - \frac{1}{4} + \frac{1}{5} + \frac{1}{6} + \frac{1}{7} - \frac{1}{8} + \frac{1}{9} + \frac{1}{10} + \frac{1}{11} - \frac{1}{12} + \cdots$

6.4 Rearrangements and Products

When adding a finite number of real numbers, the order in which the terms are added together makes no difference in the sum; this is a consequence of the commutative and associative properties of addition. However, this property of addition does not extend to infinite sums of real numbers. Consider the two series

$$1 - 1 + 1 - 1 + 1 - 1 + \cdots \qquad \text{and} \qquad 1 + 1 - 1 + 1 + 1 - 1 + \cdots,$$

which consist of the same numbers, but in different orders. Since the first series has bounded partial sums and the second series has unbounded partial sums, it is clear that the order in which the terms appear in an infinite series may affect the sum. Although one could argue that this is not a fair example since both series diverge, it turns out that even for convergent series, the order of the terms can make a difference in the sum. In this regard, the distinction between series that converge absolutely and those that converge nonabsolutely is quite dramatic.

To begin this discussion, it is necessary to precisely define what it means to add up an infinite number of terms in a different order. The concept of a permutation is useful in this situation. A **permutation** σ of the positive integers is a one-to-one function from $\mathbb{Z}^+$ onto $\mathbb{Z}^+$ (that is, $\sigma\colon \mathbb{Z}^+ \to \mathbb{Z}^+$ is a bijection). In other words, a permutation is a sequence of positive integers such that each positive integer appears exactly once. Most permutations are best expressed by writing out enough terms to indicate the pattern. For example,

$$1, 2, 4, 3, 6, 8, 5, 10, 12, 7, 14, 16, 9, 18, 20, \ldots$$

is a permutation of the positive integers consisting of one odd integer followed by two even integers, with the orders of the odd and even integers preserved.

DEFINITION 6.15 Let $\sum_{k=1}^{\infty} a_k$ be an arbitrary series and let σ be a permutation of the positive integers. The series $\sum_{k=1}^{\infty} a_{\sigma(k)}$ is a **rearrangement** of the series $\sum_{k=1}^{\infty} a_k$.

To illustrate this idea, let σ be the permutation mentioned prior to the definition and consider the series

$$1 - \frac{1}{2} + \frac{1}{3} - \frac{1}{4} + \frac{1}{5} - \frac{1}{6} + \frac{1}{7} - \cdots.$$

Applying σ to this series yields the rearrangement

$$1 - \frac{1}{2} - \frac{1}{4} + \frac{1}{3} - \frac{1}{6} - \frac{1}{8} + \frac{1}{5} - \frac{1}{10} - \frac{1}{12} + \cdots.$$

This series and its relationship to the original series are considered in the exercises at the end of this section.

Suppose that $\sum_{k=1}^{\infty} a_{\sigma(k)}$ is a rearrangement of the convergent series $\sum_{k=1}^{\infty} a_k$. Does it follow that $\sum_{k=1}^{\infty} a_{\sigma(k)}$ converges and that the two series have the same sum? The answer is yes for series that converge absolutely.

THEOREM 6.16 Every rearrangement of an absolutely convergent series converges and its sum is that of the original series.

Proof. Let $\sum_{k=1}^{\infty} a_k$ be an absolutely convergent series and let $\sum_{k=1}^{\infty} a_{\sigma(k)}$ be any rearrangement of this series. For each positive integer n, define

$$s_n = \sum_{k=1}^{n} a_k \quad \text{and} \quad t_n = \sum_{k=1}^{n} a_{\sigma(k)}.$$

We must show that the sequence $\{t_n\}$ converges and has the same limit as the convergent sequence $\{s_n\}$. Let $\epsilon > 0$. Since the series $\sum_{k=1}^{\infty} a_k$ converges absolutely, there exists an integer p such that $\sum_{k=p+1}^{\infty} |a_k| < \epsilon$. Now choose a positive integer N such that

$$\{1, 2, \ldots, p\} \subseteq \{\sigma(1), \sigma(2), \ldots, \sigma(N)\}.$$

(To be completely precise, $N = \max\{\sigma_{\text{inv}}(k) : 1 \leq k \leq p\}$.) For each $n \geq N$, the terms $a_1, a_2, \ldots, a_p$ cancel in the expression $t_n - s_n$. (This is a finite sum so the commutative and associative properties of addition are valid.) Hence,

$$|t_n - s_n| \leq \sum_{k=p+1}^{\infty} |a_k| < \epsilon$$

for all $n \geq N$. This shows that $\{t_n - s_n\}$ converges to 0. Since $t_n = (t_n - s_n) + s_n$, the sequence $\{t_n\}$ converges and has the same limit as the sequence $\{s_n\}$. This completes the proof. ∎

This theorem fails dramatically for series that do not converge absolutely: the terms of a nonabsolutely convergent series can be rearranged to add up to any given real number. This is a truly surprising result! The main idea behind the proof of this fact is not difficult, but putting the idea into a formal proof is a bit tedious and leads to some intimidating notation. It is a good idea to go through the proof once and seek out the main idea, then come back and read the details.

THEOREM 6.17 Let $\sum_{k=1}^{\infty} a_k$ be a nonabsolutely convergent series. If L is any real number, then there exists a permutation σ of the positive integers such that $\sum_{k=1}^{\infty} a_{\sigma(k)} = L$.

Proof. Without loss of generality, we may assume that $a_k \neq 0$ for all k. Let $\{p_k\}$ be the subsequence of all the positive terms of $\{a_k\}$ and let $\{q_k\}$ be the subsequence of all the negative terms of $\{a_k\}$. Since $\sum_{k=1}^{\infty} a_k$ is a nonabsolutely convergent series, the sequences $\{p_k\}$ and $\{q_k\}$ converge to 0, but the series $\sum_{k=1}^{\infty} p_k$ and $\sum_{k=1}^{\infty} q_k$ both diverge. (See Exercise 26 in the previous section.) Let $L \geq 0$; the proof for $L < 0$ is similar.

Let $p_0 = 0$ and $q_0 = 0$. Since the partial sums of $\sum_{k=1}^{\infty} p_k$ are not bounded above, there exists an index i_1 (which may be 1) such that

$$p_1 + \cdots + p_{i_1 - 1} \leq L < p_1 + \cdots + p_{i_1} \equiv x_1.$$

(The symbol $\equiv$ is used to represent a definition; x_1 is defined to be the indicated sum.) Since the partial sums of $\sum_{k=1}^{\infty} q_k$ are not bounded below, there exists an index j_1 such that

$$x_1 + q_1 + \cdots + q_{j_1 - 1} \geq L > x_1 + q_1 + \cdots + q_{j_1} \equiv y_1.$$

Similarly, there exists an index $i_2 > i_1$ such that

$$y_1 + p_{i_1+1} + \cdots + p_{i_2 - 1} \leq L < y_1 + p_{i_1+1} + \cdots + p_{i_2} \equiv x_2$$

and an index $j_2 > j_1$ such that

$$x_2 + q_{j_1+1} + \cdots + q_{j_2 - 1} \geq L > x_2 + q_{j_1+1} + \cdots + q_{j_2} \equiv y_2.$$

Continue this process and generate four sequences $\{i_n\}$, $\{j_n\}$, $\{x_n\}$, and $\{y_n\}$. The sequences $\{i_n\}$ and $\{j_n\}$ are strictly increasing sequences of positive integers, and the sequences $\{x_n\}$ and $\{y_n\}$ satisfy

$$x_n - p_{i_n} \leq L < x_n \quad \text{and} \quad y_n < L \leq y_n - q_{j_n}$$

for all n. The series

$$p_1 + \cdots + p_{i_1} + q_1 + \cdots + q_{j_1} + p_{i_1+1} + \cdots + p_{i_2} + q_{j_1+1} + \cdots + q_{j_2} + \cdots$$

is a rearrangement of the series $\sum_{k=1}^{\infty} a_k$ and will be denoted by $\sum_{k=1}^{\infty} a_{\sigma(k)}$. We claim that this series converges to L.

Let $\{s_n\}$ be the sequence of partial sums of the series $\sum_{k=1}^{\infty} a_{\sigma(k)}$. Let $\{n_k\}$ be the strictly increasing sequence of positive integers that satisfies

$$s_{n_1} = x_1,\ s_{n_2} = y_1,\ s_{n_3} = x_2,\ s_{n_4} = y_2,\ s_{n_5} = x_3,\ s_{n_6} = y_3,\ \text{ and so on.}$$

We make the following observations based upon the construction of the sequences $\{x_n\}$ and $\{y_n\}$.

If $n_1 \le n < n_2$, then $y_1 < s_n \le x_1$ and $|s_n - L| \le \max\{p_{i_1}, q_{j_1}\}$.
If $n_2 \le n < n_3$, then $y_1 \le s_n < x_2$ and $|s_n - L| \le \max\{p_{i_2}, q_{j_1}\}$.
If $n_3 \le n < n_4$, then $y_2 < s_n \le x_2$ and $|s_n - L| \le \max\{p_{i_2}, q_{j_2}\}$.
If $n_4 \le n < n_5$, then $y_2 \le s_n < x_3$ and $|s_n - L| \le \max\{p_{i_3}, q_{j_2}\}$.
If $n_{2k-1} \le n < n_{2k}$, then $y_k < s_n \le x_k$ and $|s_n - L| \le \max\{p_{i_k}, q_{j_k}\}$.
If $n_{2k} \le n < n_{2k+1}$, then $y_k \le s_n < x_k$ and $|s_n - L| \le \max\{p_{i_{k+1}}, q_{j_k}\}$.

Since both the sequences $\{p_k\}$ and $\{q_k\}$ converge to 0, it follows that $\{s_n\}$ converges to L. This completes the proof. ■

It takes a moment for the full impact of this theorem to sink in; the terms of a nonabsolutely convergent series can be rearranged to add up to any number. This result contrasts completely with the situation for finite sums and for absolutely convergent series (Theorem 6.16). In fact, this property of nonabsolutely convergent series lies behind the term conditional convergence. A series is said to **converge conditionally** if the sum of the series (as well as its convergence) depends on the arrangement of the terms. As the last two results indicate, nonabsolutely convergent series are conditionally convergent, while absolutely convergent series are unconditionally convergent. Several of the exercises at the end of this section show how rearrangements of nonabsolutely convergent series can give different sums.

Since some of the consequences of the next result are more interesting in the context of complex numbers, we take a short digression to remind the reader of some of the basic properties of complex numbers. A **complex number** is a number of the form $x + iy$, where x and y are real numbers and $i = \sqrt{-1}$. Addition and multiplication of complex numbers are defined just as they would be under the assumption that i operates like any other square root:

$$(a+ib)+(c+id) = (a+c)+i(b+d) \text{ and } (a+ib)(c+id) = (ac-bd)+i(ad+bc).$$

The set of complex numbers with these two operations can be defined formally and shown to satisfy all of the axioms for a field. The **absolute value** of a complex number is defined by $|x + iy| = \sqrt{x^2 + y^2}$. Note that

$$|x| \le |x + iy|,\ |y| \le |x + iy|,\ \text{ and } |x + iy| \le |x| + |y|.$$

An important and rather fascinating result is the relationship $e^{ix} = \cos x + i \sin x$, which is valid for all real numbers x. One way to discover this relationship is discussed in the supplementary exercises of the next chapter. Finally, the laws of exponents are valid for complex numbers; in particular, $e^{ix} = (e^i)^x$ for all real numbers x.

Given two sequences $\{a_k\}$ and $\{b_k\}$, an interesting problem is to find conditions that will guarantee that the series $\sum_{k=1}^{\infty} a_k b_k$ converges. The following theorem provides one answer to this question; another can be found in the exercises at the end of the section.

THEOREM 6.18 Let $\sum_{k=1}^{\infty} a_k$ be a series of complex numbers and let $\{b_k\}$ be a sequence of real numbers. If the sequence of partial sums of $\sum_{k=1}^{\infty} a_k$ is bounded and $\{b_k\}$ is a decreasing sequence that converges to 0, then the series $\sum_{k=1}^{\infty} a_k b_k$ converges.

Proof. Let $\{s_n\}$ be the sequence of partial sums for the series $\sum_{k=1}^{\infty} a_k$ and let M be a bound for $\{s_n\}$. Let $\epsilon > 0$ and choose a positive integer $N > 1$ such that $b_N < \epsilon/2M$. For $n > m \geq N$, we have

$$\begin{aligned}
\left|\sum_{k=m}^{n} a_k b_k\right| &= \left|\sum_{k=m}^{n} (s_k - s_{k-1}) b_k\right| \\
&= \left|\sum_{k=m}^{n} s_k b_k - \sum_{k=m-1}^{n-1} s_k b_{k+1}\right| \\
&= \left|\sum_{k=m}^{n-1} s_k (b_k - b_{k+1}) + s_n b_n - s_{m-1} b_m\right| \\
&\leq \sum_{k=m}^{n-1} |s_k|\,|b_k - b_{k+1}| + |s_n|\,|b_n| + |s_{m-1}|\,|b_m| \\
&\leq M\left(\sum_{k=m}^{n-1} |b_k - b_{k+1}| + |b_n| + |b_m|\right) \\
&= M\left(\sum_{k=m}^{n-1} (b_k - b_{k+1}) + b_n + b_m\right) \\
&= 2Mb_m \leq 2Mb_N < \epsilon.
\end{aligned}$$

By Theorem 6.3 (the statement and proof of this result easily extend to the context of complex numbers), the series $\sum_{k=1}^{\infty} a_k b_k$ converges. ■

This theorem can be used to give another proof of the Alternating Series Test; see the exercises in this section. Further comments and results concerning the product of infinite series will be considered in the next chapter.

Exercises

1. Let $\{a_k\}$ be a sequence of positive numbers that converges to 0. Prove that there exists a permutation σ of the positive integers such that $\{a_{\sigma(k)}\}$ is a decreasing sequence.

2. Let σ be a permutation of the positive integers. Find the sum of the series $\sum_{k=1}^{\infty} \frac{1}{(\sigma(k))^2}$.

3. Show that the series $\sum_{k=1}^{\infty} \frac{(-1)^{k+1}}{\left(3+(-1)^k\right)^k}$ converges and find its sum. Is Theorem 6.16 necessary here?

4. Is it necessary to essentially mimic the details of the proof of Theorem 6.17 for the case in which $L < 0$ or is there an easier way?

5. Let $\sum_{k=1}^{\infty} a_k$ be a nonabsolutely convergent series.

a) Show that there is a rearrangement of this series so that the sequence of partial sums of the rearranged series converges to ∞.

b) Show that there is a rearrangement of this series so that the sequence of partial sums of the rearranged series is bounded but does not converge.

6. Consider the following rearrangement of the series $\sum_{k=1}^{\infty} (-1)^{k+1}/k$:

$$1 - \frac{1}{2} - \frac{1}{4} + \frac{1}{3} - \frac{1}{6} - \frac{1}{8} + \frac{1}{5} - \frac{1}{10} - \frac{1}{12} + \cdots.$$

Let $\{s_n\}$ be the sequence of partial sums for the original series and let $\{t_n\}$ be the sequence of partial sums for the rearranged series. Show that $t_{3n} = s_{2n}/2$ for each n. What does this imply about the sums of these two series?

7. Let $\{a_k\}$ be a sequence that converges to 0, let $\{s_n\}$ be the sequence of partial sums for the series $\sum_{k=1}^{\infty} a_k$, and let p be a positive integer. Suppose that the subsequence $\{s_{pn}\}$ converges. Prove that $\sum_{k=1}^{\infty} a_k$ converges.

Remark. The next six exercises involve, either explicitly or implicitly, the convergent sequence $\{\gamma_n\}$ defined by

$$\gamma_n = 1 + \frac{1}{2} + \cdots + \frac{1}{n} - \int_1^n \frac{1}{x}\,dx.$$

This sequence was first considered in Exercise 18 in Section 5.2.

8. The purpose of this exercise is to find the sum of the series $\sum_{k=1}^{\infty} (-1)^{k+1}/k$. Let $\{s_n\}$ be the sequence of partial sums of this series and show that $s_{2n} = \gamma_{2n} - \gamma_n + \ln 2$. What then is the sum of the series?

9. Consider the series $\sum_{k=1}^{\infty}(-1)^{k+1}/k$. Using the method in the proof of Theorem 6.17, find the first 20 terms of a rearrangement of this series that converges to 0. Then make and prove a conjecture for a rearrangement of the series that converges to 0.

10. Let p and q be positive integers and consider the rearrangement of the series

$$1 - 1 + \frac{1}{2} - \frac{1}{2} + \frac{1}{3} - \frac{1}{3} + \frac{1}{4} - \frac{1}{4} + \frac{1}{5} - \frac{1}{5} + \cdots$$

that consists of p positive terms followed by q negative terms, where terms of the same sign are taken in the same relative order as in the original series. For example, if $p = 3$ and $q = 2$, the rearrangement would be

$$1 + \frac{1}{2} + \frac{1}{3} - 1 - \frac{1}{2} + \frac{1}{4} + \frac{1}{5} + \frac{1}{6} - \frac{1}{3} - \frac{1}{4} + \cdots.$$

Show that this rearrangement converges and find a simple formula for its sum.

11. Let p and q be positive integers and consider the rearrangement of the series

$$1 - \frac{1}{2} + \frac{1}{3} - \frac{1}{4} + \frac{1}{5} - \frac{1}{6} + \frac{1}{7} - \frac{1}{8} + \cdots$$

that consists of p positive terms followed by q negative terms, where terms of the same sign are taken in the same relative order as in the original series. (See the example in the previous exercise.) Show that this rearrangement converges and find a simple formula for its sum. It may help to initially consider the case in which q is 1.

12. Find an explicit patterned rearrangement of the series

$$1 - 1 + \frac{1}{2} - \frac{1}{2} + \frac{1}{3} - \frac{1}{3} + \frac{1}{4} - \frac{1}{4} + \frac{1}{5} - \frac{1}{5} + \cdots$$

for which the partial sums of the series converge to infinity.

13. Let $p \geq 2$ be a fixed positive integer. For each positive integer k, let $a_k = 1/k$ and let

$$b_k = \begin{cases} 1, & \text{if } k \text{ is not a multiple of } p; \\ 1 - p, & \text{if } k \text{ is a multiple of } p. \end{cases}$$

Prove that the series $\sum_{k=1}^{\infty} a_k b_k$ converges and find its sum.

14. Use Theorem 6.18 to prove the Alternating Series Test.

15. Suppose that $\sum_{k=1}^{\infty} a_k^2$ and $\sum_{k=1}^{\infty} b_k^2$ converge. Prove that $\sum_{k=1}^{\infty} a_k b_k$ converges absolutely.

Remark. The rest of the exercises in this section require the use of complex numbers. Exercise 17 gives an indication of the fact that complex numbers can provide a nice way to prove results in real analysis.

16. Let $z_k = x_k + iy_k$ for all positive integers k, where x_k and y_k are real numbers.

a) Prove that $\sum_{k=1}^{\infty} z_k$ converges if and only if $\sum_{k=1}^{\infty} x_k$ and $\sum_{k=1}^{\infty} y_k$ converge.

b) Prove that $\sum_{k=1}^{\infty} z_k$ converges absolutely if and only if $\sum_{k=1}^{\infty} x_k$ and $\sum_{k=1}^{\infty} y_k$ converge absolutely.

17. For this exercise, use the facts $e^{ik} = (e^i)^k$ and $e^{ik} = \cos k + i \sin k$, for any positive integer k.

a) Show that the sequence of partial sums for the series $\sum_{k=1}^{\infty} e^{ik}$ is bounded.

b) Prove that the series $\sum_{k=1}^{\infty} \frac{e^{ik}}{k}$ converges.

c) Prove that the series $\sum_{k=1}^{\infty} \frac{\sin k}{k}$ and $\sum_{k=1}^{\infty} \frac{\cos k}{k}$ converge.

18. Let $z \neq 1$ be a complex number that satisfies $|z| \leq 1$. Prove that the series $\sum_{k=1}^{\infty} z^k/k$ converges. (Use the fact that $z = e^{i\theta}$ for some $0 \leq \theta < 2\pi$ if $|z| = 1$.)

6.5 Supplementary Exercises

1. Suppose that a certain ball has the property that when dropped from a height h it rebounds to a height rh, where $0 < r < 1$. How far does the ball travel before coming to rest? How long does the ball bounce? Express your answers in terms of r and h.

2. Let $\sum_{k=1}^{\infty} a_k$ be a series of real numbers. Suppose that $\sum_{k=1}^{\infty} a_k b_k$ converges for every bounded sequence $\{b_k\}$. Prove that $\sum_{k=1}^{\infty} a_k$ converges absolutely.

3. Suppose that $\sum_{k=1}^{\infty} a_k$ is a convergent series of positive numbers. Prove that the series $\sum_{k=1}^{\infty} \sqrt{a_k a_{k+1}}$ converges.

4. Let $\sum_{k=1}^{\infty} a_k$ be an infinite series and let $\{s_n\}$ be its corresponding sequence of partial sums. Suppose that $s_n = (2n+5)/(3n-4)$ for all $n \geq 1$. Find a_1, a_3, and the sum of the series.

5. Let $\sum_{k=1}^{\infty} a_k$ be a convergent series of positive terms and let $t_n = \sum_{k=n}^{\infty} a_k$ for each positive integer n.

a) Prove that $\sum_{k=m}^{n} \frac{a_k}{t_k} > 1 - \frac{t_n}{t_m}$ for all positive integers m and n with $m < n$.

b) Use part (a) to prove that the series $\sum_{k=1}^{\infty} a_k/t_k$ diverges.

c) Prove that $\frac{a_k}{\sqrt{t_k}} < 2\left(\sqrt{t_k} - \sqrt{t_{k+1}}\right)$ for all positive integers k.

d) Use part (c) to prove that the series $\sum_{k=1}^{\infty} a_k/\sqrt{t_k}$ converges.

e) To see how the series defined in parts (b) and (d) compare to the original series, find both of these series when $a_k = 2^{-k}$ and also when $a_k = (k^2 + k)^{-1}$.

6. Let $\sum_{k=1}^{\infty} a_k$ be a divergent series of positive terms and let $\{s_n\}$ be the sequence of partial sums of this series.

a) Prove that the series $\sum_{k=1}^{\infty} a_k/s_k$ diverges.

b) Prove that the series $\sum_{k=1}^{\infty} a_k/s_k^2$ converges.

c) To see how the series defined in parts (a) and (b) compare to the original series, find both of these series when $a_k = 1$ and also when $a_k = \sqrt{k+1} - \sqrt{k}$.

7. Find the sum of the series $\displaystyle\sum_{k=1}^{\infty} \frac{3k}{(k+1)!}$.

8. Show that the series $\displaystyle\sum_{k=1}^{\infty} k^{-3+(-1)^{k+1}}$ converges and find its sum. Justify any algebraic manipulations that you make.

9. Let $\sum_{k=1}^{\infty} a_k$ be a series with positive terms. Show that $\sum_{k=1}^{\infty} a_k$ and $\sum_{k=1}^{\infty} a_k/(1+a_k)$ either both converge or both diverge.

10. Determine whether or not the series $\sum_{k=1}^{\infty} a_k$ converges, where a_k is given by

a) $\ln(1 - 1/k^2)$ **b)** $\displaystyle\frac{(-1)^k}{k\ln(1+1/k)}$ **c)** $\displaystyle\frac{1}{k} - \ln\bigl(1 + 1/k\bigr)$

d) $k^{-1-1/k}$ **e)** $\displaystyle\frac{\sqrt{k+1} - \sqrt{k}}{\sqrt{k}}$ **f)** $\displaystyle\frac{k!}{1 \cdot 3 \cdot 5 \cdot \cdots \cdot (2k-1)}$

11. For each positive integer k, let $a_k = \int_k^{k+1} x^{-1} \sin(\pi x)\, dx$. Prove that the series $\sum_{k=1}^{\infty} a_k$ converges nonabsolutely.

12. Let $\{d_k\}$ be the strictly increasing sequence of positive integers that do not contain the digit 0. (For example, 47 is in the sequence and 105 is not.)

a) Prove that the series $\sum_{k=1}^{\infty} 1/d_k$ converges and has a sum less than 90.

b) Find, to the nearest integer, the sum of this series.

13. Determine whether or not the series $\sum_{k=1}^{\infty} a_k$ converges, where

a) $\displaystyle a_k = \frac{2 \cdot 4 \cdot 6 \cdot \cdots \cdot (2k)}{3 \cdot 5 \cdot 7 \cdot \cdots \cdot (2k+1)}$ **b)** $\displaystyle a_k = \frac{2 \cdot 4 \cdot 6 \cdot \cdots \cdot (2k)}{3 \cdot 5 \cdot 7 \cdot \cdots \cdot (2k+5)}$

14. Let P and Q be polynomials of degrees p and q, respectively, and suppose that $Q(k) \neq 0$ for all positive integers k. Prove that the series $\sum_{k=1}^{\infty} (-1)^k P(k)/Q(k)$ converges if and only if $p < q$.

15. Determine what the Root Test and the Ratio Test say about the convergence of the series $\sum_{k=1}^{\infty} a_k$, where

a) $a_k = \begin{cases} 2^{-k}, & \text{if } k \text{ is odd;} \\ 2^{-k+2}, & \text{if } k \text{ is even;} \end{cases}$ **b)** $a_k = \begin{cases} 2^{-(k+1)/2}, & \text{if } k \text{ is odd;} \\ 3^{-k/2}, & \text{if } k \text{ is even.} \end{cases}$

16. For what values of p does the series $\sum_{k=1}^{\infty} \frac{k^p}{p^k}$ converge?

17. Find the values of r for which the series $\sum_{k=0}^{\infty} (r - r^2)^k$ converges and the sum of the series for these values of r. What is the maximum possible sum?

18. Find the values of x for which the series $\sum_{k=1}^{\infty} \cos^{2k} x$ converges and the sum of the series for these values of x.

19. The purpose of this exercise is to prove a portion of **Stirling's Formula**. Let $f: [1, \infty) \to \mathbb{R}$ be a nonnegative differentiable function whose derivative f' is positive and decreasing on $(1, \infty)$. Examples of such functions include $\sqrt{x}$, $\ln x$, $\sqrt{x} + 2x$, and $\sqrt[3]{x}$. For each positive integer k, let A_k be the area of the region bounded by the graph of $y = f(x)$ and the line passing through the points $(k, f(k))$ and $(k+1, f(k+1))$. It may be helpful to sketch a graph of a function f with the given properties and see what the region whose area is A_k looks like. Other quantities and functions defined in this exercise can also be given geometrical interpretations.

a) For each positive integer $k \geq 2$, let L_k and U_k be the functions defined by

$$L_k(x) = \big(f(k+1) - f(k)\big)(x - k) + f(k) \quad \text{and}$$
$$U_k(x) = \big(f(k) - f(k-1)\big)(x - k) + f(k).$$

Prove that $L_k(x) \leq f(x) \leq U_k(x)$ for all $x \in [k, k+1]$. (The techniques used in the proofs of Theorems 4.23 and 4.24 may be useful here.)

b) For each positive integer $k \geq 2$, let $D_k = f(k+1) - f(k)$. Prove that

$$A_k = \int_k^{k+1} (f - L_k) \leq \int_k^{k+1} (U_k - L_k) = \frac{1}{2}(D_{k-1} - D_k).$$

c) Prove that the sequence $\{D_k\}_{k=2}^{\infty}$ converges.

d) Prove that the series $\sum_{k=1}^{\infty} A_k$ converges.

e) For the specific case in which $f(x) = \ln x$, prove that

$$\sum_{k=1}^{n-1} A_k = 1 + \ln\left(\frac{n^n \sqrt{n}}{n!\, e^n}\right).$$

f) Use part (e) to prove that the sequence $\left\{\frac{n!}{(n/e)^n \sqrt{n}}\right\}$ converges.

g) Stirling's Formula includes the fact that the limit of the sequence in part (f) is $\sqrt{2\pi}$. There are many proofs of this fact; see Olmsted [19] or Stromberg [24].

20. Suppose that $\{a_k\}$ converges to 0. Prove that there exists a subsequence $\{a_{p_k}\}$ of $\{a_k\}$ such that $\sum_{k=1}^{\infty} a_{p_k}$ converges absolutely.

21. Give an example of each of the following:

a) a divergent series $\sum_{k=1}^{\infty} a_k$ for which $\sum_{k=1}^{\infty} a_{3k}$ converges;

b) a convergent series $\sum_{k=1}^{\infty} b_k$ for which $\sum_{k=1}^{\infty} b_{3k}$ diverges.

22. Let a and b be positive numbers.

a) Prove that the series $\sum_{k=1}^{\infty}\big(\ln(ak+b) - \ln(ak)\big)$ diverges.

b) Prove that the series $\sum_{k=1}^{\infty}\big(\ln(ak^2+b) - \ln(ak^2)\big)$ converges.

23. Let p and q be positive integers with $p > q$ and consider the rearrangement of the series

$$1 - 1 + \frac{1}{2} - \frac{1}{2} + \frac{1}{3} - \frac{1}{3} + \frac{1}{4} - \frac{1}{4} + \frac{1}{5} - \frac{1}{5} + \cdots$$

that consists of p positive terms followed by q negative terms, where terms of the same sign are taken in the same relative order as in the original series. Let $\{s_n\}$ be the sequence of partial sums for this series. For each positive integer n, let ${}^t\mathcal{P}_n$ be the tagged partition of the interval $[q, p]$ that consists of n subintervals of equal length, with the right endpoint of each subinterval chosen as tag. Prove that

$$S(1/x, {}^t\mathcal{P}_{(p-q)n}) = s_{(p+q)n}$$

for each positive integer n.

Remark. Another convergence test for infinite series, which often plays a prominent role in the discussion of infinite series in calculus books, is the Integral Test. Two nice features of this test are that it provides a geometrical interpretation for the convergence or divergence of a series and that it gives an estimate for how accurately the partial sums of a convergent series approximate the sum of the series. Its major disadvantage is that it can only be used when the appropriate function has a simple antiderivative. In addition, from a theoretical point of view, it seems unfortunate to require the elaborate and difficult theory of integration to prove that a series converges. The next few exercises consider the Integral Test and some of its applications.

> **Integral Test:** Let $\sum_{k=1}^{\infty} a_k$ be a series with positive terms and define a function f on $[1, \infty)$ so that $f(k) = a_k$ for each positive integer k. Suppose that f is continuous and decreasing on $[1, \infty)$ and that $\lim_{x\to\infty} f(x) = 0$. Then the series $\sum_{k=1}^{\infty} a_k$ converges if and only if the improper integral $\int_1^{\infty} f$ converges. Furthermore, if S is the sum of the series and s_n is the nth partial sum of the series, then $0 < S - s_n < \int_n^{\infty} f$ for each positive integer n.

24. Prove the Integral Test.

25. Use the area interpretation of the integral to give a geometrical interpretation for the Integral Test.

26. Use the Integral Test to find an integer n so that the nth partial sum of the series $\sum_{k=1}^{\infty} 1/k^4$ approximates its sum to 10^{-8} accuracy.

27. Use the Integral Test to prove Theorem 6.9.

28. Adopting the notation in the statement of the Integral Test, prove that

$$\int_1^{n+1} f < \sum_{k=1}^{n} a_k < a_1 + \int_1^{n} f.$$

29. This exercise shows that computers offer little assistance in determining whether or not a series converges.

a) Use the previous exercise to determine a value of n for which $\sum_{k=1}^{n} 1/k > 100$.

b) Use the Integral Test to prove that the series $\sum_{k=2}^{\infty} 1/(k \ln k)$ diverges.

c) Use the previous exercise to determine as best you can a value of n for which $\sum_{k=2}^{\infty} 1/(k \ln k) > 5$.

d) Try to find a series similar to the series in part (b) that diverges even more slowly.

Remark. Another interesting test for the convergence of an infinite series is known as the Cauchy Condensation Test. The name for this test arises from the fact that the terms of the series are regrouped or condensed. The next few exercises consider the Cauchy Condensation Test and explore some of its uses.

Cauchy Condensation Test: Let $\{a_k\}$ be a decreasing sequence of positive numbers that converges to 0. Then the series $\sum_{k=1}^{\infty} a_k$ converges if and only if the series $\sum_{k=1}^{\infty} 2^k a_{2^k}$ converges.

30. Prove the Cauchy Condensation Test.

31. Use the Cauchy Condensation Test to prove Theorem 6.9.

32. Find all values of p for which the series $\sum_{k=2}^{\infty} 1/(k(\ln k)^p)$ converges.

33. Give an example to show that the condition that $\{a_k\}$ is decreasing is necessary in the Cauchy Condensation Test.

34. Let $\{a_k\}$ be a decreasing sequence of positive numbers that converges to 0 and suppose that $\sum_{k=1}^{\infty} a_k$ converges. Prove that $\lim_{k\to\infty} ka_k = 0$. Give an example to show that this result may be false for an arbitrary convergent series of positive terms.

7

Sequences and Series of Functions

One of the advantages of a geometric series is that there is a simple formula for the sum of the series. If $|r| < 1$ and a is a real number, then

$$\sum_{k=0}^{\infty} ar^k = \frac{a}{1-r}.$$

Letting $a = 1$ and $r = x$ leads to the equation

$$\frac{1}{1-x} = \sum_{k=0}^{\infty} x^k = 1 + x + x^2 + x^3 + x^4 + x^5 + \cdots,$$

which is valid for $|x| < 1$. The function $(1-x)^{-1}$ has been expressed as an infinite degree polynomial for all x in the interval $(-1, 1)$. Since the function $(1-x)^{-1}$ does not behave at all like a polynomial, this is a rather surprising result. Is this an isolated incident, perhaps a unique feature of the geometric series, or is it possible to represent other functions in this way? Naive symbolic manipulation of the equation for a geometric series leads to other interesting results: differentiating and integrating both sides yields the equations

$$\frac{1}{(1-x)^2} = 1 + 2x + 3x^2 + 4x^3 + 5x^4 + 6x^5 + \cdots;$$

$$-\ln|1-x| = x + \frac{1}{2}x^2 + \frac{1}{3}x^3 + \frac{1}{4}x^4 + \frac{1}{5}x^5 + \frac{1}{6}x^6 + \cdots.$$

As another example to indicate the extent to which the formula for the sum of a geometric series can be manipulated, substitute $-x^2$ for x in the equation for a geometric series, then integrate both sides to obtain the equations

$$\frac{1}{1+x^2} = 1 - x^2 + x^4 - x^6 + x^8 - x^{10} + \cdots;$$

$$\arctan x = x - \frac{1}{3}x^3 + \frac{1}{5}x^5 - \frac{1}{7}x^7 + \frac{1}{9}x^9 - \frac{1}{11}x^{11} + \cdots.$$

Are the last four equations valid? Is it possible to represent every function as an infinite degree polynomial? Can the operations of integration and differentiation be performed on infinite degree polynomials in the simple way illustrated by these examples?

To answer such questions, it is necessary to first introduce the concept of an infinite series of functions. As with series of real numbers, an infinite series of functions can be considered as a sequence of functions. The deeper questions lie in the behavior of sequences and series of functions with respect to continuity, integration, and differentiation. One of the primary goals of this chapter is to determine when it is permissible to integrate and differentiate infinite series of functions in the "obvious" way—that is, term by term just like finite sums of functions. We will also consider the problem of expressing functions as polynomials with an infinite number of terms.

Since sequences and series of functions are more abstract than sequences and series of real numbers, a cautionary note is appropriate. In this chapter (and in analysis in general), it is necessary to learn to think of a function as a single entity. For example, the function f defined by $f(x) = x^2$ is the squaring function, and there is a distinction (sometimes a subtle distinction) between the function f and the formula x^2. In general, the symbol f represents the function and the symbol $f(x)$ represents the value of the function f at x. Since one symbol represents a function and the other a real number, the symbols f and $f(x)$ are not interchangeable. In order to define a function f, it must be shown how to compute $f(x)$ for each x in the domain of f. To make progress in this chapter, it is important to understand this distinction between f and $f(x)$.

7.1 Pointwise Convergence

As with real numbers, it is easiest to begin with sequences of functions before considering series of functions. According to Definition 2.1, a sequence is a function whose domain is the set of positive integers. To obtain a sequence of functions, there must be a function associated with each positive integer. A sequence of functions is normally written as $\{f_n\}$, where f_n is a function for each positive integer n. As an example of a sequence of functions, for each positive integer n, let f_n be the function defined by $f_n(x) = x^n$ for all real numbers x. Another example is the sequence $\{g_n\}$, where g_n is the function defined by $g_n(x) = \sin(nx)$. As in these two examples, the integer n usually appears as a parameter in the functions that appear in the sequence.

In most of the situations involving a sequence $\{f_n\}$, the functions f_n will have a common domain. Let I be an interval and let f_n be a function defined on I for each positive integer n. For each $x \in I$, the sequence $\{f_n(x)\}$ is a sequence of real numbers. If this sequence converges for some $a \in I$ (that is, if the sequence $\{f_n(a)\}$ of real numbers converges), then the sequence $\{f_n\}$ is said to converge at a. In most of the cases that are relevant to this discussion, the sequence $\{f_n\}$ will converge at each point of the interval I. This type of convergence is known as pointwise convergence.

DEFINITION 7.1 Let $\{f_n\}$ be a sequence of functions defined on an interval I and let f be a function defined on I. The sequence $\{f_n\}$ **converges pointwise** to f on I if the sequence $\{f_n(x)\}$ converges to $f(x)$ for each $x \in I$. In other words, $f(x) = \lim_{n\to\infty} f_n(x)$ for each $x \in I$.

It should be clear that the limit function f is unique—it is not possible for a sequence of functions to converge pointwise to different functions. In many cases, the determination of the limit function is not difficult; a number of examples are presented below. It is recommended that the reader sketch the graphs of several functions in each of the following sequences.

1. For each positive integer n, define $f_n\colon [0, 1] \to \mathbb{R}$ by $f_n(x) = x^n$. If $0 \le x < 1$, then the sequence $\{x^n\}$ converges to 0. If $x = 1$, then the sequence $\{x^n\}$ converges to 1. Therefore, the sequence $\{f_n\}$ converges pointwise on $[0, 1]$ to the function f defined by

$$f(x) = \begin{cases} 0, & \text{if } 0 \le x < 1; \\ 1, & \text{if } x = 1. \end{cases}$$

It is interesting to note that each of the functions f_n is continuous on $[0, 1]$, but the limit function f is not continuous on $[0, 1]$.

2. For each positive integer n, define $g_n\colon \mathbb{R} \to \mathbb{R}$ by $g_n(x) = x/n$. If x is any real number, then the sequence $\{x/n\}$ converges to 0. Therefore, the sequence $\{g_n\}$ converges pointwise on $\mathbb{R}$ to the function g defined by $g(x) = 0$ for all $x \in \mathbb{R}$.

3. For each positive integer n, define $h_n\colon (-1, \infty) \to \mathbb{R}$ by

$$h_n(x) = \frac{x^{2n+1}}{1 + x^{2n+1}}.$$

It is easy to verify that the sequence $\{h_n\}$ converges pointwise on $(-1, \infty)$ to the function h defined by

$$h(x) = \begin{cases} 0, & \text{if } -1 < x < 1; \\ 0.5, & \text{if } x = 1; \\ 1, & \text{if } x > 1. \end{cases}$$

Note that each of the functions h_n is unbounded on $(-1, \infty)$, but the limit function h is bounded on $(-1, \infty)$.

4. For each positive integer n, define $\phi_n\colon [0, 1] \to \mathbb{R}$ by

$$\phi_n(x) = \begin{cases} 0, & \text{if } 0 \le x < 1/n; \\ \sin(\pi/x), & \text{if } 1/n \le x \le 1. \end{cases}$$

The sequence $\{\phi_n\}$ converges pointwise on $[0, 1]$ to the function ϕ defined by

$$\phi(x) = \begin{cases} \sin(\pi/x), & \text{if } x > 0; \\ 0, & \text{if } x = 0. \end{cases}$$

To prove this, suppose that $0 < x \leq 1$. Then there exists a positive integer q such that $1/q < x$. For each $n \geq q$, we find that $\phi_n(x) = \sin(\pi/x)$, so $\{\phi_n(x)\}$ converges to $\sin(\pi/x)$. The convergence at 0 is obvious. In this example, each of the functions ϕ_n is continuous on $[0, 1]$, but $\lim_{x \to 0^+} \phi(x)$ does not exist.

5. For each positive integer n, define $F_n\colon [0, 1] \to \mathbb{R}$ by

$$F_n(x) = \begin{cases} 0, & \text{if } x = 0; \\ n, & \text{if } 0 < x < 1/n; \\ 0, & \text{if } 1/n \leq x \leq 1. \end{cases}$$

As in the previous example, the sequence $\{F_n\}$ converges pointwise on $[0, 1]$ to the function F defined by $F(x) = 0$ for all $x \in [0, 1]$. Each of the functions F_n is Riemann integrable on $[0, 1]$ and the limit function F is Riemann integrable on $[0, 1]$. However,

$$\int_0^1 F = 0 \neq 1 = \lim_{n \to \infty} \int_0^1 F_n.$$

In this case, the limit function is Riemann integrable, but the value of the integral is not the expected value.

6. Let $\{r_n\}$ be a sequence consisting of all the rational numbers in $[0, 1]$ with each rational number listed exactly once. Define functions G_n for each positive integer n, and a function G by

$$G_n(x) = \begin{cases} 1, & \text{if } x = r_1, r_2, \ldots, r_n; \\ 0, & \text{otherwise}; \end{cases} \quad \text{and} \quad G(x) = \begin{cases} 1, & \text{if } x \in \mathbb{Q}; \\ 0, & \text{if } x \notin \mathbb{Q}. \end{cases}$$

The sequence $\{G_n\}$ converges pointwise on $[0, 1]$ to the function G. Since each G_n has a finite number of discontinuities in the interval $[0, 1]$, each G_n is Riemann integrable on $[0, 1]$. However, the limit function G is a bounded function that is not Riemann integrable on $[0, 1]$.

7. For each positive integer n, define $H_n\colon [-2, 2] \to \mathbb{R}$ by

$$H_n(x) = \begin{cases} |x|, & \text{if } |x| \geq 1/n; \\ 1.5nx^2 - 0.5n^3x^4, & \text{if } |x| < 1/n. \end{cases}$$

A proof that each H_n is differentiable on $[-2, 2]$ will be left as an exercise. However, the limit function $H(x) = |x|$ is not differentiable at 0.

Suppose that the sequence $\{f_n\}$ converges pointwise to the function f on an interval I and that all of the functions f_n have some property on I. An important question in many applications is whether or not the limit function f inherits this property. The following questions are representative of this situation.

1) If each f_n is bounded on I, is f bounded on I?

2) If each f_n is continuous on I, is f continuous on I?

3) If each f_n is Riemann integrable on I, is f Riemann integrable on I?

4) If each f_n is differentiable on I, is f differentiable on I?

The preceding examples indicate that the limit function may not inherit the properties of continuity, integrability, or differentiability. An example to illustrate that the pointwise limit of a sequence of bounded functions may be unbounded is requested in the exercises. It appears that pointwise convergence does not preserve any useful properties of functions. Since it is often necessary that the limit function inherit some properties, we will introduce a type of convergence that is stronger than pointwise convergence in the next section.

As discussed in the introduction to this chapter, we will be interested in series of functions as well as sequences of functions. An **infinite series of functions** is an expression of the form $\sum_{k=1}^{\infty} f_k$, where $\{f_k\}$ is a sequence of functions defined on an interval I. (It will sometimes be useful to let the first value of k be 0 or some other integer different from 1.) As with infinite series of real numbers, this expression is defined in terms of a sequence by letting s_n be the function defined by $s_n = \sum_{k=1}^{n} f_k$ for each positive integer n. If the sequence $\{s_n\}$ converges for some value of $x \in I$, then the series converges for that value of x, and if the sequence $\{s_n\}$ converges pointwise on I, then the series converges pointwise on I. This discussion is formalized in the following definition.

DEFINITION 7.2 Let $\{f_k\}$ be a sequence of functions defined on an interval I and let f be a function defined on I. The series $\sum_{k=1}^{\infty} f_k$ **converges pointwise** to f on I if the corresponding sequence $\{s_n\} = \left\{\sum_{k=1}^{n} f_k\right\}$ of partial sums converges pointwise to f on I.

Since a series of functions is essentially a sequence of functions, the questions posed earlier concerning the properties of the limit of a sequence of functions are also relevant for the sum of a series of functions. Because of the relationship between sequences and series, every result that is valid for a sequence of functions can be translated into a result that is valid for series of functions. As with series of real numbers, it can be more difficult to find the limit of a series of functions than it is to find the limit of a sequence of functions. The problem lies in the fact that it may not be possible to find a useful representation for the sequence of partial sums. As one example of a series of functions, consider the series

$$\sum_{k=1}^{\infty} \frac{(x-1)^k}{5^{k+1}}.$$

This series can be represented as $\sum_{k=1}^{\infty} f_k$, where for each k the function f_k is defined by $f_k(x) = (x-1)^k/5^{k+1}$. Rather than find the sequence of partial sums in this case, we can determine the limit function by recognizing that the series is a geometric series with $r = (x-1)/5$. The series converges for those values of x that satisfy $|x-1|/5 < 1$. In other words, the series $\sum_{k=1}^{\infty} f_k$ converges on the interval $(-4, 6)$. Since

$$\sum_{k=1}^{\infty} \frac{(x-1)^k}{5^{k+1}} = \frac{(x-1)/25}{1-\big((x-1)/5\big)} = \frac{x-1}{30-5x},$$

the series $\sum_{k=1}^{\infty} f_k$ converges pointwise on $(-4, 6)$ to the function f defined by $f(x) = (x-1)/(30-5x)$.

Exercises

1. Show that each sequence converges pointwise on the given interval and find its limit function.

a) Define $f_n\colon [0,\infty) \to \mathbb{R}$ by $f_n(x) = \dfrac{x+1}{2+nx}$.

b) Define $g_n\colon (-1,1) \to \mathbb{R}$ by $g_n(x) = \arctan(nx)$.

c) Define $h_n\colon [0,1] \to \mathbb{R}$ by $h_n(x) = (1+n^2x^2)^{-1}$.

2. For each positive integer n, define $g_n\colon [0,1] \to \mathbb{R}$ by

$$g_n(x) = \begin{cases} 1/x, & \text{if } x \geq 1/n; \\ n^2x, & \text{if } x < 1/n. \end{cases}$$

Show that this sequence converges pointwise on the interval $[0,1]$ and find its limit function. Sketch the graphs of g_2 andg_5.

3. For each positive integer n, define $h_n\colon \mathbb{R} \to \mathbb{R}$ by $h_n(x) = ne^{x/n} - n$. Show that this sequence converges pointwise on $\mathbb{R}$ and find its limit function.

4. Consider the function f defined by $f(x) = x^2$. Find at least four sequences of functions that converge pointwise to f on $[0,1]$. Try to find sequences that are qualitatively different, not just four sequences based on one idea.

5. Let $f\colon [a,b] \to \mathbb{R}$ be differentiable on $[a,b]$. Use the definition of a derivative to construct a sequence of continuous functions that converges pointwise to f' on $[a,b]$. Recall that f' may not be a continuous function.

6. Let I be an interval. Suppose that $\{f_n\}$ converges pointwise to f on I and that $\{g_n\}$ converges pointwise to g on I.

a) Prove that $\{f_n + g_n\}$ converges pointwise to $f+g$ on I.

b) Prove that $\{f_n g_n\}$ converges pointwise to fg on I.

7. Let $\{f_n\}$ be a sequence of functions defined on an interval I. Suppose that $\{f_n\}$ converges pointwise to f on I and that each of the functions f_n is bounded by the same number M. Prove that f is bounded by M.

8. Give an example of a sequence of bounded functions on $[0,1]$ that converges pointwise on $[0,1]$ to an unbounded function.

9. Give an example of a sequence $\{f_n\}$ of continuous functions defined on $[0,1]$ such that $\{f_n\}$ converges pointwise to the zero function on $[0,1]$, but the sequence $\left\{\int_0^1 f_n\right\}$ is unbounded.

10. Fill in the details behind the properties of the sequence $\{H_n\}$ in Example (7).

11. For each positive integer n, define $g_n\colon [0, 1] \to \mathbb{R}$ by $g_n(x) = \sin(nx)/\sqrt{n}$. Show that this sequence converges pointwise on $[0, 1]$ to a differentiable function g, but that $\{g_n'(0)\}$ does not converge to $g'(0)$.

12. State and prove the analogue of part (a) of Exercise 6 for series of functions.

13. Show that the series $\sum_{k=1}^{\infty} \sin^{2k} x$ converges pointwise on the interval $(-\pi/2, \pi/2)$ and find the limit function.

14. Let $\{a_k\}$ be a sequence of real numbers and consider the series of functions $\sum_{k=1}^{\infty} a_k x^k$.

a) Suppose that this series converges for some $x_1 \neq 0$. Show that the series converges for all x that satisfy $|x| < |x_1|$.

b) Suppose that this series diverges for some $x_2 \neq 0$. Show that the series diverges for all x that satisfy $|x| > |x_2|$.

15. For each of the following sequences $\{f_k\}$, the series $\sum_{k=1}^{\infty} f_k$ converges pointwise on an interval I. Find the largest interval I with this property and find the limit function f.

a) $f_k(x) = \left(\dfrac{x}{3}\right)^k$ **b)** $f_k(x) = x^k(1 - x)$

c) $f_k(x) = \dfrac{x^2}{(1 + x^2)^k}$ **d)** $f_k(x) = \dfrac{x}{(kx + 1)(kx + x + 1)}$

16. For what values of x does the series $\displaystyle\sum_{k=1}^{\infty} \frac{1}{k + k^2 x}$ converge?

7.2 Uniform Convergence

As indicated in the last section, pointwise convergence does not guarantee that the limit function possesses the same properties as the functions in the sequence. Fortunately, there is a stronger type of convergence that is sufficient in almost every case to ensure that the limit function f inherits the properties of the functions f_n. This type of convergence is known as uniform convergence and is the topic for this section.

To motivate the definition of this stronger form of convergence, we will look more closely at the issue of continuity. Suppose that each f_n is continuous on an interval I and that the sequence $\{f_n\}$ converges pointwise to f on I. We would like to prove that f is continuous on I or, to say the same thing, that f is continuous at each point c in I. The continuity of f at c should somehow be related to the continuity of the f_n's at c. Some thoughtful scribbling leads to the inequality

$$|f(x) - f(c)| \leq |f(x) - f_n(x)| + |f_n(x) - f_n(c)| + |f_n(c) - f(c)|.$$

The first and last terms on the right can be made small if n is chosen large enough; this follows from the pointwise convergence of the sequence $\{f_n\}$ to f. Since f_n is continuous at c, the middle term can be made small for an appropriate choice of δ. Hence, the continuity of f at c follows from the continuity of the f_n's at c. What is the problem with this proof? Since counterexamples have already been discussed (see the previous section), there must be an error. The problem—it is a big problem,

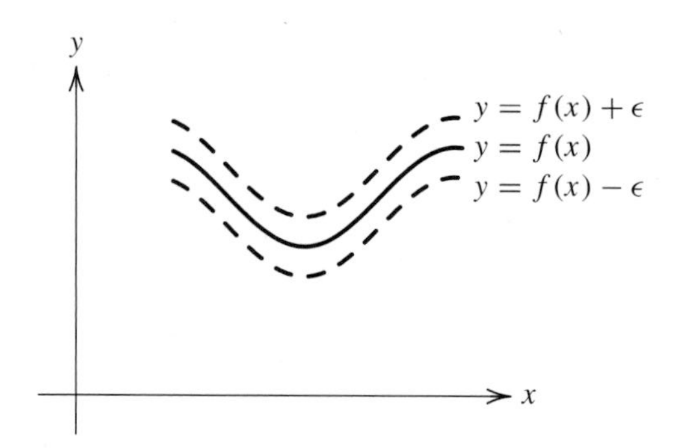

Figure 7.1 The graph of f_n must lie between the dashed lines

but also a subtle one—lies in the fact that there are many different values of x to contend with. The term $|f(x) - f_n(x)|$ must be small for all values of x near c. The problem is that the choice of n may vary with the value of x and as n varies, so does the value of δ needed for the continuity of f_n. We need to know that $|f(x) - f_n(x)|$ is small for all x using the same value of n, then choose $\delta > 0$ from the definition of continuity for that particular function f_n. The key to the concept of uniform convergence is making the quantity $|f(x) - f_n(x)|$ small for all x using the same value of n.

Suppose that $\{f_n\}$ converges pointwise to f on I. This means that for each $x \in I$, the sequence $\{f_n(x)\}$ converges to $f(x)$. Let $x \in I$ and let $\epsilon > 0$. Then there exists a positive integer $N(\epsilon, x)$ such that $|f_n(x) - f(x)| < \epsilon$ for all $n \geq N(\epsilon, x)$. The integer $N(\epsilon, x)$, as indicated, generally depends on both ϵ and x. For a fixed ϵ, the integer N will depend on x. If for a fixed but arbitrary $\epsilon > 0$, the integer N can be chosen independently of x on an interval I (so that the same N works for all x), then the sequence $\{f_n\}$ is said to converge uniformly to f on I. (This discussion should be compared with the discussion preceding the definition of uniform continuity.)

DEFINITION 7.3 Let $\{f_n\}$ be a sequence of functions defined on an interval I and let f be a function defined on I. The sequence $\{f_n\}$ **converges uniformly** to f on I if for each $\epsilon > 0$ there exists a positive integer N such that $|f_n(x) - f(x)| < \epsilon$ for all $x \in I$ and for all $n \geq N$. A sequence $\{f_n\}$ is said to converge uniformly on I if there exists a function f to which $\{f_n\}$ converges uniformly on I.

Although not explicitly stated in the definition, it should be clear that $\{f_n\}$ converges pointwise to f on I whenever $\{f_n\}$ converges uniformly to f on I. In other words, uniform convergence is pointwise convergence with an extra property: for each $\epsilon > 0$, the choice of N is independent of $x \in I$. It should also be clear that if $\{f_n\}$ converges uniformly on I, then it converges uniformly on every subinterval of I. The inequality $|f_n(x) - f(x)| < \epsilon$ for all $x \in I$ can be written as

$$f(x) - \epsilon < f_n(x) < f(x) + \epsilon$$

for all $x \in I$. Thus, for large enough n, the graph of f_n lies within a narrow tube around the graph of f; see Figure 7.1.

Consider the sequence $\{f_n\}$, where $f_n(x) = \sin(nx)/\sqrt{n}$ for each n. This sequence converges uniformly on $\mathbb{R}$ to the function f defined by $f(x) = 0$ for all $x \in \mathbb{R}$. To see this, let $\epsilon > 0$ and choose a positive integer N such that $N > 1/\epsilon^2$. For each $n \geq N$ and each $x \in \mathbb{R}$,

$$|f_n(x) - f(x)| = \frac{|\sin(nx)|}{\sqrt{n}} \leq \frac{1}{\sqrt{n}} \leq \frac{1}{\sqrt{N}} < \epsilon.$$

Therefore, the sequence $\{f_n\}$ converges uniformly to f on $\mathbb{R}$.

To prove that a sequence of functions does not converge uniformly on I, we must negate the definition of uniform convergence. Since there are a number of quantifiers, the negation of this definition requires some care (see Appendix A). Suppose that $\{f_n\}$ converges pointwise to f on I. (If the sequence does not converge pointwise on I, then it certainly does not converge uniformly on I.) The sequence $\{f_n\}$ does not converge uniformly to f on I if there exists an $\epsilon > 0$ such that for every positive integer N there is a point $x \in I$ and an integer $n \geq N$ such that $|f_n(x) - f(x)| \geq \epsilon$. Read this statement carefully to make certain that you understand exactly what it says.

As an example, we will show that the sequence $\{x^n\}$ does not converge uniformly to the zero function on the interval $(0, 1)$. As x gets closer to 1, the sequence $\{x_n\}$ "takes longer" to get close to 0. Let $\epsilon = 0.5$ and let N be any positive integer. The number $x = \sqrt[N]{0.5}$ belongs to the interval $(0, 1)$ (note that the value for x approaches 1 as N increases) and $|f_N(x)| = 0.5$. This is sufficient to show that the convergence to the zero function is not uniform on $(0, 1)$.

As you can probably imagine, working with the definition of uniform convergence can be difficult. It is therefore helpful to have some methods for determining uniform convergence other than appealing directly to the definition. The next theorem provides one such method. The proof, a simple consequence of the definitions, will be left as an exercise.

THEOREM 7.4 Suppose that the sequence $\{f_n\}$ converges pointwise to f on an interval I and let $M_n = \sup\{|f_n(x) - f(x)| : x \in I\}$ for each n. Then $\{f_n\}$ converges uniformly to f on I if and only if $\{M_n\}$ converges to 0. ■

To illustrate the use of this theorem, consider again the sequence $\{x^n\}$, which converges pointwise to the zero function on $(0, 1)$. Since

$$M_n = \sup\{|x^n - 0| : x \in (0, 1)\} = \sup\{x^n : x \in (0, 1)\} = 1,$$

the sequence $\{x^n\}$ does not converge uniformly to the zero function on the interval $(0, 1)$. Conceptually, this is much easier than negating the definition of uniform convergence.

As a second example, for each positive integer n, define $f_n\colon [0, 1] \to \mathbb{R}$ by

$$f_n(x) = \frac{x}{1 + nx^2}.$$

It is easy to verify that this sequence converges pointwise to f, where $f(x) = 0$ for each $x \in [0, 1]$. Since $f(x) = 0$ for all $x \in [0, 1]$ and since each f_n is nonnegative, the number M_n from Theorem 7.4 is the maximum value of the function f_n on the interval [0, 1]. To find M_n, compute f_n' and find the critical points of f_n. Since

$$f_n'(x) = \frac{1 - nx^2}{(1 + nx^2)^2},$$

the only relevant critical point is $1/\sqrt{n}$. Evaluating f_n at the critical point and the endpoints yields

$$f_n(0) = 0, \quad f_n(1/\sqrt{n}) = \frac{1}{2\sqrt{n}}, \quad \text{and} \quad f_n(1) = \frac{1}{n+1}.$$

Hence, the maximum value of f_n on [0, 1] is $1/(2\sqrt{n})$. Since $\{1/(2\sqrt{n})\}$ converges to 0, the sequence $\{f_n\}$ converges uniformly to f on [0, 1] by Theorem 7.4.

As with sequences of real numbers, there is also a Cauchy criterion for uniform convergence. It has the advantage of being useful when the limit function cannot be found explicitly.

THEOREM 7.5 Cauchy Criterion for Uniform Convergence A sequence $\{f_n\}$ of functions defined on an interval I converges uniformly on I if and only if for each $\epsilon > 0$ there exists a positive integer N such that $|f_n(x) - f_m(x)| < \epsilon$ for all $x \in I$ and for all $m, n \geq N$.

Proof. Suppose that for each $\epsilon > 0$ there exists a positive integer N such that $|f_n(x) - f_m(x)| < \epsilon$ for all $x \in I$ and for all $m, n \geq N$. The first step is to find a function f such that $\{f_n\}$ converges pointwise to f on I. By hypothesis, the sequence $\{f_n(x)\}$ is a Cauchy sequence of real numbers for each $x \in I$. Define a function f on I by $f(x) = \lim_{n\to\infty} f_n(x)$. To show that $\{f_n\}$ converges uniformly to f on I, let $\epsilon > 0$ and choose a positive integer N such that $|f_n(x) - f_m(x)| < \epsilon$ for all $x \in I$ and for all $m, n \geq N$. Fix $n \geq N$ and $x \in I$. For each $m \geq N$, the inequality $|f_n(x) - f_m(x)| < \epsilon$ is valid, and the sequence $\{|f_n(x) - f_m(x)|\}_{m=N}^{\infty}$ converges to $|f_n(x) - f(x)|$. It follows that $|f_n(x) - f(x)| \leq \epsilon$. Since $n \geq N$ and $x \in I$ were arbitrary, the sequence $\{f_n\}$ converges uniformly to f on I.

The proof of the converse will be left as an exercise. ∎

Given the relationship between sequences and series, the definition of uniform convergence for a series of functions should come as no surprise. After the definition, a convenient test for uniform convergence of series is stated and proved.

DEFINITION 7.6 Let $\{f_k\}$ be a sequence of functions defined on an interval I. The series $\sum_{k=1}^{\infty} f_k$ **converges uniformly** on I if its corresponding sequence of partial sums converges uniformly on I.

THEOREM 7.7 Weierstrass M-test Let $\{f_k\}$ be a sequence of functions defined on an interval I and suppose that for each positive integer k, there exists a positive number M_k such that $|f_k(x)| \leq M_k$ for all $x \in I$. If the series $\sum_{k=1}^{\infty} M_k$ converges, then the series $\sum_{k=1}^{\infty} f_k$ converges uniformly on I.

Proof. Let $\{s_n\}$ be the sequence of partial sums for the series of functions and let $\epsilon > 0$. Since $\sum_{k=1}^{\infty} M_k$ converges, there exists a positive integer N such that $\sum_{k=m+1}^{n} M_k < \epsilon$ for all $n > m \geq N$ (see Theorem 6.3). It follows that

$$|s_n(x) - s_m(x)| = \left| \sum_{k=m+1}^{n} f_k(x) \right| \leq \sum_{k=m+1}^{n} |f_k(x)| \leq \sum_{k=m+1}^{n} M_k < \epsilon$$

for each $x \in I$ and for all $n > m \geq N$. Hence, the sequence $\{s_n\}$ converges uniformly on I by the Cauchy criterion. This completes the proof. ■

As a simple illustration of the Weierstrass M-test, let g be a bounded function defined on $\mathbb{R}$ and consider the series $\sum_{k=1}^{\infty} g(kx)/k^2$. If M is a bound for g on $\mathbb{R}$, then for each positive integer k, the terms of this series satisfy

$$\left| \frac{g(kx)}{k^2} \right| \leq \frac{M}{k^2}$$

for all $x \in \mathbb{R}$. Since the series $\sum_{k=1}^{\infty} M/k^2$ converges, the series $\sum_{k=1}^{\infty} g(kx)/k^2$ converges uniformly on $\mathbb{R}$ by the Weierstrass M-test.

Exercises

1. For each real number x, the sequence $\{x/n\}$ converges to 0. Given $\epsilon > 0$ and $x \in \mathbb{R}$, find a positive integer $N(\epsilon, x)$ such that $|x/n| < \epsilon$ for all $n \geq N(\epsilon, x)$.
2. For each $x \in (0, 1)$, the sequence $\{x^n\}$ converges to 0. Given $\epsilon > 0$ and $x \in (0, 1)$, find a positive integer $N(\epsilon, x)$ such that $|x^n| < \epsilon$ for all $n \geq N(\epsilon, x)$.
3. Use the definition of uniform convergence to prove each of the following.
   a) The sequence in Example (5) from Section 7.1 does not converge uniformly on the interval [0, 1].
   b) The sequence in Example (6) from Section 7.1 does not converge uniformly on the interval [0, 1].
4. Let I be an interval. Suppose that $\{f_n\}$ converges uniformly to f on I and that $\{g_n\}$ converges uniformly to g on I.
   a) Prove that $\{f_n + g_n\}$ converges uniformly to $f + g$ on I.
   b) Give an example to show that $\{f_n g_n\}$ may not converge uniformly to fg on I.
   c) Suppose that f and g are bounded on I. Prove that $\{f_n g_n\}$ converges uniformly to fg on I.
5. Prove Theorem 7.4.

6. Use Theorem 7.4 to prove that the sequence in Example (4) from Section 7.1 does not converge uniformly on $[0, 1]$.

7. Determine whether or not the given sequence converges uniformly on $[0, 1]$.

a) The sequence $\{f_n\}$, where $f_n(x) = nx(1 - x^2)^n$.

b) The sequence $\{g_n\}$, where $g_n(x) = x(1 - x)^n$.

c) The sequence $\{h_n\}$, where $h_n(x) = nx(1 - x)^n$.

8. Does the sequence in Example (7) from Section 7.1 converge uniformly on $[0, 1]$?

9. Finish the proof of Theorem 7.5.

10. Let g be a bounded function defined on $\mathbb{R}$ and consider the series $\sum_{k=1}^{\infty} g(k^2x)/2^k$. Prove that this series converges uniformly on $\mathbb{R}$.

11. Prove that the series $\sum_{k=1}^{\infty} \sin^{2k} x$ converges uniformly on the interval $[0, \pi/3]$.

12. Show that the series $\sum_{k=1}^{\infty} x^k$ does not converge uniformly on the interval $(0, 1)$.

13. For which intervals I does the series $\sum_{k=1}^{\infty} (x/3)^k$ converge uniformly on I?

14. Prove that the series $\displaystyle\sum_{k=1}^{\infty} (-1)^{k+1} \frac{x + \sqrt{k}}{k}$ converges uniformly on $[-1, 1]$ but does not converge absolutely for any value of x in this interval.

15. Consider the series $\displaystyle\sum_{k=1}^{\infty} \frac{1}{1 + k^2 x}$. Prove that this series

a) converges for each $x > 0$;

b) converges uniformly on the interval $[1, 2]$;

c) does not converge uniformly on the interval $(0, 1)$.

7.3 Uniform Convergence and Inherited Properties

Uniform convergence is sufficient in most cases to guarantee that the limit function f inherits the properties of the functions in the sequence $\{f_n\}$. In this section, we will consider the properties of continuity, Riemann integrability, and differentiability. Since the property of continuity was used to motivate the definition of uniform convergence (see the previous section), this property will be considered first.

Suppose that $\{f_n\}$ converges pointwise to f on an interval I and that each f_n is continuous at $c \in I$. In order for f to be continuous at c, we must have $\lim_{x \to c} f(x) = f(c)$. In this context, the last equality can be rewritten as

$$\lim_{x \to c} \lim_{n \to \infty} f_n(x) = \lim_{n \to \infty} \lim_{x \to c} f_n(x).$$

The continuity of f at c is dependent upon the validity of interchanging the order of two limit operations. In general, limit operations cannot be interchanged. A simple example involving sequences is

$$\lim_{n\to\infty}\lim_{m\to\infty}\frac{n}{m+n}=0\neq 1=\lim_{m\to\infty}\lim_{n\to\infty}\frac{n}{m+n}.$$

As the results in this section will show, uniform convergence guarantees that two limits can be interchanged without affecting the value of the limit.

THEOREM 7.8 Suppose that the sequence $\{f_n\}$ converges pointwise to a function f on an interval I and that each f_n is continuous at $c \in I$. If $\{f_n\}$ converges uniformly to f on I, then f is continuous at c.

Proof. Let $\epsilon > 0$. Since $\{f_n\}$ converges uniformly to f on I, there exists a positive integer p such that $|f_p(x) - f(x)| < \epsilon$ for all $x \in I$. Since the function f_p is continuous at c, there exists $\delta > 0$ such that $|f_p(x) - f_p(c)| < \epsilon$ for all $x \in I$ that satisfy $|x - c| < \delta$. For these same values of x,

$$\begin{aligned}|f(x) - f(c)| &\leq |f(x) - f_p(x)| + |f_p(x) - f_p(c)| + |f_p(c) - f(c)|\\ &< \epsilon + \epsilon + \epsilon = 3\epsilon.\end{aligned}$$

Hence, the function f is continuous at c. ■

COROLLARY 7.9 Suppose that the sequence $\{f_n\}$ converges pointwise to a function f on an interval I and that each f_n is continuous on I. If $\{f_n\}$ converges uniformly to f on I, then f is continuous on I. ■

Corollary 7.9 provides a quick way of determining that a sequence does not converge uniformly. If each of the functions f_n is continuous on I and the limit function f is not continuous on I, then the convergence is not uniform on I. For example, the sequences $\{h_n\}$ and $\{\phi_n\}$ defined in Examples (3) and (4) in Section 7.1 do not converge uniformly on their given intervals.

Uniform convergence is not a necessary condition for the limit function to be continuous; that is, it is possible for the limit function to be continuous even if the convergence is not uniform. For a simple example, the sequence $\{x/n\}$ converges pointwise but not uniformly on $\mathbb{R}$ to the zero function. It is even possible for a limit function to be continuous at a point where none of the functions in the sequence are continuous. The sequence defined in Example (5) in Section 7.1 illustrates this situation.

Although the converse to Corollary 7.9 is false, a partial converse to this result was discovered by Ulisse Dini (1845–1918). It requires that the interval be closed and bounded and that the sequence of functions be increasing. A sequence $\{f_n\}$ is increasing on $[a, b]$ if $f_n(x) \leq f_{n+1}(x)$ for all n and for all $x \in [a, b]$. The proof of Dini's Theorem will be left as an exercise.

THEOREM 7.10 Dini's Theorem Let $\{f_n\}$ be an increasing sequence of continuous functions that converges pointwise to a function f on an interval $[a, b]$. If f is continuous on $[a, b]$, then $\{f_n\}$ converges uniformly to f on $[a, b]$. ■

Uniform convergence is also sufficient to guarantee that the limit of a sequence of Riemann integrable functions is Riemann integrable. In this case, there is a further matter to check. Suppose that each f_n is Riemann integrable on $[a, b]$, that $\{f_n\}$ converges pointwise to f on $[a, b]$, and that f is Riemann integrable on $[a, b]$. Does it follow that the equation

$$\int_a^b f = \lim_{n\to\infty} \int_a^b f_n$$

is valid? A counterexample was given in Example (5) of Section 7.1. This equation can be restated as

$$\lim_{\|{}^tP\|\to 0} \lim_{n\to\infty} S(f_n, {}^tP) = \lim_{n\to\infty} \lim_{\|{}^tP\|\to 0} S(f_n, {}^tP),$$

which shows that the question again reduces to the interchange of limit operations.

THEOREM 7.11 Let $\{f_n\}$ be a sequence of Riemann integrable functions defined on $[a, b]$ and let f be a function defined on $[a, b]$. If $\{f_n\}$ converges uniformly to f on $[a, b]$, then f is Riemann integrable on $[a, b]$ and $\int_a^b f = \lim_{n\to\infty} \int_a^b f_n$.

Proof. The first step in the proof is to show that the sequence $\{\int_a^b f_n\}$ converges. Let $\eta > 0$. Since $\{f_n\}$ converges uniformly to f on $[a, b]$, there exists a positive integer K such that $|f_n(x) - f_m(x)| < \eta$ for all $x \in [a, b]$ and for all $m, n \geq K$. It follows that

$$\left|\int_a^b f_n - \int_a^b f_m\right| = \left|\int_a^b (f_n - f_m)\right| \leq \int_a^b |f_n - f_m| \leq \eta\,(b - a)$$

for all $m, n \geq K$. Hence, the sequence $\{\int_a^b f_n\}$ is Cauchy. Let L be the limit of this sequence. We will prove that f is Riemann integrable on $[a, b]$ and that $\int_a^b f = L$.

Let $\epsilon > 0$. Since $\{f_n\}$ converges uniformly to f on $[a, b]$, there exists a positive integer N such that $|f_n(x) - f(x)| < \epsilon$ for all $x \in [a, b]$ and for all $n \geq N$. Since $\{\int_a^b f_n\}$ converges to L, there exists an integer $q \geq N$ such that $\left|\int_a^b f_q - L\right| < \epsilon$. Since f_q is Riemann integrable on $[a, b]$, there exists a positive number δ such that $|S(f_q, {}^tP) - \int_a^b f_q| < \epsilon$ whenever ${}^tP$ is a tagged partition of $[a, b]$ that satisfies $\|{}^tP\| < \delta$. Then $|S(f, {}^tP) - S(f_q, {}^tP)| < \epsilon(b - a)$ for any such tagged partition ${}^tP$ (see the exercises) and

$$\begin{aligned} |S(f, {}^tP) - L| &\leq |S(f, {}^tP) - S(f_q, {}^tP)| + \left|S(f_q, {}^tP) - \int_a^b f_q\right| + \left|\int_a^b f_q - L\right| \\ &< \epsilon\,(b - a) + \epsilon + \epsilon. \end{aligned}$$

Hence, the function f is Riemann integrable on $[a, b]$ and $\int_a^b f = \lim_{n\to\infty} \int_a^b f_n$. ■

The last inherited property that we will discuss is differentiability. As with the other two properties considered in this section, the key question involves the interchange of limit operations. Suppose that a sequence $\{f_n\}$ converges pointwise to f on an interval I, that each f_n is differentiable at $c \in I$, and that the sequence

$\{f_n'(c)\}$ converges. The best case scenario is that f is differentiable at c and that

$$f'(c) = \lim_{n\to\infty} f_n'(c).$$

In terms of limits, this becomes

$$\lim_{x\to c}\lim_{n\to\infty} \frac{f_n(x) - f_n(c)}{x - c} = \lim_{n\to\infty}\lim_{x\to c} \frac{f_n(x) - f_n(c)}{x - c}.$$

After studying the last two theorems, one might anticipate that this result is valid if $\{f_n\}$ converges uniformly to f on I. However, a counterexample to this statement can be found in Example (7) of Section 7.1. Therefore, the hypotheses needed to guarantee the desired result must be even stronger than uniform convergence. The result that will be most useful in this textbook involves continuously differentiable functions. A function f is said to be **continuously differentiable** on an interval I if it is differentiable on I and the function f' is continuous on I. The proof of the following theorem will be left as an exercise.

THEOREM 7.12 Let $\{f_n\}$ be a sequence of continuously differentiable functions defined on an interval I, let f and g be functions defined on I, and suppose that $\{f_n\}$ converges pointwise to f on I. If $\{f_n'\}$ converges uniformly to g on I, then f is differentiable on I and $f' = g$. ■

It should be pointed out that all of the results in this section can be improved in the sense that the hypotheses of each result can be weakened without changing the conclusion. The tradeoff, of course, is that the proof of a theorem with weaker hypotheses generally involves more work. The versions stated in this section are suitable for the needs of this book, but some further results on this topic are discussed in the exercises at the end of this section and in the supplementary exercises.

Since series of functions are sequences of functions expressed in a different form, the results in this section can be applied to series of functions as well. The formal statements for series are given in the following theorem; the proof will be left as an exercise.

THEOREM 7.13 Let $\{f_k\}$ be a sequence of functions defined on an interval I and suppose that $\sum_{k=1}^{\infty} f_k$ converges uniformly to a function f defined on I.

a) If each f_k is continuous on I, then f is continuous on I.

b) If each f_k is Riemann integrable on an interval $[a, b] \subseteq I$, then f is Riemann integrable on $[a, b]$ and $\int_a^b f = \sum_{k=1}^{\infty} \int_a^b f_k$.

c) If each f_k is continuously differentiable on I and the series $\sum_{k=1}^{\infty} f_k'$ converges uniformly on I, then f is differentiable on I and $f'(x) = \sum_{k=1}^{\infty} f_k'(x)$ for each $x \in I$. ■

Exercises

1. Suppose that the sequence $\{f_n\}$ converges uniformly to a function f on an interval I and that each of the functions f_n is bounded on I. Prove that f is bounded on I.

2. Suppose that the sequence $\{f_n\}$ converges uniformly to a function f on $[0, 1]$ and that the equation $f_n(x) = 0$ has exactly three solutions in $[0, 1]$ for each positive integer n. Give an example to show that the equation $f(x) = 0$ may not have exactly three solutions in $[0, 1]$.

3. Give an example of a sequence $\{f_n\}$ of functions defined on $[0, 1]$ such that each f_n is nowhere continuous but $\{f_n\}$ converges uniformly on $[0, 1]$ to a function f that is continuous on $[0, 1]$.

4. Show that the hypotheses of Theorem 7.8 can be weakened without affecting the proof very much by considering the following questions:

a) Does the convergence need to be uniform on the entire interval I?

b) Do all of the functions f_n need to be continuous at c?

c) Does considering subsequences make any difference?

5. A sequence $\{f_n\}$ of continuous functions defined on an interval I is said to be **equicontinuous** on I if for each $\epsilon > 0$ there exists $\delta > 0$ such that $|f_n(t) - f_n(s)| < \epsilon$ for all n and for all $s, t \in I$ that satisfy $|t - s| < \delta$.

a) Prove that the sequence $\{\sin(n\pi x)\}$ is not equicontinuous on $[0, 1]$.

b) Suppose that $\{f_n\}$ is a sequence of equicontinuous functions that converges pointwise to f on I. Prove that f is continuous on I.

c) Suppose that $\{f_n\}$ is a sequence of continuous functions that converges uniformly on I. Prove that $\{f_n\}$ is equicontinuous on I.

6. Prove Dini's Theorem by completing the following steps.

a) For each positive integer n, let $g_n = f - f_n$. Prove that $\{g_n\}$ is a decreasing sequence of continuous functions that converges pointwise to the zero function on $[a, b]$.

b) For each positive integer n, let M_n be the maximum value of g_n on $[a, b]$. Show that the proof is complete if $\{M_n\}$ has a subsequence that converges to 0.

c) Prove that for each positive integer n, there exists a point $x_n \in [a, b]$ such that $g_n(x_n) = M_n$.

d) By the Bolzano-Weierstrass Theorem, there exists a subsequence $\{x_{p_n}\}$ of $\{x_n\}$ that converges to a point $z \in [a, b]$. Prove that $\{M_{p_n}\}$ converges to 0.

7. Let $\{f_n\}$ be a sequence of nonnegative continuous functions that converge pointwise to the zero function on an interval $[a, b]$. For each positive integer n, define a function g_n on $[a, b]$ by $g_n(x) = \min\{f_1(x), f_2(x), \ldots, f_n(x)\}$. Prove that the sequence $\{g_n\}$ converges uniformly to the zero function on $[a, b]$.

8. Adopting the notation of the last part of the proof of Theorem 7.11, prove that

$$|S(f, {}'P) - S(f_q, {}'P)| < \epsilon\,(b - a).$$

9. For the given sequence of functions, examine both the hypotheses and the conclusion of Theorem 7.11 on the interval $[0, 2]$.

a) The sequence $\{f_n\}$, where $f_n(x) = \begin{cases} 1 - |1 - nx|, & \text{if } 0 \le x \le 2/n; \\ 0, & \text{if } x \ge 2/n. \end{cases}$

b) The sequence $\{g_n\}$, where $g_n(x) = 2nxe^{-nx^2}$.

10. Give some examples to show that Theorem 7.11 does not extend to improper integrals of the form $\int_a^\infty f$ without some additional hypotheses.

11. Let $\{f_n\}$ be a sequence of Riemann integrable functions defined on $[a, b]$ and suppose that $\{f_n\}$ converges uniformly to f on $[a, b]$. For each n, define $F_n\colon [a, b] \to \mathbb{R}$ by $F_n(x) = \int_a^x f_n$. Prove that the sequence $\{F_n\}$ converges uniformly on $[a, b]$.

12. Adopting the notation and hypotheses of Theorem 7.12, use the result of the previous exercise to prove that $\{f_n\}$ converges uniformly to f on each closed and bounded subinterval of I.

13. Use the Fundamental Theorem of Calculus and Theorem 7.11 to prove Theorem 7.12. (You first need to come up with an appropriate interval $[a, b]$.) How must your proof be modified if the functions f_n' are only assumed to be Riemann integrable?

14. For the given sequence of functions, examine both the hypotheses and the conclusion of Theorem 7.12 on the interval $[0, 1]$.

a) The sequence $\{f_n\}$, where $f_n(x) = \sin(nx)/n^2$.

b) The sequence $\{g_n\}$, where $g_n(x) = x^n/n$.

15. Prove Theorem 7.13.

16. Let $\sum_{k=1}^{\infty} a_k$ be an absolutely convergent series.

a) Prove that $\sum_{k=1}^{\infty} a_k \sin(kx)$ and $\sum_{k=1}^{\infty} a_k \cos(kx)$ converge uniformly on $\mathbb{R}$.

b) Let f be the function defined by $f(x) = \sum_{k=1}^{\infty} a_k \sin(kx)$. Prove that

$$\int_0^{\pi/2} f = \sum_{k=1}^{\infty} \frac{a_{2k-1} + a_{4k-2}}{2k - 1}.$$

c) For the function f defined in part (b), express $\int_0^\pi f$ as an infinite series of real numbers.

17. Explain why the function f defined by $f(x) = \sum_{k=1}^{\infty} 1/k^x$ is defined for all $x > 1$, then determine whether or not f is continuous and differentiable on $(1, \infty)$.

18. Prove that the series $\sum_{k=1}^{\infty} x^k(1 - x)$ does not converge uniformly on $[0, 1]$.

7.4 Power Series

Power series are a particularly simple and useful type of series of functions. The terms of a power series are powers of x or, more generally, powers of $x - c$ for some constant c. In essence, a power series is an infinite degree polynomial; as we will see, such series possess some nice properties that the typical series of functions does not possess.

DEFINITION 7.14 Given a real number c and a sequence $\{a_k\}_{k=0}^{\infty}$ of real numbers, the expression $\sum_{k=0}^{\infty} a_k(x-c)^k$ is called a **power series** centered at c. The numbers a_k are the **coefficients** of the power series and the number c is the **center** of the power series.

Writing out the terms of a general power series, we obtain

$$a_0 + a_1(x-c) + a_2(x-c)^2 + a_3(x-c)^3 + a_4(x-c)^4 + \cdots.$$

For the special case in which $c = 0$, a power series assumes the form

$$a_0 + a_1 x + a_2 x^2 + a_3 x^3 + a_4 x^4 + \cdots.$$

(For the record, when $x = c$, the first term of a power series becomes $a_0 \cdot 0^0$. Since 0^0 is an indeterminate form, this represents an unfortunate notational problem. The shorthand summation notation for power series is quite convenient and will be used throughout this chapter; just remember that a_0 is the first term of the series.) Since this expression looks so much like a polynomial, it is tempting to differentiate and integrate it just like a polynomial. Such manipulations of power series are indeed valid, but a proof is required (see Theorem 7.17).

Given a power series, the first thing to determine is the set of values of x for which it converges. By now, the reader should have enough insight to realize that this set will depend on the coefficients of the power series. As the next theorem indicates, the set of values of x for which a power series converges is an interval whose length can be found by a simple formula. As a reminder, the limit superior of a sequence was defined in Section 2.3.

THEOREM 7.15 Let $\sum_{k=0}^{\infty} a_k(x-c)^k$ be a power series, let $s = \limsup_{k\to\infty} \sqrt[k]{|a_k|}$, and let

$$\rho = \begin{cases} 0, & \text{if } s = \infty; \\ 1/s, & \text{if } 0 < s < \infty; \\ \infty, & \text{if } s = 0. \end{cases}$$

Then $\sum_{k=0}^{\infty} a_k(x-c)^k$ converges absolutely for all values of x that satisfy $|x-c| < \rho$ and diverges for all values of x that satisfy $|x-c| > \rho$. Furthermore, the convergence of the series is uniform on the interval $[c-r, c+r]$ for each r that satisfies $0 < r < \rho$.

Proof. Suppose first that $0 < \rho < \infty$ and let x be any real number. We must show that the series $\sum_{k=0}^{\infty} a_k(x-c)^k$ converges absolutely if $|x-c| < \rho$ and diverges if $|x-c| > \rho$. Note that

$$\limsup_{k\to\infty} \sqrt[k]{|a_k|\,|x-c|^k} = |x-c| \limsup_{k\to\infty} \sqrt[k]{|a_k|} = |x-c|/\rho.$$

By the Root Test, the series does indeed converge absolutely if $|x-c| < \rho$ and diverge if $|x-c| > \rho$. Now let r be a real number that satisfies $0 < r < \rho$. The series $\sum_{k=0}^{\infty} |a_k| r^k$ converges according to the Root Test since

$$\limsup_{k\to\infty} \sqrt[k]{|a_k| r^k} = r \limsup_{k\to\infty} \sqrt[k]{|a_k|} = r/\rho < 1.$$

Since $|a_k| |x - c|^k \leq |a_k| r^k$ for all x in $[c - r, c + r]$, the series $\sum_{k=0}^{\infty} a_k (x - c)^k$ converges uniformly on $[c - r, c + r]$ by the Weierstrass M-test. This completes the proof for the $0 < \rho < \infty$ case. The case in which $\rho = \infty$ is similar; the details will be left as an exercise.

Now suppose that $\rho = 0$. This means that $\limsup_{k\to\infty} \sqrt[k]{|a_k|} = \infty$. If $x \neq c$, then

$$\limsup_{k\to\infty} \sqrt[k]{|a_k| |x - c|^k} = |x - c| \limsup_{k\to\infty} \sqrt[k]{|a_k|} = \infty.$$

By the Root Test, this series does not converge, so the power series converges only for $x = c$. This completes the proof. ■

Thus, the set of values of x for which a power series converges is an interval centered at c. This interval is known as the **interval of convergence** for the power series. If $0 < \rho < \infty$, then the power series converges for all x in the interval $(c - \rho, c + \rho)$ and diverges for all x outside the interval $[c - \rho, c + \rho]$. The convergence at the endpoints $c - \rho$ and $c + \rho$ must be checked separately since these two points fall between the cracks of the Root Test. If $\rho = \infty$, then the interval of convergence consists of all real numbers. If $\rho = 0$, the power series converges only at the point c. (A single point can be considered as a degenerate interval; the point c can be written as the interval $[c, c]$, which has length 0. Thus, the term interval of convergence applies to all power series.) For the record, if a power series centered at c has radius of convergence ρ, then it is not technically correct to say the series converges on the interval $(c - \rho, c + \rho)$; a problem occurs when $\rho = \infty$. To avoid dealing with two separate cases, $\rho < \infty$ and $\rho = \infty$, we will sometimes write $\{x : |x - c| < \rho\}$ to represent this interval.

The number ρ defined in Theorem 7.15 is called the **radius of convergence** of the power series. The term "radius" is usually reserved for circles and spheres, so its use here may seem odd. A better setting for the study of power series is actually the set of complex numbers. The complex number $a + bi$ can be associated with the point (a, b) in the complex plane. It can be shown that the set of complex numbers in the complex plane for which a power series converges includes all of the points in the interior of a circle and perhaps some of its boundary points. The radius of this circle is the radius of convergence of the power series. The use of this term has been carried over to the real number line; a one-dimensional circle is an interval.

Since the quantity $\sqrt[k]{|a_k|}$ can be difficult to compute, it is sometimes easier to use the Ratio Test to determine the radius of convergence of a power series. If the number $\lim_{k\to\infty} |a_k|/|a_{k+1}|$ exists or is ∞, then this value equals the number ρ defined in Theorem 7.15. (A justification of this fact is provided by Exercise 15 in Section 6.3). Note that the ratio used here is the reciprocal of the ratio that appears in the Ratio Test and that the resulting limit is the radius of convergence.

As an example of a power series, consider $\sum_{k=0}^{\infty}(k+1)(x-4)^k/2^k$. This series is centered at 4 and its coefficients are $(k+1)/2^k$. Since

$$\limsup_{k\to\infty}\sqrt[k]{\frac{k+1}{2^k}} = \frac{1}{2}\lim_{k\to\infty}\sqrt[k]{k+1} = \frac{1}{2},$$

the radius of convergence of the power series is 2. Alternatively, we could have computed

$$\lim_{k\to\infty}\left(\frac{k+1}{2^k}\cdot\frac{2^{k+1}}{k+2}\right) = 2$$

and reached the same conclusion. In any event, the series converges for all x in the interval $(2, 6)$. For $x = 2$ and $x = 6$, the resulting series are

$$\sum_{k=0}^{\infty}(-1)^k(k+1) \quad \text{and} \quad \sum_{k=0}^{\infty}(k+1),$$

respectively. Since each of these series diverges, the interval of convergence for this power series is $(2, 6)$.

Let $\sum_{k=0}^{\infty} a_k(x-c)^k$ be a power series with interval of convergence I. A function f can be defined on I by letting $f(x)$ be the sum of the series for each $x \in I$. In other words,

$$f(x) = a_0 + a_1(x-c) + a_2(x-c)^2 + a_3(x-c)^3 + \cdots.$$

Suppose that the function f is differentiable and integrable and that the series can be differentiated and integrated term by term. Then

$$f'(x) = a_1 + 2a_2(x-c) + 3a_3(x-c)^2 + \cdots \quad \text{and}$$

$$\int_c^x f = a_0(x-c) + \frac{a_1}{2}(x-c)^2 + \frac{a_2}{3}(x-c)^3 + \frac{a_3}{4}(x-c)^4 + \cdots.$$

As we will see in a moment, these equations are valid for all x in the interval of convergence, except possibly at the endpoints. An important observation is the fact that f' and $\int f$ are also power series centered at c. In addition, these power series have the same radius of convergence as the original power series. A proof of this fact, which is recorded in the following lemma, will be left as an exercise.

LEMMA 7.16 If $\{a_k\}_{k=0}^{\infty}$ is a sequence of real numbers and c is a real number, then the power series

$$\sum_{k=0}^{\infty} a_k(x-c)^k, \quad \sum_{k=1}^{\infty} ka_k(x-c)^{k-1}, \quad \text{and} \quad \sum_{k=0}^{\infty}\frac{a_k}{k+1}(x-c)^{k+1}$$

all have the same radius of convergence. ■

Although the power series for both f' and $\int f$ have the same radius of convergence as the power series for f, the interval of convergence for these power series may vary. In other words, convergence at the endpoints should be checked carefully. An example of a power series for which the series listed in Lemma 7.16 have different intervals of convergence will be requested in the exercises.

THEOREM 7.17 Let $\sum_{k=0}^{\infty} a_k(x-c)^k$ be a power series with radius of convergence $\rho > 0$, let I be the interval $\{x : |x-c| < \rho\}$, and let $f(x) = \sum_{k=0}^{\infty} a_k(x-c)^k$ for each $x \in I$. Then the function f is continuous and differentiable on I, and

$$f'(x) = \sum_{k=1}^{\infty} ka_k(x-c)^{k-1} \quad \text{and} \quad \int_c^x f = \sum_{k=0}^{\infty} \frac{a_k}{k+1}(x-c)^{k+1}$$

for each $x \in I$.

Proof. We will prove that f is differentiable on I and that f' has the indicated power series representation. The continuity of f on I then follows immediately. A proof of the result for integration will be left as an exercise.

Let $x \in I$. Choose r so that $0 < r < \rho$ and $x \in [c-r, c+r]$. By Theorem 7.15 and Lemma 7.16, the series $\sum_{k=0}^{\infty} a_k(x-c)^k$ and $\sum_{k=1}^{\infty} ka_k(x-c)^{k-1}$ converge uniformly on $[c-r, c+r]$. By Theorem 7.13, the function f is differentiable on $[c-r, c+r]$ and

$$f'(t) = \sum_{k=1}^{\infty} ka_k(t-c)^{k-1}$$

for all $t \in [c-r, c+r]$. In particular, the function f is differentiable at x and $f'(x) = \sum_{k=1}^{\infty} ka_k(x-c)^{k-1}$. This completes the proof. ■

To illustrate Theorem 7.17, consider the series $\sum_{k=0}^{\infty} (k+1)(x-4)^k/2^k$ discussed earlier in this section. This power series represents the function f defined by

$$f(x) = 1 + (x-4) + \frac{3}{4}(x-4)^2 + \frac{1}{2}(x-4)^3 + \frac{5}{16}(x-4)^4 + \cdots$$

for all $x \in (2, 6)$. By Theorem 7.17, we find that

$$f'(x) = 1 + \frac{3}{2}(x-4) + \frac{3}{2}(x-4)^2 + \frac{5}{4}(x-4)^3 + \cdots \quad \text{and}$$

$$\int_4^x f = (x-4) + \frac{1}{2}(x-4)^2 + \frac{1}{4}(x-4)^3 + \frac{1}{8}(x-4)^4 + \cdots$$

for all x in the interval $(2, 6)$.

Since the power series representation for f' has the same radius of convergence as that of f, Theorem 7.17 shows that f'' exists on I and that

$$f''(x) = \sum_{k=2}^{\infty} k(k-1)a_k(x-c)^{k-2}$$

for all $x \in I$. This power series also has the same radius of convergence as the power series for f and f'. It follows that f''' exists on I and that

$$f'''(x) = \sum_{k=3}^{\infty} k(k-1)(k-2)a_k(x-c)^{k-3}$$

for all $x \in I$. Similarly, the function f has derivatives of all orders on I, and $f^{(k)}$ (the kth derivative of f) has a power series representation for each positive integer k. This observation leads to the following theorem; a proof of the corollary will be left as an exercise.

THEOREM 7.18 Let $\sum_{k=0}^{\infty} a_k(x-c)^k$ be a power series with radius of convergence $\rho > 0$, let I be the interval $\{x : |x-c| < \rho\}$, and let $f(x) = \sum_{k=0}^{\infty} a_k(x-c)^k$ for each $x \in I$. Then the function f has derivatives of all orders on I and $f^{(k)}(c) = k!a_k$ for all positive integers k.

Proof. By the discussion preceding the theorem, the function f has derivatives of all orders on I. Using Theorem 7.17 repeatedly, it follows that for each positive integer n,

$$f^{(n)}(x) = \sum_{k=n}^{\infty} k(k-1)(k-2)\cdots(k-n+1)a_k(x-c)^{k-n}$$

for all $x \in I$. Substituting c for x yields $f^{(n)}(c) = n!\,a_n$. ■

COROLLARY 7.19 Let $\sum_{k=0}^{\infty} a_k(x-c)^k$ and $\sum_{k=0}^{\infty} b_k(x-c)^k$ be two power series, let $r > 0$, and suppose that both series converge on the interval $(c-r, c+r)$. If

$$\sum_{k=0}^{\infty} a_k(x-c)^k = \sum_{k=0}^{\infty} b_k(x-c)^k$$

for all x in the interval $(c-r, c+r)$, then $a_k = b_k$ for all k. ■

In other words, the corollary states that a power series representation of a function for a given center c is unique. A function with a power series representation at a point c is infinitely differentiable and, as indicated by Theorem 7.18, the coefficients of the power series are determined by the derivatives of the function at c. Consequently, for a function expressed as a power series, the values of its derivatives at a single point completely determine the function. The fact that all of the values of the function in the interval of convergence are determined by information at a single point is rather surprising.

Suppose that f is a function (such as a trigonometric or exponential function) that is infinitely differentiable at a point c. Referring to Theorem 7.18, it is natural to consider the power series

$$\sum_{k=0}^{\infty} \frac{f^{(k)}(c)}{k!}(x-c)^k.$$

A series of this form is called the **Taylor series** of f at c in honor of Brook Taylor (1685–1731). In the special but common case in which $c = 0$, the series is known as a **Maclaurin series**, named after Colin Maclaurin (1698–1746). A natural question to ask is whether or not the equation

$$f(x) = \sum_{k=0}^{\infty} \frac{f^{(k)}(c)}{k!}(x-c)^k$$

is valid for those values of x for which both the function is defined and the series converges. The general answer to this question is no; one example is provided in the exercises at the end of this section. The existence of such functions means that some further conditions must be satisfied by a function in order to guarantee that the function and its Taylor series are equal.

Since a power series representation of a function is unique, a power series for a function found by any method must be the Taylor series for the function. For example, from the formula for the sum of a geometric series,

$$\frac{1}{1+x} = \frac{1}{1-(-x)} = \sum_{k=0}^{\infty}(-x)^k = \sum_{k=0}^{\infty}(-1)^k x^k$$

for $|x| < 1$. By the Fundamental Theorem of Calculus and Theorem 7.17,

$$\begin{aligned}
\ln|1+x| = \int_0^x \frac{1}{1+t}\,dt &= \int_0^x \sum_{k=0}^{\infty}(-1)^k t^k\,dt \\
&= \sum_{k=0}^{\infty} \int_0^x (-1)^k t^k\,dt \\
&= \sum_{k=0}^{\infty} \frac{(-1)^k}{k+1} x^{k+1} = \sum_{k=1}^{\infty} \frac{(-1)^{k+1}}{k} x^k
\end{aligned}$$

for $|x| < 1$. Since this is a power series representation of $\ln|1+x|$ that is centered at 0, it must also be the Maclaurin series for $\ln|1+x|$. In a few short steps, we have found the Maclaurin series for $\ln|1+x|$ and shown that the two are equal for $|x| < 1$. In the next section, a general method for determining when a function equals its Taylor series will be discussed.

As a final example to indicate how algebra can be used to determine Taylor series representations, note that

$$\frac{1}{x} = \frac{1/2}{1-\dfrac{2-x}{2}} = \frac{1}{2}\sum_{k=0}^{\infty}\left(\frac{2-x}{2}\right)^k = \sum_{k=0}^{\infty} \frac{(-1)^k}{2^{k+1}}(x-2)^k.$$

This equation gives the Taylor series centered at 2 for the function $f(x) = 1/x$ and shows that the two are equal on the interval $(0, 4)$, which is the interval of convergence of the series.

Exercises

1. Find the radius of convergence for the given power series.

   a) $\sum_{k=1}^{\infty} k^k (x+4)^k$ **b)** $\sum_{k=0}^{\infty} (3+(-1)^k)^k x^k$ **c)** $\sum_{k=0}^{\infty} \frac{(k!)^2}{(2k)!}(x-2)^k$

2. Find the interval of convergence for the given power series.

   a) $\sum_{k=1}^{\infty} \frac{(x+3)^k}{k2^k}$ **b)** $\sum_{k=0}^{\infty} \frac{10^k}{k!}(x+2)^k$ **c)** $\sum_{k=0}^{\infty} \frac{(2k+1)!}{2^k (k!)^2} x^k$

3. Find the sum of the series in Exercise 1 (b).
4. For each of the intervals (a) $(-1, 1)$, (b) $[-2, 2)$, (c) $(-3, 3]$, and (d) $[-4, 4]$, give an example of a power series that has the interval as its interval of convergence.
5. The interval of convergence of a power series is $(-1, 10]$. What is its center and radius of convergence?
6. Finish the proof of Theorem 7.17 by considering the case in which $\rho = \infty$.
7. Let $\sum_{k=0}^{\infty} a_k (x-c)^k$ be a power series with radius of convergence $\rho > 0$, let I be the interval $\{x : |x-c| < \rho\}$, and let $[u, v] \subseteq I$. Prove that the power series converges uniformly on $[u, v]$.
8. Suppose that the power series $\sum_{k=0}^{\infty} a_k x^k$ has radius of convergence $\rho > 0$ and define a function f by letting $f(x)$ be the sum of this series for $|x| < \rho$. Suppose that $f(x) = f(-x)$ for all such x. Prove that $a_k = 0$ for every odd integer k.
9. Let $\{b_k\}$ be a sequence of positive numbers that converges to 1, and let $\{c_k\}$ be a sequence of nonnegative numbers.

   a) Suppose that $\limsup_{k\to\infty} c_k = 0$. Prove that $\limsup_{k\to\infty} b_k c_k = 0$.

   b) Suppose that $\limsup_{k\to\infty} c_k = \infty$. Prove that $\limsup_{k\to\infty} b_k c_k = \infty$.

   c) Suppose that $\limsup_{k\to\infty} c_k = r$, where $0 < r < \infty$. Prove that $\limsup_{k\to\infty} b_k c_k = r$.

10. Let $\{a_k\}_{k=0}^{\infty}$ be a sequence of real numbers.

    a) Use the previous exercise to prove that the series $\sum_{k=0}^{\infty} a_k (x-c)^k$ and $\sum_{k=1}^{\infty} k a_k (x-c)^k$ have the same radius of convergence.

    b) Prove that the series $\sum_{k=1}^{\infty} k a_k (x-c)^{k-1}$ and $\sum_{k=1}^{\infty} k a_k (x-c)^k$ converge for the same values of x.

    c) Use parts (a) and (b) to prove one part of Lemma 7.16.

    d) Using a similar analysis, prove the other part of Lemma 7.16.

11. Let $\{a_k\}_{k=0}^{\infty}$ be a sequence of real numbers and let p be a positive integer. Prove that the power series $\sum_{k=0}^{\infty} a_k(x-c)^k$ and $\sum_{k=0}^{\infty} a_{k+p}(x-c)^k$ have the same radius of convergence.

12. Prove the integration portion of Theorem 7.17.

13. Give an example of a power series $f(x) = \sum_{k=0}^{\infty} a_k x^k$ for which the domain of f is different than the domain of f'.

14. Use the Maclaurin series for $\ln|1+x|$ to find the sum of the series $\sum_{k=1}^{\infty} 3^{-k}/k$.

15. Find the sum of the series in Exercise 2 (a).

16. Show that the series $\sum_{k=0}^{\infty}(k+1)(x-4)^k/2^k$ is the derivative of a geometric series. Then find the sum of this series by differentiating the formula for the sum of the geometric series. (The series in this exercise appeared as an example following Theorem 7.15).

17. Consider the function f defined by $f(x) = \arctan x$.

a) Integrate an appropriate geometric series to determine the Maclaurin series for f.

b) What is the interval of convergence of the series found in part (a)? How does it compare with the interval of convergence of the original geometric series?

c) Find $f^{(201)}(0)$.

d) Explain how to use the Maclaurin series for f to obtain the equation

$$\frac{\pi}{4} = 1 - \frac{1}{3} + \frac{1}{5} - \frac{1}{7} + \frac{1}{9} - \frac{1}{11} + \cdots.$$

Can you verify that this equation is valid? There is more going on here than simple substitution; look again at your answer to part (b).

18. Find the Maclaurin series for the function g defined by $g(x) = (1+x)^{-2}$. What is the interval of convergence for this series?

19. Suppose that a function f is defined by $f(x) = \sum_{k=0}^{\infty} \frac{k+3}{4^k}(x-1)^k$. Find the domain of f and find $f^{(97)}(1)$.

20. Consider the function f defined by $f(x) = (2x+3)^{-1}$. Use the formula for a geometric series and some appropriate algebra to find the Taylor series for f centered at -1, 0, and 4. (That is, find three different series.) Determine the radius of convergence for each of these series.

21. Consider the statement: the values of a function expressed as a power series are determined by information at a single point. Explain the sense in which this statement is true, but also discuss why the values of the function at some other points are relevant.

22. Let $f(x) = e^{-1/x^2}$ for $x \neq 0$ and $f(0) = 0$ for $x = 0$. Show that $f^{(k)}(0) = 0$ for all positive integers k. It follows that the function f does not equal its Taylor series centered at 0 for any $x \neq 0$.

23. Define a sequence of real numbers by $a_0 = 1$, $a_1 = 1$, and $a_{k+1} = a_k + a_{k-1}$ for all $k \geq 1$. (This recursively defined sequence generates the Fibonacci numbers.) Define a function f by $f(x) = \sum_{k=0}^{\infty} a_k x^k$ for each value of x for which the series converges.

a) Prove that $a_k \leq 2^k$ for all k and conclude that the radius of convergence of the power series defining f is at least $1/2$.

b) Prove that $xf(x) + x^2 f(x) = f(x) - 1$ for each x in the domain of f. Use this fact to show that $f(x) = 1/(1 - x - x^2)$.

c) Let α and β be the roots of $1 - x - x^2$ and use partial fractions to write

$$f(x) = \frac{A}{\alpha - x} + \frac{B}{\beta - x},$$

where A and B are constants that depend on α and β.

d) Use the formula for a geometric series to find the Maclaurin series for the functions $A/(\alpha - x)$ and $B/(\beta - x)$.

e) Use the fact that the Maclaurin series representation of a function is unique to conclude that

$$a_k = \frac{1}{\sqrt{5}}\left(\left(\frac{1+\sqrt{5}}{2}\right)^{k+1} - \left(\frac{1-\sqrt{5}}{2}\right)^{k+1}\right)$$

for each integer $k \geq 0$.

7.5 TAYLOR'S FORMULA

Near the end of the last section, we used a geometric series to show that the function $\ln|1 + x|$ and its Maclaurin series are equal on the interval $(-1, 1)$. Since it is not always possible to find a simple method to show that a function and its Taylor series are equal, a general technique is needed. Suppose that f is a function that is infinitely differentiable at a point c. For each positive integer n, let

$$P_n(x) = \sum_{k=0}^{n} \frac{f^{(k)}(c)}{k!}(x - c)^k \quad \text{and} \quad R_n(x) = f(x) - P_n(x).$$

The polynomial P_n is called the nth **Taylor polynomial** of f at c and the remainder R_n measures the difference between $f(x)$ and $P_n(x)$. A function and its Taylor series are equal at a point x if and only if $\lim_{n\to\infty} R_n(x) = 0$. To determine whether or not this limit is 0, a formula or estimate for the remainder term $R_n(x)$ is required.

As we will see, there are several different ways to express the remainder term R_n. The formula for $R_n(x)$ given in the next theorem involves an integral; it is therefore known as Taylor's formula with integral remainder. The proof uses integration by parts (see Theorem 5.18); $\int u\,dv = uv - \int v\,du$. The choice for the function v involves an arbitrary constant, which is usually chosen to be 0, but, as in the following proof, sometimes a different constant is useful.

THEOREM 7.20 Taylor's Formula with Integral Remainder If f is infinitely differentiable on some open interval I that contains the point c, then for each $x \in I$

and for each positive integer n,

$$f(x) = \sum_{k=0}^{n} \frac{f^{(k)}(c)}{k!}(x-c)^k + \frac{1}{n!}\int_c^x (x-t)^n f^{(n+1)}(t)\,dt.$$

Proof. Adopting the notation from the introduction to this section, the theorem asserts that the remainder term R_n, which is defined by $R_n(x) = f(x) - P_n(x)$, satisfies

$$R_n(x) = \frac{1}{n!}\int_c^x (x-t)^n f^{(n+1)}(t)\,dt$$

for each $x \in I$ and for each positive integer n. Fix $x \in I$. We will use induction to prove that the formula for $R_n(x)$ is valid for each positive integer n. Using the Fundamental Theorem of Calculus and integration by parts (let $u(t) = f'(t) - f'(c)$ and $v(t) = t - x$),

$$\begin{aligned}
R_1(x) &= f(x) - f(c) - f'(c)(x-c) \\
&= \int_c^x \big(f'(t) - f'(c)\big)\,dt \\
&= \big(f'(t) - f'(c)\big)(t-x)\Big|_c^x - \int_c^x (t-x)f''(t)\,dt \\
&= \int_c^x (x-t)f''(t)\,dt.
\end{aligned}$$

This shows that $R_1(x)$ has the desired form. Now suppose that

$$R_n(x) = \frac{1}{n!}\int_c^x (x-t)^n f^{(n+1)}(t)\,dt$$

for some positive integer n. Using the same approach as in the base case,

$$\begin{aligned}
R_{n+1}(x) &= R_n(x) - \frac{f^{(n+1)}(c)}{(n+1)!}(x-c)^{n+1} \\
&= \frac{1}{n!}\int_c^x (x-t)^n f^{(n+1)}(t)\,dt - \frac{1}{n!}\int_c^x f^{(n+1)}(c)(x-t)^n\,dt \\
&= \frac{1}{n!}\int_c^x \big(f^{(n+1)}(t) - f^{(n+1)}(c)\big)(x-t)^n\,dt \\
&= \frac{1}{n!}\Big[-\frac{(x-t)^{n+1}}{n+1}\big(f^{(n+1)}(t) - f^{(n+1)}(c)\big)\Big|_c^x \\
&\qquad + \int_c^x \frac{(x-t)^{n+1}}{n+1} f^{(n+2)}(t)\,dt\Big] \\
&= \frac{1}{(n+1)!}\int_c^x (x-t)^{n+1} f^{(n+2)}(t)\,dt.
\end{aligned}$$

This shows that the remainder term $R_{n+1}(x)$ has the desired form. Since $x \in I$ was arbitrary, the proof is complete. ∎

The integrals involved in the integral form of the remainder are usually too complicated to evaluate by finding an antiderivative. Fortunately, proving that the limit of the sequence $\{R_n(x)\}$ is 0 is more important than actually finding the values of the terms $R_n(x)$, so estimation is a viable option. In some cases, all of the derivatives of a function are bounded by the same number. When this occurs, it is fairly easy to prove that $\{R_n(x)\}$ converges to 0.

THEOREM 7.21 Suppose that f is infinitely differentiable on an open interval I that contains the point c. If there exists a positive number M such that $|f^{(n)}(t)| \leq M$ for all $t \in I$ and for all positive integers n, then the function f and its Taylor series centered at c are equal for all $x \in I$.

Proof. As in the introduction to this section, let

$$R_n(x) = f(x) - \sum_{k=0}^{n} \frac{f^{(k)}(c)}{k!}(x-c)^k.$$

It is sufficient to prove that $\lim_{n\to\infty} R_n(x) = 0$ for all $x \in I$. Let $x \in I$ and suppose that $x > c$; the proof for the case $x < c$ is almost identical. Since all of the derivatives of f are bounded by M on the interval $[c, x]$, Taylor's Formula with integral remainder yields

$$\begin{aligned} |R_n(x)| &= \left|\frac{1}{n!}\int_c^x (x-t)^n f^{(n+1)}(t)\,dt\right| \\ &\leq \frac{M}{n!}\int_c^x |x-t|^n\,dt \\ &= \frac{M}{(n+1)!}|x-c|^{n+1}. \end{aligned}$$

Since the sequence $\{(x-c)^n/n!\}$ converges to 0 (see the exercises), the sequence $\{R_n(x)\}$ converges to 0 by the squeeze theorem. This completes the proof. ∎

To illustrate this theorem, we will prove that the function $f(x) = e^x$ equals its Maclaurin series for all values of x. Since all of the derivatives of e^x are e^x, it is clear that $f^{(k)}(0) = 1$ for all k. Thus $\sum_{k=0}^{\infty} x^k/k!$ is the Maclaurin series for e^x. Using the Ratio Test, it is easy to verify that this series converges for all real numbers x. Let p be any positive integer. On the interval $(-\infty, p)$, all of the derivatives of e^x are bounded by e^p. By Theorem 7.21, the function e^x and its Maclaurin series are equal for all $x \in (-\infty, p)$. Since this is true for all positive integers p, it follows that $e^x = \sum_{k=0}^{\infty} x^k/k!$ for all real numbers x. Substituting 1 for x yields the interesting formula

$$e = 1 + 1 + \frac{1}{2!} + \frac{1}{3!} + \frac{1}{4!} + \frac{1}{5!} + \frac{1}{6!} + \cdots,$$

which was used as the definition of e in Exercise 35 of Section 2.2.

In addition to establishing the equality of a function and its Taylor series, the remainder term can determine how accurately a Taylor polynomial approximates

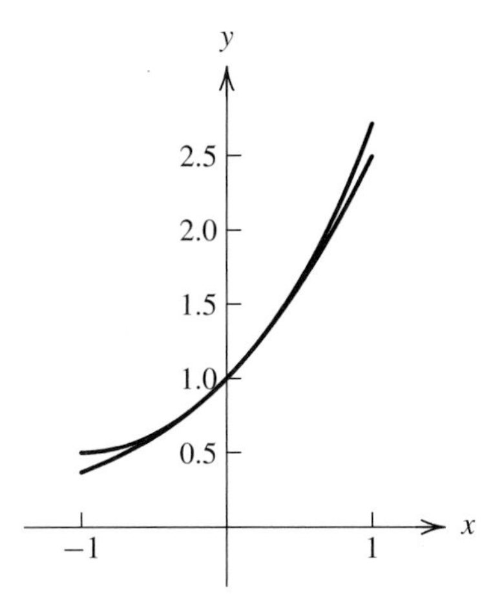

Figure 7.2 The graphs of $y = e^x$ (lower curve) and $y = 1 + x + x^2/2$

the given function. Let P_n and R_n be the Taylor polynomial for e^x centered at 0 and the corresponding remainder, respectively, for each positive integer n. Let b be a positive number. Using ideas similar to those in the proof of Theorem 7.21 (let $M = e^b$ and $c = 0$), we find that

$$|e^x - P_n(x)| \leq \frac{e^b}{(n+1)!}|x|^{n+1} \leq \frac{e^b b^{n+1}}{(n+1)!}$$

for each $x \in [-b, b]$ and for each $n \in \mathbb{Z}^+$. For example, with $b = 1$ and $n = 2$,

$$|e^x - P_2(x)| \leq \frac{e^1 \cdot 1^3}{3!} = \frac{e}{6} < 0.46$$

for all x in the interval $[-1, 1]$. The graphs of the equations $y = e^x$ and $y = P_2(x)$ on the interval $[-1, 1]$ are shown in Figure 7.2. As is evident from the graph, this is a conservative estimate since the functions are actually much closer than 0.46 units on this interval.

If the derivatives of a function are not all bounded by the same number, then Theorem 7.21 does not apply and some other method to estimate the integral must be found. An important example of this situation is the function $(1 + x)^r$, where r is a real number. The next theorem states that for each real number r, the Maclaurin series for $(1 + x)^r$ converges to $(1 + x)^r$ for $|x| < 1$. This result is an extension of the familiar Binomial Theorem (see Appendix C). In keeping with the notation for binomial coefficients, let

$$\binom{r}{k} = \frac{r(r-1)(r-2)\cdots(r-k+1)}{k!}$$

for all real numbers r and all positive integers k. The proof of the following lemma will be left as an exercise.

LEMMA 7.22 If r is a real number and $|x| < 1$, then $\lim_{n\to\infty} n \binom{r}{n} |x|^n = 0$. ■

THEOREM 7.23 Binomial Series For each real number r, the formula

$$(1+x)^r = 1 + \sum_{k=1}^{\infty} \binom{r}{k} x^k$$

is valid for all x that satisfy $|x| < 1$.

Proof. A proof that the given series is actually the Maclaurin series for $(1+x)^r$ will be left as an exercise. Suppose that $r > 1$ and fix $x \in (-1, 1)$. Using Taylor's formula with integral remainder, we must show that $\lim_{n\to\infty} R_n(x) = 0$, where

$$\begin{aligned} R_n(x) &= \int_0^x \frac{(x-t)^n}{n!} r(r-1)\cdots(r-n)(1+t)^{r-n-1}\,dt \\ &= (n+1)\binom{r}{n+1} \int_0^x \left(\frac{x-t}{1+t}\right)^n (1+t)^{r-1}\,dt. \end{aligned}$$

Note that

if $x > 0$ and $0 \le t \le x$, then $0 \le \dfrac{x-t}{1+t} \le x$ and $0 \le 1+t \le 1+|x|$, and

if $x < 0$ and $x \le t \le 0$, then $x \le \dfrac{x-t}{1+t} \le 0$ and $0 \le 1+t \le 1+|x|$.

It follows that

$$\begin{aligned} |R_n(x)| &\le (n+1)\left|\binom{r}{n+1}\right| \left|\int_0^x |x|^n (1+|x|)^{r-1}\,dt\right| \\ &= (n+1)\left|\binom{r}{n+1}\right| (1+|x|)^{r-1}|x|^{n+1}. \end{aligned}$$

Hence, the sequence $\{R_n(x)\}$ converges to 0 by Lemma 7.22. This completes the proof for the case $r > 1$. The proof for the case $r \le 1$ is similar and will be left as an exercise. ■

We conclude this section with another representation for the remainder term R_n. This form of the remainder involves a derivative, namely,

$$R_n(x) = \frac{f^{(n+1)}(z)}{(n+1)!}(x-c)^{n+1},$$

where z is some point between x and c. It was discovered by Joseph-Louis Lagrange (1736–1813) and represents an extension of the Mean Value Theorem to higher derivatives. There are several ways to prove this result; the proof presented here is fairly simple to follow, but the motivating idea behind it is difficult to grasp.

THEOREM 7.24 Taylor's Formula with Derivative Remainder Suppose that f is defined on some open interval I that contains the point c and let n be a positive integer. If $f^{(n+1)}$ exists on I, then for each $x \in I$ there exists a point z between x

and c such that

$$f(x) = \sum_{k=0}^{n} \frac{f^{(k)}(c)}{k!}(x-c)^k + \frac{f^{(n+1)}(z)}{(n+1)!}(x-c)^{n+1}.$$

Proof. Fix $x \in I$ with $x \neq c$ and define a function F on I by

$$F(t) = \sum_{k=0}^{n} \frac{f^{(k)}(t)}{k!}(x-t)^k + A(x-t)^{n+1}$$

for each $t \in I$, where the constant A is chosen so that $F(c) = f(x)$. By the hypotheses, the function F is continuous and differentiable on the interval with endpoints x and c, and $F(x) = f(x) = F(c)$. By the Mean Value Theorem, there exists a point z between x and c such that $F'(z) = 0$. Differentiating F and solving the equation $F'(z) = 0$ yields $A = \dfrac{f^{(n+1)}(z)}{(n+1)!}$. (The computations to find A are left as an exercise.) Hence,

$$f(x) = F(c) = \sum_{k=0}^{n} \frac{f^{(k)}(c)}{k!}(x-c)^k + \frac{f^{(n+1)}(z)}{(n+1)!}(x-c)^{n+1}.$$

This completes the proof. ∎

Exercises

1. Let x and c be real numbers. Prove that the sequence $\{(x-c)^n/n!\}$ converges to 0. This result is used in the proof of Theorem 7.21.

2. Let n be a positive integer, let P_n be the nth Taylor polynomial for e^x centered at 0, and let $R_n(x) = e^x - P_n(x)$. Prove that R_n is increasing on $[0, \infty)$.

3. The following problems involve the Maclaurin series for $\sin x$.

a) Show that the Maclaurin series for $\sin x$ is $\displaystyle\sum_{k=0}^{\infty} \frac{(-1)^k}{(2k+1)!} x^{2k+1}$.

b) Prove that the Maclaurin series for $\sin x$ equals $\sin x$ for all values of x.

c) Determine the accuracy of the third degree Taylor polynomial $x - x^3/6$ to $\sin x$ on the interval $[0, 1]$. Give two answers: one from the theory and one from a careful sketch of the two graphs.

d) Using the theory in this section, find the smallest degree Taylor polynomial centered at 0 that approximates $\sin x$ to within 10^{-5} on the interval $[-3, 3]$.

e) By differentiating the Maclaurin series for $\sin x$, find the Maclaurin series for $\cos x$.

f) Find the Maclaurin series for $\sin^2 x$.

g) Let $f(x) = \sin x^2$ and find $f^{(81)}(0)$.

4. Use the theory in this section to find the smallest degree Taylor polynomial centered at 0 that approximates e^x to within 10^{-3} on the interval $[-2, 2]$.

5. Express each of the integrals $\displaystyle\int_0^1 e^{-x^2}\,dx$ and $\displaystyle\int_0^{0.5} \frac{\sin x}{x}\,dx$ as infinite series.

6. Find the Maclaurin series for $\sinh x$ and $\cosh x$. (See Exercise 15 in Section 1.5 for the definitions of these functions.)

7. Use the Maclaurin series for one of the functions $\sin x$, $\cos x$, e^x, $\sinh x$, or $\cosh x$ to find the sum of the given series.

a) $\displaystyle\sum_{k=1}^{\infty} \frac{2^{k+1}}{k!}$ **b)** $\displaystyle\sum_{k=0}^{\infty} \frac{(-2)^{k+1}}{(2k+1)!}$ **c)** $\displaystyle\sum_{k=0}^{\infty} \frac{3}{2^k k!}$

d) $\displaystyle\sum_{k=0}^{\infty} \frac{(-6)^k}{(2k)!}$ **e)** $\displaystyle\sum_{k=0}^{\infty} \frac{6^k}{(2k+1)!}$ **f)** $\displaystyle\sum_{k=1}^{\infty} \frac{(-1)^k 4^k}{(2k+1)!}$

8. Prove Lemma 7.22.

9. Verify that the series in Theorem 7.23 is the Maclaurin series for the function $(1+x)^r$.

10. Find $\binom{3/2}{2}$, $\binom{-3}{5}$, and $\binom{-1/2}{4}$.

11. Let r be a real number. Find $\binom{r}{1}$. Based upon the binomial series, what is a reasonable value to assign to $\binom{r}{0}$?

12. Use the binomial series to prove each of the following.

a) $$\frac{1}{(1+x)^4} = \sum_{k=0}^{\infty} (-1)^k \cdot \frac{(k+1)(k+2)(k+3)}{6} x^k$$

b) $$\frac{1}{\sqrt{1-x}} = \sum_{k=0}^{\infty} \frac{(2k)!}{4^k (k!)^2} x^k$$

13. Prove the binomial series result for the case in which $r < 1$. Begin by determining where the fact that r is greater than 1 is used in the proof in the text.

14. This exercise outlines a derivation of the binomial series that involves the use of differential equations.

a) Show that the radius of convergence of the power series $\sum_{k=1}^{\infty} \binom{r}{k} x^k$ is 1.

b) Let f be the function defined by $f(x) = 1 + \sum_{k=1}^{\infty} \binom{r}{k} x^k$ for each $x \in (-1, 1)$. Show that $(1+x) f'(x) = r f(x)$.

c) Solve the differential equation in part (b) to find the function f.

15. Verify the value of A that appears in the proof of Theorem 7.24.

16. Write out the conclusion of Theorem 7.24 for the cases $n = 1$ and $n = 2$ without using summation notation.

17. Write the conclusion of the Mean Value Theorem so that it has the same form as Theorem 7.24.

18. Use Theorem 7.24 to prove that e^x equals its Maclaurin series for all real numbers.

19. Show that Theorem 7.24 can be used to prove the binomial series result for x in the interval $(-0.5, 1)$.

7.6 Several Miscellaneous Results

As is evident from its title, this section contains a variety of results concerning infinite series, both of numbers and functions. The first few results are related to each other and involve the product of two series, but the last two results in the section are independent of these and can be studied in any order. One of these results shows that every continuous function on $[a, b]$ is the limit of a uniformly convergent sequence of polynomials; the other result establishes the existence of continuous functions that are nowhere differentiable.

The distributive law makes it possible to write the product

$$(a_1 + a_2 + \cdots + a_m)(b_1 + b_2 + \cdots + b_n)$$

as a sum of mn products of the form $a_i b_j$ without affecting the value of the expression. This raises the question of whether or not the distributive law extends to the product

$$(a_0 + a_1 + a_2 + \cdots)(b_0 + b_1 + b_2 + \cdots)$$

of two convergent series. Since the collection of all ordered pairs of positive integers is countably infinite (see Section 1.4), the set $\{a_i b_j : i, j \geq 0\}$ can be expressed as a sequence $\{p_k\}_{k=0}^{\infty}$. The series $\sum_{k=0}^{\infty} p_k$ represents the distributed sum of the product of the two original series. We want to know whether or not the series $\sum_{k=0}^{\infty} p_k$ converges and satisfies

$$(a_0 + a_1 + a_2 + \cdots)(b_0 + b_1 + b_2 + \cdots) = p_0 + p_1 + p_2 + p_3 + \cdots.$$

Based upon previous work, you might guess that absolute convergence plays a role in the answer to this question. This is indeed the case; the following result is valid when both series are absolutely convergent.

THEOREM 7.25 Let $\sum_{k=0}^{\infty} a_k$ and $\sum_{k=0}^{\infty} b_k$ be two series and let $\{p_k\}_{k=0}^{\infty}$ be a listing of all possible products $a_i b_j$ for $i, j \geq 0$. If $\sum_{k=0}^{\infty} a_k$ and $\sum_{k=0}^{\infty} b_k$ are absolutely convergent, then $\sum_{k=0}^{\infty} p_k$ is absolutely convergent and $\sum_{k=0}^{\infty} p_k = \left(\sum_{k=0}^{\infty} a_k\right)\left(\sum_{k=0}^{\infty} b_k\right)$.

Proof. For each positive integer n, it is possible to find a positive integer q so that the terms $\{a_k : k > q\}$ and $\{b_k : k > q\}$ do not appear in any of the products $\{p_k : 0 \leq k \leq n\}$. (In other words, if $p_k = a_i b_j$ and $0 \leq k \leq n$, then $i \leq q$ and $j \leq q$.) Then

$$\sum_{k=0}^{n} |p_k| \leq \left(\sum_{k=0}^{q} |a_k|\right)\left(\sum_{k=0}^{q} |b_k|\right) \leq \left(\sum_{k=0}^{\infty} |a_k|\right)\left(\sum_{k=0}^{\infty} |b_k|\right).$$

Because this inequality is valid for each positive integer n, the series of products is absolutely convergent. Since the sum of an absolutely convergent series is independent of the order of the terms (see Theorem 6.16), the sum of the series of products can be found using any arrangement of the products. At least one arrangement of the series of products satisfies

$$\sum_{k=0}^{n^2-1} p_k = \Big(\sum_{k=0}^{n-1} a_k\Big)\Big(\sum_{k=0}^{n-1} b_k\Big)$$

for each positive integer n. Since all three sequences of partial sums converge, letting $n \to \infty$ yields the desired result. ■

If the two given series are not absolutely convergent, then the series of products may not be absolutely convergent. In this case, the order of the terms of the product series becomes important. One common way to order the terms of the product series is known as the Cauchy product. The **Cauchy product** of two series $\sum_{k=0}^{\infty} a_k$ and $\sum_{k=0}^{\infty} b_k$ is the series $\sum_{k=0}^{\infty} c_k$, where $c_k = \sum_{i=0}^{k} a_i b_{k-i}$. The motivation for this particular product is the formal expansion of the product of two power series

$$(a_0 + a_1 x + a_2 x^2 + \cdots)(b_0 + b_1 x + b_2 x^2 + \cdots).$$

Multiplying these two series and combining like terms yields

$$a_0 b_0 + (a_0 b_1 + a_1 b_0)\, x + (a_0 b_2 + a_1 b_1 + a_2 b_0)\, x^2 + \cdots.$$

It is easy to verify that the Cauchy product term c_k is the coefficient of x^k in this expansion. As the next theorem indicates, the Cauchy product of two convergent series converges as long as at least one of the series forming the product is absolutely convergent. The lemma preceding the theorem provides a condition that guarantees the terms of the Cauchy product converge to zero.

LEMMA 7.26 Let $\sum_{k=0}^{\infty} a_k$ and $\sum_{k=0}^{\infty} b_k$ be two series and let $\sum_{k=0}^{\infty} c_k$ be their Cauchy product. If $\sum_{k=0}^{\infty} a_k$ is absolutely convergent and $\lim_{k\to\infty} b_k = 0$, then $\lim_{k\to\infty} c_k = 0$.

Proof. Let $S = \sum_{k=0}^{\infty} |a_k|$, let M be a bound for the sequence $\{b_k\}_{k=0}^{\infty}$, and let $\epsilon > 0$. By hypothesis, there exists a positive integer K such that $\sum_{i=K+1}^{\infty} |a_i| < \epsilon/M$ and $|b_i| < \epsilon/S$ for all $i \geq K$. For each $k \geq 2K$,

$$\begin{aligned} |c_k| = \Big|\sum_{i=0}^{k} a_i b_{k-i}\Big| &\leq \sum_{i=0}^{K} |a_i|\,|b_{k-i}| + \sum_{i=K+1}^{k} |a_i|\,|b_{k-i}| \\ &< \frac{\epsilon}{S} \sum_{i=0}^{K} |a_i| + M \sum_{i=K+1}^{k} |a_i| < \epsilon + \epsilon = 2\epsilon. \end{aligned}$$

This shows that the sequence $\{c_k\}$ converges to 0. ■

THEOREM 7.27 Let $\sum_{k=0}^{\infty} a_k$ and $\sum_{k=0}^{\infty} b_k$ be two convergent series and let $\sum_{k=0}^{\infty} c_k$ be their Cauchy product. If $\sum_{k=0}^{\infty} a_k$ is absolutely convergent, then $\sum_{k=0}^{\infty} c_k$ converges and $\sum_{k=0}^{\infty} c_k = \left(\sum_{k=0}^{\infty} a_k\right)\left(\sum_{k=0}^{\infty} b_k\right)$.

Proof. For each nonnegative integer n, let

$$A_n = \sum_{k=0}^{n} a_k, \quad B_n = \sum_{k=0}^{n} b_k, \quad C_n = \sum_{k=0}^{n} c_k, \quad \text{and } D_n = \sum_{k=n+1}^{\infty} b_k.$$

In addition, let $A = \sum_{k=0}^{\infty} a_k$ and $B = \sum_{k=0}^{\infty} b_k$ and let $\sum_{k=0}^{\infty} p_k$ be the Cauchy product of the series $\sum_{k=0}^{\infty} a_k$ and $\sum_{k=0}^{\infty} D_k$. Then

$$\begin{aligned}
C_n &= a_0 b_0 + \left(a_0 b_1 + a_1 b_0\right) + \cdots + \left(a_0 b_n + a_1 b_{n-1} + \cdots + a_{n-1} b_1 + a_n b_0\right) \\
&= a_0 B_n + a_1 B_{n-1} + \cdots + a_n B_0 \\
&= a_0 (B - D_n) + a_1 (B - D_{n-1}) + \cdots + a_n (B - D_0) \\
&= A_n B - \left(a_0 D_n + a_1 D_{n-1} + \cdots + a_n D_0\right) \\
&= A_n B - p_n.
\end{aligned}$$

Since $\sum_{k=0}^{\infty} a_k$ converges absolutely and $\{D_k\}$ converges to 0, the sequence $\{p_n\}$ converges to 0 by the previous lemma. Hence,

$$\sum_{k=0}^{\infty} c_k = \lim_{n \to \infty} C_n = \lim_{n \to \infty} (A_n B - p_n) = AB = \left(\sum_{k=0}^{\infty} a_k\right)\left(\sum_{k=0}^{\infty} b_k\right),$$

the desired result. ■

Near the end of Section 7.4, it was shown that

$$\ln|1 + x| = \sum_{k=1}^{\infty} \frac{(-1)^{k+1}}{k} x^k$$

for all values of x in the open interval $(-1, 1)$. In this case, both the function and the series are defined at 1, but the derivation of the formula, which involved a geometric series, offers no guarantee that the two values are equal. Such a guarantee is provided by the following theorem, which was first proved by Niels Henrik Abel (1802–1829). The proof uses the fact that the radius of convergence of any power series whose coefficients form a bounded sequence is at least 1; see the exercises.

THEOREM 7.28 **Abel's Theorem** If the series $\sum_{k=1}^{\infty} a_k$ converges and a function f is defined by $f(x) = \sum_{k=1}^{\infty} a_k x^k$, then $\lim_{x \to 1^-} f(x) = \sum_{k=1}^{\infty} a_k$.

Proof. Since the sequence $\{a_k\}$ is bounded, the radius of convergence of the power series defining f is at least 1. It follows that the domain of the function f includes the interval $(-1, 1]$. Let $\{s_n\}$ be the sequence of partial sums of the series $\sum_{k=1}^{\infty} a_k$, let s be the sum of the series, and let M be a bound for the sequence $\{s_n\}$. It will be left as an exercise to prove that

$$f(x) = (1-x)\sum_{k=1}^{\infty} s_k x^k$$

for all x in the interval $(-1, 1)$. Let $0 < \epsilon < 1$. Choose a positive integer N so that $|s_n - s| < \epsilon$ for all $n > N$ and let $\delta = 1 - \sqrt[N]{1-\epsilon}$. It is easy to check that $0 < \delta < \epsilon$. If $1 - \delta < x < 1$, then

$$\begin{aligned}
|f(x) - s| &= \left|(1-x)\sum_{k=1}^{\infty} s_k x^k - (1-x)\sum_{k=0}^{\infty} s x^k\right| \\
&\le (1-x)\sum_{k=1}^{N} |s_k - s| x^k + (1-x)\sum_{k=N+1}^{\infty} |s_k - s| x^k + |s|(1-x) \\
&< (1-x)\sum_{k=1}^{N} 2M\, x^k + (1-x)\sum_{k=N+1}^{\infty} \epsilon\, x^k + M\delta \\
&= 2M(x - x^{N+1}) + \epsilon\, x^{N+1} + M\delta \\
&< 2M\epsilon + \epsilon + M\epsilon = (3M+1)\epsilon.
\end{aligned}$$

This shows that $\lim_{x \to 1^-} f(x) = s$, as desired. ■

To illustrate Abel's Theorem, consider once again the Maclaurin series for $\ln|1+x|$. The function g defined by

$$g(x) = \sum_{k=1}^{\infty} \frac{(-1)^{k+1}}{k} x^k$$

has the interval $(-1, 1]$ as its domain and is continuous on $(-1, 1)$ by Theorem 7.17. Abel's Theorem asserts that g is actually continuous on the interval $(-1, 1]$. Since $g(x) = \ln|1+x|$ for $x \in (-1, 1)$ and since both functions are continuous at 1, it follows that $g(1) = \ln 2$. In other words,

$$\ln 2 = \sum_{k=1}^{\infty} \frac{(-1)^{k+1}}{k} = 1 - \frac{1}{2} + \frac{1}{3} - \frac{1}{4} + \frac{1}{5} - \frac{1}{6} + \cdots.$$

As indicated by this example, the conclusion of Abel's Theorem shows that power series are continuous on their interval of convergence. A proof of this result for the general case (as opposed to the case when the radius of convergence is 1) will be left as an exercise.

Although Abel's Theorem deals with power series, it can be used to prove another result concerning the Cauchy product of two convergent series. The proof of the following result will be left as an exercise.

THEOREM 7.29 Let $\sum_{k=0}^{\infty} a_k$ and $\sum_{k=0}^{\infty} b_k$ be two convergent series and let $\sum_{k=0}^{\infty} c_k$ be their Cauchy product. If $\sum_{k=0}^{\infty} c_k$ converges, then $\sum_{k=0}^{\infty} c_k = \left(\sum_{k=0}^{\infty} a_k\right)\left(\sum_{k=0}^{\infty} b_k\right)$. ■

Suppose that a function f can be represented by its Maclaurin series with radius of convergence $\rho > 0$. If $0 < r < \rho$, then the sequence $\{P_n\}$ of polynomials defined by

$$P_n(x) = \sum_{k=0}^{n} \frac{f^{(k)}(0)}{k!} x^k$$

converges uniformly to f on the interval $[-r, r]$. This shows that any function f which is represented by its Maclaurin series (or, more generally, by a Taylor series) is the limit of a uniformly convergent sequence of polynomials. In order for f to have a Maclaurin series, it must be infinitely differentiable on an open interval containing 0. However, it turns out that every continuous function defined on a closed and bounded interval is the limit of a uniformly convergent sequence of polynomials. This is an interesting result since, in contrast with arbitrary continuous functions, polynomials are easy to evaluate, graph, differentiate, etc. In addition, since polynomials are a particularly simple class of continuous functions and since uniform convergence preserves many properties of functions, it is a useful theoretical result. There are several different ways to prove this fact; the method used here involves integrals. Some of the computations in the proof will be left to the reader.

THEOREM 7.30 Weierstrass Approximation Theorem If f is a continuous function on an interval $[a, b]$, then there exists a sequence $\{P_n\}$ of polynomials such that $\{P_n\}$ converges uniformly to f on $[a, b]$.

Proof. We will prove the theorem for the case in which f is a continuous function on $[0, 1]$ with $f(0) = 0 = f(1)$. The proof that the general case follows from this result will be left as an exercise.

Extend the function f to $\mathbb{R}$ by letting $f(x) = 0$ for $x \in \mathbb{R} \setminus [0, 1]$. The function f is then continuous for all real numbers. For each positive integer n, let Q_n be the polynomial defined by $Q_n(x) = c_n(1 - x^2)^n$, where c_n is a constant chosen so that $\int_{-1}^{1} Q_n = 1$. To estimate the magnitude of c_n, compute

$$\begin{aligned}\int_{-1}^{1} (1 - x^2)^n \, dx = 2\int_0^1 (1 - x^2)^n \, dx &\geq 2\int_0^{1/\sqrt{n}} (1 - x^2)^n \, dx \\ &\geq 2\int_0^{1/\sqrt{n}} (1 - nx^2) \, dx = \frac{4}{3\sqrt{n}} > \frac{1}{\sqrt{n}}.\end{aligned}$$

This shows that $c_n < \sqrt{n}$ for each value of n. The graphs of several Q_n's are shown in Figure 7.3. Define a function $P_n \colon [0, 1] \to \mathbb{R}$ by

$$P_n(x) = \int_{-1}^{1} f(x + t)Q_n(t) \, dt.$$

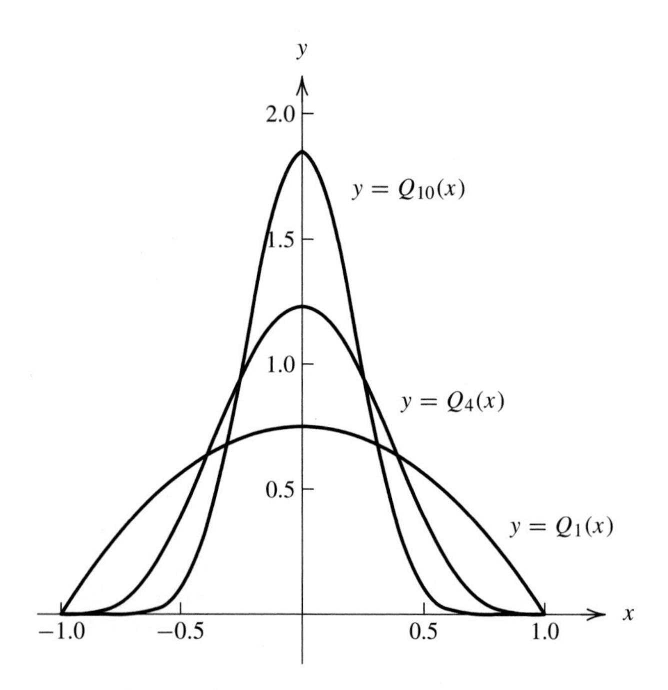

Figure 7.3 The graphs of Q_1, Q_4, and Q_{10}

Noting that $f(t) = 0$ for all t outside of the interval $[0, 1]$ and making a substitution of variables in the integral yields

$$P_n(x) = \int_{-x}^{1-x} f(x+t)Q_n(t)\,dt = \int_0^1 f(s)Q_n(s-x)\,ds.$$

Since the expression $f(s)Q_n(s-x)$ can be written as a polynomial in powers of x with coefficients that are functions of s (see the exercises), integrating this expression with respect to s yields a polynomial in x. Hence, the function P_n is a polynomial for each positive integer n. We will prove that the sequence $\{P_n\}$ converges uniformly to f on $[0, 1]$.

Let M be a bound for f and let $\epsilon > 0$. Since f is uniformly continuous on $\mathbb{R}$, there exists a positive number $\delta < 1$ such that $|f(y) - f(x)| < \epsilon/2$ for all x, y that satisfy $|y - x| \leq \delta$. For $x \in [\delta, 1]$,

$$0 \leq Q_n(x) = c_n(1-x^2)^n < \sqrt{n}(1-\delta^2)^n.$$

Since the sequence $\{\sqrt{n}(1-\delta^2)^n\}$ converges to 0 (see Theorem 2.14), the sequence $\{Q_n\}$ converges uniformly to 0 on $[\delta, 1]$. It then follows from Theorem 7.11 that $\lim\limits_{n\to\infty} \int_\delta^1 Q_n = 0$. Consequently, there exists a positive integer N such that

$\int_\delta^1 Q_n < \epsilon/(8M)$ for all $n \geq N$. If $x \in [0, 1]$ and $n \geq N$, then

$$\begin{aligned}
|P_n(x) - f(x)| &= \left| \int_{-1}^{1} f(x+t) Q_n(t)\, dt - \int_{-1}^{1} f(x) Q_n(t)\, dt \right| \\
&\leq \int_{-1}^{1} |f(x+t) - f(x)| Q_n(t)\, dt \\
&\leq \int_{-1}^{-\delta} 2M\, Q_n + \int_{-\delta}^{\delta} (\epsilon/2) Q_n + \int_{\delta}^{1} 2M\, Q_n \\
&\leq \int_{\delta}^{1} 2M\, Q_n + \int_{-1}^{1} (\epsilon/2) Q_n + \int_{\delta}^{1} 2M\, Q_n \\
&< \epsilon/4 + \epsilon/2 + \epsilon/4 = \epsilon.
\end{aligned}$$

Hence, the sequence $\{P_n\}$ converges uniformly to f on $[0, 1]$. This completes the proof. ∎

This proof of the Weierstrass Approximation Theorem does not involve any deep ideas, but it is not all that enlightening. The origin of the polynomials is not clear and the convergence of the polynomials to the function is difficult to visualize. In addition, it is not easy to actually find the polynomials that approximate the function. Nevertheless, a sequence of polynomials that converges uniformly to a continuous function f on an interval $[a, b]$ does exist.

The final result in this section is the construction of a continuous nowhere differentiable function. An example of such a function was first published by Weierstrass in 1872. The existence of such a function created quite a stir among mathematicians. It had been taken for granted that continuous functions were differentiable at most points; think about the type of graph you normally draw to represent a continuous function. After rigorous definitions for continuity of functions and convergence of series were given, it was possible to see where these definitions led. It is imperative that a simple mental picture of a continuous function be set aside and replaced with any function that satisfies the definition of continuity. Results contrary to intuition sometimes appear and catch mathematicians by surprise. When this occurs, either the definition has been poorly formulated (and thus needs to be altered) or intuition needs to be expanded to include new possibilities. In this case, since the definitions of continuity and convergence are well established, it is the intuition that must adapt.

The example of a continuous nowhere differentiable function given in the proof of the next theorem is different than the one originally published by Weierstrass. Weierstrass' example is easier to write down (see the exercises), but proving that it is nowhere differentiable is more difficult. The construction of the function in the proof uses a common method for creating continuous functions: express the function as the sum of a uniformly convergent series of continuous functions.

THEOREM 7.31 There exists a continuous function that is not differentiable at any point.

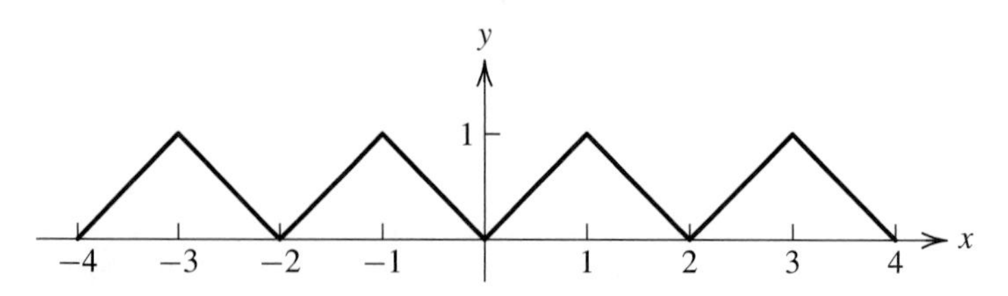

Figure 7.4 The graph of g as defined in the proof of Theorem 7.31

Proof. Define a function $g: \mathbb{R} \to \mathbb{R}$ by letting $g(x) = |x|$ for $-1 \le x \le 1$ and $g(x+2) = g(x)$ for all other values of x. (The graph of g can be found in Figure 7.4.) By definition, the function g is continuous on $\mathbb{R}$, $0 \le g(x) \le 1$ for all x, and $|g(y) - g(x)| \le |y - x|$ for all x and y. The function f defined by

$$f(x) = \sum_{k=0}^{\infty} \left(\frac{3}{4}\right)^k g(4^k x)$$

is continuous on $\mathbb{R}$ since the convergence of the series is uniform (by the Weierstrass M-test) and each term of the series is continuous. We will show that for each real number x, there exists a sequence $\{\delta_n\}$ of nonzero real numbers such that $\{\delta_n\}$ converges to 0, but the sequence

$$\left\{\frac{f(x+\delta_n) - f(x)}{\delta_n}\right\}$$

does not converge. This will show that f is not differentiable at x.

Fix $x \in \mathbb{R}$ and let n be a positive integer. Choose $\delta_n = \pm 4^{-n}/2$ so that there are no integers between $4^n x$ and $4^n(x + \delta_n)$. Note that

$$\begin{aligned} \left|g(4^k x + 4^k \delta_n) - g(4^k x)\right| &= 0 \text{ if } k > n, \\ \left|g(4^k x + 4^k \delta_n) - g(4^k x)\right| &= 1/2 \text{ if } k = n, \text{ and} \\ \left|g(4^k x + 4^k \delta_n) - g(4^k x)\right| &\le |4^k \delta_n| \text{ if } k < n. \end{aligned}$$

This assertion requires some justification. For $k > n$, the value is 0 since $4^k \delta_n$ is a multiple of 2 and g is a periodic function with period 2. For $k = n$, the value is $1/2$ since $|4^n \delta_n| = 1/2$ and the function g is linear with a slope of 1 or -1 on the interval whose endpoints are $4^k x$ and $4^k x + 4^k \delta_n$. (This is where the choice of δ_n is used.) For $k < n$, the inequality follows from the fact that $|g(y) - g(x)| \le |y - x|$. These results and the Reverse Triangle Inequality yield

$$\begin{aligned} \left|\frac{f(x+\delta_n) - f(x)}{\delta_n}\right| &= \left|\sum_{k=0}^{n} \left(\frac{3}{4}\right)^k \cdot \frac{g(4^k x + 4^k \delta_n) - g(4^k x)}{\delta_n}\right| \\ &\ge \left(\frac{3}{4}\right)^n 4^n - \sum_{k=0}^{n-1} \left(\frac{3}{4}\right)^k \cdot \left|\frac{g(4^k x + 4^k \delta_n) - g(4^k x)}{\delta_n}\right| \\ &\ge 3^n - \sum_{k=0}^{n-1} 3^k = 3^n - \frac{3^n - 1}{3 - 1} > 3^n/2. \end{aligned}$$

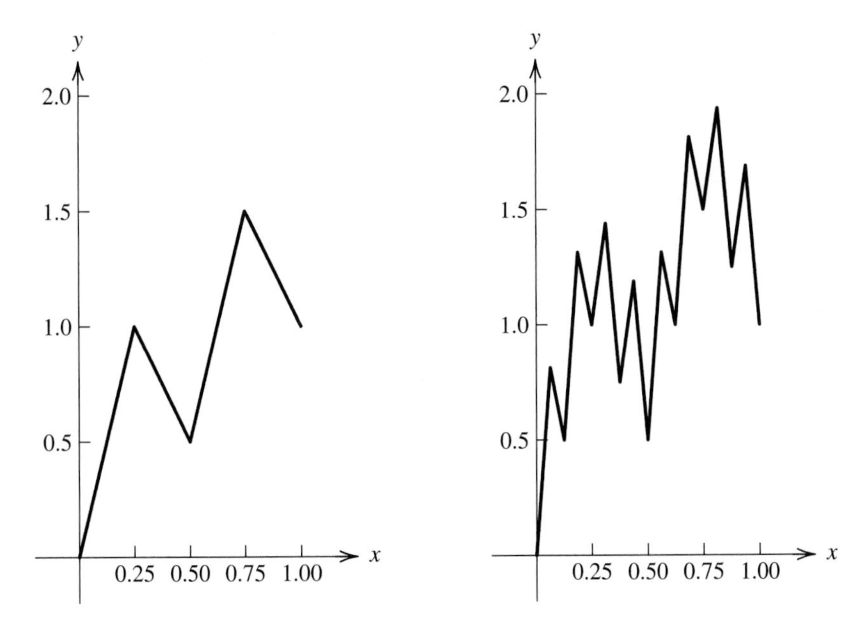

Figure 7.5 The graphs of $\sum_{k=0}^{n} (\frac{3}{4})^k g(4^k x)$ for $n = 1$ (left) and $n = 2$

The sequence $\{\delta_n\}$ converges to 0, but the sequence

$$\left\{ \frac{f(x+\delta_n) - f(x)}{\delta_n} \right\}$$

is unbounded and hence does not converge. It follows that f is not differentiable at x. This completes the proof. ∎

Each of the continuous functions $g(4^k x)$ has points of nondifferentiability. As k increases, the number of points of nondifferentiability increases. The sum of all of these functions turns out to be a continuous function with no points of differentiability. Two of the partial sums of the series defining f are graphed in Figure 7.5; for $n = 1$ and $n = 2$, the erratic behavior of this function is already evident. However, it is impossible to sketch the graph of a nowhere differentiable function because the graph oscillates an infinite number of times in every interval.

Although the functions used to define f were not differentiable functions, a nowhere differentiable function can be the limit of a uniformly convergent sequence of differentiable functions. By Theorem 7.30, any continuous nowhere differentiable function defined on $[a, b]$ is the limit of a uniformly convergent sequence of polynomials on the interval $[a, b]$. The actual construction of these polynomials is very difficult, but such a sequence does exist. The original example by Weierstrass (see the exercises) also involves a limit of differentiable functions. Consequently, uniform convergence does not preserve differentiability at all.

Exercises

1. Let $\{a_k\}$ and $\{b_k\}$ be two sequences and let $\{p_k\}$ be a listing of all the terms in the set $\{a_i b_j : i, j \geq 1\}$. Suppose that $\{a_k\}$ converges to 0 and that $\{b_k\}$ is bounded.

a) Give an example to show that the sequence $\{p_k\}$ may not converge to 0.

b) Suppose further that $\{b_k\}$ converges to 0. Does it follow that $\{p_k\}$ converges to 0?

2. In the proof of Theorem 7.25, a certain arrangement of the products p_k is required. Write out the terms $p_0, p_1, p_2, \ldots, p_{24}$ for such an arrangement.

3. Let $a_0 = 0$ and $a_n = (-1)^{n+1}/\sqrt{n}$ for each $n \geq 1$. Show that the Cauchy product of the series $\sum_{k=0}^{\infty} a_k$ with itself does not converge.

4. Suppose that f and g are infinitely differentiable functions.

a) Find and prove a formula for the higher derivatives of fg. That is, find a formula for $(fg)^{(n)}$ in terms of the derivatives of f and g.

b) Assume that f and g have Maclaurin series. Write down the Maclaurin series for fg based upon your answer to part (a).

c) Show that the Maclaurin series from part (b) is the same as the Cauchy product of the Maclaurin series for f and g.

5. Use the Cauchy product of appropriate series to find the first 5 nonzero terms of the Maclaurin series of the given function.

a) $e^x \sin x$ **b)** $e^x \cos x$ **c)** $\sin x \cosh x$

6. Prove that the radius of convergence of any power series whose coefficients form a bounded sequence is at least 1.

7. Adopting the notation of the proof of Abel's Theorem, verify that the function f satisfies $f(x) = (1 - x) \sum_{k=1}^{\infty} s_k x^k$.

8. Give a simpler proof of Abel's Theorem under the condition that $\sum_{k=1}^{\infty} a_k$ is absolutely convergent.

9. Let $\{a_n\}$ be the sequence defined by

$$a_n = \frac{1}{n+1} + \frac{1}{n+2} + \cdots + \frac{1}{2n}$$

for each positive integer n. Prove that this sequence converges to $\ln 2$ by showing that a_n is related to the partial sums of the series $\sum_{k=1}^{\infty} (-1)^{k+1}/k$.

10. Use Abel's Theorem and the Maclaurin series for $\arctan x$ to find a series representation for π. (Compare this result with Exercise 17 in Section 7.4.)

11. Suppose that the radius of convergence of the power series $\sum_{k=0}^{\infty} a_k x^k$ is $\rho > 0$ and that $\sum_{k=0}^{\infty} a_k \rho^k$ converges. Use Abel's Theorem to prove that $\lim_{x \to \rho^-} \left(\sum_{k=0}^{\infty} a_k x^k \right) = \sum_{k=0}^{\infty} a_k \rho^k$.

12. Use the result of the previous exercise to prove that a power series is continuous at each point in its interval of convergence.

13. Prove Theorem 7.29 by considering the product of the functions f and g defined by $f(x) = \sum_{k=0}^{\infty} a_k x^k$ and $g(x) = \sum_{k=0}^{\infty} b_k x^k$, respectively.

14. Prove that $(1 - x^2)^n \geq 1 - nx^2$ for all $x \in [0, 1]$ and for all positive integers n. This result is used in the proof of Theorem 7.30.

15. Adopting the notation of the proof of Theorem 7.30, find c_1, c_2, c_3, and c_4.

16. Prove that the function P_n, as defined in the proof of Theorem 7.30, is a polynomial for each positive integer n.

17. For the function $\sin(\pi x)$ on the interval $[0, 1]$, find the polynomials P_1, P_2, and P_3 defined in Theorem 7.30. (A calculus software package makes this exercise much easier.) The Maclaurin series for $\sin(\pi x)$ also converges uniformly to $\sin(\pi x)$ on the interval $[0, 1]$ and is therefore another sequence of polynomials that converges uniformly to $\sin(\pi x)$ on $[0, 1]$. How do these two sequences compare?

18. Finish the proof of Theorem 7.30 by considering a general continuous function defined on $[a, b]$.

19. Let f be a continuous function defined on $[a, b]$ and let $\epsilon > 0$. Prove that there exists a polynomial P such that $|P(x) - f(x)| < \epsilon$ for all $x \in [a, b]$.

20. Let f be a continuous function defined on $[a, b]$ and let $\epsilon > 0$.

a) Prove that there exists a polynomial P such that $\int_a^b |f - P| < \epsilon$.

b) Show that the polynomial in part (a) can be chosen so that all of its coefficients are rational numbers.

21. Prove that the function $\sin x$ cannot be the limit on $\mathbb{R}$ of a uniformly convergent sequence of polynomials.

22. Suppose that f is continuous on $[0, 1]$ and that $\int_0^1 f(x)x^n \, dx = 0$ for each nonnegative integer n. Prove that $f(x) = 0$ for all $x \in [0, 1]$.

23. Consider the function g defined in the proof of Theorem 7.31. Show that $g(x)$ represents the distance from x to the nearest even integer.

24. Consider the function $f: \mathbb{R} \to \mathbb{R}$ defined by

$$f(x) = \sum_{k=0}^{\infty} \frac{1}{2^k} \cos(3^k x).$$

Each term of this series (an example of a Fourier series) is differentiable at every point.

a) Prove that f is continuous on $\mathbb{R}$.

b) Use the sequence $\{\pi/(2 \cdot 3^n)\}$ to prove that f is not differentiable at 0.

c) Use the fact that f is periodic to prove that f is not differentiable at any integer multiple of π.

d) Prove that $f(x) - \cos(x)$ is not differentiable at 0.

e) Prove that f is not differentiable at any integer multiple of $\pi/3$.

f) Prove that f is not differentiable at any number of the form $n\pi/3^k$, where n is an integer and k is a positive integer. How "common" are these numbers?

g) The proofs of the preceding facts are not that difficult, but the proof that this function, which is of the form of the original examples of Weierstrass, is nowhere differentiable is a bit more involved. The interested reader can consult the work by Titchmarsh [26].

7.7 SUPPLEMENTARY EXERCISES

1. For the sequence $\{f_n\}$ of functions defined on $[0, 1]$ by $f_n(x) = n^2 x e^{-nx}$, examine both the hypotheses and the conclusion of Theorem 7.12.

2. Consider the series $\sum_{k=1}^{\infty} (|x| - k)^{-2}$ and define a function f by letting $f(x)$ be the sum of this series for each value of x for which the series converges.

a) For what values of x does the series converge?

b) On what intervals does the series converge uniformly?

c) Is f continuous at each point for which it is defined?

d) Is the function f bounded?

3. Show that the set of all polynomials with rational coefficients can be expressed as a sequence $\{P_n\}$. Then prove that every continuous function on $[0, 1]$ is the uniform limit of some subsequence of $\{P_n\}$.

4. Let P be the polynomial $P(x) = x^5 - 4x^2 + 3x - 2$. Find the Taylor series centered at 2 for P and the Taylor series centered at -1 for P. Is it possible to find these series using algebra only?

5. Complete the following steps to prove that e is an irrational number.

a) Suppose that $e = p/q$, where p and q are positive integers. Prove that

$$p(q-1)! = \sum_{k=0}^{\infty} \frac{q!}{k!}.$$

b) Use the result from (a) to prove that $\sum_{k=q+1}^{\infty} (q!/k!)$ is an integer.

c) Prove that $\sum_{k=q+1}^{\infty} (q!/k!)$ belongs to the interval $(0, 1)$ and obtain a contradiction.

6. Complete the following steps to prove that π is an irrational number. (This exercise has been adapted from the article by Niven [16].)

a) Suppose that $\pi = p/q$, where p and q are positive integers. For each positive integer n, define f_n by

$$f_n(x) = \frac{x^n (p - qx)^n}{n!}.$$

Prove that $f_n^{(k)}(0)$ and $f_n^{(k)}(\pi)$ are integers for $k = 0, 1, \ldots, 2n$.

b) Prove that $\{f_n\}$ converges uniformly to the zero function on $[0, \pi]$.

c) Prove that $\lim_{n\to\infty} \int_0^{\pi} f_n(x) \sin x \, dx = 0$.

d) For each positive integer n, define F_n by $F_n(x) = \sum_{k=0}^{n} (-1)^k f_n^{(2k)}(x)$. Prove that $F_n(0)$ and $F_n(\pi)$ are integers for each positive integer n.

e) Verify the following equality:

$$\int_0^{\pi} f_n(x) \sin x \, dx = \Big(F_n'(x) \sin x - F_n(x) \cos x\Big)\Big|_0^{\pi} = F_n(0) + F_n(\pi).$$

f) Use parts (c), (d), and (e) to obtain a contradiction.

7. For each positive integer n, define a function f_n by $f_n(x) = n\cos(2x/n) - n$. Determine whether or not $\{f_n\}$ converges pointwise on any intervals. If so, find those intervals on which the convergence is uniform.

8. Let $f: [a, b] \to \mathbb{R}$ be continuous on $[a, b]$. Prove that there exists a sequence $\{\phi_n\}$ of step functions that converges uniformly to f on $[a, b]$. Does this result extend to bounded functions?

9. Find the first four nonzero terms of the Maclaurin series for $\tan x$ in two different ways.

a) Compute derivatives and use the formula for the coefficients.

b) Perform long division on the Maclaurin series for $\sin x$ and $\cos x$.

10. Suppose that the sequence $\{f_n\}$ converges pointwise to f on an interval I and that f is bounded by M on I. For each positive integer n, define g_n on I by

$$g_n(x) = \begin{cases} -M, & \text{if } f_n(x) < -M; \\ f_n(x), & \text{if } -M \le f_n(x) \le M; \\ M, & \text{if } f_n(x) > M. \end{cases}$$

Prove that the sequence $\{g_n\}$ converges pointwise to f on I.

11. Let $f: [0, 1] \to \mathbb{R}$ be a continuous function with $f(1) = 0$. For each positive integer n, define $g_n: [0, 1] \to \mathbb{R}$ by $g_n(x) = f(x)x^n$. Prove that the sequence $\{g_n\}$ converges uniformly on the interval $[0, 1]$.

12. For each positive integer n, define f_n by $f_n(x) = \sqrt{2n+1}\,x(1 - x^2)^n$. Determine whether or not the sequence $\{f_n\}$ converges uniformly on $[0, 1]$.

13. For each positive integer n, define f_n by $f_n(x) = x + 1/n$. Show that $\{f_n\}$ converges uniformly on $\mathbb{R}$, but that $\{f_n^2\}$ does not.

14. We have seen that the set of points at which a power series converges is an interval. This is not the case for a Fourier series. Prove that the series $\sum_{k=1}^{\infty} \sin(kx)$ converges for $x = 0$ and $x = \pi$ and diverges for $x = \pi/2$.

15. Prove that $\int_0^1 x^x\,dx = \sum_{k=1}^{\infty} (-1)^{k+1}/k^k$.

16. For each real number x, define $f(x) = \sum_{k=1}^{\infty} \dfrac{kx - \lfloor kx \rfloor}{k^2}$, where, as usual, $\lfloor x \rfloor$ is the greatest integer less than or equal to x.

a) Show that the domain of f is all real numbers.

b) Show that f has a point of discontinuity in each interval.

c) Show that f is Riemann integrable on an arbitrary interval $[a, b]$.

d) Express $\int_0^1 f$ as an infinite series.

17. Evaluate $\lim_{x\to 0} \dfrac{e^{x^3} - 1}{x - \sin x}$ by first finding the Maclaurin series for the numerator and denominator.

18. Find the Maclaurin series for the function $\ln\left|\dfrac{1+x}{1-x}\right|$. Use this series to find a series whose sum is $\ln 10$.

19. Let $\{f_n\}$ be a sequence of convex functions (see Definition 4.25) defined on an interval I. Suppose that $\{f_n\}$ converges uniformly on I to a function f. Prove that f is convex on I.

Remark. As mentioned in this chapter, the set of complex numbers is better suited for the study of power series than the set of real numbers. While a knowledge of complex numbers is not necessary for the study of real analysis, the following exercise gives a tantalizing hint of the interesting results and fascinating connections that are contained in complex analysis.

20. Consider the Maclaurin series for e^x, $\sin x$, and $\cos x$. Assume that $i = \sqrt{-1}$ behaves just like any other constant.

a) Use some algebraic manipulation to show that $e^{ix} = \cos x + i \sin x$.

b) Use the result in part (a) to derive the remarkable equation $e^{i\pi} + 1 = 0$, which links the mathematical constants 0, 1, π, e, and i.

c) By part (a), it follows that $i = e^{i\pi/2}$. Use this fact and the laws of exponents to compute the rather bizarre quantity i^i.

d) Use the equation from part (a) to express $\sin x$ and $\cos x$ in terms of e, i, and x.

e) Use part (d) to find relationships between the functions $\sin x$ and $\sinh x$ and the functions $\cos x$ and $\cosh x$. (See Exercise 15 in Section 1.5 for the definitions of $\sinh x$ and $\cosh x$.)

Remark. One of the many applications of power series is as a tool to find solutions to certain differential equations. The first step is to assume that the solution has a power series representation centered at some point c; for the exercises here, we will always use $c = 0$. Substituting the series $\sum_{k=0}^{\infty} a_k x^k$ into the differential equation leads to one or more equations involving the coefficients a_k of the power series. (The uniqueness of power series representations is used to determine these equations.) Solving these equations for the coefficients yields a power series representation of the solution. The following exercises provide simple illustrations of this method.

21. Find a function f that satisfies $f'(x) = f(x)$ for all x by assuming that f has a power series representation of the form $\sum_{k=0}^{\infty} a_k x^k$, substituting this expression into the differential equation, and solving for the a_k's. The answer should look familiar.

22. Use a power series centered at 0 to find a function f that satisfies $f'(x) = -xf(x)$.

23. Use a power series centered at 0 to find a solution to the second order differential equation $y'' - x^2y' - 2xy = 0$. Your answer should contain two arbitrary constants.

Remark. The next few exercises discuss conditions that guarantee that the limit of a sequence of continuous functions is continuous. We will make extensive use of the following definition.

Let $\{f_k\}$ be a sequence of functions defined on an interval I and suppose that $\{f_k\}$ converges pointwise on I to a function f.

a) The sequence $\{f_k\}$ converges **quasi-uniformly** to f at the point $c \in I$ if for each $\epsilon > 0$ and positive integer K there exist $\delta > 0$ and a positive integer $m \geq K$ such that $|f_m(x) - f(x)| < \epsilon$ for all $x \in I$ that satisfy $|x - c| < \delta$.

b) The sequence $\{f_k\}$ converges **quasi-uniformly** to f on I if for each $\epsilon > 0$ and positive integer K there exist a finite number of positive integers $K_1, K_2, \ldots, K_q$ such that $K_i \geq K$ for all i and $\min\{|f_{K_i}(x) - f(x)| : 1 \leq i \leq q\} < \epsilon$ for all $x \in I$.

24. Suppose that $\{f_k\}$ converges pointwise to a function f on an interval I, let $c \in I$, and assume that each f_k is continuous at c. Prove that $\{f_k\}$ converges quasi-uniformly to f at c if and only if f is continuous at c.

25. Suppose that $\{f_k\}$ converges pointwise to a function f on an interval $[a, b]$ and that $\{f_k\}$ converges quasi-uniformly to f at each point of $[a, b]$. Prove that $\{f_k\}$ converges quasi-uniformly to f on $[a, b]$.

26. Suppose that $\{f_k\}$ converges pointwise to a function f on an interval $[a, b]$ and assume that each f_k is continuous on $[a, b]$. Prove that $\{f_k\}$ converges quasi-uniformly to f on $[a, b]$ if and only if f is continuous on $[a, b]$.

Remark. The next set of exercises discusses conditions that guarantee that the pointwise limit of a sequence of Riemann integrable functions is Riemann integrable and the integral of the limit function is the limit of the integrals of the functions in the sequence. A sequence $\{f_n\}$ is **uniformly bounded** on $[a, b]$ if there is a number M such that $|f_n(x)| \leq M$ for all $x \in [a, b]$ and for all positive integers n.

27. For each positive integer n, let f_n be a bounded function defined on an interval I. Suppose that $\{f_n\}$ converges uniformly on I. Prove that $\{f_n\}$ is uniformly bounded.

28. Let $\{c_n\}$ be any sequence of real numbers and define f_n by $f_n(x) = c_n \sin(n\pi x)$ for $0 \leq x \leq 1/n$ and $f_n(x) = 0$ for all other values of x. Find general conditions on the sequence $\{c_n\}$ to guarantee each of the following.

a) The sequence $\{f_n\}$ converges uniformly on $[0, 1]$ to the zero function.

b) $\lim_{n\to\infty} \int_0^1 f_n = 0$.

29. Let $\{f_n\}$ be a sequence of Riemann integrable functions that converges pointwise to a function f on $[a, b]$. Suppose that $\{f_n\}$ converges uniformly to f on each closed subinterval of (a, b) and that $\{f_n\}$ is uniformly bounded on $[a, b]$. Prove that f is Riemann integrable on $[a, b]$ and $\int_a^b f = \lim_{n\to\infty} \int_a^b f_n$.

30. Let $\{f_n\}$ be a sequence of Riemann integrable functions that converges pointwise to a function f on $[a, b]$ and let $P = \{x_i : 0 \leq i \leq q\}$ be a partition of $[a, b]$. Suppose that $\{f_n\}$ converges uniformly to f on each closed subinterval of (x_{i-1}, x_i) for $1 \leq i \leq q$ and that $\{f_n\}$ is uniformly bounded on $[a, b]$. Prove that f is Riemann integrable on $[a, b]$ and $\int_a^b f = \lim_{n\to\infty} \int_a^b f_n$.

31. Let p be a positive integer and let $g_n(x) = \cos^{2n}(p!\pi x)$ for each positive integer n. The sequence $\{g_n\}$ converges pointwise on $[0, 1]$ to a function g.

a) Find the limit function g.

b) Show that $\{g_n\}$ satisfies the hypotheses of the previous exercise.

c) Find $\int_0^1 g_n$ for each positive integer n.

d) By the previous exercise, the sequence $\left\{\int_0^1 g_n\right\}$ converges to 0. Try to prove this fact using the formula for $\int_0^1 g_n$ from part (c).

32. Prove that the following two statements are equivalent. Statement (i) is known as the **Bounded Convergence Theorem** for the Riemann integral. Its proof is a bit beyond the level of this text; the interested reader can consult the articles by Lewin [13] and Luxemburg [14].

i. If $\{f_n\}$ is a uniformly bounded sequence of Riemann integrable functions that converges pointwise on an interval $[a, b]$ to a Riemann integrable function f, then $\int_a^b f = \lim_{n\to\infty} \int_a^b f_n$.

ii. If $\{f_n\}$ is a uniformly bounded sequence of nonnegative Riemann integrable functions that converges pointwise on an interval $[a, b]$ to the zero function, then $\lim_{n\to\infty} \int_a^b f_n = 0$.

Remark. As mentioned in the text, it is possible to weaken the hypotheses of Theorem 7.12 and still obtain positive results. A proof of a result that does not require the functions to be continuously differentiable is outlined in the next exercise.

33. Let $\{f_n\}$ be a sequence of differentiable functions defined on an interval $[a, b]$, let $c \in [a, b]$, and suppose that the sequence $\{f_n(c)\}$ converges. Suppose further that $\{f_n'\}$ converges uniformly to g on $[a, b]$. Then $\{f_n\}$ converges uniformly to a function f on $[a, b]$, the function f is differentiable on $[a, b]$, and $f' = g$. Complete the following steps to prove this result.

a) Use the Cauchy criterion for uniform convergence and the Mean Value Theorem to show that $\{f_n\}$ converges uniformly on $[a, b]$. Call the limit function f.

b) Fix $x \in [a, b]$ and define a sequence $\{D_n\}$ on $[a, b]$ by

$$D_n(v) = \begin{cases} \dfrac{f_n(v) - f_n(x)}{v - x}, & \text{if } v \neq x; \\ f_n'(x), & \text{if } v = x. \end{cases}$$

Note that D_n is continuous on $[a, b]$ for each n. Prove that $\{D_n\}$ converges uniformly on $[a, b]$. Find a formula for the limit function D.

c) Complete the proof by explaining the following equality.

$$\begin{aligned} f'(x) &= \lim_{v\to x} \frac{f(v) - f(x)}{v - x} = \lim_{v\to x} D(v) = D(x) \\ &= \lim_{n\to\infty} D_n(x) = \lim_{n\to\infty} f_n'(x) = g(x). \end{aligned}$$

Remark. It is an interesting problem to find the exact value of the sum of a convergent infinite series. The last exercises in this section address this problem for some particularly simple convergent series. Except for the last problem, all of the series are related to a geometric series in some way.

34. Prove that $x\left(\sum_{k=1}^{\infty} k^m x^k\right)' = \sum_{k=1}^{\infty} k^{m+1} x^k$ for each positive integer m.

35. Use the previous exercise to show that for each positive integer m,

$$\sum_{k=1}^{\infty} k^m x^k = \frac{P_m(x)}{(1-x)^{m+1}},$$

where P_m is a polynomial of degree m.

36. Find P_1, P_2, P_3, P_4, and P_5, then use these polynomials to find the following sums.

a) $\sum_{k=1}^{\infty} \frac{k}{5^k}$ **b)** $\sum_{k=1}^{\infty} \frac{k^2}{4^k}$ **c)** $\sum_{k=1}^{\infty} \frac{k^3 2^k}{3^k}$

d) $\displaystyle\sum_{k=1}^{\infty} \frac{k3^k}{4^{k+1}}$ **e)** $\displaystyle\sum_{k=1}^{\infty} \frac{k^4}{(-2)^k}$ **f)** $\displaystyle\sum_{k=1}^{\infty} \frac{(-1)^{k+1}k^5}{3^k}$

37. Prove the following properties of P_m.

a) $P_m(0) = 0$ for all positive integers m.

b) The coefficients of x^m and x in P_m are 1.

c) The sum of all the coefficients of P_m is $m!$.

38. Let $\left\{ {m \atop k} \right\}$ be the coefficient of x^k in P_m and adopt the convention that $\left\{ {m \atop k} \right\} = 0$ if $k < 0$ or $k > m$. Prove that

$$\left\{ {m+1 \atop k} \right\} = k\left\{ {m \atop k} \right\} + (m+2-k)\left\{ {m \atop k-1} \right\}$$

for all positive integers m and integers k.

39. Prove that $\left\{ {m \atop k} \right\} = \left\{ {m \atop m+1-k} \right\}$ for all positive integers m and integers k.

40. Show that $P_m(1/x) = P_m(x)/x^{m+1}$. Conclude that $1/x$ is a root of P_m if $x \neq 0$ is a root of P_m.

41. Prove that P_m has $m-1$ negative roots. (For further information on the behavior of the roots of P_m, see the article by Dubeau and Savoie [5].)

42. For a given positive integer m, determine the number of values of x that satisfy the equation $\sum_{k=1}^{\infty} k^m x^k = 0$.

43. Fix a positive integer m. Express both of the functions $\dfrac{P_m(x)}{x(1-x)^{m+1}}$ and $(1-x)^{m+1}$ as power series of the form $\sum_{k=0}^{\infty} a_k x^k$, then find their Cauchy product. Use the result to prove that

$$\left\{ {m \atop k} \right\} = \sum_{i=0}^{k-1} (-1)^i (k-i)^m \binom{m+1}{i}$$

for all positive integers k.

44. Find the values of x for which $\sum_{k=1}^{\infty} x^k/k$ converges and find the sum of the series.

45. The purpose of this exercise is to find the sum of the series $\sum_{n=1}^{\infty} 1/n^2$.

a) Use integration by parts and mathematical induction to prove that

$$\int_0^{\pi/2} \sin^{2n+1} x\, dx = \frac{2}{3} \cdot \frac{4}{5} \cdot \frac{6}{7} \cdot \cdots \cdot \frac{2n}{2n+1} = \frac{4^n (n!)^2}{(2n+1)!}$$

for each positive integer n.

b) Use the result in part (a) and a trigonometric substitution to show that

$$\int_0^1 \frac{x^{2n+1}}{\sqrt{1-x^2}}\, dx = \frac{4^n (n!)^2}{(2n+1)!}$$

for each positive integer n.

c) Use the binomial series to show that

$$\frac{1}{\sqrt{1-t^2}} = \sum_{n=0}^{\infty} \frac{(2n)!}{4^n (n!)^2} t^{2n}$$

for $|t| < 1$ and conclude that

$$\arcsin x = \sum_{n=0}^{\infty} \frac{(2n)!}{(2n+1)4^n (n!)^2} x^{2n+1}$$

for $|x| < 1$.

d) Use term by term integration to show that

$$\int_0^1 \frac{\arcsin x}{\sqrt{1-x^2}}\, dx = \sum_{n=1}^{\infty} \frac{1}{(2n-1)^2}.$$

e) Prove that $\sum_{n=1}^{\infty} 1/n^2 = \pi^2/6$.

8

Point-Set Topology

The first seven chapters of this book have been devoted primarily to the theory behind single variable calculus. The functions have been, for the most part, of a type familiar to the reader from calculus, and their domains have almost exclusively been intervals. It is possible, and sometimes necessary, to leave this familiar setting. In order to discuss more unusual functions, it is helpful to be aware of subsets of $\mathbb{R}$ that are not intervals. Although it is not immediately apparent, real numbers can form sets much more general than intervals. Before discussing the properties of various types of general sets of real numbers, it is a good idea to at least mention why such a study is useful.

The elements of the types of sets that occur in analysis are often called points. A point in $\mathbb{R}$ is a real number, a point in $\mathbb{R}^2$ is an ordered pair of real numbers, and a point in $\mathbb{R}^3$ is an ordered triple of real numbers. In a differential equations course, a set such as $\{f \in \mathcal{C}([0, 1]) : f''(x) + f(x) = x\}$, which represents the collection of all continuous functions f defined on $[0, 1]$ that satisfy the given differential equation, might be of interest. In this context, a point would refer to one of the functions in this set. In linear algebra, a point can refer to a vector in $\mathbb{R}^p$ or an element of an arbitrary vector space. Finally, if B represents the set of all bounded sequences of real numbers, then a point in B represents a sequence. Typically, a collection of elements such as those mentioned in the last few sentences is known as a **space**. The elements in a space are not just random objects. For example, in many spaces the addition of two elements can be defined and there is often a notion of distance between two elements. Examples include $\mathbb{R}^p$, the set of all bounded sequences, and the set of all continuous functions. In each of these cases, a **point** represents one element of the given space. The point might be a p-tuple, a sequence, or a function. It then becomes necessary to describe sets of points in these general spaces and, except for $\mathbb{R}$, the notion of an interval makes no sense. (The definition of an interval requires the concept of an order—most spaces cannot be ordered in a natural way.) **Point-set topology** is a branch of mathematics which deals with points and sets of points in an abstract space as well as various relationships between points and sets. It provides a framework for discussing points and sets that is independent of the objects that the points represent.

As indicated in the last paragraph, one of the motivating forces behind the development of point-set topology was a search for common descriptions of the relationships between sets and points in all different kinds of spaces. However, point-set topology is also a useful tool in real analysis because certain subsets of real numbers are not easily described in terms of intervals. For example, the following two questions give rise to sets that may not be expressible using intervals.

1. Let $f\colon [a, b] \to \mathbb{R}$ be an increasing function. What is the range of f?
2. Let $f\colon \mathbb{R} \to \mathbb{R}$ and let C be the set of all points x such that f is continuous at x. What type of set is C?

The descriptions of sets of real numbers such as these require concepts from point-set topology. The first four sections of this chapter introduce some of the ideas from point-set topology in the context of the set of real numbers. The last section of the chapter outlines these ideas in the more general setting of metric spaces.

8.1 Open and Closed Sets

The functions in this text have usually been defined on intervals. As the reader may recall, some of the theorems concerning continuous functions require that the interval be closed and bounded. (For one example, see the Extreme Value Theorem.) What property does a closed, bounded interval possess that is not possessed by an open, bounded interval? How can you determine if an arbitrary set of real numbers has this property? Furthermore, is it possible to express this property in a way that makes sense for spaces other than $\mathbb{R}$? We will begin to answer these questions in this section.

In what follows, we will be working with arbitrary subsets of real numbers. It is important to realize that there are sets of real numbers other than intervals or finite unions of intervals. This is easier said than done since sets other than intervals are hard to imagine. Familiar examples are the set of rational numbers and the set of irrational numbers. Two less familiar examples are the sets

$$A = \{x \in \mathbb{R} : 0.1 \le \sin(1/x) < 0.3\} \text{ and}$$
$$B = \{x \in \mathbb{R} : x \text{ has only even digits in its decimal expansion}\}.$$

With a little effort, you should be able to convince yourself that the set B contains no intervals at all. In other words, if $x, y \in B$ with $x < y$, then there exists a point z such that $z \notin B$ and $x < z < y$. It is important to overcome the natural tendency to think in terms of intervals and to realize that there are many varieties of sets of real numbers.

As mentioned in the introduction to this chapter, point-set topology considers the relationship between points and sets in an arbitrary space. The next definition considers some possible relationships between points and sets of real numbers.

DEFINITION 8.1 Let E be a set of real numbers.

a) A point x is an **interior point** of E if there exists a positive number r such that $(x - r, x + r) \subseteq E$.

b) A point x is an **isolated point** of E if there exists a positive number r such that $(x - r, x + r) \cap E = \{x\}$.

c) A point x is a **limit point** of E if for each positive number r, the set $(x - r, x + r) \cap E$ contains a point of E other than x.

d) The set E is **open** if all of its points are interior points.

e) The set E is **closed** if it contains all of its limit points.

To illustrate the definition, let $E = (-1, 1) \cup \{3\}$. If $x \in (-1, 1)$ and r is $\min\{x + 1, 1 - x\}$, then $r > 0$ and $(x - r, x + r) \subseteq (-1, 1) \subseteq E$. Hence, each point $x \in (-1, 1)$ is an interior point of E. If $r = 1$, then $(3 - 1, 3 + 1) \cap E = \{3\}$, so 3 is an isolated point of E. Finally, if $x \in [-1, 1]$ and $0 < r < 1$, then the set $(x - r, x + r) \cap E$ is actually an interval and thus contains infinitely many points of E. Therefore, each point $x \in [-1, 1]$ is a limit point of E. Based upon these observations, the set E is neither open (the point 3 is not an interior point) nor closed (it does not contain 1, which is one of its limit points).

The following observations involving the terms defined in Definition 8.1 are valid; the proofs of some of these observations will be requested in the exercises.

1. Every point of $\mathbb{Z}$ is an isolated point of $\mathbb{Z}$.
2. Every real number is a limit point of $\mathbb{Q}$ and a limit point of $\mathbb{R} \setminus \mathbb{Q}$.
3. Every real number is an interior point of $\mathbb{R}$.
4. The set $\mathbb{Q}$ has no interior points.
5. An isolated point of a set E must belong to E.
6. A limit point of a set E may or may not belong to E.
7. Every point in a set E is either an isolated point of E or a limit point of E.
8. A point x is a limit point of a set E if and only if for each positive number r, the set $(x - r, x + r) \cap E$ is infinite.
9. A point x is a limit point of a set E if and only if $O \cap (E \setminus \{x\}) \neq \emptyset$ for each open set O that contains x.
10. The sets $\emptyset$ and $\mathbb{R}$ are both open and closed.

There is no standard terminology for the concept of a limit point. In some texts, limit points are called **points of accumulation**, while other texts refer to them as **cluster points**. The term "point of accumulation" is a bit lengthy to say or write, the word "cluster" is a bit unfamiliar, and the adjective "limit" has already been used to describe limits of sequences and functions. Thus, none of these terms is ideally suited for this concept. However, the next theorem provides an explanation for the use of the term "limit point": a limit point of E is the limit of a sequence in E.

THEOREM 8.2 Let E be a set of real numbers. A point x is a limit point of the set E if and only if there exists a sequence of points in $E \setminus \{x\}$ that converges to x.

Proof. Suppose first that x is a limit point of E. By the definition of a limit point, for each positive integer n, there exists a point $x_n \neq x$ that belongs to the set $(x - 1/n, x + 1/n) \cap E$. Since $|x_n - x| < 1/n$ for all n, the sequence $\{x_n\}$ converges to x.

Now suppose that there exists a sequence $\{x_n\}$ of points in $E \setminus \{x\}$ that converges to x and let r be a positive number. Since $\{x_n\}$ converges to x, there exists a positive integer p such that $|x_n - x| < r$ for all $n \geq p$. It follows that x_p is an element of the set $(x - r, x + r) \cap E$ that is different from x. Hence, the point x is a limit point of E. ■

There is a simple connection between open and closed sets. As indicated in the following theorem, the complement of one type of set yields the other.

THEOREM 8.3 Let E be a set of real numbers.

a) The set E is open if and only if $\mathbb{R} \setminus E$ is closed.

b) The set E is closed if and only if $\mathbb{R} \setminus E$ is open.

Proof. To prove part (a), suppose first that E is open. We must show that $\mathbb{R} \setminus E$ contains all of its limit points. Let $x \in E$. Since E is an open set, there exists $r > 0$ such that $(x - r, x + r) \subseteq E$. This set inclusion is equivalent to the set equality $(x - r, x + r) \cap (\mathbb{R} \setminus E) = \emptyset$, so x is not a limit point of $\mathbb{R} \setminus E$. Therefore, every limit point of $\mathbb{R} \setminus E$ must belong to $\mathbb{R} \setminus E$.

To finish the proof of part (a), suppose that $\mathbb{R} \setminus E$ is closed. We must show that every point of E is an interior point of E. Let $x \in E$. Since $x \notin \mathbb{R} \setminus E$ and $\mathbb{R} \setminus E$ contains all of its limit points, the point x is not a limit point of $\mathbb{R} \setminus E$. Hence, there exists a positive number r such that $(x - r, x + r) \cap (\mathbb{R} \setminus E) = \emptyset$. It follows that $(x - r, x + r) \subseteq E$, so x is an interior point of E.

Part (b) follows from part (a) and the observation that $E = \mathbb{R} \setminus (\mathbb{R} \setminus E)$. This completes the proof. ■

Since the terms "open" and "closed" refer to sets, it is natural to determine how set operations affect these types of sets. The next theorem considers unions and intersections of open and closed sets.

THEOREM 8.4

a) The union of any collection of open sets is open and the intersection of any finite collection of open sets is open.

b) The intersection of any collection of closed sets is closed and the union of any finite collection of closed sets is closed.

Proof. Let $\{E_\alpha : \alpha \in A\}$ be an arbitrary collection of open sets (that is, for each element α of some set A there is an associated set E_α) and let $E = \bigcup_{\alpha \in A} E_\alpha$. We must show that each point of E is an interior point of E. Let $x \in E$. By the definition of union, there exists an index $\alpha_0 \in A$ such that $x \in E_{\alpha_0}$. Since E_{α_0} is an open set, there exists $r > 0$ such that $(x - r, x + r) \subseteq E_{\alpha_0}$. Since $E_{\alpha_0} \subseteq E$, it follows that x is an interior point of E.

Now let $\{E_i : 1 \le i \le n\}$ be a finite collection of open sets and let E be the set $\bigcap_{i=1}^{n} E_i$. Suppose that $x \in E$. By the definition of intersection, the point x belongs to each of the sets E_i. Consequently, for each appropriate integer i, there exists $r_i > 0$ such that $(x - r_i, x + r_i) \subseteq E_i$. Since there are only a finite number of sets, we can let $r = \min\{r_i : 1 \le i \le n\}$. Then $r > 0$ and

$$(x - r, x + r) \subseteq (x - r_i, x + r_i) \subseteq E_i$$

for each i. It follows that $(x - r, x + r) \subseteq E$, which shows that the set E is open. This proves part (a).

Part (b) follows from part (a) and DeMorgan's Laws (see Appendix B). The details will be left as an exercise. ■

In different terminology, Theorem 8.4 states that the collection of open sets is closed under arbitrary unions and finite intersections, and the collection of closed sets is closed under arbitrary intersections and finite unions. (It is perhaps unfortunate that the term "closed" has two different meanings in the previous sentence. One of its uses is consistent with the phrase, "the set of rational numbers is closed under addition and multiplication", which was discussed in Chapter 1. The context provides a clue to the intended meaning.) To illustrate the importance of a finite collection for each part of the theorem, note that

$$\bigcap_{n=1}^{\infty} \left(-\frac{1}{n}, \frac{1}{n}\right) = \{0\} \quad \text{and} \quad \bigcup_{n=1}^{\infty} \left[\frac{1}{n}, 2\right] = (0, 2].$$

The first equality shows that a countably infinite intersection of open sets may not be open and the second equality shows that a countably infinite union of closed sets may not be closed.

In Chapter 1, it was shown that an interval had one of nine possible forms (see Theorem 1.9). Some of these forms were called open intervals and others were called closed intervals. At the time, the important distinction was whether or not the interval contained its endpoints. However, the use of these adjectives to describe intervals is consistent with the definitions of open and closed sets.

THEOREM 8.5 Every open interval is an open set and every closed interval is a closed set.

Proof. A proof that every open interval is an open set will be left as an exercise. Let I be a closed interval. If $I = \mathbb{R}$, then I is closed by observation (10). If $I \neq \mathbb{R}$, then the interval I can assume one of three possible forms: $[a, b]$, $(-\infty, b]$, or $[a, \infty)$. In each of these cases, the set $\mathbb{R} \setminus I$ is either an open interval or the union of two open intervals. By the first part of the theorem and Theorem 8.4, the set $\mathbb{R} \setminus I$ is open. It then follows from Theorem 8.3 that the interval I is a closed set. ■

The converse of Theorem 8.5 is false: it is an easy matter to write down an open set that is not an open interval and a closed set that is not a closed interval.

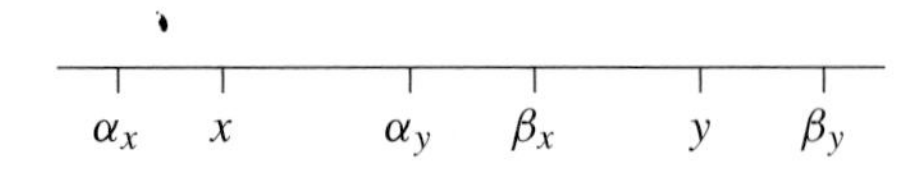

Figure 8.1 A possible arrangement of the points in the proof of Theorem 8.6

However, open sets are related to open intervals in a simple way. This important relationship is given in the following theorem.

THEOREM 8.6 Every nonempty open set of real numbers can be expressed as a countable union of disjoint open intervals.

Proof. Let O be a nonempty open set. For each $x \in O$, let

$$\alpha_x = \inf\{a : (a, x) \subseteq O\} \text{ and } \beta_x = \sup\{b : (x, b) \subseteq O\},$$

and let $I_x = (\alpha_x, \beta_x)$. (If, for a given x, the sets in question are unbounded, then define $\alpha_x = -\infty$ or $\beta_x = \infty$ as appropriate.) It will be left as an exercise to verify that $x \in I_x \subseteq O$ and that $\alpha_x, \beta_x \notin O$ for each $x \in O$. We will prove that for each $x, y \in O$ either $I_x \cap I_y = \emptyset$ or $I_x = I_y$. Let $x, y \in O$ with $x < y$ and suppose that $I_x \cap I_y \neq \emptyset$. Since $I_x \cap I_y \neq \emptyset$, it follows that $\alpha_y < \beta_x$. (See Figure 8.1 for one possible arrangement of these points.) Now

if $\alpha_y > \alpha_x$, then $(\alpha_x, y) \subseteq (\alpha_x, \beta_x) \cup (\alpha_y, y) \subseteq O$, contradicting the definition of α_y;

if $\alpha_y < \alpha_x$, then $(\alpha_y, x) \subseteq (\alpha_y, y) \subseteq O$, contradicting the definition of α_x.

It follows that $\alpha_x = \alpha_y$. Using the fact that $\alpha_x = \alpha_y$,

if $\beta_x < \beta_y$, then $(x, \beta_y) \subseteq (\alpha_y, \beta_y) \subseteq O$, contradicting the definition of β_x;

if $\beta_y < \beta_x$, then $(y, \beta_x) \subseteq (\alpha_x, \beta_x) \subseteq O$, contradicting the definition of β_y.

It follows that $\beta_x = \beta_y$. Therefore, the intervals I_x and I_y are the same.

To complete the proof, let $\mathcal{G}$ be the collection of all distinct intervals I_x chosen from the collection $\{I_x : x \in O\}$. Then $\mathcal{G}$ is a collection of disjoint open intervals, and O is the union of all these intervals. Since each of the intervals in $\mathcal{G}$ must contain a rational number and no two intervals in $\mathcal{G}$ contain the same rational number, there are a countable number of intervals in $\mathcal{G}$. This completes the proof. ■

Since open intervals are very familiar sets and, as shown in Theorem 8.6, an open set is just a countable union of open intervals, an open set is fairly easy to visualize. Since the complement of a closed set is an open set, the complement of a closed set is also easy to visualize. However, there is no simple characterization of a closed set in terms of closed intervals. This is an important point, but one that is difficult to accept without some experience with closed sets. It is tempting to think of a closed set E as a countable union of closed intervals, where the closed intervals form the gaps left by the countable union of open intervals comprising the open set $\mathbb{R} \setminus E$. To illustrate the fact that an arbitrary closed set cannot be written in this way, we will construct a closed set that contains no rational numbers. Let $\{r_n\}$ be a sequence of distinct numbers such that $\{r_n : n \in \mathbb{Z}^+\} = [0, 1] \cap \mathbb{Q}$. Define sets G and H by $G = \bigcup_{n=1}^{\infty} (r_n - 4^{-n}, r_n + 4^{-n})$ and $H = [0, 1] \setminus G$. The set G is open, the set H is closed, and $\{r_n : n \in \mathbb{Z}^+\} \subseteq G$. The sum of the lengths of all the intervals in G is

$$\sum_{n=1}^{\infty} \frac{2}{4^n} = \frac{2/4}{1-(1/4)} = \frac{2}{3}.$$

Since the interval $[0, 1]$ has length 1, there must be quite a few points in H. (This is not a proof, but hopefully it is a convincing argument.) The set H does not contain any rational numbers and, as a result, it contains no intervals. In other words, the set H is a rather "big" closed set that contains no rational numbers. This is very surprising given the fact that there is a rational number between any two real numbers. The set H is typical of the kind of set that appears frequently in higher analysis, but it is difficult to become comfortable with sets of this type. For now, let this example serve as a warning to avoid simplistic thinking about closed sets. An open set is a countable union of open intervals, but a closed set may not be a simple combination of closed intervals.

Given a set E, it is sometimes useful to construct sets that are related to E. The next definition provides the necessary terminology.

DEFINITION 8.7 Let E be a set of real numbers.

a) The **interior** of E, denoted E°, is the set of all interior points of E.

b) The **derived set** of E, denoted E', is the set of all limit points of E.

c) The **closure** of E, denoted $\overline{E}$, is the set $E \cup E'$.

For the set $E = (-1, 1) \cup \{3\}$ that was discussed after Definition 8.1, we find that $E^\circ = (-1, 1)$, $E' = [-1, 1]$, and $\overline{E} = [-1, 1] \cup \{3\}$. Since every real number is a limit point of $\mathbb{Q}$, the closure of $\mathbb{Q}$ is $\mathbb{R}$, that is, $\overline{\mathbb{Q}} = \mathbb{R}$. Since every point of $\mathbb{Z}$ is an isolated point of $\mathbb{Z}$, it follows that $\mathbb{Z}' = \emptyset$ and $\overline{\mathbb{Z}} = \mathbb{Z}$.

As indicated by the next theorem, for an arbitrary set E, the set E° is open and the sets E' and $\overline{E}$ are closed. In the proof, we will use observation (8) that was listed after Definition 8.1.

THEOREM 8.8 If E is a set of real numbers, then the set E° is open and the sets E' and $\overline{E}$ are closed.

Proof. We will prove that E' is a closed set and leave the proofs for the other two sets as exercises. To prove that E' is a closed set, it is sufficient to prove that E' contains all of its limit points. Let x be a limit point of E'. We must show that x is a limit point of E. Let $r > 0$. Since x is a limit point of E', the set $(x-r, x+r) \cap E'$ contains a point $y \neq x$. Since $(x-r, x+r)$ is an open set that contains y, there exists a positive number s such that $(y-s, y+s) \subseteq (x-r, x+r)$. Since y is a limit point of E, the set $(y-s, y+s) \cap E$ is infinite. It follows that the set $(x-r, x+r) \cap E$ is infinite as well. This shows that x is a limit point of E, which means that $x \in E'$. Since E' contains all of its limit points, it is a closed set. ■

Using the notation of Definition 8.7, a set E is closed if and only if $E' \subseteq E$. Since $E' \subseteq E$ if and only if $E \cup E' = E$, it follows that a set E is closed if and only if $\overline{E} = E$. It should also be clear that a set E is open if and only if $E = E^\circ$.

Exercises

1. Consider the set $E = (1, 2] \cup \{3\} \cup [5, 7)$. Find the interior points of E, the isolated points of E, and the limit points of E.
2. Let $A = \{1/n : n \in \mathbb{Z}^+\}$.
   a) Prove that every point of A is an isolated point of A.
   b) Prove that 0 is a limit point of A.
3. Give an example of a set E and a point $x \in E$ such that x is neither an interior point of E nor an isolated point of E.
4. Prove observations (2) and (4) listed in the discussion following Definition 8.1.
5. Prove observation (8) listed in the discussion following Definition 8.1.
6. Prove that every finite set is closed.
7. Prove observation (9) listed in the discussion following Definition 8.1.
8. Prove that x is a limit point of a set E if and only if $E \cap O$ is infinite for each open set O that contains x.
9. Prove that x is a limit point of a set E if and only if $\{y : 0 < |y - x| < 1/n\} \cap E \neq \emptyset$ for each positive integer n.
10. Give three examples of sets that are neither open nor closed. Make the examples as distinctly different as possible.
11. Let E be a closed and bounded set. Prove that E contains its infimum and supremum.
12. Let E be a closed set, let $\{x_n\}$ be a sequence in E, and suppose that $\{x_n\}$ converges to x. Prove that $x \in E$.
13. Prove part (b) of Theorem 8.4.
14. Prove that every set of real numbers can be expressed as a union of closed sets and as an intersection of open sets.
15. Suppose A is an open set and B is a closed set. Prove that $A \setminus B$ is an open set and that $B \setminus A$ is a closed set.
16. Prove the portion of Theorem 8.5 that states that every open interval is an open set.
17. Use Theorem 8.2 to prove that the interval $[a, b]$ is a closed set.
18. Give an example of each of the following.
    a) An open set that is not an open interval.
    b) A closed set that is not a closed interval.
    c) A bounded, closed, uncountable set that is not an interval.
19. Express the set $[0, 1) \cup \{4\}$ as a countably infinite union of closed sets.
20. Express the set $[0, 1] \cup [3, 5]$ as a countably infinite intersection of open sets.
21. Adopting the notation of the proof of Theorem 8.6, prove that $x \in I_x \subseteq O$ and that α_x, β_x do not belong to O.
22. This exercise assumes knowledge of equivalence relations (see any elementary text in abstract algebra) and offers a different proof of Theorem 8.6. Let O be a nonempty open set. Define a relation $\sim$ on O by $x \sim y$ if and only if the closed interval with endpoints x and y is a subset of O. (If $x = y$, the interval is degenerate; it consists of a single point.) Prove that $\sim$ is an equivalence relation on O and that the equivalence classes of $\sim$ form the desired decomposition of O.

23. Determine whether each statement is true or false. If it is true, find an example. If it is false, prove that no such example can be found.

a) There exists a countably infinite closed set.

b) There exists a countably infinite set that is not closed.

c) There exists a countably infinite open set.

24. Let B be the set of all numbers in $[0, 1]$ that have a decimal expansion that contains only even digits. Prove that every point of B is a limit point of B and that no point of B is an interior point of B.

25. For the set $E = (-5, 3] \cup \mathbb{Z}^+$, find E°, E', and $\overline{E}$.

26. Give an example of a set E with the indicated property.

a) The derived set of E is $\{1, 2, 3, 4\}$.

b) The derived set of E is $\mathbb{Z}$.

c) The derived set of E is $\{1/n : n \in \mathbb{Z}^+\} \cup \{0\}$.

d) The set E is open, is not an interval, and $\overline{E} = [0, 2]$.

e) The set E is a proper subset of E'.

f) The set E is nonempty and $E' = E$.

g) The sets $(E')'$ and E' are different.

27. Prove that the set E° is open for every set E.

28. Let A and B be sets of real numbers and suppose that x is a limit point of $A \cup B$. Prove that x is a limit point of A or a limit point of B (or a limit point of both A and B).

29. Prove that $\overline{E}$ is a closed set for every set E. There are several ways to do this, but try to find a proof that uses the previous exercise.

30. Explain why there is no set E such that $E' = \{1/n : n \in \mathbb{Z}^+\}$.

31. Let E be a set. Prove that $x \in \overline{E}$ if and only if there exists a sequence $\{x_n\}$ in E that converges to x.

32. Let E be a bounded set. Prove that $\overline{E}$ is a bounded set.

33. Let E be a set of real numbers. Prove that $\overline{\overline{E}} = \overline{E}$.

34. Let E be a set of real numbers. Prove that E and $\overline{E}$ have the same limit points.

35. Let A and B be sets of real numbers.

a) Prove that $\overline{A \cup B} = \overline{A} \cup \overline{B}$.

b) Prove that $\overline{A \cap B} \subseteq \overline{A} \cap \overline{B}$.

c) Give an example to show that the inclusion in part (b) may be proper.

36. Let A and B be sets of real numbers.

a) Suppose that $A \subseteq B$. Prove that $\overline{A} \subseteq \overline{B}$.

b) Suppose that $\overline{A} \subseteq \overline{B}$. Does it follow that $A \subseteq B$?

Remark. The rest of the exercises in this section are in no particular order.

37. Let E be a set of real numbers. Prove that $\overline{E} \cap (\mathbb{R} \setminus E) = \overline{E} \setminus E^\circ$.

38. Let E be a set of real numbers. Prove that $\overline{E} = \displaystyle\bigcap_{n=1}^{\infty} \left(\bigcup_{x \in E} \left(x - \frac{1}{n}, x + \frac{1}{n} \right) \right)$.

39. Let E be an arbitrary set of real numbers and let E^i denote the set of all isolated points of E. Prove that E^i is a countable set.

40. The Bolzano-Weierstrass Theorem (see Theorem 2.19) states that every bounded sequence has a convergent subsequence. A more general statement of this result is the following: every bounded infinite set of real numbers has a limit point.

a) Give an example of an infinite unbounded set that has no limit points.

b) Use the Nested Intervals Theorem (see Theorem 2.15) to prove that every bounded infinite set of real numbers has a limit point.

c) Show how Theorem 2.19 is a special case of the result in part (b).

41. Let $\{a_n\}$ be a sequence of real numbers. Prove that $\{a_n\}$ converges to a if and only if for each open set O containing a, there exists a positive integer N such that $a_n \in O$ for all $n \geq N$.

42. The point x is a **boundary point** of E if each open set that contains x contains points of both E and $\mathbb{R} \setminus E$. The **boundary** of E, denoted ∂E, is the set of all boundary points of E.

a) Prove that $\partial E = \overline{E} \cap (\overline{\mathbb{R} \setminus E})$.

b) Prove that ∂E is a closed set.

c) Consider the statement: a point x is a boundary point of E if and only if x is a limit point of E. Determine whether or not either portion of this biconditional statement is valid. Give proofs and/or find counterexamples to support your answer.

8.2 Compact Sets

The introduction of point-set topology into real analysis is intended to clarify certain concepts. To make this point clear, consider the following two results, which were discussed in Chapter 3.

1. If K is a closed and bounded interval and if $f: K \to \mathbb{R}$ is continuous on K, then $f(K)$ is a closed and bounded interval.
2. If K is a closed and bounded interval and if $f: K \to \mathbb{R}$ is continuous on K, then f is uniformly continuous on K.

These results may not be valid if K is a bounded, open interval or if K is an unbounded interval. What is the property possessed by closed and bounded intervals that is crucial to these theorems? After studying the first section of this chapter, it should seem plausible that these results are valid if the word "interval" is replaced by "set". The resulting statements are indeed theorems, but there is a slight problem. The modified statements are not necessarily valid in other abstract spaces. Consequently, there must be some property inherent in closed and bounded intervals that is lacking in arbitrary closed and bounded sets. This property is known as compactness. The following two theorems are valid in any space.

1. If K is a compact set and if $f: K \to \mathbb{R}$ is continuous on K, then $f(K)$ is a compact set.
2. If K is a compact set and if $f: K \to \mathbb{R}$ is continuous on K, then f is uniformly continuous on K.

The purpose of this section is to define and study compact sets; the two theorems used here as illustrations will be proved in Section 8.3.

The definition of a compact set is more abstract than the other concepts defined thus far in this chapter. For sets of real numbers, it will follow that a set is compact if and only if it is closed and bounded. However, since this result is not valid in arbitrary spaces and since compact sets play an important role in analysis, it is necessary to learn the general definition of a compact set. Although the definition will seem foreign at first, with time and practice its usefulness should be appreciated.

In order to give a general definition of a compact set, we need to introduce the concept of an open cover. The symbol $\mathcal{G}$ is a script version of the letter G.

DEFINITION 8.9 Let E be a set of real numbers. A collection $\mathcal{G}$ of sets is an **open cover** of E if each set in $\mathcal{G}$ is open and E is contained in the union of all the sets in $\mathcal{G}$. The open cover $\mathcal{G}$ has a **finite subcover** if E is contained in the union of a finite number of sets in $\mathcal{G}$.

In other words, a collection $\mathcal{G}$ of open sets is an open cover of E if $E \subseteq \bigcup_{G \in \mathcal{G}} G$. The open cover $\mathcal{G}$ has a finite subcover if there exist sets $G_1, G_2, \ldots, G_n$ in $\mathcal{G}$ such that $E \subseteq \bigcup_{i=1}^{n} G_i$. To say it another way, $\mathcal{G}_0$ is a finite subcover if $\mathcal{G}_0 \subseteq \mathcal{G}$, the set $\mathcal{G}_0$ is finite, and $E \subseteq \bigcup_{G \in \mathcal{G}_0} G$. It is important to keep the notation straight. The set $\mathcal{G}$ is a collection of open sets (an element G of $\mathcal{G}$ is an open set) and the set $\mathcal{G}_0$ is a subset of $\mathcal{G}$, that is, each of the open sets that belongs to $\mathcal{G}_0$ also belongs to $\mathcal{G}$. The union of all of the elements of $\mathcal{G}$ is an open set that contains E. If it is possible to find a finite number of sets in $\mathcal{G}$ whose union contains E, then $\mathcal{G}$ has a finite subcover.

There are many possible open covers for any given set of real numbers. For the interval $(0, 1)$, all of the following collections are open covers:

$$\begin{aligned}
\mathcal{G}_1 &= \left\{\left(\frac{1}{n}, 1\right) : n \in \mathbb{Z}^+ \setminus \{1\}\right\}, \\
\mathcal{G}_2 &= \left\{\left(\frac{1}{n}, n\right) : n \in \mathbb{Z}^+ \setminus \{1\}\right\}, \\
\mathcal{G}_3 &= \left\{\left(\frac{1}{2n}, \frac{2n-1}{2n}\right) : n \in \mathbb{Z}^+ \setminus \{1\}\right\}, \\
\mathcal{G}_4 &= \left\{\left(\frac{n}{4}, \frac{n+2}{4}\right) : n \in \mathbb{Z}\right\}, \\
\mathcal{G}_5 &= \left\{\left(\frac{1}{r}, r\right) : r > 1\right\}, \\
\mathcal{G}_6 &= \left\{(-r, 1-r) : 0 < r < 1\right\}, \\
\mathcal{G}_7 &= \left\{(-0.2, 0.3), (0.1, 0.4), (0.3, 0.9), (0.7, 1.4)\right\}, \quad \text{and} \\
\mathcal{G}_8 &= \left\{(0, 1)\right\}.
\end{aligned}$$

The reader should verify that in each case, the union of all of the elements of $\mathcal{G}_i$ contains $(0, 1)$. Note that $\mathcal{G}_2 \subseteq \mathcal{G}_5$, but $\mathcal{G}_2$ is a countably infinite subcover rather than a finite subcover. The open cover $\mathcal{G}_4$ contains a finite subcover—namely,

$$\{(0, 0.5), (0.25, 0.75), (0.5, 1.0)\}.$$

Although these examples provide a variety of open covers of $(0, 1)$, there are many other possibilities.

DEFINITION 8.10 A set E is **compact** if every open cover of E has a finite subcover.

Using only the definition, it is usually easier to prove that a set is not compact than it is to prove that a set is compact. To show that a set is not compact, it is necessary to find one open cover that does not have a finite subcover. For example, the set $\mathbb{Z}^+$ is not compact since the open cover $\{(n-1, n+1) : n \in \mathbb{Z}^+\}$ has no finite subcover. In fact, if any open set is removed from this collection, the new collection is no longer a cover of $\mathbb{Z}^+$. As an aside, note that $\{(r, \infty) : r > 0\}$ is an open cover of $\mathbb{Z}^+$ that does have a finite subcover—namely, the one element set $\{(0.5, \infty)\}$. However, in order for a set to be compact, every open cover must have a finite subcover. Since there is at least one open cover of $\mathbb{Z}^+$ that does not have a finite subcover, the set $\mathbb{Z}^+$ is not compact.

The open cover $\{(-n, n) : n \in \mathbb{Z}^+\}$ of $\mathbb{R}$ illustrates that $\mathbb{R}$ is not a compact set. This follows from the fact that the union of the elements of any finite subset of this cover is a bounded set. The collection $\mathcal{G}_1$ is an open cover of $(0, 1)$ that has no finite subcover. To verify this, suppose that $\{(1/n_i, 1) : 1 \le i \le p\}$ is a finite subset of $\mathcal{G}_1$. Let $q = \max\{n_i : 1 \le i \le p\}$ and note that

$$\bigcup_{i=1}^{p} \left(\frac{1}{n_i}, 1\right) = \left(\frac{1}{q}, 1\right).$$

It follows that no finite subset of $\mathcal{G}_1$ can cover $(0, 1)$. Since there is an open cover of $(0, 1)$ that contains no finite subcover, the interval $(0, 1)$ is not a compact set.

To prove that a set is compact, it must be shown that every possible open cover contains a finite subcover. As an example, we will prove that the set

$$E = \{1/n : n \in \mathbb{Z}^+\} \cup \{0\}$$

is compact. Let $\mathcal{G}$ be an open cover of E. Then there exists a set $G_0 \in \mathcal{G}$ such that $0 \in G_0$. Since G_0 is an open set, there exists $r > 0$ such that $(-r, r) \subseteq G_0$. Choose a positive integer p such that $1/n < r$ for all $n > p$. Using the fact that $\mathcal{G}$ is an open cover once again, there exist sets $G_1, G_2, \ldots, G_p$ in $\mathcal{G}$ such that $1/n \in G_n$ for $n = 1, 2, \ldots, p$. It follows that the collection $\{G_n : 0 \le n \le p\}$ is a finite subcover of E. Since every open cover of E has a finite subcover, the set E is compact.

It is generally tedious to prove that a set is compact using only the definition. In practice, it is often easier to find simpler properties that characterize compact sets, then show that a set has these other properties. Recording some of the properties of compact sets is the goal of the next few theorems. Any proof that is omitted will be left as an exercise.

THEOREM 8.11 A compact set is closed and bounded.

Proof. Let K be a compact set. We will prove that K is closed and leave the proof that K is bounded as an exercise. To prove that K is closed, it is sufficient to prove that $\mathbb{R} \setminus K$ is open. Let $z \in \mathbb{R} \setminus K$. For each positive integer n, define $G_n = \{x \in \mathbb{R} : |x - z| > 1/n\}$. Since each G_n is the union of two open intervals, it is an open set. The collection $\{G_n : n \in \mathbb{Z}^+\}$ is an open cover of $\mathbb{R} \setminus \{z\}$ and thus an open cover of K. Since K is compact, this collection has a finite subcover $\{G_{n_i} : 1 \le i \le p\}$. Let $q = \max\{n_i : 1 \le i \le p\}$ and note that

$$K \subseteq \bigcup_{i=1}^{p} G_{n_i} = \{x \in \mathbb{R} : |x - z| > 1/q\}.$$

It follows that $(z - 1/q, z + 1/q) \subseteq \mathbb{R} \setminus K$, so z is an interior point of $\mathbb{R} \setminus K$. Since every point of $\mathbb{R} \setminus K$ is an interior point, the set $\mathbb{R} \setminus K$ is open. This completes the proof. ■

THEOREM 8.12 A closed subset of a compact set is compact.

Proof. Let K be a compact set and let $E \subseteq K$ be closed. Suppose that $\mathcal{G}$ is an open cover of E. The collection $\mathcal{G}^* = \mathcal{G} \cup \{\mathbb{R} \setminus E\}$ is an open cover of K. Since K is compact, this collection contains a finite subcover $\mathcal{G}_1^*$ of K. The collection $\mathcal{G}_1 = \mathcal{G}_1^* \setminus \{\mathbb{R} \setminus E\}$ is a finite subcover of E from the collection $\mathcal{G}$. Since every open cover of E contains a finite subcover, the set E is compact. ■

THEOREM 8.13 An arbitrary intersection of compact sets is compact and a finite union of compact sets is compact. ■

A **nested sequence of sets** is a sequence $\{A_n\}$ of sets such that $A_{n+1} \subseteq A_n$ for every n. Each set in the nested sequence contains the next one and all of the sets thereafter. Are there points that are common to all of the sets—that is, is the intersection of all the sets nonempty? In general, the answer is no; the intersection of a nested sequence of sets may be empty. One example is the nested sequence of intervals $\{[n, \infty)\}$. However, the intersection is nonempty if all of the sets in the nested sequence are nonempty and compact.

THEOREM 8.14 If $\{K_n\}$ is a nested sequence of nonempty compact sets, then there exists a point z that belongs to all of the sets K_n.

Proof. This will be a proof by contradiction. Suppose that there is no point z that belongs to all of the sets K_n. By taking complements and using one of DeMorgan's Laws, we find that

$$\mathbb{R} = \mathbb{R} \setminus \left(\bigcap_{n=1}^{\infty} K_n \right) = \bigcup_{n=1}^{\infty} (\mathbb{R} \setminus K_n).$$

In particular, the collection $\{\mathbb{R} \setminus K_n : n \in \mathbb{Z}^+\}$ is an open cover of K_1. Since K_1 is a compact set, this collection has a finite subcover $\{\mathbb{R} \setminus K_{n_i} : 1 \le i \le p\}$. Let $q = \max\{n_i : 1 \le i \le p\}$ and note that

$$K_1 \subseteq \bigcup_{i=1}^{p} (\mathbb{R} \setminus K_{n_i}) = \mathbb{R} \setminus K_q.$$

It follows that $K_1 \cap K_q = \emptyset$. However, $K_1 \cap K_q = K_q$ since the sequence $\{K_n\}$ is nested, and this contradicts the fact that K_q is nonempty. Hence, there is a point z that belongs to all of the sets K_n. ■

The next theorem gives a simple characterization of compact sets of real numbers: a set of real numbers is compact if and only if it is closed and bounded. This result is known as the Heine-Borel Theorem in honor of the mathematicians H. E. Heine (1821–1881) and Emile Borel (1871–1956). The Heine-Borel Theorem is also valid in $\mathbb{R}^n$, but there are spaces (such as the set of all continuous functions defined on the interval $[0, 1]$) in which there exist closed and bounded sets that are not compact. In other words, the proof of the next theorem involves special properties of the set of real numbers.

THEOREM 8.15 Heine-Borel Theorem A set of real numbers is compact if and only if it is closed and bounded.

Proof. Since every compact set is closed and bounded (see Theorem 8.11), it remains to be shown that every closed and bounded set of real numbers is compact. We first show that every closed and bounded interval $[a, b]$ is compact. Let $\mathcal{G}$ be an open cover of $[a, b]$. Let E be the set of all points x in $(a, b]$ such that the interval $[a, x]$ can be covered by a finite number of elements of $\mathcal{G}$. Since $\mathcal{G}$ is an open cover of $[a, b]$, there exists a set G_a in $\mathcal{G}$ such that $a \in G_a$. Since G_a is an open set, there exists a point $x \in (a, b]$ such that $[a, x] \subseteq G_a$. It follows that the set E is nonempty. Since E is bounded above by b, the Completeness Axiom states that E has a supremum, call it z. To show that $\mathcal{G}$ has a finite subcover, it is sufficient to prove that $z \in E$ and $z = b$.

Since $\mathcal{G}$ is an open cover of $[a, b]$, there exists a set $G_z \in \mathcal{G}$ such that $z \in G_z$. Since G_z is an open set, there exists a point $c \in (a, z)$ such that $[c, z] \subseteq G_z$. Since $z = \sup E$ and $c < z$, there exists a point $x \in (c, z] \cap E$. By the definition of the set E, the interval $[a, x]$ can be covered by a finite number of elements of $\mathcal{G}$. Since $[x, z] \subseteq G_z$, it follows that $[a, z] = [a, x] \cup [x, z]$ can also be covered by a finite number of elements of $\mathcal{G}$. This shows that $z \in E$. To show that $z = b$, we will assume that $z < b$ and obtain a contradiction. Since $z < b$ and G_z is an open set containing z, there exists a point $d \in (z, b)$ such that $[z, d] \subseteq G_z$. Now $[a, d] = [a, z] \cup [z, d]$ can be covered by a finite number of elements of $\mathcal{G}$. This is a contradiction to the fact that $z = \sup E$. It follows that $z = b$. Since $b \in E$, the open cover $\mathcal{G}$ contains a finite subcover. As every open cover of $[a, b]$ contains a finite subcover, the interval $[a, b]$ is a compact set.

Now let K be any closed and bounded set. Since K is bounded, there exists a number $M > 0$ such that $K \subseteq [-M, M]$. By the first part of the proof, the interval $[-M, M]$ is compact. By Theorem 8.12, the set K is compact. Hence, every closed and bounded set of real numbers is compact. This completes the proof. ■

The Heine-Borel Theorem makes it much easier to prove that a set is compact. It is usually difficult to prove that every open cover has a finite subcover, but relatively easy to show that a set is closed and bounded. For example, the set

$$E = \{n/(n+1) : n \in \mathbb{Z}^+\} \cup [1, 2]$$

is closed (it contains all of its limit points) and bounded. Therefore, the set E is compact. To prove that this set is compact using the definition would require much more effort.

The most important particular case of the Heine-Borel Theorem is the fact that every closed and bounded interval is compact. It is crucial to remember, however, that compact sets have no simple description in terms of closed intervals. Since compact sets are closed sets, an arbitrary compact set may be difficult to visualize.

We have related many of the basic concepts in real analysis to sequences. As indicated by the next result, it is also possible to do this with compact sets.

THEOREM 8.16 A set K of real numbers is compact if and only if every sequence in K has a subsequence that converges to a point in K.

Proof. Suppose that K is a compact set. By the Heine-Borel Theorem, the set K is closed and bounded. Let $\{x_n\}$ be a sequence in K. Since K is bounded, the sequence $\{x_n\}$ is bounded. By the Bolzano-Weierstrass Theorem, this sequence has a convergent subsequence $\{x_{p_n}\}$. Let z be the limit of this subsequence. Since z is the limit of a sequence in K, either $z \in K$ or z is a limit point of K. Since K is closed, it contains all of its limit points. It follows that $z \in K$. Hence, every sequence in K has a subsequence that converges to a point in K.

The proof of the converse will be left as an exercise. ■

A set K with the property that every sequence in K has a subsequence that converges to a point in K is sometimes said to be **sequentially compact**. The previous theorem thus states that a set of real numbers is compact if and only if it is sequentially compact. As we will see in Section 8.5, this result is valid in any metric space.

Exercises

1. Let E be a set of real numbers.
   - **a)** Prove that the collection $\{(x-1, x+1) : x \in E\}$ is an open cover of E.
   - **b)** Prove that the collection $\{(-n, n) : n \in \mathbb{Z}^+\}$ is an open cover of E.
   - **c)** Prove that E is bounded if and only if the collection in part (b) has a finite subcover.
2. Find an open cover of $[1, 2)$ that does not have a finite subcover.
3. Find an open cover $\mathcal{G}$ of the set $E = \{1/n : n \in \mathbb{Z}^+\}$ such that any proper subset of $\mathcal{G}$ is not an open cover of E.
4. Consider the following two open covers of $(0, 1)$:
$$\mathcal{G}_1 = \left\{\left(\frac{x}{2}, \frac{x+1}{2}\right) : 0 < x < 1\right\} \quad \text{and} \quad \mathcal{G}_2 = \left\{\left(\frac{x}{4}, \frac{x+1}{4}\right) : -1 < x < 4\right\}.$$
   Show that $\mathcal{G}_1$ does not contain a finite subcover, but that $\mathcal{G}_2$ does contain a finite subcover.
5. Let $\mathcal{G} = \{G_n : n \in \mathbb{Z}^+\}$ be an open cover of a set E. Discuss the validity and/or merits of the statement "the collection $\bigcup_{n=1}^{\infty} G_n$ is an open cover of E".

6. Use the definition of a compact set to prove each of the following.

a) A finite set is compact.

b) The set $E = \{(-1)^n/n : n \in \mathbb{Z}^+\}$ is not compact.

c) The interval $[0, \infty)$ is not compact.

d) The set E that contains 1 and every term of the sequence $\{n/(n+1)\}$ is compact.

7. Suppose that the sequence $\{x_n\}$ converges to the number x_0. Use the definition of a compact set to prove that the set $\{x_n : n = 0, 1, 2, \ldots\}$ is compact.

8. Suppose that closed covers and finite closed covers are defined in the obvious way. If E is a set of real numbers with the property that every closed cover has a finite subcover, what can you conclude about the set E?

9. Finish the proof of Theorem 8.11 by proving that a compact set is bounded.

10. Prove that the set $\mathbb{Q} \cap [0, 1]$ is not compact.

11. Prove Theorem 8.13. (Do not use the Heine-Borel Theorem.)

12. Use the Nested Intervals Theorem (see Theorem 2.15) to prove that the interval $[a, b]$ is compact.

13. Prove that the set E that contains all of the points in the sequence $\{(-1)^n/\sqrt[n]{2}\}$ and all the points in the set $\{x \in \mathbb{R} : 1 \leq |x| \leq 3\}$ is a compact set.

14. Let E be the set of all points x such that either $x = 0$ or there exist positive integers m and n such that $x = (m+n)/mn$. Prove that E is a compact set.

15. Use the Heine-Borel Theorem to prove Theorem 8.13.

16. Finish the proof of Theorem 8.16.

Remark. The rest of the exercises in this section are in no particular order.

17. Prove that every open (bounded or unbounded) interval can be expressed as a countable union of compact sets. Is the same result valid for any open set?

18. Given a point $c \in \mathbb{R}$ and a set $E \subseteq \mathbb{R}$, define the distance $\rho(c, E)$ from c to E by $\rho(c, E) = \inf\{|x - c| : x \in E\}$.

a) Prove that $x \in \overline{E}$ if and only if $\rho(x, E) = 0$.

b) Let $\alpha \geq 0$. Prove that $\{x \in \mathbb{R} : \rho(x, E) \geq \alpha\}$ is a closed set.

c) Suppose that E is a compact set. Prove that there exists a point $z \in E$ such that $\rho(c, E) = |z - c|$. (Try to do this exercise without using particular properties of $\mathbb{R}$, just the properties of compact sets considered in this section.)

d) Is the result in part (c) valid if E is an arbitrary closed set?

19. Let A and B be two sets of real numbers. Define the distance from A to B by $\rho(A, B) = \inf\{|a - b| : a \in A, b \in B\}$.

a) Give an example to show that the distance between two sets can be 0 even if the two sets are disjoint.

b) Give an example to show that the distance between two closed sets can be 0 even if the two sets are disjoint.

c) Suppose that A and B are disjoint compact sets. Prove that $\rho(A, B) > 0$.

20. Let E be a set of real numbers.

a) Prove that the collection of all intervals of the form $(s - r, s + r)$, where $r, s \in \mathbb{Q}$ and $r > 0$, is countable.

b) Let x be a real number and let O be an open set that contains x. Prove that there exist rational numbers $r_x > 0$ and s_x such that $x \in (s_x - r_x, s_x + r_x) \subseteq O$.

c) Prove that every open cover of E contains a countable subcover (a subcover that contains a countable number of elements). Recall that an open cover may contain an uncountable number of sets. This result is known as the **Lindelöf covering theorem**.

21. This exercise illustrates how open covers and the Heine-Borel Theorem can be used to prove properties of real-valued functions. Let $f: [a, b] \to \mathbb{R}$ and suppose that f has a one-sided limit at each point of $[a, b]$.

a) Prove that for each $c \in [a, b]$, there exist positive numbers δ_c and M_c such that $|f(x)| \leq M_c$ for all $x \in [a, b]$ that satisfy $|x - c| < \delta_c$.

b) Prove that $\{(c - \delta_c, c + \delta_c) : c \in [a, b]\}$ is an open cover of $[a, b]$ and therefore has a finite subcover.

c) Use parts (a) and (b) to prove that f is bounded on $[a, b]$.

22. This exercise shows how open covers and the Heine-Borel Theorem can be used to prove that continuous functions are uniformly continuous on closed and bounded intervals. Let $f: [a, b] \to \mathbb{R}$ be continuous and let $\epsilon > 0$.

a) Prove that for each $c \in [a, b]$, there exists $\delta_c > 0$ such that $|f(x) - f(c)| < \epsilon$ for all $x \in [a, b]$ that satisfy $|x - c| < \delta_c$.

b) Prove that $\{(c - \delta_c, c + \delta_c) : c \in [a, b]\}$ is an open cover of $[a, b]$ and therefore has a finite subcover.

c) Use parts (a) and (b) to prove that there exists $\delta > 0$ such that $|f(y) - f(x)| < 2\epsilon$ for all $x, y \in [a, b]$ that satisfy $|x - y| < \delta$.

23. Let C be the set of all continuous functions $f: [a, b] \to \mathbb{R}$ and let $E \subseteq C$. The set E is **equicontinuous** at $z \in [a, b]$ if for each $\epsilon > 0$ there exists $\delta > 0$ such that $|f(x) - f(z)| < \epsilon$ for all $f \in E$ and for all $x \in [a, b]$ that satisfy $|x - z| < \delta$. The set E is equicontinuous on $[a, b]$ if for each $\epsilon > 0$ there exists $\delta > 0$ such that $|f(y) - f(x)| < \epsilon$ for all $f \in E$ and for all $x, y \in [a, b]$ that satisfy $|y - x| < \delta$. Prove that E is equicontinuous on $[a, b]$ if and only if E is equicontinuous at each point of $[a, b]$.

8.3 Continuous Functions

Continuous functions are particularly easy to describe using terms from point-set topology. This description, which involves open sets, will be discussed in this section after presenting a slight generalization of the notion of open and closed sets. In addition, we will consider the relationship between continuous functions and compact sets and prove the two statements made in the introduction to Section 8.2.

The definitions of open and closed sets given in Section 8.1 implicitly assume that the universe is the set of all real numbers. A set $E \subseteq \mathbb{R}$ is open or closed if it possesses certain properties in relation to the set $\mathbb{R}$. What does it mean, for example, to say that a set $E \subseteq [a, b]$ is open if the interval $[a, b]$ is the universe under consideration? There are various ways to define open and closed sets relative to another set; one approach is given in the following definition.

DEFINITION 8.17 Let A be a set of real numbers.

a) A set $E \subseteq A$ is **open in** A if there is an open set O such that $E = A \cap O$.

b) A set $E \subseteq A$ is **closed in** A if there is a closed set K such that $E = A \cap K$.

The open set O and closed set K that appear in this definition are defined according to Definition 8.1. To be completely consistent, we should say that an open set is open in $\mathbb{R}$, but we will always assume that the terms "open" and "closed" apply to sets in relationship to $\mathbb{R}$ unless another set is explicitly mentioned; that is, the universe is assumed to be $\mathbb{R}$ unless stated otherwise.

To illustrate the definition, an interval of the form $[a, c)$, where $a < c \leq b$, is open in $[a, b]$ since

$$[a, c) = [a, b] \cap (a - 1, c).$$

Similarly, an interval of the form $[c, b)$, where $a < c < b$, is closed in (a, b) since

$$[c, b) = (a, b) \cap [c, b + 2].$$

As in these two examples, there are often many choices for the open set O or the closed set K. The next theorem, whose proof will be left as an exercise, records a number of simple properties of sets that are open or closed in an arbitrary set.

THEOREM 8.18 Let A be a set of real numbers.

1. A set $E \subseteq A$ is open in A if and only if $A \setminus E$ is closed in A.
2. A set $E \subseteq A$ is closed in A if and only if $A \setminus E$ is open in A.
3. If A is an open set, then a set $E \subseteq A$ is open in A if and only if it is open.
4. If A is a closed set, then a set $E \subseteq A$ is closed in A if and only if it is closed.
5. A set $E \subseteq A$ is closed in A if and only if $E' \cap A \subseteq E$. ■

The next theorem considers a type of function that has appeared frequently in this text—namely, a function defined on a closed and bounded interval. It states that these functions are continuous if and only if each member of a particular collection of sets is open.

THEOREM 8.19 A function $f\colon [a, b] \to \mathbb{R}$ is continuous on $[a, b]$ if and only if the sets

$$\{x \in [a, b] : f(x) < r\} \quad \text{and} \quad \{x \in [a, b] : f(x) > r\}$$

are open in $[a, b]$ for each real number r.

Proof. Suppose first that $\{x \in [a, b] : f(x) < r\}$ and $\{x \in [a, b] : f(x) > r\}$ are open in $[a, b]$ for each real number r. Let $c \in [a, b]$ and let $\epsilon > 0$. By hypothesis, there exist open sets O_1 and O_2 such that

$$\{x \in [a, b] : f(x) < f(c) + \epsilon\} = [a, b] \cap O_1;$$
$$\{x \in [a, b] : f(x) > f(c) - \epsilon\} = [a, b] \cap O_2.$$

The set $O = O_1 \cap O_2$ is an open set that contains c. Choose $\delta > 0$ so that $(c - \delta, c + \delta) \subseteq O$. If $|x - c| < \delta$ and $x \in [a, b]$, then $x \in [a, b] \cap O_1$ and $x \in [a, b] \cap O_2$ and it follows that

$$f(c) - \epsilon < f(x) < f(c) + \epsilon.$$

We have thus shown that $|f(x) - f(c)| < \epsilon$ for all $x \in [a, b]$ that satisfy $|x - c| < \delta$. Hence, the function f is continuous at c. Since c was an arbitrary point in $[a, b]$, the function f is continuous on $[a, b]$.

Now suppose that f is continuous on $[a, b]$. Let r be any real number. By parts (2) and (4) of Theorem 8.18, it is sufficient to prove that the sets

$$\{x \in [a, b] : f(x) \geq r\} \quad \text{and} \quad \{x \in [a, b] : f(x) \leq r\}$$

are closed. We will prove that the first set is closed; the proof for the other set is similar. Let $E = \{x \in [a, b] : f(x) \geq r\}$ and let c be a limit point E. Note that $c \in [a, b]$ since $E \subseteq [a, b]$. By Theorem 8.2, there exists a sequence $\{x_n\}$ in E that converges to c. Since f is continuous at c, the sequence $\{f(x_n)\}$ converges to $f(c)$. Since $f(x_n) \geq r$ for all n, we must have $f(c) \geq r$ as well. Hence, the limit point c belongs to E. This completes the proof. ∎

The functions that have been considered so far in this book have been defined on intervals or perhaps defined on an interval except at a single point. It is sometimes necessary and/or interesting to consider functions defined on more general sets. Only a minor change is necessary in order to modify the definition of continuity given in Chapter 3 to allow for functions defined on arbitrary sets.

DEFINITION 8.20 Let A be a set of real numbers, let $f: A \to \mathbb{R}$, and let $a \in A$. The function f is continuous at a if for each $\epsilon > 0$ there exists $\delta > 0$ such that $|f(x) - f(a)| < \epsilon$ for all $x \in A$ that satisfy $|x - a| < \delta$. The function f is continuous on A if f is continuous at each point of A.

The only significant change in this definition is the extra condition $x \in A$, which is added to guarantee that $f(x)$ is defined. One simple consequence of this definition is that a function $f: A \to \mathbb{R}$ is automatically continuous at an isolated point of A. If a is an isolated point of A and $\delta > 0$ is chosen so that $(a - \delta, a + \delta) \cap A = \{a\}$, then the only point in A that satisfies $|x - a| < \delta$ is a, and the inequality $|f(a) - f(a)| < \epsilon$ is trivially satisfied for every $\epsilon > 0$. In particular, every function $f: \mathbb{Z} \to \mathbb{R}$ is continuous on $\mathbb{Z}$.

For a more interesting example of a function that is continuous on a set, let $I_n = \big[2/(2n + 1), 1/n\big]$ for each positive integer n, let $E = \{0\} \cup \bigcup_{n=1}^{\infty} I_n$, and define a function $g: E \to \mathbb{R}$ by $g(0) = 0$ and $g(x) = 1/n$ if $x \in I_n$. We will prove that g is continuous on E. Let $c \in E$. There are two cases to consider. Suppose first that $c \neq 0$ and choose a positive integer p such that $c \in I_p$. It can be shown that there exists $\delta > 0$ such that $x \in E$ and $|x - c| < \delta$ implies $x \in I_p$ (see the exercises). For these values of x, it is clear that $|g(x) - g(c)| = 0$; this shows that g is continuous at c. Now suppose that $c = 0$. Let $\epsilon > 0$ and choose a positive integer q so that $1/q < \epsilon$. If $x \in E$ and $0 < x < 1/q$, then $x \in I_n$ for some $n \geq q$

and $|g(x) - g(c)| = 1/n \leq 1/q < \epsilon$. Hence, the function g is continuous at 0. We have thus shown that g is continuous on E.

Since the definition of continuity on a set is essentially the same as the definition of continuity on an interval, many of the theorems from Section 3.2 concerning continuous functions on an interval are valid for continuous functions on a set. If these results are needed, we will use them without offering a new proof. However, because of its importance and its validity in many spaces, one such result is stated here. The proof will be left as an exercise.

THEOREM 8.21 Let A be a set, let $f: A \to \mathbb{R}$, and let $c \in A$. Then the function f is continuous at c if and only if the sequence $\{f(x_n)\}$ converges to $f(c)$ for each sequence $\{x_n\}$ in A that converges to c. ■

The next result extends Theorem 8.19 to the case in which the domain is an arbitrary set of real numbers. It provides two statements that are equivalent to the statement "a function f is continuous on a set A".

THEOREM 8.22 Let A be a set of real numbers and let $f: A \to \mathbb{R}$. The following statements are equivalent:

1. The function f is continuous on A.
2. For each closed set E, the set $\{x \in A : f(x) \in E\}$ is closed in A.
3. For each open set G, the set $\{x \in A : f(x) \in G\}$ is open in A.

Proof. We will prove that (1) $\Rightarrow$ (2) and (2) $\Rightarrow$ (3); a proof that (3) $\Rightarrow$ (1) will be left as an exercise.

Suppose first that the function f is continuous on A. Let E be a closed set and let $H = \{x \in A : f(x) \in E\}$. By part (5) of Theorem 8.18, it is sufficient to prove that $H' \cap A \subseteq H$. Let $z \in H' \cap A$. Since z is a limit point of H, there exists a sequence $\{x_n\}$ in H that converges to z. Since $z \in A$ and f is continuous on A, the sequence $\{f(x_n)\}$ converges to $f(z)$. Since E is closed and $\{f(x_n)\}$ is a sequence in E, the point $f(z)$ belongs to E. It follows that $z \in H$, and we conclude that H is closed in A.

Now suppose that (2) holds and let G be an open set. We first note that

$$A \setminus \{x \in A : f(x) \in \mathbb{R} \setminus G\} = \{x \in A : f(x) \in G\}.$$

Since $\mathbb{R} \setminus G$ is closed, the hypothesis implies that the set $\{x \in A : f(x) \in \mathbb{R} \setminus G\}$ is closed in A. By part (2) of Theorem 8.18, the set $\{x \in A : f(x) \in G\}$ is open in A. This proves that (2) $\Rightarrow$ (3). ■

Let $f: A \to \mathbb{R}$. Given a set $G \subseteq \mathbb{R}$, the set $\{x \in A : f(x) \in G\}$ is called the **inverse image** of G. Statements (1) and (3) of Theorem 8.22 show that a function f is continuous on A if and only if the inverse image of every open set is open in A. The equivalence of these two statements is often used as the definition of a continuous function in an arbitrary space. Since this definition of continuity depends only on open sets and does not require inequalities, absolute values, or any measure of distance, it is a very general definition.

In Section 8.2, the notion of a compact set was introduced by looking at two theorems that involve continuous functions and closed and bounded intervals. It was claimed there that these two theorems remain valid for compact sets. We are now in a position to prove these two results.

The first theorem states that the continuous image of a compact set is compact. In other words, continuous functions preserve compactness: if f is continuous on a compact set K, then $f(K)$ is also a compact set. Since there are several ways to describe compact sets in $\mathbb{R}$ (see Definition 8.10, Theorem 8.15, and Theorem 8.16), there are several ways to prove this theorem. The proof presented here uses the connection between compact sets and sequences.

THEOREM 8.23 Let K be a compact set and let $f: K \to \mathbb{R}$. If f is continuous on K, then $f(K)$ is compact.

Proof. We will use Theorem 8.16 to prove that $f(K)$ is compact. Let $\{y_n\}$ be any sequence in $f(K)$. For each positive integer n, let x_n be a point in K for which $f(x_n) = y_n$. Since K is a compact set, Theorem 8.16 asserts that the sequence $\{x_n\}$ has a subsequence $\{x_{p_n}\}$ that converges to a point $x \in K$. Since f is continuous on K, the sequence $\{f(x_{p_n})\}$ converges to $f(x)$. Note that $f(x_{p_n}) = y_{p_n}$ for each n and that $f(x) \in f(K)$. Therefore, the sequence $\{y_n\}$ has a subsequence $\{y_{p_n}\}$ that converges to a point in $f(K)$. By Theorem 8.16, the set $f(K)$ is compact. ■

The second theorem concerning compact sets states that a continuous function is uniformly continuous if its domain is a compact set. (The reader should have no trouble extending the definition of uniform continuity for functions defined on an interval to functions defined on arbitrary sets.) Recall that, using informal language, a function is uniformly continuous on a set A if for each $\epsilon > 0$, a single $\delta > 0$ can be found that works for all $x \in A$. Let $f: A \to \mathbb{R}$ be continuous on A. Given $\epsilon > 0$, for each $x \in A$ there exists $\delta_x > 0$ such that $|f(y) - f(x)| < \epsilon$ for all $y \in A$ that satisfy $|y - x| < \delta_x$. To find one $\delta > 0$ that works for all x, it is natural to take the smallest δ_x. Unless A is a finite set, there are an infinite number of δ_x's, and the number $\inf\{\delta_x : x \in K\}$ may be 0 rather than positive. The advantage of a compact set is that it makes it possible to consider only a finite number of δ_x's. This is evident in the proof of the following theorem.

THEOREM 8.24 Let K be a compact set and let $f: K \to \mathbb{R}$. If f is continuous on K, then f is uniformly continuous on K.

Proof. Let $\epsilon > 0$. For each $x \in K$, there exists a positive number δ_x such that $|f(y) - f(x)| < \epsilon/2$ for all $y \in K$ that satisfy $|y - x| < 2\delta_x$. The collection

$$\mathcal{G} = \{(x - \delta_x, x + \delta_x) : x \in K\}$$

is an open cover of K. Since K is compact, this collection contains a finite subcover

$$\{(x_i - \delta_{x_i}, x_i + \delta_{x_i}) : 1 \le i \le p\}.$$

Let $\delta = \min\{\delta_{x_i} : 1 \le i \le p\}$ and note that $\delta > 0$. Suppose that $x, y \in K$ with $|y - x| < \delta$. Choose an index k such that $x \in (x_k - \delta_{x_k}, x_k + \delta_{x_k})$ and note that

$$|y - x_k| \le |y - x| + |x - x_k| < \delta + \delta_{x_k} \le 2\delta_{x_k}.$$

It follows that

$$|f(y) - f(x)| \le |f(y) - f(x_k)| + |f(x_k) - f(x)| < \epsilon/2 + \epsilon/2 = \epsilon.$$

Therefore, the function f is uniformly continuous on K. ■

We close this section with one more result about continuous functions defined on arbitrary sets. Let A be a set of real numbers and let $f\colon A \to \mathbb{R}$. There are instances in which it is useful to extend f so that f is defined for all real numbers. In other words, we want a function $f_e\colon \mathbb{R} \to \mathbb{R}$ such that $f_e(x) = f(x)$ for all $x \in A$. One simple way to do this is to define f_e by

$$f_e(x) = \begin{cases} f(x), & \text{if } x \in A; \\ 0, & \text{if } x \notin A. \end{cases}$$

This extension of f is sometimes useful, but one of its disadvantages is that f_e may not be continuous on $\mathbb{R}$ even when f is continuous on A. As a simple illustration, define $f\colon \mathbb{Q} \to \mathbb{R}$ by $f(x) = 1$ for all $x \in \mathbb{Q}$. The function f is continuous on $\mathbb{Q}$, but its extension f_e using the preceding construction is not continuous at any point. The natural extension in this case would be to let $f_e(x) = 1$ for all $x \in \mathbb{R}$. If $g\colon [a, b] \to \mathbb{R}$ is continuous on $[a, b]$, then the function g_e defined by

$$g_e(x) = \begin{cases} g(a), & \text{if } x < a; \\ g(x), & \text{if } a \le x \le b; \\ g(b), & \text{if } x > b; \end{cases}$$

is continuous on $\mathbb{R}$. However, it is not always possible to extend a continuous function defined on a set A to a continuous function defined on $\mathbb{R}$. For example, the function $h\colon (0, 1) \to \mathbb{R}$ defined by $h(x) = 1/x$ is continuous on $(0, 1)$, but h cannot be extended to a continuous function on $\mathbb{R}$. The problem in this case is that the function $1/x$ is unbounded at 0 and the domain of h contains points that are arbitrarily close to 0. A moment's thought may reveal that such a problem can only arise if the domain of the function does not contain all of its limit points. If the domain is a closed set, then a continuous extension should be possible.

The fact that continuous functions on closed sets can be extended to continuous functions on $\mathbb{R}$ is known as the Tietze Extension Theorem (named after H. Tietze (1880-1964)). As with several other results in this chapter, the version of this theorem stated here is a specific case of a much more general result. (A more general statement can be found in Munkres [15].) The following proof uses the idea of one-sided limit points. A point x is a **right-hand limit point** of a set E if $(x, x + r) \cap E \ne \emptyset$ for all $r > 0$, and x is **isolated on the right side** of E if $x \in E$ and $[x, x + r) \cap E = \{x\}$ for some $r > 0$. The left-hand versions are defined analogously.

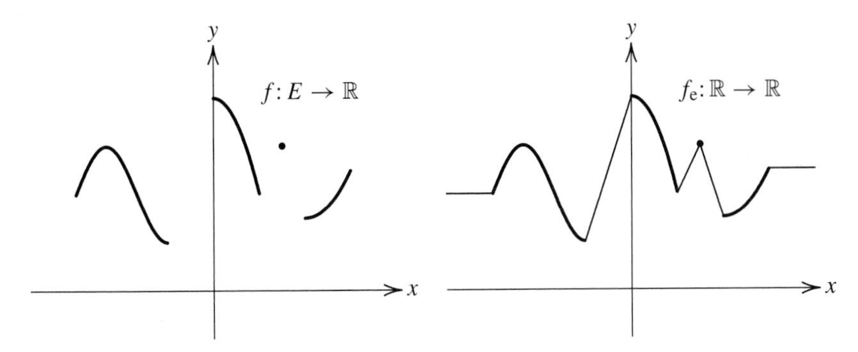

Figure 8.2 The function f_e is the linear extension of f

THEOREM 8.25 Tietze Extension Theorem Let E be a closed set. If a function $f: E \to \mathbb{R}$ is continuous on E, then there exists a continuous function $f_e: \mathbb{R} \to \mathbb{R}$ such that $f(x) = f_e(x)$ for all $x \in E$.

Proof. We will assume that E is unbounded both above and below; the other cases will be left as exercises. Since the set $\mathbb{R} \setminus E$ is open, it can be represented as a countable union of open intervals. Let $\{(a_n, b_n)\}$ be the sequence of disjoint open intervals whose union is $\mathbb{R} \setminus E$. (If there are only a finite number of open intervals, the proof is no different.) Define a function $f_e: \mathbb{R} \to \mathbb{R}$ by $f_e(x) = f(x)$ for each $x \in E$ and

$$f_e(x) = \frac{f(b_n) - f(a_n)}{b_n - a_n}(x - a_n) + f(a_n)$$

for each $x \in (a_n, b_n)$. Note that the function f_e is linear on the intervals comprising the complement of E. We will prove that f_e is continuous on $\mathbb{R}$.

Since f_e is linear on each of the intervals (a_n, b_n), the function f_e is continuous at each point of $\mathbb{R} \setminus E$. In addition, the function f_e is continuous at each isolated point of E since the two linear parts of f_e have the same value at these points. Suppose that c is a limit point of E. We will prove that $\lim_{x \to c^+} f_e(x) = f_e(c)$; the proof that $\lim_{x \to c^-} f_e(x) = f_e(c)$ is similar. If c is isolated on the right side of E, then f_e is linear on an interval having c as a left-endpoint, and the limit follows easily from the definition of f_e. Suppose that c is a right-hand limit point of E and let $\epsilon > 0$. Since f is continuous on E, there exists a point $c_1 \in E \cap (c, c+1)$ such that $|f(x) - f(c)| < \epsilon$ for all $x \in [c, c_1] \cap E$. Let $x \in [c, c_1] \setminus E$ and choose an index k such that $x \in (a_k, b_k)$. Since f_e is linear on $[a_k, b_k]$ and $a_k, b_k \in [c, c_1] \cap E$,

$$\begin{aligned} |f_e(x) - f_e(c)| &\le \max\{|f_e(a_k) - f_e(c)|, |f_e(b_k) - f_e(c)|\} \\ &= \max\{|f(a_k) - f(c)|, |f(b_k) - f(c)|\} < \epsilon. \end{aligned}$$

It follows that $\lim_{x \to c^+} f_e(x) = f_e(c)$. This completes the proof. ∎

The function f_e defined in the proof of the Tietze Extension Theorem is called the **linear extension** of f from E to $\mathbb{R}$. As indicated in Figure 8.2, the function f_e is essentially a "connect the dots" function. If the function f is bounded on E, then the function f_e constructed in the proof has the same bound as f (see the exercises). This is an important fact in some applications of this theorem.

Exercises

1. Give an example of a set E such that E is open in $[a, b]$, E is not an open set, and E is not an interval.
2. Let $A = (-\infty, 0) \cup \{1/n : n \in \mathbb{Z}^+\}$. Give an example of a set E that is closed in A but not closed in $\mathbb{R}$.
3. Let A be a set of real numbers. Prove that the sets A and $\emptyset$ are both open in A and closed in A.
4. Prove parts (1) and (2) of Theorem 8.18.
5. Prove parts (3) and (4) of Theorem 8.18.
6. Prove part (5) of Theorem 8.18.
7. Let $f: [-1, 1] \to \mathbb{R}$ be defined by $f(x) = 0$ if $-1 \le x \le 0$ and $f(x) = 1$ if $0 < x \le 1$. Use Theorem 8.19 to prove that f is not continuous on $[-1, 1]$.
8. Let $f: [-3, 1] \to \mathbb{R}$ be defined by $f(x) = x^2$. Use Theorem 8.19 to prove that f is continuous on $[-3, 1]$.
9. In the proof that the function g discussed after Definition 8.20 is continuous on E, the existence of a $\delta > 0$ with a certain property was used. The value of this δ depends on the integer p. Find an explicit formula for δ in terms of p.
10. Let $I_n = \big[2/(2n+1), 1/n\big]$ for each positive integer n, let $E = \{0\} \cup \bigcup_{n=1}^{\infty} I_n$, and define a function $f: E \to \mathbb{R}$ by $f(0) = 1$ and $f(x) = \sqrt[n]{5}$ if $x \in I_n$. Prove that f is continuous on E.
11. Let $I_n = \big[2^{-n}, 2^{-n} + 4^{-n}\big]$ for each positive integer n, let $E = \{0\} \cup \bigcup_{n=1}^{\infty} I_n$, and define a function $f: E \to \mathbb{R}$ by $f(0) = 0$ and $f(x) = (1/n)\sin(\pi/x)$ if $x \in I_n$. Prove that f is continuous on E.
12. Let $\{(a_n, b_n)\}$ be any sequence of disjoint open intervals and let $G = \bigcup_{n=1}^{\infty} (a_n, b_n)$. Given a sequence $\{v_n\}$ of real numbers, define a function $f: G \to \mathbb{R}$ by $f(x) = v_n$ if $x \in (a_n, b_n)$. Prove that f is continuous on G.
13. Prove Theorem 8.21.
14. Let f and g be continuous functions defined on a set A. Let E be the set of all real numbers $x \in A$ such that $f(x) = g(x)$. Prove that E is closed in A.
15. Finish the proof of Theorem 8.22 by proving that (3) $\Rightarrow$ (1).
16. Give an example of a nonconstant continuous function $f: \mathbb{R} \to \mathbb{R}$ and an open set G such that $f(G)$ is not open.
17. Define $f: \mathbb{R} \to \mathbb{R}$ by $f(x) = x$ if $x < 0$ and $f(x) = 1$ if $x \ge 0$. Find an open set G and a closed set E such that the set $\{x \in \mathbb{R} : f(x) \in G\}$ is not open and the set $\{x \in \mathbb{R} : f(x) \in E\}$ is not closed.
18. Give an example of a function $f: \mathbb{R} \to \mathbb{R}$ such that the set $\{x \in \mathbb{R} : f(x) = 0\}$ is neither open nor closed.

19. Consider the function $f:(-6,4] \to \mathbb{R}$ defined by $f(x) = x^2 + 1$. Find the inverse image of each set.
a) $[3, 10)$ **b)** $(-1, 2)$ **c)** $(-5, 5) \cup (50, \infty)$

20. Let A and B be nonempty subsets of $\mathbb{R}$ and let $f: A \to B$. Prove that f is continuous on A if and only if the inverse image of each set that is open in B is open in A.

21. Use the definition of a compact set to prove Theorem 8.23.

22. Let K be a compact set and let $f: K \to \mathbb{R}$ be continuous on K. Prove that there exist points $c, d \in K$ such that $f(c) \leq f(x) \leq f(d)$ for all $x \in K$. Note that this is a generalization of the Extreme Value Theorem.

23. Let K be a compact set and let $f: K \to \mathbb{R}$ be continuous on K. Suppose that f is one-to-one on K. Use Theorem 8.22 to prove that f_{inv} is continuous on $f(K)$.

24. Let A and E be nonempty sets of real numbers. Define a function $f: A \to \mathbb{R}$ by $f(x) = \inf\{|y - x| : y \in E\}$. Prove that f is uniformly continuous on A.

25. Use an argument similar to the one given in the proof of Theorem 8.24 to prove the following statement. If $f: [a, b] \to \mathbb{R}$ has a limit at each point of $[a, b]$, then f is bounded on $[a, b]$.

26. Let $E = \{1/n : n \in \mathbb{Z}^+\}$ and let $f: E \to \mathbb{R}$. Find and prove a general condition on f that guarantees that f can be extended to a continuous function on $\mathbb{R}$.

27. Finish the proof of the Tietze Extension Theorem by considering the cases in which E is bounded above, bounded below, or bounded. It is possible to do this by reducing the other cases to the case in which E is unbounded both above and below. For example, if E is bounded above, let $f(x) = f(\sup E)$ for all $x > \sup E$. Write up this informal suggestion precisely and show that the resulting set is still closed and that the new function f is still continuous on the larger set.

28. Let $E = \mathbb{Z} \cup \bigcup_{n=1}^{\infty} [n + 0.3, n + 0.7]$ and define $f: E \to \mathbb{R}$ by $f(n) = (-1)^n$ for each $n \in \mathbb{Z}$ and $f(x) = x$ for $x \in [n + 0.3, n + 0.7]$ for each $n \in \mathbb{Z}^+$.

a) Prove that E is a closed set.

b) Prove that f is continuous on E.

c) Sketch the graph of the linear extension of f on the interval $[-5, 5]$.

29. Let f be a bounded continuous function defined on a closed set E. Show that the linear extension of f has the same bound as f.

30. Let $A \subseteq \mathbb{R}$ be nonempty and let $f: A \to \mathbb{R}$ be uniformly continuous on A.

a) Suppose that $a \in \overline{A} \setminus A$ and let $\{x_n\}$ be any sequence in A that converges to a. Prove that $\{f(x_n)\}$ converges, then show that the limit of the sequence $\{f(x_n)\}$ is the same for every sequence $\{x_n\}$ in A that converges to a.

b) For each $a \in \overline{A} \setminus A$, let $f(a)$ be the common limit of the sequences $\{f(x_n)\}$ from part (a). In this way, the function f can be extended to $\overline{A}$. Prove that f is uniformly continuous on $\overline{A}$.

c) Prove that there exists a continuous function $g: \mathbb{R} \to \mathbb{R}$ such that $f(x) = g(x)$ for all $x \in A$.

Remark. The rest of the exercises in this section are in no particular order.

31. Let A and B be two nonempty disjoint compact sets. Prove that there exists a continuous function $f: \mathbb{R} \to \mathbb{R}$ such that $f(x) = 0$ for all $x \in A$ and $f(x) = 1$ for all $x \in B$.

32. Let A be a set of real numbers. Prove that a function $f: A \to \mathbb{R}$ is continuous on A if and only if the sets

$$\{x \in A : f(x) < r\} \quad \text{and} \quad \{x \in A : f(x) > r\}$$

are open in A for each real number r.

33. Let A be a set of real numbers and consider the function $f: \mathbb{R} \to \mathbb{R}$ defined by $f(x) = 1$ if $x \in A$ and $f(x) = 0$ if $x \notin A$. Determine the set of points at which f is continuous.

34. Let A be a set of real numbers, let $f: A \to \mathbb{R}$, and let $a \in A$. Prove that f is continuous at a if and only if for every open set V containing $f(a)$ there exists an open set U containing a such that $f(A \cap U) \subseteq V$.

8.4 Miscellaneous Results

The first three sections of this chapter have only scratched the surface of the topics considered in point-set topology. Some further topics in this area that are of interest in elementary real analysis are discussed in this section. One such topic is the concept of a connected set, which is the point-set topology analogue of an interval. Other topics include perfect sets, G_δ sets, and nowhere dense sets. The last two topics will help answer a question about the set of points of continuity of a function.

In Section 8.2, it was shown that compact sets are, in many ways, a topological extension of closed and bounded intervals. However, a compact set of real numbers may not be an interval, and an open interval is not a compact set. Thus, we are still lacking a point-set topology concept that generalizes the notion of an interval. The following definition can be used to extend the notion of an interval to a general space.

DEFINITION 8.26 Let E be a set of real numbers.

a) A **separation** of E is a pair of nonempty disjoint sets U and V such that U and V are each open in E and $E = U \cup V$.

b) The set E is **connected** if there exists no separation of E.

The empty set and sets containing only one point are trivial examples of connected sets. Since a set is connected if it has no separation, it is easier to prove that a set is not connected than it is to prove a set is connected. The set $A = \{0\} \cup \{1/n : n \in \mathbb{Z}^+\}$ is not connected since the sets $A \cap (-1, 0.3)$ and $A \cap (0.3, 2)$ are a separation of A. The set $B = [0, 1] \cup (4, 5)$ also has a separation since the sets $[0, 1]$ and $(4, 5)$ are each open in B. Since a set H is closed in E if and only if $E \setminus H$ is open in E, a set E is connected if and only if the only sets that are both open and closed in E are $\emptyset$ and E. It turns out that, aside from the trivial examples of the empty set and sets containing only one point, the only sets of real numbers with this property are intervals.

THEOREM 8.27 A set of real numbers that contains at least two points is connected if and only if it is an interval.

Proof. We will give a proof by contradiction that every interval is a connected set. Let I be an interval and suppose that the sets U and V form a separation of I. Let

$a \in U$ and $b \in V$ and, without loss of generality, assume that $a < b$. Define sets U_1 and V_1 by $U_1 = [a, b] \cap U$ and $V_1 = [a, b] \cap V$. A proof that the sets U_1 and V_1 form a separation of $[a, b]$ will be left as an exercise. Let $c = \sup U_1$. Since U_1 is closed in $[a, b]$ (and therefore a closed set), the point c belongs to U_1. Since $c \in U_1$ and U_1 is open in $[a, b]$, there exists a positive number r such that $[c, c + r) \subseteq U_1$. But this contradicts the fact that $c = \sup U_1$. Therefore, the interval I is connected.

A proof that every connected set is an interval will be left as an exercise. ■

One of the consequences of the Intermediate Value Theorem is the fact that a nonconstant continuous function maps intervals onto intervals (see Theorem 3.18). In other words, the range $f(I)$ of a nonconstant continuous function defined on an interval I is an interval. Combining this result with Theorem 8.27, we find that the range $f(E)$ of a continuous function f defined on a connected set E is a connected set. However, this result is valid in a more general setting (see Section 8.5); the proof given below uses only the definition of a connected set and some basic properties of continuous functions.

THEOREM 8.28 Let E be a connected set and let $f: E \to \mathbb{R}$. If f is continuous on E, then $f(E)$ is connected.

Proof. We will prove the following form of the contrapositive of this theorem: if f is continuous and $f(E)$ is not connected, then E is not connected. Suppose that $f(E)$ is not connected. By the definition of connected set, there exist open sets O_1 and O_2 such that

$$f(E) = \big(f(E) \cap O_1\big) \cup \big(f(E) \cap O_2\big),$$

where the sets $f(E) \cap O_1$ and $f(E) \cap O_2$ are disjoint and nonempty. By Theorem 8.22, the sets $U = \{x \in E : f(x) \in O_1\}$ and $V = \{x \in E : f(x) \in O_2\}$ are open in E. It is not difficult to verify that U and V are a separation of E (see the exercises). Therefore, the set E is not connected. This completes the proof. ■

We now turn to the concept of a perfect set. The difference between an open interval and a closed interval is that closed intervals contain all of their limit points. However, both open and closed intervals have the property that every point in the interval is a limit point of the interval. In other words, intervals have no isolated points. More generally, an open set has no isolated points (see the exercises), but a closed set may have isolated sets. A closed set that has no isolated points is known as a perfect set. A more common way to express this concept is given in the following definition.

DEFINITION 8.29 A set E is **perfect** if $E' = E$.

According to the definition, a set E is perfect if it contains all of its limit points ($E' \subseteq E$) and every point in E is a limit point of E ($E \subseteq E'$). Since E' is a closed set, every perfect set is closed. The empty set is a trivial example of a perfect set, and closed intervals of the form $[a, b]$ are the most familiar examples of perfect sets. Since a nonempty perfect set has limit points, every nonempty perfect set contains an infinite number of points. In fact, as with closed intervals, every nonempty perfect set has an uncountable number of points. To prove this result, we will use the following lemma, whose proof will be left as an exercise.

LEMMA 8.30 Let P be a nonempty perfect set, let r be a positive number, and let $[c, d]$ be an interval. If $x \in P$ and the set $[c, d] \cap P$ is infinite, then there exists an interval $[u, v] \subseteq [c, d]$ such that $x \notin [u, v]$, $v - u < r$, and $[u, v] \cap P$ is infinite. ■

THEOREM 8.31 A nonempty perfect set is uncountable.

Proof. Let P be a nonempty perfect set. To prove that P is uncountable, it is sufficient to prove that $P \neq \{x_n : n \in \mathbb{Z}^+\}$ for every sequence $\{x_n\}$ of distinct points in P. Let $\{x_n\}$ be any sequence of distinct points in P. Since P has a limit point, there exists an interval $[a, b]$ such that $[a, b] \cap P$ is infinite. By the lemma, there exists an interval $[a_1, b_1] \subseteq [a, b]$ such that

$$x_1 \notin [a_1, b_1],\ [a_1, b_1] \cap P \text{ is infinite, and } b_1 - a_1 < 1.$$

Similarly, there exists an interval $[a_2, b_2] \subseteq [a_1, b_1]$ such that

$$x_2 \notin [a_2, b_2],\ [a_2, b_2] \cap P \text{ is infinite, and } b_2 - a_2 < 1/2.$$

Continuing this process yields a nested sequence $\{[a_n, b_n]\}$ of closed intervals such that for each n,

$$x_n \notin [a_n, b_n],\ [a_n, b_n] \cap P \text{ is infinite, and } b_n - a_n < 1/n.$$

By the Nested Intervals Theorem, there is a unique point z that lies in the intersection of all of the intervals $[a_n, b_n]$. A proof that $z \in P$ will be left as an exercise. It follows that $P \neq \{x_n : n \in \mathbb{Z}^+\}$ since $x_n \neq z$ for each n. Therefore, the set P is uncountable. ■

We now present an interesting example of a perfect set. Construct a nested sequence $\{K_n\}$ of closed sets as follows. Begin with the closed interval $[0, 1]$ and remove the open middle third of this interval—that is, remove the interval $(1/3, 2/3)$ from $[0, 1]$. This leaves two closed intervals; call the union of these two intervals K_1. Next, remove the open middle third of each of the two intervals in K_1. Let K_2 be the union of the four remaining closed intervals. In general, to obtain the set K_{n+1} from the set K_n, remove the open middle third from each of the intervals in K_n. The first three sets in this sequence are

$$\begin{aligned}
K_1 &= \left[0, \tfrac{1}{3}\right] \cup \left[\tfrac{2}{3}, 1\right],\\
K_2 &= \left[0, \tfrac{1}{9}\right] \cup \left[\tfrac{2}{9}, \tfrac{3}{9}\right] \cup \left[\tfrac{6}{9}, \tfrac{7}{9}\right] \cup \left[\tfrac{8}{9}, 1\right],\\
K_3 &= \left[0, \tfrac{1}{27}\right] \cup \left[\tfrac{2}{27}, \tfrac{3}{27}\right] \cup \left[\tfrac{6}{27}, \tfrac{7}{27}\right] \cup \left[\tfrac{8}{27}, \tfrac{9}{27}\right] \cup \left[\tfrac{18}{27}, \tfrac{19}{27}\right] \cup \left[\tfrac{20}{27}, \tfrac{21}{27}\right]\\
&\qquad \cup \left[\tfrac{24}{27}, \tfrac{25}{27}\right] \cup \left[\tfrac{26}{27}, 1\right].
\end{aligned}$$

The set K_n is the union of 2^n disjoint closed intervals, each with length 3^{-n}. The **Cantor set**, which we will denote by T, is defined by $T = \bigcap_{n=1}^{\infty} K_n$. Since $\{K_n\}$ is a nested sequence of nonempty compact sets, the set T is a nonempty closed set (see Theorem 8.14). Note that the endpoints of all the intervals whose union forms a set

K_n belong to T. Since $[0, 1] \setminus T$ is an open set, it can be expressed as a countable union of open intervals. These open intervals are the intervals removed to form the sets K_n. Hence, the set $[0, 1] \setminus T$ consists of one interval of length $1/3$, two intervals of length $1/9$, four intervals of length $1/27$, and so on. The sum of the lengths of the intervals forming the set $[0, 1] \setminus T$ is thus $\sum_{n=1}^{\infty} 2^{n-1}/3^n = 1$. Although not a proof, this observation seems to imply that the Cantor set contains no intervals. Some of the properties of the Cantor set are summarized in the following theorem; further properties of this set are explored in the exercises.

THEOREM 8.32 The Cantor set T is a nonempty perfect set that contains no intervals.

Proof. We have already noted that the Cantor set T is nonempty and closed. Let $T = \bigcap_{n=1}^{\infty} K_n$, where the sequence $\{K_n\}$ is the sequence constructed in the previous paragraph.

To prove that T is perfect, we must show that every point of T is a limit point of T. Let $x \in T$ and let $r > 0$. Choose a positive integer n such that $3^{-n} < r$. Since $x \in K_n$, there exists a closed interval I of length 3^{-n} such that $x \in I \subseteq K_n$. Let a be an endpoint of I that is distinct from x. Then a is an element of the set $(x - r, x + r) \cap T$ that is different than x. It follows that x is a limit point of T.

To prove that T contains no intervals, let x and y be two distinct points in T with $x < y$. Choose a positive integer p such that $3^{-p} < y - x$. Since each interval in K_p has length 3^{-p}, the points x and y cannot belong to the same interval in K_p. Since the closed intervals that comprise K_p are disjoint, there must be a point $z \notin T$ such that $x < z < y$. Hence, the set T contains no intervals. This completes the proof. ■

The Cantor set and sets similar to it appear frequently in real analysis. Although they are difficult to visualize, it is important to develop some intuition about sets of this type.

One of the reasons for the development of point-set topology was a desire to find a way to express relationships between points and sets that was independent of what the individual points represented. The terms that have been defined in this chapter can be used in the diverse spaces $\mathbb{R}$, $\mathbb{R}^2$, the collection of continuous functions on $[0, 1]$, and many other spaces as well. However, the concepts of point-set topology are also needed to answer certain questions concerning the set of real numbers. As an illustration, we will examine the following question:

> Let $f: \mathbb{R} \to \mathbb{R}$ and let C be the set of all points x such that f is continuous at x. What type of set is C?

The answer to this question requires some further concepts from point-set topology.

Open sets and closed sets are useful ways to describe sets of real numbers, but there are many sets that are neither open nor closed. It is sometimes useful to have descriptions that apply to a wider variety of sets. Two possibilities are provided in the following definition.

DEFINITION 8.33 Let E be a set of real numbers.

a) The set E is a $\boldsymbol{G_\delta}$ **set** if E is a countable intersection of open sets.

b) The set E is an $\boldsymbol{F_\sigma}$ **set** if E is a countable union of closed sets.

Recall that a countable set is a set that is either finite or countably infinite. Consequently, a G_δ set is a set that is either a finite intersection of open sets or a countably infinite intersection of open sets. Since $G = G \cap G$ for any open set G, it is clear that every open set is a G_δ set. A finite intersection of open sets is still an open set, so countably infinite intersections of open sets are of greatest interest here. For example, the sets

$$[0, 1] = \bigcap_{n=1}^{\infty} \left(-\frac{1}{n}, 1 + \frac{1}{n}\right) \quad \text{and} \quad [0, 1) = \bigcap_{n=1}^{\infty} \left(-\frac{1}{n}, 1\right)$$

are G_δ sets that are not open. Every finite set is a G_δ set since

$$\{x_1, x_2, \ldots, x_p\} = \bigcap_{n=1}^{\infty} \left(\bigcup_{i=1}^{p} \left(x_i - \frac{1}{n}, x_i + \frac{1}{n}\right)\right).$$

A similar discussion applies to F_σ sets. Consequently, every closed set is an F_σ set, but countably infinite unions of closed sets are of greatest interest. The sets

$$(0, 1) = \bigcup_{n=3}^{\infty} \left[\frac{1}{n}, 1 - \frac{1}{n}\right] \quad \text{and} \quad [0, 1) = \bigcup_{n=2}^{\infty} \left[0, 1 - \frac{1}{n}\right]$$

are F_σ sets that are not closed. Every countable set is an F_σ set since a countable set is a countable union of one-point sets.

It is actually rather difficult to write down a subset of $\mathbb{R}$ that is neither a G_δ set nor an F_σ set. In fact, no example will be presented in this book since the construction of a set that is neither a G_δ set nor an F_σ set would take us too far astray from the goals of this chapter. (For the record, every nonmeasurable set is a set that is neither a G_δ set nor an F_σ set; a construction of a nonmeasurable set can be found in Gordon [8] or most any other graduate text in real analysis.)

Some general properties of F_σ and G_δ sets are recorded in the following theorem.

THEOREM 8.34 The following properties are valid for F_σ and G_δ sets.

1. The complement of an F_σ set is a G_δ set.

2. The complement of a G_δ set is an F_σ set.

3. The union of countably many F_σ sets is an F_σ set.

4. The intersection of countably many G_δ sets is a G_δ set.

5. The intersection of two F_σ sets is an F_σ set.

6. The union of two G_δ sets is a G_δ set.

7. Every open set is an F_σ set.

8. Every closed set is a G_δ set.

9. The set difference of two closed sets is an F_σ set.

Proof. Parts (1) and (2) are simple consequences of DeMorgan's Laws. Parts (3) and (4) follow from the fact that a countable union of countable sets is still countable. To prove part (5), let A and B be F_σ sets. Then $A = \bigcup_{i=1}^{\infty} A_i$ and $B = \bigcup_{j=1}^{\infty} B_j$, where the A_i's and B_j's are closed sets. It follows that

$$A \cap B = \left(\bigcup_{i=1}^{\infty} A_i\right) \cap \left(\bigcup_{j=1}^{\infty} B_j\right) = \bigcup_{i=1}^{\infty}\bigcup_{j=1}^{\infty}(A_i \cap B_j).$$

Since $A_i \cap B_j$ is a closed set for all i and j, we find that $A \cap B$ is a countable union of closed sets. This shows that $A \cap B$ is an F_σ set. The proof of part (6) is similar. The proofs of the remaining three parts will be left as exercises. ■

In order to discuss the set of points of continuity of a function, we will need to consider the oscillation of a function at a point. This concept was defined in the supplementary exercises of Chapter 5, but its definition will be repeated here. The reader may wish to review the definition of the oscillation of a function on an interval (Definition 5.6) before proceeding.

DEFINITION 8.35 Let $f:\mathbb{R} \to \mathbb{R}$ and let $c \in \mathbb{R}$.

a) Suppose that f is bounded on some open interval containing c. Then the **oscillation** of the function f at the point c is defined by

$$\omega(f, c) = \lim_{r \to 0^+} \omega(f, [c-r, c+r]).$$

b) Suppose that f is unbounded on every open interval containing c. Then the **oscillation** $\omega(f, c)$ of the function f at the point c is ∞.

For a given function f that is bounded on an open interval containing the point c, the expression $\omega(f, [c-r, c+r])$ is an increasing function of r for $r > 0$, so the one-sided limit in Definition 8.35 is guaranteed to exist and

$$\omega(f, c) = \inf\{\omega(f, [c-r, c+r]) : r > 0\}$$

(see Theorem 3.31). The oscillation of a function at a point is related to the continuity of the function at that point in a simple way: a function $f:\mathbb{R} \to \mathbb{R}$ is continuous at c if and only if $\omega(f, c) = 0$. A proof of this fact (which happens to also be an exercise in Chapter 5) will be left as an exercise. The next theorem considers the set of points at which the oscillation of a function has a certain magnitude.

THEOREM 8.36 Let $f:\mathbb{R} \to \mathbb{R}$. If α is a positive real number, then the set $\{x \in \mathbb{R} : \omega(f, x) \geq \alpha\}$ is closed.

Proof. Let $E = \{x \in \mathbb{R} : \omega(f, x) \geq \alpha\}$ and let z be a limit point of E. If f is unbounded on every open interval containing z, then $\omega(f, z) = \infty$ and z certainly belongs to E. Suppose that f is bounded on some open interval containing z. For each $r > 0$, the set $E \cap (z-r, z+r)$ contains a point other than z. Let x be such a point, then choose $\delta > 0$ so that $[x-\delta, x+\delta] \subseteq [z-r, z+r]$. Since $x \in E$,

$$\omega(f, [z-r, z+r]) \geq \omega(f, [x-\delta, x+\delta]) \geq \omega(f, x) \geq \alpha.$$

Since this is valid for all $r > 0$, it follows that $\omega(f, z) \geq \alpha$. This shows that $z \in E$. Since E contains all of its limit points, it is a closed set. ■

It is now possible to classify the types of sets that can be the set of points of continuity of a function. As indicated by the following theorem, the set of points of continuity of $f: \mathbb{R} \to \mathbb{R}$ is a G_δ set.

THEOREM 8.37 The set of points at which a function $f: \mathbb{R} \to \mathbb{R}$ is continuous is a G_δ set.

Proof. It is sufficient to prove that the set D of all points at which f is not continuous is an F_σ set. Note that $x \in D$ if and only if $\omega(f, x) > 0$. For each positive integer n, let $E_n = \{x \in \mathbb{R} : \omega(f, x) \geq 1/n\}$. By Theorem 8.36, each E_n is a closed set. Since $D = \bigcup_{n=1}^{\infty} E_n$, the set D is an F_σ set. ■

Since there are sets of real numbers that are not G_δ sets, not every set of real numbers can be the set of points of continuity of a function $f: \mathbb{R} \to \mathbb{R}$. This result is not all that satisfying since we have not yet given examples of sets of this type. However, as we will prove shortly, the set of rational numbers is a familiar set that is not a G_δ set. Therefore, there is no function $f: \mathbb{R} \to \mathbb{R}$ whose set of points of continuity is $\mathbb{Q}$. In order to prove this, we need some more concepts from point-set topology.

DEFINITION 8.38 Let E be a set of real numbers.

a) The set E is **dense** if $\overline{E} = \mathbb{R}$.

b) The set E is **nowhere dense** if $\overline{E}$ contains no open intervals.

The simplest nontrivial example of a dense set is the set $\mathbb{Q}$ of rational numbers. Although the concept of a dense set is important in point-set topology, we will have no occasion to use it. Its definition is included as a contrast to the notion of a nowhere dense set. The simplest examples of nowhere dense sets are finite sets of real numbers, but some infinite sets are nowhere dense as well. Two familiar examples are the sets $\mathbb{Z}$ and $\{1/n : n \in \mathbb{Z}^+\}$. The Cantor set is an example of an uncountable, nowhere dense set.

It follows easily from the definition that a set E is nowhere dense if and only if $\overline{E}$ is nowhere dense. Another useful property of nowhere dense sets is recorded in the next theorem.

THEOREM 8.39 If E is a closed, nowhere dense set and $[u, v]$ is a closed, bounded interval, then there exists an interval $[c, d] \subseteq [u, v]$ such that $[c, d] \cap E = \emptyset$.

Proof. Since E is a nowhere dense set, it contains no open intervals. In particular, there is a point $z \in (u, v) \setminus E$. Since E is a closed set and $z \notin E$, there exists a positive number δ such that $(z - \delta, z + \delta) \cap E = \emptyset$. By taking δ to be a smaller positive number if necessary, we may assume that $[z - \delta, z + \delta] \subseteq [u, v]$. The interval $[z - \delta, z + \delta]$ has the desired properties. ■

It will be left as an exercise to prove that the union of two nowhere dense sets is nowhere dense. By the Principle of Mathematical Induction, a finite union of nowhere dense sets is still a nowhere dense set, but this result does not extend to a countably infinite union of nowhere dense sets: the set $\mathbb{Q}$ is a countably infinite union of one point sets. However, as shown by the next theorem, a countably infinite union of nowhere dense sets has many "holes" in it. (This theorem is actually a special case of a result known as the **Baire Category Theorem**; a more general statement can be found in Munkres [15]. For some interesting elementary consequences of this theorem, see Boas [3].)

THEOREM 8.40 If $\{E_n\}$ is any sequence of nowhere dense sets, then every closed interval $[a, b]$ contains a point x that does not belong to any of the sets E_n.

Proof. Without loss of generality, we may assume that each E_n is closed. By Theorem 8.39, there exists an interval $[a_1, b_1] \subseteq [a, b]$ such that $[a_1, b_1] \cap E_1 = \emptyset$. Using Theorem 8.39 once again, there exists an interval $[a_2, b_2] \subseteq [a_1, b_1]$ such that $[a_2, b_2] \cap E_2 = \emptyset$. Continuing this process yields a nested sequence $\{[a_n, b_n]\}$ of closed intervals in $[a, b]$ such that $[a_n, b_n] \cap E_n = \emptyset$ for all n. By the Nested Intervals Theorem, there exists a point x that belongs to all of the intervals $[a_n, b_n]$. By the method used to determine the intervals $[a_n, b_n]$, the point x does not belong to any of the sets E_n. ∎

One simple consequence of the preceding theorem is the fact that the set of all real numbers cannot be expressed as a countably infinite union of nowhere dense sets. We can use this fact to prove that the set of rational numbers is not a G_δ set. Therefore, there is no function $f: \mathbb{R} \to \mathbb{R}$ that is continuous at each rational number and not continuous at each irrational number.

THEOREM 8.41 The set of rational numbers is not a G_δ set.

Proof. This will be a proof by contradiction. Suppose that the set of rational numbers is a G_δ set. Then the set of irrational numbers is an F_σ set. Let

$$\mathbb{R} \setminus \mathbb{Q} = \bigcup_{n=1}^{\infty} E_n,$$

where each E_n is a closed set. Since $E_n = \overline{E_n}$ contains no rational numbers, it contains no open intervals. Therefore, each E_n is a nowhere dense set. Let $\mathbb{Q} = \{r_n : n \in \mathbb{Z}^+\}$. Then the equation

$$\mathbb{R} = (\mathbb{R} \setminus \mathbb{Q}) \cup \mathbb{Q} = \left(\bigcup_{n=1}^{\infty} E_n\right) \bigcup \left(\bigcup_{n=1}^{\infty} \{r_n\}\right)$$

(where $\{r_n\}$ represents a set containing the single point r_n) shows that $\mathbb{R}$ can be expressed as a countably infinite union of nowhere dense sets. This contradicts Theorem 8.40, so the set of rational numbers is not a G_δ set. ∎

Exercises

1. Let E be a finite set with at least two points. Find a separation of E.

2. Find a separation of the set $\mathbb{Q}$.
3. Use only the definition of connected set and other results in point-set topology to prove each of the following.
   a) Let E be a set that is not connected and suppose that the sets U and V form a separation of E. Suppose that A is a connected subset of E. Prove that either $A \subseteq U$ or $A \subseteq V$.
   b) Let A be a connected set and suppose that B is a set that satisfies $A \subseteq B \subseteq \overline{A}$. Prove that B is connected. In particular, the closure of a connected set is connected.
4. Adopting the notation of the proof of Theorem 8.27, prove that the sets U_1 and V_1 form a separation of $[a, b]$. Then prove that U_1 is open and closed in $[a, b]$.
5. Finish the proof of Theorem 8.27 by proving that every connected set that contains more than one point is an interval.
6. Prove that every connected set that contains more than one point contains an uncountable number of points.
7. Finish the proof of Theorem 8.28 by showing that U and V form a separation of E.
8. Let G be an open set that is neither $\emptyset$ nor $\mathbb{R}$. Prove that $G \subseteq G'$ and that $G' \setminus G \neq \emptyset$.
9. Give an example of a closed and bounded set that has a countably infinite number of isolated points.
10. Give a simple example of a nonempty perfect set that is not an interval.
11. Prove Lemma 8.30.
12. Adopting the notation of the proof of Theorem 8.31, prove that $z \in P$.
13. Prove that every countably infinite closed set has an isolated point.
14. Suppose that a set A is constructed in the same way as the Cantor set except that at each stage of the construction, an open interval one-fifth the size of the original interval is removed from the center of the original interval. That is, the interval

$$\left(\frac{a+b}{2} - \frac{b-a}{10}, \frac{a+b}{2} + \frac{b-a}{10}\right)$$

   is removed from $[a, b]$.
   a) Prove that A is a nonempty perfect set that contains no intervals.
   b) Find the sum of the lengths of the intervals that form the set $[0, 1] \setminus A$.
15. Prove that the Cantor set is an uncountable set of measure zero. (See the supplementary exercises of Chapter 5 for the definition of a set of measure zero).
16. Prove that the Cantor set consists of all points $x \in [0, 1]$ such that x has a base three expansion using only 0's and 2's. (See Exercise 46 in Section 1.5).
17. Let E be the set of all real numbers in $[0, 1]$ that have a decimal expansion that uses only the digits 0 and 5. Prove that E is a perfect set that contains no intervals.
18. A point x is a **condensation point** of a set E if $E \cap O$ is uncountable for every open set O that contains x. Let E^c denote the set of all condensation points of E.
   a) Prove that $E \setminus E^c$ is countable.
   b) Prove that E^c is a closed set.
   c) Prove that E^c is a perfect set.
   d) Prove that every uncountable set has a condensation point.

e) Prove that every closed set can be written as the union of a perfect set and a countable set.

19. Prove that the set of rational numbers is an F_σ set.

20. Find a sequence $\{G_n\}$ of open sets such that $\mathbb{Z} = \bigcap_{n=1}^{\infty} G_n$.

21. Find a sequence $\{K_n\}$ of closed sets such that $(0, 1) \cup (2, 3) = \bigcup_{n=1}^{\infty} K_n$.

22. Prove part (6) of Theorem 8.34.

23. Prove parts (7) and (8) of Theorem 8.34.

24. Prove part (9) of Theorem 8.34.

25. Let $f: \mathbb{R} \to \mathbb{R}$ and let $c \in \mathbb{R}$. Prove that the function f is continuous at c if and only if $\omega(f, c) = 0$.

26. Let $f: \mathbb{R} \to \mathbb{R}$ and $\beta > \alpha > 0$. Prove that $\{x \in \mathbb{R} : \alpha \leq \omega(f, x) < \beta\}$ is an F_σ set.

27. Fill in the details of the following alternate proof of Theorem 8.37. Let C be the set of all points at which f is continuous. For each $c \in C$ and $n \in \mathbb{Z}^+$, choose $\delta_n^c > 0$ such that $|f(x) - f(c)| < 1/n$ for all x that satisfy $|x - c| < \delta_n^c$. Prove that

$$C = \bigcap_{n=1}^{\infty} \Big(\bigcup_{c \in C} (c - \delta_n^c, c + \delta_n^c) \Big)$$

and conclude that C is a G_δ set.

28. Let $f: [a, b] \to \mathbb{R}$. Prove that the set of points in $[a, b]$ at which f is continuous is a G_δ set. Is the set of points in (a, b) at which an arbitrary function $f: (a, b) \to \mathbb{R}$ is continuous a G_δ set?

29. Let K represent the set of irrational numbers.

a) Prove that K is a G_δ set.

b) Find a function $f: \mathbb{R} \to \mathbb{R}$ whose set of points of continuity is K.

30. Prove that the set of irrational numbers is a dense set.

31. Let E be the set of all real numbers that can be expressed in the form $n/2^p$, where n is an integer and p is a positive integer. Prove that E is dense in $\mathbb{R}$.

32. Show that the intersection of two dense sets may not be a dense set.

33. Prove that a closed set that is countably infinite is nowhere dense.

34. Prove the following properties of nowhere dense sets.

a) A subset of a nowhere dense set is a nowhere dense set.

b) The intersection of two nowhere dense sets is a nowhere dense set.

c) The union of two nowhere dense sets is a nowhere dense set.

35. Prove that a set E is dense if and only if every interval contains a point of E.

36. Let $\{O_n\}$ be a sequence of open dense sets. Use the previous exercise to prove that $\bigcap_{n=1}^{\infty} O_n$ is a dense set.

37. Let E be a closed nowhere dense set and define $f: \mathbb{R} \to \mathbb{R}$ by $f(x) = 1$ if $x \in E$ and $f(x) = 0$ if $x \notin E$. Prove that the set of points at which f is not continuous is E.

38. Let E be a set that can be expressed as $E = \bigcup_{n=1}^{\infty} E_n$, where each E_n is closed and nowhere dense. (Note that E may not be nowhere dense.) Define $f: \mathbb{R} \to \mathbb{R}$ by

$$f(x) = \begin{cases} 0, & \text{if } x \notin E; \\ 1/p, & \text{if } x \in E, \text{ where } p = \min\{n : x \in E_n\}. \end{cases}$$

a) Prove that the set of points at which f is continuous is $\mathbb{R} \setminus E$.

b) Prove that $\{x \in \mathbb{R} : f(x) \geq r\}$ is closed for every real number r.

39. Let E be a nonempty, bounded, perfect, nowhere dense set, let $a = \inf E$, let $b = \sup E$, and let $[a, b] \setminus E = \bigcup_{k=1}^{\infty} (a_k, b_k)$, where the intervals (a_k, b_k) are disjoint. Let $\{y_k\}$ be a sequence of real numbers, and for each positive integer k, let c_k be the midpoint of $[a_k, b_k]$. Define a function $f: [a, b] \to \mathbb{R}$ by $f(x) = 0$ for $x \in E$, $f(c_k) = y_k$ for each k, and letting f be linear on each of the intervals $[a_k, c_k]$ and $[c_k, b_k]$. That is, the graph of f joins the points $(a_k, 0)$ and (c_k, y_k) with a straight line and the points (c_k, y_k) and $(b_k, 0)$ with a straight line.

a) Prove that f has the intermediate value property.

b) Prove that there exists a sequence $\{f_n\}$ of continuous functions that converges pointwise to f on $[a, b]$.

c) Prove that f is continuous on $[a, b]$ if and only if $\{y_k\}$ converges to 0.

d) Suppose that there exists a positive number m such that $|y_k| > m$ for all k. Find the set of points of continuity of f.

8.5 METRIC SPACES

As we have mentioned several times in this chapter, one of the goals of point-set topology is to provide a common framework for a discussion of the relationships between points and sets in an arbitrary space. One of the most important properties of the spaces that appear in analysis is the existence of a measure of the distance between two points. The quantity $|x - y|$ represents the distance between the real numbers x and y; a moment's reflection will reveal how often this measure of distance (as well as its important property, the triangle inequality) have appeared in this text. Given two points $\mathbf{x} = (x_1, x_2, x_3)$ and $\mathbf{y} = (y_1, y_2, y_3)$ in $\mathbb{R}^3$, the formula

$$\sqrt{(x_1 - y_1)^2 + (x_2 - y_2)^2 + (x_3 - y_3)^2}$$

represents the Euclidean distance from $\mathbf{x}$ to $\mathbf{y}$. Given two continuous functions f and g defined on $[0, 1]$, the number $\sup\{|f(t) - g(t)| : 0 \leq t \leq 1\}$ can be used to define the distance between the functions f and g. These three spaces ($\mathbb{R}$, $\mathbb{R}^3$, and the continuous functions defined on $[0, 1]$) have different types of elements, but the distance function has the same properties in every case. If all the special properties of a space are removed, leaving only points and a measure of the distance between points, the space that remains is known as a metric space. In this section, we will explore some of the properties of general metric spaces and look at some of the unique features of particular metric spaces.

For the record, this section will be written in a style different than the other sections in the text. Except for a few brief comments, it will merely consist of a list of definitions, theorems, and exercises. Many of the proofs of theorems in this section are adaptations of a corresponding proof for $\mathbb{R}$ found earlier in the chapter. The exercises, which parallel the results in the section, will provide examples of the concepts and suggestions for some of the proofs. They also include problems that my students have found helpful and/or I think are interesting. Because the discussion is limited, this section is essentially a guided self-study. As you proceed, be patient and have confidence in your ability to learn these concepts.

We begin with the definition of a metric space. Let X be a set. Recall that the Cartesian product $X \times X$ represents the set of all ordered pairs of elements from the set X.

DEFINITION 8.42 A **metric space** (X, d) consists of a set X and a function $d: X \times X \to \mathbb{R}$ that satisfies the following four properties.

1. $d(x, y) \geq 0$ for all $x, y \in X$.
2. $d(x, y) = 0$ if and only if $x = y$.
3. $d(x, y) = d(y, x)$ for all $x, y \in X$.
4. $d(x, y) \leq d(x, z) + d(z, y)$ for all $x, y, z \in X$.

The function d, which gives the distance between two points in X, is known as a **metric**.

Property (1) simply asserts that the distance between two points in (X, d) is nonnegative, while property (2) guarantees that the distance between distinct points is positive. Property (3) is a symmetry condition; the distance between points should not depend on the order the points appear. Finally, property (4) is the triangle inequality. This property, as you may imagine, will be very important in the discussion of metric spaces.

There are many familiar examples of metric spaces. To list some of these, we will adopt the following notation for some specific sets. (The sequences mentioned below are assumed to be sequences of real numbers.)

$\mathbb{R}^p$ is the set of all p-tuples of real numbers.

$C([a, b])$ is the set of all continuous functions defined on $[a, b]$.

c_0 is the set of all sequences that converge to 0.

c_{00} is the set of all sequences that have a finite number of nonzero terms.

ℓ_1 is the set of all sequences $\{a_i\}$ for which $\sum_{i=1}^{\infty} |a_i|$ converges.

ℓ_∞ is the set of all bounded sequences.

All of the following sets, along with the given distance function, are metric spaces. The verification of the four properties of a metric space will be left to the reader. It should be noted that properties (1), (2), and (3) are trivial in every case. In proving the triangle inequality for the metric space $(\mathbb{R}^p, d_2)$, the reader might find the Cauchy-Schwarz Inequality and the exercises near the end of Section 2.4 helpful.

1. $(\mathbb{Q}, d_1)$, where $d_1(x, y) = |x - y|$.
2. $(\mathbb{R}, d_1)$, where $d_1(x, y) = |x - y|$.

3. $(\mathbb{R}^p, d_1)$, where $d_1(x, y) = \sum_{i=1}^{p} |x_i - y_i|$.

4. $(\mathbb{R}^p, d_2)$, where $d_2(x, y) = \left(\sum_{i=1}^{p} (x_i - y_i)^2\right)^{1/2}$.

5. $(\mathbb{R}^p, d_\infty)$, where $d_\infty(x, y) = \max\{|x_i - y_i| : 1 \le i \le p\}$.

6. $(\mathcal{C}([a, b]), d_\infty)$, where $d_\infty(x, y) = \sup\{|x(t) - y(t)| : t \in [a, b]\}$.

7. $(\mathcal{C}([a, b]), d_1)$, where $d_1(x, y) = \int_a^b |x(t) - y(t)|\, dt$.

8. (c_0, d_∞), where $d_\infty(x, y) = \sup\{|x_i - y_i| : 1 \le i < \infty\}$.

9. (c_{00}, d_∞), where $d_\infty(x, y) = \sup\{|x_i - y_i| : 1 \le i < \infty\}$.

10. (ℓ_1, d_1), where $d_1(x, y) = \sum_{i=1}^{\infty} |x_i - y_i|$.

11. (ℓ_∞, d_∞), where $d_\infty(x, y) = \sup\{|x_i - y_i| : 1 \le i < \infty\}$.

12. (X, d_0), where X is any set and $d_0(x, y) = \begin{cases} 0, & \text{if } x = y; \\ 1, & \text{if } x \ne y. \end{cases}$

If (X, d) is a metric space and $E \subseteq X$, then (E, d) is also a metric space. Unless stated otherwise, the symbol (X, d) will represent a fixed but arbitrary metric space. In order for a theorem concerning metric spaces to be valid, it must be true for every metric space. In this regard, the reader should keep Example (12), which is known as the **discrete metric space**, in mind, as it can be a source of counterexamples.

DEFINITION 8.43 Let (X, d) be a metric space.

a) Let $v \in X$ and let $r > 0$. The **open ball** centered at v with radius r is defined by $B_d(v, r) = \{x \in X : d(x, v) < r\}$.

b) Let $E \subseteq X$. The set E is **bounded** in (X, d) if there exist $x \in X$ and $M > 0$ such that $E \subseteq B_d(x, M)$.

DEFINITION 8.44 Let $\{x_n\}$ be a sequence in a metric space (X, d).

a) The sequence $\{x_n\}$ **converges** to $x \in X$ if for each $\epsilon > 0$ there exists a positive integer N such that $d(x_n, x) < \epsilon$ for all $n \ge N$.

b) The sequence $\{x_n\}$ converges or is **convergent** if there exists a point $x \in X$ such that $\{x_n\}$ converges to x.

c) The sequence $\{x_n\}$ is a **Cauchy sequence** if for each $\epsilon > 0$ there exists a positive integer N such that $d(x_n, x_m) < \epsilon$ for all $m, n \ge N$.

d) Let $\{p_n\}$ be a strictly increasing sequence of positive integers. The sequence $\{x_{p_n}\}$ is a **subsequence** of $\{x_n\}$.

THEOREM 8.45 Let $\{x_n\}$ be a sequence in a metric space (X, d). If $\{x_n\}$ converges, then the set $\{x_n : n \in \mathbb{Z}^+\}$ is bounded and $\{x_n\}$ is a Cauchy sequence. ■

THEOREM 8.46 If $\{x_n\}$ is a Cauchy sequence in a metric space (X, d), then the set $\{x_n : n \in \mathbb{Z}^+\}$ is bounded. Furthermore, if a subsequence of $\{x_n\}$ converges, then $\{x_n\}$ converges. ■

DEFINITION 8.47 A metric space (X, d) is **complete** if every Cauchy sequence in (X, d) converges to a point in X.

DEFINITION 8.48 Let (X, d) be a metric space, let $E \subseteq X$, and let $x \in X$.

a) The point x is an **isolated point** of E if there exists $r > 0$ such that $B_d(x, r) \cap E = \{x\}$.

b) The point x is an **interior point** of E if there exists $r > 0$ such that $B_d(x, r) \subseteq E$.

c) The point x is a **limit point** of E if for each $r > 0$, the set $E \cap B_d(x, r)$ contains a point of E other than x.

d) The point x is a **boundary point** of E if for each $r > 0$, the ball $B_d(x, r)$ contains at least one point of E and one point of $X \setminus E$.

e) The set E is **open** in (X, d) if each point of E is an interior point of E.

f) The set E is **closed** in (X, d) if E contains all of its limit points.

THEOREM 8.49 Let (X, d) be a metric space, let $E \subseteq X$, and let $x \in X$. The point x is a limit point of the set E if and only if there exists a sequence of points in $E \setminus \{x\}$ that converges to x. ■

THEOREM 8.50 Let (X, d) be a metric space and let $E \subseteq X$.

a) The set E is open in (X, d) if and only if $X \setminus E$ is closed in (X, d).

b) The set E is closed in (X, d) if and only if $X \setminus E$ is open in (X, d). ■

THEOREM 8.51 Let (X, d) be a metric space.

a) The union of any collection of open sets in (X, d) is open in (X, d) and the intersection of any finite collection of open sets in (X, d) is open in (X, d).

b) The intersection of any collection of closed sets in (X, d) is closed in (X, d) and the union of any finite collection of closed sets in (X, d) is closed in (X, d). ■

DEFINITION 8.52 Let (X, d) be a metric space and let $E \subseteq X$.

a) The **interior** of E, denoted E°, is the set of all interior points of E.

b) The **derived set** of E, denoted E', is the set of all limit points of E.

c) The **closure** of E, denoted $\overline{E}$, is the set $E \cup E'$.

d) The **boundary** of E, denoted ∂E, is the set of all boundary points of E.

THEOREM 8.53 If E is a subset of a metric space (X, d), then the sets E', $\overline{E}$, and ∂E are closed in (X, d), and the set E° is open in (X, d). ■

DEFINITION 8.54 Let (X, d) be a metric space and let $K \subseteq X$.

a) A collection $\mathcal{G}$ of subsets of X is an **open cover** of K if each set in $\mathcal{G}$ is open in (X, d) and K is contained in the union of all the sets in $\mathcal{G}$. An open cover $\mathcal{G}$ of K has a **finite subcover** if K is contained in the union of a finite number of the sets in $\mathcal{G}$.

b) The set K is **compact** in (X, d) if every open cover of K has a finite subcover.

c) The metric space (X, d) is compact if X is compact in (X, d).

THEOREM 8.55 Let (X, d) be a metric space.

a) If $K \subseteq X$ is compact in (X, d), then K is closed in (X, d) and bounded in (X, d).

b) If $K \subseteq X$ is compact in (X, d) and $E \subseteq K$ is closed in (X, d), then E is compact in (X, d).

c) An arbitrary intersection of sets that are compact in (X, d) is compact in (X, d).

d) A finite union of sets that are compact in (X, d) is compact in (X, d). ■

THEOREM 8.56 Let (X, d) be a metric space. If $\{K_n\}$ is a nested sequence of nonempty compact sets in (X, d), then the set $\bigcap_{n=1}^{\infty} K_n$ is nonempty. ■

DEFINITION 8.57 Let (X, d) be a metric space and let $K \subseteq X$. The set K is **sequentially compact** in (X, d) if each sequence in K has a subsequence that converges to a point in K.

DEFINITION 8.58 Let (X, d) be a metric space and let $K \subseteq X$. The set K is **totally bounded** in (X, d) if for each $\epsilon > 0$ there is a finite subset $\{x_i : 1 \leq i \leq n\}$ of K such that $K \subseteq \bigcup_{i=1}^{n} B_d(x_i, \epsilon)$.

THEOREM 8.59 Let K be a set in a metric space (X, d). The set K is compact in (X, d) if and only if it is sequentially compact in (X, d). ■

THEOREM 8.60 Let K be a set in a metric space (X, d). The set K is compact in (X, d) if and only if (K, d) is complete and K is totally bounded in (X, d). ■

THEOREM 8.61 Heine-Borel Theorem A set is compact in $(\mathbb{R}^p, d_2)$ if and only if it is closed in $(\mathbb{R}^p, d_2)$ and bounded in $(\mathbb{R}^p, d_2)$. ■

DEFINITION 8.62 Let (X, d) be a metric space.

a) A **separation** of X is a pair of nonempty, disjoint open sets U and V such that $X = U \cup V$.

b) The metric space (X, d) is **connected** if there exists no separation of X.

c) A set $E \subseteq X$ is connected in (X, d) if the metric space (E, d) is connected.

THEOREM 8.63 Let (X, d) be a metric space and suppose that the sets U and V are a separation of X. If A is connected in (X, d), then either $A \subseteq U$ or $A \subseteq V$. ■

THEOREM 8.64 Let (X, d) be a metric space and let $A \subseteq X$ be connected in (X, d). If B is a set that satisfies $A \subseteq B \subseteq \overline{A}$, then B is connected in (X, d). ■

THEOREM 8.65 Let (X, d) be a metric space and let $\mathcal{F}$ be a collection of sets that are connected in (X, d). If $\bigcap_{F \in \mathcal{F}} F \neq \emptyset$, then $\bigcup_{F \in \mathcal{F}} F$ is connected in (X, d). ■

DEFINITION 8.66 Let (X, d) and (Y, ρ) be metric spaces.

a) A function $f: X \to Y$ is **continuous** at $x_0 \in X$ if for each $\epsilon > 0$ there exists $\delta > 0$ such that $f(x) \in B_\rho(f(x_0), \epsilon)$ for all $x \in B_d(x_0, \delta)$.

b) A function $f: X \to Y$ is continuous on X if it is continuous at each point of X.

c) A function $f: X \to Y$ is **uniformly continuous** on X if for each $\epsilon > 0$ there exists $\delta > 0$ such that $\rho(f(x_1), f(x_2)) < \epsilon$ for all $x_1, x_2 \in X$ that satisfy $d(x_1, x_2) < \delta$.

THEOREM 8.67 Let (X, d) and (Y, ρ) be metric spaces, let $f: X \to Y$, and let $x_0 \in X$. The function f is continuous at x_0 if and only if the sequence $\{f(x_n)\}$ converges to $f(x_0)$ for each sequence $\{x_n\}$ in X that converges to x_0. ■

THEOREM 8.68 Let (X, d) and (Y, ρ) be metric spaces and let $f: X \to Y$. The following statements are equivalent:

1. The function f is continuous on X.
2. For each set $A \subseteq X$, $f(\overline{A}) \subseteq \overline{f(A)}$.
3. For each set K closed in (Y, ρ), $\{x \in X : f(x) \in K\}$ is closed in (X, d).
4. For each set G open in (Y, ρ), $\{x \in X : f(x) \in G\}$ is open in (X, d). ■

THEOREM 8.69 Let (X, d) and (Y, ρ) be metric spaces and let $f: X \to Y$ be continuous. If $K \subseteq X$ is compact in (X, d), then $f(K)$ is compact in (Y, ρ). ■

THEOREM 8.70 Let (X, d) be a compact metric space, let (Y, ρ) be a metric space, and let $f: X \to Y$. If f is continuous on X, then f is uniformly continuous on X. ■

THEOREM 8.71 Let (X, d) and (Y, ρ) be metric spaces and let $f: X \to Y$ be continuous. If $E \subseteq X$ is connected in (X, d), then $f(E)$ is connected in (Y, ρ). ■

Exercises

1. Let d be a metric on a set X. Prove that the following functions are metrics on X.

a) $d_1(x, y) = 2d(x, y)$

b) $d_2(x, y) = \min\{1, d_1(x, y)\}$

c) $d_3(x, y) = \sqrt{d(x, y)}$

d) $d_4(x, y) = \dfrac{d(x, y)}{1 + d(x, y)}$

2. Suppose that d_1 and d_2 are two metrics on a set X. Prove that the following functions are metrics on X.

   a) $d_3(x, y) = d_1(x, y) + d_2(x, y)$

   b) $d_4(x, y) = \max\{d_1(x, y), d_2(x, y)\}$

3. Show that the function d defined by $d(x, y) = \left|\dfrac{1}{x} - \dfrac{1}{y}\right|$ is a metric on the set of positive real numbers.

4. Let $R([a, b])$ be the collection of all Riemann integrable functions defined on $[a, b]$. Explain why the function

$$d(x, y) = \int_a^b |x(t) - y(t)|\, dt$$

   does not define a metric on $R([a, b])$.

5. Prove that $c_{00} \subseteq \ell_1 \subseteq c_0 \subseteq \ell_\infty$ and that all of the inclusions are proper.

6. Give a geometric description of the open balls in the metric spaces $(\mathbb{R}, d_1)$, $(\mathbb{R}^2, d_2)$, and $(\mathbb{R}^3, d_2)$.

7. Consider the set $\mathbb{R}^2$ and let $\Theta = (0, 0)$. Sketch a graph of the set of points that belong to the given ball.

   a) $B_{d_1}(\Theta, 1)$ b) $B_{d_2}(\Theta, 1)$ c) $B_{d_\infty}(\Theta, 1)$

8. Let X be any set. Describe the sets that are open balls in (X, d_0).

9. Consider the set $\mathcal{C}([0, 1])$ and let Θ be the zero function. Give a geometric description of the balls $B_{d_1}(\Theta, 1)$ and $B_{d_\infty}(\Theta, 1)$.

10. Let (X, d) be a metric space, let E be a nonempty subset of X, and consider the set $\{d(x, y) : x, y \in E\}$. If this set is bounded, then the **diameter** of the set E is defined by $\operatorname{diam} E = \sup\{d(x, y) : x, y \in E\}$; otherwise $\operatorname{diam} E = \infty$. The diameter of the empty set is defined to be 0.

    a) Find the diameter of the set $\{x \in \mathbb{R} : 1 \leq x^2 < 6\}$ in $(\mathbb{R}, d_1)$.

    b) Find the diameter of $\mathbb{Z}^+$ in $(\mathbb{R}, d_1)$.

    c) What are the possible values for the diameter of a set in (X, d_0)?

    d) Find the diameter of the set $\{(x, x) : |x| < 1\}$ in both $(\mathbb{R}^2, d_2)$ and $(\mathbb{R}^2, d_\infty)$.

    e) Consider the functions $x(t) = t$, $y(t) = -t$, and $z(t) = \sin t$. Find the diameter of the set $\{x, y, z\}$ in both $(\mathcal{C}([a, b]), d_\infty)$ and $(\mathcal{C}([a, b]), d_1)$.

    f) Let $x \in X$ and let $r > 0$. Prove that the diameter of the ball $B_d(x, r)$ is at most $2r$.

    g) Give an example of a metric space (X, d) and a ball $B_d(x, r)$ in (X, d) such that $\operatorname{diam} B_d(x, r) \neq 2r$.

    h) Prove that a set E is bounded in (X, d) if and only if $\operatorname{diam} E < \infty$.

11. Prove that a sequence $\{x_n\}$ converges in $(\mathcal{C}([a, b]), d_\infty)$ if and only if the sequence $\{x_n\}$ converges uniformly on $[a, b]$.

12. For each positive integer n, let x_n be the sequence

$$1, \tfrac{1}{4}, \tfrac{1}{9}, \ldots, \tfrac{1}{n^2}, 0, 0, 0, \ldots.$$

    Prove that $\{x_n\}$ converges to the sequence $\{1/i^2\}$ in (ℓ_1, d_1).

13. Let $\{x_n\}$ be a sequence in a metric space (X, d). Prove that $\{x_n\}$ converges to $x \in X$ if and only if the sequence $\{d(x_n, x)\}$ converges to 0.

14. Let $\{x_n\}$ and $\{y_n\}$ be sequences in a metric space (X, d). Suppose that $\{x_n\}$ converges to x and $\{y_n\}$ converges to y. Prove that $\{d(x_n, y_n)\}$ converges to $d(x, y)$.

15. Prove Theorem 8.45.

16. Prove Theorem 8.46.

17. Prove that $(\mathbb{R}, d_1)$ is a complete metric space and that $(\mathbb{Q}, d_1)$ is not a complete metric space.

18. Let $p > 1$ be a positive integer.

a) Prove that $(\mathbb{R}^p, d_1)$ is a complete metric space.

b) Prove that $d_\infty(x, y) \le d_1(x, y) \le p\, d_\infty(x, y)$ for all $x, y \in \mathbb{R}^p$.

c) Prove that a sequence is convergent in $(\mathbb{R}^p, d_\infty)$ if and only if it is convergent in $(\mathbb{R}^p, d_1)$. Also prove that a sequence is Cauchy in $(\mathbb{R}^p, d_\infty)$ if and only if it is Cauchy in $(\mathbb{R}^p, d_1)$.

d) Prove that the metric space $(\mathbb{R}^p, d_\infty)$ is complete.

e) Use an argument similar to the one outlined in parts (b), (c), and (d) to prove that the metric space $(\mathbb{R}^p, d_2)$ is complete.

19. Prove that (ℓ_∞, d_∞) is a complete metric space.

20. Prove that $(\mathcal{C}([a, b]), d_\infty)$ is a complete metric space.

21. For each positive integer n, define $x_n\colon [-1, 1] \to \mathbb{R}$ by

$$x_n(t) = \begin{cases} -1, & \text{if } -1 \le t \le -1/n; \\ nt, & \text{if } -1/n < t < 1/n; \\ 1, & \text{if } 1/n \le t \le 1. \end{cases}$$

a) Prove that $\{x_n\}$ is a Cauchy sequence in $(\mathcal{C}([-1, 1]), d_1)$.

b) Let x be any continuous function defined on $[-1, 1]$. Prove that $\{x_n\}$ does not converge to x in $(\mathcal{C}([-1, 1]), d_1)$. (You cannot simply find the pointwise limit of $\{x_n\}$ and arrive at this conclusion!) Conclude that $(\mathcal{C}([-1, 1]), d_1)$ is not a complete metric space.

c) Let $[a, b]$ be any interval. Prove that $(\mathcal{C}([a, b]), d_1)$ is not a complete metric space.

22. Determine whether (c_{00}, d_∞) is a complete metric space.

23. Determine whether (X, d_0) is a complete metric space.

24. Prove that the isolated points and the interior points of a set must belong to the set, but give examples to show that the limit points and boundary points of a set may or may not belong to the set.

25. Consider the set $E = \{(x, y) : x > 0\} \cup \{(-1, 0), (-2, 0)\}$ as a subset in the metric space $(\mathbb{R}^2, d_2)$. Find the isolated points of E, the interior points of E, the limit points of E, and the boundary points of E.

26. Prove that in any metric space (X, d), the sets X and $\emptyset$ are both open and closed.

27. Let (X, d) be a metric space and let $E \subseteq X$. Prove that a point $x \in X$ is a limit point of E if and only if $E \cap O$ is infinite for each open set O in (X, d) that contains x.

28. Determine whether each of the following statements is true or false. If it is true, give a proof; if it is false, provide a counterexample. Assume that all points and sets are in a metric space (X, d).

a) Each point x of a set E is either an isolated point of E or a limit point of E.

b) If x is a boundary point of E, then x is a limit point of E.

c) If x is a limit point of E, then x is a boundary point of E.

d) An open set contains no isolated points.

29. Prove Theorem 8.49.

30. Prove Theorem 8.50.

31. Give some examples of sets that are neither open nor closed in $(\mathbb{R}^3, d_2)$.

32. Prove that open balls are open in any metric space.

33. Let (X, d) be a metric space, let $v \in X$, and let r be a positive number. Prove that the set $\{x \in X : d(x, v) \leq r\}$ is closed in (X, d).

34. Prove that every open set can be written as a union of closed sets.

35. Let X be any nonempty set. Determine which subsets of X are open in (X, d_0).

36. Let (X, d) be a metric space and let x and y be distinct points in X. Prove that there exist open sets U and V in (X, d) such that $x \in U$, $y \in V$, and $U \cap V = \emptyset$.

37. Prove Theorem 8.51. Show that the result may fail for arbitrary intersections of open sets and for arbitrary unions of closed sets.

38. Prove that $\partial E = \overline{E} \cap \overline{(X \setminus E)}$ for any set E in a metric space (X, d).

39. Prove that $\overline{\overline{E}} = \overline{E}$ for any set E in a metric space (X, d).

40. Let A and B be subsets of a metric space (X, d).

a) Prove that $\overline{A \cup B} = \overline{A} \cup \overline{B}$.

b) Prove that $\overline{A \cap B} \subseteq \overline{A} \cap \overline{B}$. Give an example for which the inclusion is proper.

c) Suppose that $A \subseteq B$. Prove that $\overline{A} \subseteq \overline{B}$.

d) Suppose that $\overline{A} \subseteq \overline{B}$. Can you conclude anything about the relationship between A and B?

41. Let v be a point in a metric space (X, d), let $r > 0$, and let $E = \{x \in X : d(x, v) \leq r\}$. (This set is sometimes referred to as a closed ball.) Determine the relationship between the sets E and $\overline{B_d(x, r)}$.

42. Prove Theorem 8.53.

43. Let (X, d) be a metric space. Given a point $y \in X$ and a set $E \subseteq X$, define the distance from y to E by $\rho(y, E) = \inf\{d(x, y) : x \in E\}$.

a) Prove that $y \in \overline{E}$ if and only if $\rho(y, E) = 0$.

b) Let $\alpha \geq 0$. Prove that the set $\{x \in X : \rho(x, E) \geq \alpha\}$ is closed in (X, d).

44. Let (X, d) be a metric space and let $E \subseteq X$.

a) Suppose that (E, d) is a complete metric space. Prove that E is closed in (X, d).

b) Suppose that (X, d) is a complete metric space and that E is closed in (X, d). Prove that (E, d) is a complete metric space.

45. Let (X, d) be a metric space and let $E \subseteq X$. (Recall that (E, d) is itself a metric space.) Prove that a set $U \subseteq E$ is open in (E, d) if and only if there exists a set G open in (X, d) such that $U = E \cap G$.

46. Let E be the set of all continuous functions x defined on $[a, b]$ with the property that there exists a point $t \in [a, b]$ such that $x(t) = 0$. Prove that E is closed in $(\mathcal{C}([a, b]), d_\infty)$.

47. Let (X, d) be a metric space. Prove that every finite subset of X is compact in (X, d).

48. Prove that the set $E = \{(x, y) : x \geq 0\}$ is not compact in $(\mathbb{R}^2, d_2)$.

49. Prove Theorem 8.55.

50. Prove Theorem 8.56.

51. Let $\mathcal{F}$ be a collection of sets in a metric space (X, d). The collection $\mathcal{F}$ has the **finite intersection property** if the intersection of any finite collection of sets in $\mathcal{F}$ is nonempty. Prove that (X, d) is compact if and only if $\bigcap_{F \in \mathcal{F}} F$ is nonempty for every collection $\mathcal{F}$ of closed sets in (X, d) with the finite intersection property.

52. Prove that a set is compact in (X, d_0) if and only if it is finite.

53. Prove that a subset of a totally bounded set is totally bounded.

54. Show that a totally bounded set is bounded.

55. Give an example of a set in (ℓ_∞, d_∞) that is bounded, but not totally bounded.

56. Show that every bounded set in $(\mathbb{R}, d_1)$ is totally bounded.

57. Let (X, d) be a metric space and suppose that K is a compact set in (X, d). Prove that K is sequentially compact in (X, d). (This is one part of the proof of Theorem 8.59.)

58. Here is an outline for a proof of Theorem 8.59. (See the previous exercise for one half of the proof.) We will make use of the following definition. Let E be a subset of a metric space (X, d) and let $\mathcal{G}$ be an open cover of E. A number $\delta > 0$ is called a **Lebesgue number** of $\mathcal{G}$ if each subset of E with diameter less than δ is a subset of some set in $\mathcal{G}$. Now let (X, d) be a metric space and let K be a set that is sequentially compact in (X, d).

a) Prove that every open cover of K has a Lebesgue number.

b) Prove that K is totally bounded.

c) Use parts (a) and (b) to prove that K is compact in (X, d).

59. Let K be a set in a metric space (X, d). The purpose of this exercise is to prove that K is totally bounded in (X, d) if and only if every sequence in K has a Cauchy subsequence.

a) Suppose first that K is not totally bounded. Prove that there is a sequence in K that has no Cauchy subsequence.

b) Now suppose that K is totally bounded and let $\{x_n\}$ be a sequence in K. Prove that there is a sequence $\{y_k\}$ in K and a doubly-indexed sequence $\{x_{kn}\}$, represented as

$$\begin{array}{l} x_{11}, x_{12}, x_{13}, x_{14}, x_{15}, \ldots \\ x_{21}, x_{22}, x_{23}, x_{24}, x_{25}, \ldots \\ x_{31}, x_{32}, x_{33}, x_{34}, x_{35}, \ldots \\ x_{41}, x_{42}, x_{43}, x_{44}, x_{45}, \ldots \\ \quad\quad\quad\vdots \end{array}$$

of subsequences of $\{x_n\}$ such that $\{x_{kn} : n \in \mathbb{Z}^+\} \subseteq B_d(y_k, \frac{1}{k})$ and $\{x_{(k+1)n}\}_{n=1}^\infty$ is a subsequence of $\{x_{kn}\}_{n=1}^\infty$ for each positive integer k. Then prove that the sequence $\{x_{nn}\}$ is a subsequence of $\{x_n\}$ that is Cauchy.

60. Use the previous exercise to prove Theorem 8.60.

61. Prove the Heine-Borel Theorem.

62. Give an example of a set in (ℓ_∞, d_∞) that is closed and bounded, but not compact.

63. As indicated by the theorems and exercises thus far, for an arbitrary metric space, every compact set is closed and bounded, but not every closed and bounded set is compact. The purpose of this exercise is to characterize the compact sets in $(\mathcal{C}([a, b]), d_\infty)$. We will make use of the following definition. A set E of functions in $\mathcal{C}([a, b])$ is **equicontinuous** if for each $\epsilon > 0$ there exists $\delta > 0$ such that $|x(s) - x(t)| < \epsilon$ for all $x \in E$ and for all $s, t \in [a, b]$ that satisfy $|s - t| < \delta$. As shown in this exercise, a set is compact in $(\mathcal{C}([a, b]), d_\infty)$ if and only if it is closed, bounded, and equicontinuous.

a) Let $\{x_n\}$ be a sequence of continuous functions defined on $[a, b]$ and suppose that $\{x_n\}$ converges uniformly on $[a, b]$. Prove that $\{x_n : n \in \mathbb{Z}^+\}$ is equicontinuous on $[a, b]$.

b) Suppose that K is a compact set in $(\mathcal{C}([a, b]), d_\infty)$. Prove that K is equicontinuous.

c) Prove that the following subsets of $\mathcal{C}([0, 1])$ are not equicontinuous:

$$\{\sin(n\pi t) : n \in \mathbb{Z}^+\},\ \ \{t^n : n \in \mathbb{Z}^+\}, \text{ and } \{t^2/(t^2 + (1 - nt)^2) : n \in \mathbb{Z}^+\}.$$

d) The sets listed in part (c) can be interpreted as sequences. Prove that these sequences have no subsequences that converge in $(\mathcal{C}([0, 1]), d_\infty)$.

e) For each $a > 0$, let x_a be the function defined by $x_a(t) = (a + t^2)^{-1}$. Prove that the set $\{x_a : 1 \le a \le 10\}$ is equicontinuous on $[0, 1]$ but that the set $\{x_a : 0 < a \le 10\}$ is not equicontinuous on $[0, 1]$.

f) Let $\{x_n\}$ be a bounded sequence in $(\mathcal{C}([a, b]), d_\infty)$ and let C be a countably infinite subset of $[a, b]$. Write C as a sequence $\{t_n\}$ of distinct numbers, then prove that there exists a doubly-indexed sequence $\{x_{kn}\}$, represented as

$$\begin{array}{l} x_{11}, x_{12}, x_{13}, x_{14}, x_{15}, \ldots \\ x_{21}, x_{22}, x_{23}, x_{24}, x_{25}, \ldots \\ x_{31}, x_{32}, x_{33}, x_{34}, x_{35}, \ldots \\ x_{41}, x_{42}, x_{43}, x_{44}, x_{45}, \ldots \\ \qquad\quad \vdots \end{array}$$

of subsequences of $\{x_n\}$ such that for each positive integer k, the sequence $\{x_{kn}(t_k)\}_{n=1}^{\infty}$ converges and the sequence $\{x_{(k+1)n}\}_{n=1}^{\infty}$ is a subsequence of $\{x_{kn}\}_{n=1}^{\infty}$. Prove that $\{x_{nn}\}$ is a subsequence of $\{x_n\}$ that converges pointwise on C.

g) Let $\{x_n\}$ be a sequence of continuous functions defined on $[a, b]$ that converge pointwise to x on $[a, b]$. Suppose that $\{x_n : n \in \mathbb{Z}^+\}$ is equicontinuous. Prove that x is continuous on $[a, b]$ and that $\{x_n\}$ converges uniformly to x on $[a, b]$.

h) Suppose that K is closed and bounded in $(\mathcal{C}([a, b]), d_\infty)$ and that K is equicontinuous. To prove that K is compact, it is sufficient to prove that K is sequentially compact. Let $\{x_n\}$ be a sequence in K. By part (f), there exists a subsequence $\{y_n\}$ of $\{x_n\}$ that converges pointwise on $[a, b] \cap \mathbb{Q}$. Prove that $\{y_n\}$ converges pointwise on $[a, b]$ to a function y, then show that $\{y_n\}$ converges to y in $(\mathcal{C}([a, b]), d_\infty)$ and that $y \in K$.

i) Another way to prove the result obtained in part (h) is the following. Suppose that K is closed and bounded in $(\mathcal{C}([a, b]), d_\infty)$ and that K is equicontinuous. It is sufficient to prove that K is totally bounded. Since K is bounded, there exists a positive number M such that $|x(t)| \le M$ for all $x \in K$ and for all $t \in [a, b]$. Use the fact that the intervals $[a, b]$ and $[-M, M]$ are totally bounded to show that K is totally bounded.

64. Prove that a metric space (X, d) is connected if and only if X and $\emptyset$ are the only subsets of X that are both open and closed in (X, d).

65. Prove that any subset of (X, d_0) that contains more than one point is not connected.

66. Let (X, d) be a metric space and let $E \subseteq X$ be a set that contains more than one point.

a) Suppose that E contains an isolated point. Prove that E is not connected.

b) Suppose that E is a finite set. Prove that E is not connected.

67. Let A and B be connected subsets of $(\mathbb{R}^2, d_2)$. Find examples to show that the sets $A \cap B$, $A \cup B$, $A \setminus B$, and A° may not be connected.

68. Prove that every subset of $(\mathbb{Q}, d_1)$ with more than one point is not connected.

69. Prove Theorem 8.63.

70. Prove Theorem 8.64.

71. Let $D_1 = \{(x, y) : (x-1)^2 + y^2 \le 1\}$ and $D_2 = \{(x, y) : (x+1)^2 + y^2 < 1\}$. Prove that $D_1 \cup D_2$ is a connected subset of $(\mathbb{R}^2, d_2)$.

72. Prove Theorem 8.65.

73. Let p be a positive integer. Prove that every open ball in $(\mathbb{R}^p, d_2)$ is a connected set.

74. Prove that a metric space (X, d) is connected if and only if every proper nonempty subset of X has nonempty boundary.

75. The sets U and V are said to be **separated** if $U \cap \overline{V} = \emptyset = \overline{U} \cap V$. Show that X is connected if and only if X is not the union of two nonempty separated sets.

76. Let A and C be subsets of a metric space (X, d). Suppose that C is connected in (X, d) and that both of the sets $C \cap A$ and $C \cap (X \setminus A)$ are nonempty. Prove that C intersects the boundary of A.

77. Consider the function $f: \mathcal{C}([0, 1]) \to \ell_\infty$ defined by $f(x) = \{x(1/k)\}_{k=1}^{\infty}$. Assume that the metric on ℓ_∞ is the usual d_∞ metric.

a) Prove that f is uniformly continuous on $(\mathcal{C}([0, 1]), d_\infty)$.

b) Prove that f is not continuous on $(\mathcal{C}([0, 1]), d_1)$.

78. Prove that any function is continuous at an isolated point of a metric space (X, d). Consequently, every function on (X, d_0) is continuous.

79. Define $f: \mathcal{C}([0, 1]) \to \mathbb{R}$ by $f(x) = x(0)$. Assume that the metric on $\mathbb{R}$ is the usual d_1 metric.

a) Assume that the metric on $\mathcal{C}([0, 1])$ is the d_∞ metric. Prove that f is uniformly continuous on $\mathcal{C}([0, 1])$.

b) Assume that the metric on $\mathcal{C}([0, 1])$ is the d_1 metric. Prove that f is not continuous on $\mathcal{C}([0, 1])$.

80. Prove that the composition of two continuous functions is continuous. Be certain to formulate the necessary hypotheses for the statement of this result.

81. Prove Theorem 8.67.

82. Let (X, d) and (Y, ρ) be two metric spaces and let $f: X \to Y$ be uniformly continuous on X. Suppose that $\{x_n\}$ is a Cauchy sequence in (X, d). Prove that $\{f(x_n)\}$ is a Cauchy sequence in (Y, ρ).

83. Define $f: \mathcal{C}([0,1]) \to \mathcal{C}([0,1])$ by $f(x)(t) = \int_0^t x(s)\,ds$. Prove that this function is uniformly continuous if the d_∞ metric is used for both the domain and the codomain.

84. Let d_1 and d_2 be two metrics on a set X. Suppose that the identity function (the function f defined by $f(x) = x$ for all $x \in X$) from (X, d_1) to (X, d_2) is continuous. Let $\mathcal{G}_1$ be the collection of all sets open in (X, d_1) and let $\mathcal{G}_2$ be the collection of all sets open in (X, d_2). How are $\mathcal{G}_1$ and $\mathcal{G}_2$ related? Is there a converse to this result?

85. Prove Theorem 8.68.

86. Prove Theorem 8.69.

87. Prove Theorem 8.70.

88. Prove Theorem 8.71.

89. Let (X, d) and (Y, ρ) be metric spaces, let E be a compact subset of X, and let $f: X \to Y$ be continuous and one-to-one. Prove that $f_{\text{inv}}: f(E) \to X$ is continuous.

90. Let (X, d) be a metric space, let $E \subseteq X$ be connected in (X, d), and let $f: X \to \mathbb{R}$. Suppose that f is continuous on E, that x_1 and x_2 are distinct points in E, and that v is any number between $f(x_1)$ and $f(x_2)$. Prove that there exists a point $x \in E$ such that $f(x) = v$. (Note that this is an extension of the Intermediate Value Theorem.)

91. Let x and y be two points in a metric space (X, d) and let $E \subseteq X$. A **path** from x to y is a continuous function $f: [0,1] \to X$ such that $f(0) = x$ and $f(1) = y$. If, in addition, $f([0,1]) \subseteq E$, then we say that there is a path in E from x to y. The set E is **path connected** if for each pair of points $x, y \in E$ there exists a path in E from x to y. The following set of exercises involves path connected sets.

a) Prove that every path connected set is connected.

b) (Ignore this exercise if you do not know what an equivalence relation is.) Let E be a subset of a metric space (X, d). Define a relation $\sim$ on E by $x \sim y$ if there exists a path from x to y. Prove that $\sim$ is an equivalence relation on E.

c) Let E be a subset of a metric space (X, d). For each $x \in E$, let C_x be the set of all points v in E such that there exists a path in E from x to v. Suppose that x and y are distinct points in E. Prove that the sets C_x and C_y are either disjoint or equal.

d) Let $f: [a,b] \to \mathbb{R}$ be continuous on $[a,b]$. Prove that the graph of f (that is, the set $\{(x, f(x)) : x \in [a,b]\}$) is a path connected (and hence connected) subset of $(\mathbb{R}^2, d_2)$.

e) Consider the following two subsets of $(\mathbb{R}^2, d_2)$:

$$C = \{(0,y) : 0 \le y \le 1\} \cup \{(x,0) : 0 \le x \le 1\} \cup \left(\bigcup_{n=1}^{\infty} \{(1/n, y) : 0 \le y \le 1\}\right);$$

$$D = (0,1) \cup \{(x,0) : 0 \le x \le 1\} \cup \left(\bigcup_{n=1}^{\infty} \{(1/n, y) : 0 \le y \le 1\}\right).$$

The reader should sketch the graphs of these two sets; the sets C and D are sometimes called the **comb space** and the **deleted comb space**, respectively. Prove that the set C is path connected (and hence connected) and that the set D is connected, but not path connected.

f) Prove that every open ball in $(\mathbb{R}^p, d_2)$ is path connected.

g) Let E be a countable set in $\mathbb{R}^2$. Prove that the set $\mathbb{R}^2 \setminus E$ is path connected.

h) Let (X, d) be a metric space in which every open ball is path connected. Prove that every open connected set is path connected.

Remark. It is possible to define more than one metric on a given set X. Since the definitions of convergent sequence, open set, etc. depend on the metric, the metric plays a crucial role in determining the properties of a metric space. This is why it is important to write (X, d) rather than simply X. (However, when it is clear that a certain specific metric is associated with a given set, it is common practice to simply let X denote a metric space.) Let d_1 and d_2 be two metrics on a set X. The metrics d_1 and d_2 are **equivalent** if there exist constants m and M such that $md_1(x, y) \leq d_2(x, y) \leq Md_1(x, y)$ for all $x, y \in X$.

92. Let d_1 and d_2 be equivalent metrics on a set X. Prove each of the following statements.

a) A set $E \subseteq X$ is bounded in (X, d_1) if and only if it is bounded in (X, d_2).

b) A sequence $\{x_n\}$ in X is Cauchy in (X, d_1) if and only if it is Cauchy in (X, d_2).

c) A set $E \subseteq X$ is open in (X, d_1) if and only if it is open in (X, d_2).

93. Let d_1, d_2, and d_3 be metrics on a set X. Suppose that d_1 is equivalent to d_2 and that d_2 is equivalent to d_3. Prove that d_1 is equivalent to d_3.

94. Consider the three metrics d_1, d_2, and d_∞ that have been defined on the set $\mathbb{R}^p$. Show that each of these metrics is equivalent to the other two.

95. Let d_1, d_2 be metrics on a set X and let $\mathcal{G}_1, \mathcal{G}_2$ be the collection of all sets open in (X, d_1) and open in (X, d_2), respectively. Prove that $\mathcal{G}_1 = \mathcal{G}_2$ if and only if the following property holds: for each $x \in X$ and for each $r > 0$, there exist $r_1, r_2 > 0$ such that $B_{d_1}(x, r_1) \subseteq B_{d_2}(x, r)$ and $B_{d_2}(x, r_2) \subseteq B_{d_1}(x, r)$.

96. Consider the metric spaces $(\mathbb{R}, d_1)$ and $(\mathbb{R}, d_b)$, where d_1 is the usual metric on $\mathbb{R}$ and $d_b(x, y) = \min\{d_1(x, y), 1\}$. (The subscript '$b$' represents the term "bounded"; d_b is a bounded metric on $\mathbb{R}$.)

a) Prove that the collections of open sets in $(\mathbb{R}, d_1)$ and $(\mathbb{R}, d_b)$ are identical.

b) Prove that the metrics d_1 and d_b are not equivalent.

97. Are the d_1 and d_∞ metrics defined on the set $\mathcal{C}([a, b])$ equivalent?

Remark. Let (X, d_X) and (Y, d_Y) be metric spaces. There are several ways to define a metric on the Cartesian product $X \times Y$; we will adopt the following definition. The metric space $(X \times Y, d_c)$ is the metric space with set $X \times Y$ and metric d_c defined by $d_c((x_1, y_1), (x_2, y_2)) = d_X(x_1, x_2) + d_Y(y_1, y_2)$.

98. Verify that $(X \times Y, d_c)$ is a metric space.

99. Let (X, d_X) and (Y, d_Y) be metric spaces, let $A \subseteq X$, and let $B \subseteq Y$. Prove each of the following.

a) Let $\{z_n\}$ be a sequence in $X \times Y$ and suppose that $z_n = (x_n, y_n)$ for each n. The sequence $\{z_n\}$ converges in $(X \times Y, d_c)$ if and only if the sequence $\{x_n\}$ converges in (X, d_X) and the sequence $\{y_n\}$ converges in (Y, d_Y).

b) The set $A \times B$ is closed in $(X \times Y, d_c)$ if and only if the sets A and B are closed in (X, d_X) and (Y, d_Y), respectively.

c) The set $A \times B$ is open in $(X \times Y, d_c)$ if and only if the sets A and B are open in (X, d_X) and (Y, d_Y), respectively.

d) The set $A \times B$ is totally bounded in $(X \times Y, d_c)$ if and only if the sets A and B are totally bounded in (X, d_X) and (Y, d_Y), respectively.

100. Use part (d) of the previous exercise to give an alternate proof of the Heine-Borel Theorem for $\mathbb{R}^p$.

101. Let (X, d_X) and (Y, d_Y) be metric spaces.

a) Suppose that the metric spaces (X, d_X) and (Y, d_Y) are compact. Prove that the metric space $(X \times Y, d_c)$ is compact.

b) Suppose that the metric spaces (X, d_X) and (Y, d_Y) are connected. Prove that the metric space $(X \times Y, d_c)$ is connected.

Remark. The last set of exercises for this section considers a basic existence/uniqueness theorem in the theory of differential equations.

102. Let (X, d) be a metric space. A function $T: X \to X$ is a **contraction** on X if there exists $0 < \beta < 1$ such that $d(T(x), T(y)) \leq \beta\, d(x, y)$ for all $x, y \in X$. Prove that a contraction on X is uniformly continuous on X.

103. Let (X, d) be a complete metric space and suppose that $T: X \to X$ is a contraction on X. Prove that T has a unique fixed point, that is, that there exists a unique point $z \in X$ such that $T(z) = z$. (This result is known as the **Banach Fixed Point Theorem.**) To get started on this problem, let x_0 be any point in X and consider the sequence $\{x_n\}$ defined recursively by $x_n = T(x_{n-1})$.

104. We will assume that $\mathbb{R}$ and $\mathbb{R}^2$ denote the metric spaces $(\mathbb{R}, d_1)$ and $(\mathbb{R}^2, d_2)$, respectively. Let $D \subseteq \mathbb{R}^2$ be an open and bounded set, let $(t_0, x_0) \in D$, and let $f: D \to \mathbb{R}$. Suppose that f is continuous and bounded on D, and that f is Lipschitz on D with respect to the second variable. This last phrase means that there exists $K > 0$ such that $|f(t, x) - f(t, y)| \leq K|x - y|$ for all points $(t, x), (t, y) \in D$. Consider the initial value problem

$$\frac{dx}{dt} = f(t, x) \quad \text{and} \quad x(t_0) = x_0.$$

A solution to this problem is a function x defined on some interval containing the point t_0 so that $x(t_0) = x_0$ and $x'(t) = f(t, x(t))$ for all t in the interval. The following exercises outline a proof that there exists a unique solution to this initial value problem given the conditions on D and f.

a) Let M be a bound for f on D and let K be the constant that appears in the Lipschitz condition. Explain why there is a number $r > 0$ such that $Kr < 1$ and the set

$$\{(t, x) : |t - t_0| \leq r \text{ and } |x - x_0| \leq Mr\}$$

is contained in D.

b) Let I_r be the closed interval $[t_0 - r, t_0 + r]$ and let

$$E = \{x \in \mathcal{C}(I_r) : x(t_0) = x_0 \text{ and } |x(t) - x_0| \leq Mr \text{ for all } t \in I_r\}.$$

Prove that E is a closed subset of the metric space $(\mathcal{C}(I_r), d_\infty)$ and conclude that the metric space (E, d_∞) is complete.

c) Define a function $T: E \to E$ by

$$T(x)(t) = x_0 + \int_{t_0}^{t} f(s, x(s))\, ds.$$

Prove that $T(x)$ actually belongs to the set E and that T is a contraction on E.

d) By the Banach Fixed Point Theorem, the function T has a unique fixed point. Show that this fixed point satisfies the differential equation.

A

Mathematical Logic

Although the topics in real analysis will be familiar to you from calculus, this book will be quite different than your calculus book. In those courses, the focus was on techniques and applications whereas in real analysis, the focus is on theorems and proofs. For example, the Mean Value Theorem is an extremely important theorem in differential calculus. Do you remember what this theorem says? Most students do not. However, you probably remember how to find the intervals on which a function is increasing—the intervals on which the derivative is positive. This result is one of the consequences of the Mean Value Theorem. In this book, we will be more concerned with the statement and proof of the Mean Value Theorem than with finding the intervals on which a specific function is increasing. A similar change in focus applies to most of the topics in first semester calculus. For example, rather than finding derivatives of functions such as $f(x) = x^2 \sin(3x^4)$, we will carefully prove the product rule and the chain rule. Furthermore, many of the exercises in this book ask for a proof of some statement. In a nutshell, it is impossible to avoid reading and writing proofs in a course in real analysis. The purpose of this appendix is to provide some assistance in reading, writing, and understanding proofs.

A.1 Mathematical Theories

The material in this book constitutes a small fraction of a mathematical theory known as real analysis. A mathematical theory is a collection of related statements that are known or accepted to be true. The theory consists of definitions, axioms, and derived results. The derived results are usually called theorems. We begin this appendix with a general discussion of definitions, axioms, and theorems.

Definitions represent a mathematical shorthand. A word or short phrase is used to represent some concept. For example, a prime number is a positive integer p such that $p > 1$ and the only positive divisors of p are p and 1. The term "prime number" replaces the longer phrase. It is much easier to write or say "prime number" than it is to write or say "a positive integer greater than 1 whose only positive divisors are itself and 1". The tradeoff, of course, is that you must learn what is meant by a prime number. Although the longer version is not written, it must be known. Notice that the definition of a prime number requires knowledge of positive integers and the notion of divisibility of integers. New terms are defined using previously defined terms and concepts. This process cannot go on indefinitely. In order to avoid circular definitions, some terms must remain undefined. In geometry, points and lines are undefined terms. Other objects, such as triangles and squares, are defined in terms of points and lines. Although most people are comfortable with these concepts, it is not possible to give them a definition in terms of simpler concepts. Another undefined term in mathematics is the term "set". Attempts to define a set result in a list of synonyms (such as collection, group, or aggregate) that do not define the term. In summary, certain terms in a mathematical theory must remain undefined. New terms may be defined using the undefined terms or previously defined terms.

A mathematical theory cannot get off the ground with definitions only. It is necessary to know something about the terms and/or how they are related to each other. Basic information about the terms and their relationships is provided by axioms. An **axiom** is a statement that is assumed to be true. Most axioms are statements that are easy to believe. Turning to geometry once again, one axiom states that two distinct points determine exactly one line. This statement certainly makes sense. The important point, however, is that this statement cannot be proved. It is simply a statement that is assumed to be true. Although the axioms are generally chosen by intuition, the only real requirement for a list of axioms is that they be consistent. This means that the axioms do not lead to contradictions. In this book, the axioms deal with the properties of the set of real numbers; these can be found in Chapter 1.

For clarity, for aesthetics, and for ease of checking for consistency, the number of undefined terms and axioms is kept to a bare minimum. A short list of undefined terms and axioms lies at the foundation of every branch of mathematics. In fact, most branches of mathematics share a common foundation. This common base involves properties of sets and properties of positive integers. However, most mathematics courses do not start at this level. A typical mathematics course generally assumes knowledge of other aspects of mathematics. For instance, the set of positive integers can be used to define the set of real numbers. For this course in real analysis, it is assumed that the reader already has a working knowledge of the set of real numbers. It is taken for granted that a rigorous definition of the set of real numbers using more basic concepts exists. At a different level, a graduate course in real analysis would assume a working knowledge of the contents of this book.

A **theorem** is a true statement that follows from the axioms, definitions, and previously derived results. An example from calculus is the following theorem:

If f is differentiable at c, then f is continuous at c.

This result follows from the definitions of continuity and differentiability, and from previous results on limits. The bulk of a mathematical theory is made up of theorems. Most of this book is made up of theorems and their corresponding proofs. Some authors refer to derived results as propositions, but the use of the word "theorem" is much more common.

One other comment on terminology is worth mentioning. A common sequence of derived results is lemma, theorem, corollary. A **lemma** is a derived result whose only real purpose is as an aid in the proof of a theorem. The lemma is usually only referred to in the proof of the associated theorem—it is not of interest in and of itself. A lemma is often used to shorten a proof or to make a proof read more easily. If part of the proof of a theorem involves some technical details that divert the reader's attention from the main points, then this result is pulled out and called a lemma. The technical details in the proof of the theorem are replaced by a phrase such as "by the lemma". A proof that requires a number of fairly long steps is sometimes split into parts, each of which becomes a lemma. A **corollary** is a result that follows almost immediately from a theorem. It is a simple consequence of the result recorded in the theorem. None of these labels (lemma, theorem, proposition, corollary) has an exact meaning and their use may vary from author to author. The common theme is that each represents a derived result.

Another important aspect of a mathematical theory are examples. **Examples** are objects that illustrate definitions and other concepts. Examples give the mind some specific content to ponder when thinking about a definition or a concept. For instance, after defining a prime number, it is helpful to note that 7 is prime and $6 = 2 \cdot 3$ is not. Abstract mathematics is brought to life by examples. It is possible to create all kinds of definitions, but unless there are some examples that satisfy a given definition, the definition is not very useful. Consider the following "artificial" definition:

> A positive integer n is called a **neat prime** if both n and $n + 100$ are prime numbers and there are no prime numbers between n and $n + 100$.

Before proving theorems about neat primes, an example of a neat prime should be found. If there are no neat primes, there is no need to study the concept. For the calculus theorem stated earlier in this appendix, an example of a function f and a point c such that f is continuous at c but f is not differentiable at c would be interesting and enlightening. The study of and search for examples can lead to conjectures about possible theorems and/or indicate that proposed theorems are false. With every new definition and concept, you should always generate a number of specific examples.

After the axioms and definitions have been recorded, how are derived results generated? The discovery of a derived result involves hard work, intuition, and, on occasion, creative insight. The new result must then be proved. The validity of the axioms and previous results must be used to establish the validity of the new result. This is where logic enters the picture. The rules of logic make it possible to move from one true statement to another. To understand a mathematical theory, it is necessary to understand the logic that establishes the validity of derived results.

A.2 Statements and Connectives

In mathematical logic, a **statement** is a sentence that is either true or false. The following sentences are examples of statements:

1. The square root of 25 is 5.
2. The integer 39 is a prime number.
3. The function $f(x) = x^2$ is continuous.

Statements (1) and (3) are true and statement (2) is false. Not every sentence is a statement—some sentences are neither true nor false. The following sentences are not statements.

4. What is the sine of 30°?
5. This sentence is false.
6. The number x is an integer.

Since questions are neither true nor false, sentence (4) is not a statement. Sentence (5) is rather odd, but assigning it a value of either true or false leads to a contradiction. The problem with sentence (6) is that its truth value cannot be determined until a value for x has been substituted. Such sentences are very important in mathematics and will be considered later in this appendix.

Given certain statements that are assumed to be true, logic is concerned with the deductions that can be made from these statements. In other words, logic is most concerned with the connections between statements. It should be emphasized that logic does not determine absolute truth. Assuming that certain statements are true, logic helps to determine what other statements are true.

As an example of a statement that involves connections between other statements, consider the following:

If 23471 is a prime number, then 23471 is not a perfect square.

This statement represents the statements "23471 is a prime number" and "23471 is not a perfect square" connected by the "if, then" form that is so common in mathematics. Although statements in mathematics are often in the "if, then" form, there are other ways to connect statements. There are four connectives that occur frequently in mathematics. Since we are not yet concerned with particular statements, it is best to use letters to represent generic statements. It is standard practice to use capital letters such as P and Q to denote statements. The four connectives, along with the negation of a statement, are defined below.

1. The **negation** of the statement P is another statement denoted by $\neg P$ and read as "not P". The statement $\neg P$ has the opposite truth value of the statement P.

2. The **conjunction** of the statements P, Q is another statement denoted by $P \wedge Q$ and read as "P and Q". The statement $P \wedge Q$ is true when both P and Q are true.

3. The **disjunction** of the statements P, Q is another statement denoted by $P \vee Q$ and read as "P or Q". The statement $P \vee Q$ is true when either P or Q or both P and Q are true.

4. The **conditional** of the statements P, Q is another statement denoted by $P \Rightarrow Q$, which is read as "P implies Q" or "if P, then Q". The statement $P \Rightarrow Q$ is false only when P is true and Q is false.
5. The **biconditional** of the statements P, Q is another statement denoted by $P \Leftrightarrow Q$ and read as "P if and only if Q". The statement $P \Leftrightarrow Q$ is true when P and Q have the same truth value.

A statement such as $P \wedge Q$ is known as a **compound statement**; it is made up of the simpler statements P and Q. A **truth table** for a compound statement is an organized listing of its truth values in terms of the truth values of its simpler components. The truth tables for the compound statements defined thus far appear in the following table.

P	Q	$P \wedge Q$	$P \vee Q$	$P \Rightarrow Q$	$P \Leftrightarrow Q$
T	T	T	T	T	T
T	F	F	T	F	F
F	T	F	T	T	F
F	F	F	F	T	T

Two compound statements are said to be **logically equivalent** if they have the same truth table; the two statements have the same truth value in all possible cases. For example, the statements $\neg(P \wedge Q)$ and $\neg P \vee \neg Q$ are logically equivalent since the last two columns of the following truth table are identical:

P	Q	$\neg(P \wedge Q)$	$\neg P \vee \neg Q$
T	T	F	F
T	F	T	T
F	T	T	T
F	F	T	T

The biconditional connective can be used to represent this situation. Hence,

$$\neg(P \wedge Q) \Leftrightarrow (\neg P \vee \neg Q).$$

It is not necessary to use a truth table to prove that two statements are logically equivalent. For the preceding example,

$$\begin{aligned}(P \wedge Q \text{ is false}) &\Leftrightarrow (\text{either } P \text{ is false or } Q \text{ is false})\\ &\Leftrightarrow (\text{either } \neg P \text{ is true or } \neg Q \text{ is true})\\ &\Leftrightarrow (\neg P \vee \neg Q).\end{aligned}$$

Logically equivalent statements represent two ways of saying the same thing. An example from calculus is

$$(f \text{ is continuous at } c) \Leftrightarrow \lim_{x \to c} f(x) = f(c).$$

There are times when one form is preferred over the other, but since the two statements always have the same truth value, either one can be used.

Some useful logically equivalent statements are listed below. The reader should verify each one.

1. $\neg(\neg P) \Leftrightarrow P$
2. $\neg(P \wedge Q) \Leftrightarrow (\neg P \vee \neg Q)$
3. $\neg(P \vee Q) \Leftrightarrow (\neg P \wedge \neg Q)$
4. $\neg(P \Rightarrow Q) \Leftrightarrow (P \wedge \neg Q)$
5. $(P \Rightarrow Q) \Leftrightarrow (\neg P \vee Q)$
6. $(P \Rightarrow Q) \Leftrightarrow (\neg Q \Rightarrow \neg P)$
7. $(P \Leftrightarrow Q) \Leftrightarrow ((P \Rightarrow Q) \wedge (Q \Rightarrow P))$

It is important to understand what each of these statements means. Some of the proof strategies in mathematics discussed in this appendix are based on these logical equivalences.

A.3 Open Statements and Quantifiers

Theorems in mathematics are usually of the form $P \Rightarrow Q$ or $P \Leftrightarrow Q$. In addition, most statements in mathematics involve variables. In fact, it is hard to think of a statement in mathematics that does not involve one or more variables. A sentence that involves one or more variables and that becomes a statement when values are substituted for the variables is called an **open statement**. An open statement that involves one variable is denoted by $P(x)$. An open statement such as this is not a statement because it is neither true nor false until x is assigned a value. (Its truth value is "open" until a value for x is substituted.) The set of values that can be assigned to x is referred to as the **universe of discourse** (or simply the universe) and often denoted by the letter U. As an example, let $P(x)$ denote the open statement $x^2 \geq x$. It is usually necessary to determine the intended universe from the context. The choice for the universe is sometimes not unique and in ambiguous cases should be carefully specified. For this example, we will let the universe be the set of real numbers. Although $P(x)$ is not a statement, the following three sentences, all of which are based upon $P(x)$, are statements:

(i) $5^2 \geq 5$, (ii) $x^2 \geq x$ for all x, and (iii) there exists an x such that $x^2 \geq x$.

The first sentence simply assigns a value to the variable. The second sentence asserts that $P(x)$ is true for every x in the universe. The third sentence asserts that for at least one value of x in the universe, the sentence $P(x)$ true. These statements are true, false, and true, respectively.

These examples illustrate that it is possible to turn open statements into statements. (These examples also illustrate the importance of the universe of discourse. If U is the set of integers, then all three statements are true.) In general, let $P(x)$ be an open statement, let U denote the universe of discourse, and let $a \in U$. Then the following are all statements:

(i) $P(a)$, (ii) $(\forall x)(P(x))$, and (iii) $(\exists x)(P(x))$.

The first statement is just a substitution of one element from the universe. The second is the statement that $P(x)$ is true for all choices of x from the universe. The symbol $\forall$ is known as the **universal quantifier** and $(\forall x)$ is read as "for every x", "for each x", or "for all x". The third statement says that there is at least one value of x for which $P(x)$ is true. The symbol $\exists$ is known as the **existential quantifier** and $(\exists x)$ is read as "there exists x" or "for some x". These three statements represent the three ways to turn open statements into statements.

It is important to note that the statement

$$(\forall x)(P(x)) \Rightarrow (\exists x)(P(x))$$

is true, but that the statement

$$(\exists x)(P(x)) \Rightarrow (\forall x)(P(x))$$

is false. The first implication is true since an open statement that is true for all values of x in the universe is certainly true for some value of x in the universe (as long as the universe is not the empty set). The second implication does not hold since an open statement can be true for some values of x in the universe and false for other values of x in the universe. Although it is clear that the second implication is false, many students make this error when writing proofs. Students often give one or two examples and claim that this provides a proof of some statement. As a simple illustration, consider the statement

$$n^2 + n + 41 \text{ is a prime number for every positive integer } n.$$

A student may write

$$1^2 + 1 + 41 = 43 \text{ is prime;}$$
$$2^2 + 2 + 41 = 47 \text{ is prime;}$$
$$3^2 + 3 + 41 = 53 \text{ is prime;}$$

and then state that the statement is true for all positive integers n. The student is asserting that

$$(\exists n)(P(n)) \Rightarrow (\forall n)(P(n)),$$

where $P(n)$ is the open statement "$n^2 + n + 41$ is a prime number". This is not correct. Searching for examples is helpful, but examples alone cannot prove a general result like this. (We will discuss methods of proof shortly.) However, a single example can prove that a general statement is false; the example is called a **counterexample**. If there exists one value of n for which $n^2 + n + 41$ is not a prime number, then the above statement is false. The integer 40 is a counterexample in this case;

$$40^2 + 40 + 41 = 40(40 + 1) + 41 = 41 \cdot 41$$

is not a prime number. Since $P(40)$ is false, the statement $(\forall n)(P(n))$ is false. In other words,

$$n^2 + n + 41 \text{ is a prime number for every positive integer } n$$

is a false statement. When reading or writing mathematical statements, be certain to distinguish between the universal and existential quantifiers.

Let the universe be the set of all real numbers. The following simple examples illustrate existential and universal quantifiers.

1. $(\forall x)(|x| > 0)$ is false since the statement is false when $x = 0$.
2. $(\exists x)(|x| \leq 0)$ is true since $|0| \leq 0$.
3. $(\forall x)(x + 1 \geq x)$ is true since $(x + 1) - x = 1 \geq 0$ for all x.
4. $(\exists x)(2x \geq x^2 + 1)$ is true since the statement is true for $x = 1$.

To prove that the statement $(\exists x)(P(x))$ is true, it is necessary to find only one value of x in the universe for which $P(x)$ is true. In the above examples, this was easy to do. However, there are some situations in which finding even one example can be extremely difficult. For instance, an example of a neat prime, as defined previously, would be difficult to find. To prove that the statement $(\forall x)(P(x))$ is true, it is necessary to prove that $P(x)$ is true for every value of x in the universe. This requires a generic proof—a number of test cases is not sufficient unless the universe has only a finite number of elements.

A statement such as $(\forall x)(x \geq 1 \Rightarrow x^2 \geq x)$ is often written in the shorter form $(\forall x \geq 1)(x^2 \geq x)$. This simplification is done when the first part of the conditional simply limits the universe. As another example, the statement

$$(\forall x)(x < 0 \Rightarrow |x| = -x)$$

would be written as $(\forall x < 0)(|x| = -x)$. In general, if S is a subset of the universe U, the statement "every x in S satisfies $P(x)$" is written as $(\forall x \in S)(P(x))$ rather than in the longer form $(\forall x)((x \in S) \Rightarrow P(x))$.

An open statement may contain more than one variable. For example, the open statement "f is differentiable at c" contains two variables, f and c. An open statement like this can be represented by $P(f, c)$. (As in this example, it is usually necessary to determine the intended universe from the context. Here f is a function and c is a number in the domain of f, but more information is needed to determine the type of function and/or the type of number. For instance, can c be a complex number? In ambiguous cases, the universe should be carefully specified.) Open statements with more than two variables are treated similarly. The following statements are familiar facts about real numbers written in the language of quantifiers.

1. $(\forall x)(\forall y)(x + y = y + x)$ (commutative law of addition)
2. $(\forall x)(\forall y)(\forall z)((x + y) + z = x + (y + z))$ (associative law of addition)
3. $(\forall x)(\forall y)(xy = yx)$ (commutative law of multiplication)
4. $(\forall x)(\forall y)(\forall z)((x < y) \Rightarrow (x + z < y + z))$ (additive property of inequalities)
5. $(\forall x \neq 0)(\exists y)(xy = 1)$ (existence of multiplicative inverses)

Note the limiting of the universe in the last example.

In many mathematical formulas and definitions, some variables are universally quantified and others are existentially quantified. It is important to be careful in these cases since the order of the quantifiers is crucial. The statement $(\exists x)(\forall y)(x+y = y)$ requires a fixed value of x that satisfies the equation $x + y = y$ regardless of the value of y. It is easy to see that $x = 0$ satisfies this condition. For the statement, $(\forall y)(\exists x)(x + y = 0)$, the value of x depends upon y. For each value of y, it is necessary to find a value of x for which $x + y = 0$. In this case, we can choose $x = -y$. Note how the value of x depends on the value of y. Now compare the statements (with the universe being the set of real numbers)

$$(\forall x)(\exists y)(x < y) \text{ and } (\exists y)(\forall x)(x < y).$$

The first statement is true—it states that there is no largest number. The second statement is false since it requires a fixed value of y that is larger than every value of x. In general, if you compare the statements $(\exists y)(\forall x)(P(x, y))$ and $(\forall x)(\exists y)(P(x, y))$, it is clear that the first will imply the second. As in the last example, it is usually the case that this implication cannot be reversed.

A.4 Conditional Statements and Quantifiers

There are many statements in mathematics that involve more than one open statement. In fact, many theorems in mathematics are of the form "if $P(x)$, then $Q(x)$", which is equivalent to the symbolic statement $(\forall x)(P(x) \Rightarrow Q(x))$. Notice that the universal quantifier is the intended quantifier in such statements. Some simple examples are given in the following list.

1. If $x \geq 1$, then $x^2 \geq x$.

2. If f is differentiable at c, then f is continuous at c.

3. If f is continuous on $[a, b]$, then f is Riemann integrable on $[a, b]$.

4. If $\{a_n\}$ is convergent, then $\{a_n\}$ is bounded.

The universal quantifier is the quantifier that is implicitly assumed in each of these cases. The universe of discourse would be obtained from the context.

Since quantifiers are so prevalent in mathematics, a few more examples may be helpful. Let U represent the set of all functions $f : \mathbb{R} \to \mathbb{R}$, let $C(f)$ be the statement that f is continuous on $\mathbb{R}$, and let $D(f)$ be the statement that f is differentiable on $\mathbb{R}$. All differentiable functions are continuous is symbolized as

$$(\forall f)(D(f) \Rightarrow C(f)),$$

and some continuous functions are not differentiable is symbolized as

$$(\exists f)(C(f) \wedge \neg D(f)).$$

You may wonder why the first example is not

$$(\forall f)(C(f) \wedge D(f)).$$

This statement actually says that every function is both continuous and differentiable, a much stronger statement than intended, and it is false in this case.

Now consider the following very simple theorem (the universe is the set of real numbers):

$$\text{if } x \geq 1, \text{ then } x^2 \geq x.$$

Let $P(x)$ be the open statement $x \geq 1$ and let $Q(x)$ be the open statement $x^2 \geq x$. This theorem connects these two open statements with the conditional connective. In addition, the theorem contains an implicit quantifier, the universal quantifier. In symbols, this theorem is represented by

$$(\forall x)(P(x) \Rightarrow Q(x)).$$

How do you prove a general result such as this? Or, to rephrase the question, what would convince you that this result is false? Suppose that someone points out that $(0.5)^2 < 0.5$. You would probably say that 0.5 is not greater than or equal to 1 so is out of consideration. The answer would be the same if the person mentions that $0^2 \geq 0$. In both cases, the statement $P(x)$ is false, but we are not willing to conclude that $P(x) \Rightarrow Q(x)$ is false. The only evidence that should convince you that the statement is false is a number greater than 1 whose square is less than itself. This helps to explain why the conditional statement $P(x) \Rightarrow Q(x)$ is false only when $P(x)$ is true and $Q(x)$ is false.

Based on the discussion in the previous paragraph, to prove that $P(x) \Rightarrow Q(x)$, we begin with a value of x for which $P(x)$ is true. You can certainly check the result for $x = 1, 2, 3$, and 4 and even for $x = (\pi + 1)/\sqrt{2}$. However, it is obviously not possible to check this result for every real number $x \geq 1$. Plugging in some values for x can help convince you that the statement is true, but it can never prove the result. To construct a proof, it is necessary to take a generic real number that makes $P(x)$ true. Using only the given properties and previous results, you must prove that $Q(x)$ is true. The results that are considered "previous results" are a little vague here. The statement we are trying to prove may seem so simple that it should be considered a previous result. This is where the axioms enter the picture. The axioms are assumed to be true, then you begin proving other results. The axioms in this case are the field axioms listed in Chapter 1. However, you do not need to read these axioms now in order to understand the following proof. Using a two column statement/reason format, a proof of "if $x \geq 1$, then $x^2 \geq x$" might look like the following.

Statement	Reason
$x \geq 1$	given
$x - 1 \geq 0$	additive property of inequalities
$x > 0$	since $x \geq 1 > 0$
$x(x - 1) \geq 0$	multiplicative property of nonnegative numbers
$x^2 - x \geq 0$	distributive property
$x^2 \geq x$	additive property of inequalities

Note that each of the reasons is more "basic" than the result that we are proving. These reasons are therefore previous results. (We have not used the multiplicative property of inequalities; the above argument essentially proves this property.)

The advantage of a two column proof like this is that it forces you to think about each step and to find a reason for each step. When the proof is written out in sentences, some of the reasons are omitted. The omission of reasons depends upon the background of the audience; reasons that are very basic and familiar to the audience are generally left out. A proof of "if $x \geq 1$, then $x^2 \geq x$" using sentences might go like this.

> Suppose that $x \geq 1$. Then $x - 1 \geq 0$. Since the product of two nonnegative numbers is nonnegative, it follows that
>
> $$x^2 - x = x(x-1) \geq 0.$$
>
> This implies that $x^2 \geq x$.

The important aspect of this proof is the fact that x is a generic but fixed element of the universe that makes $P(x)$ a true statement. Using only the given properties of x and previous results, you must prove that $Q(x)$ is a true statement.

In the conditional statement $P(x) \Rightarrow Q(x)$, the statement $P(x)$ is usually referred to as the **hypothesis** and the statement $Q(x)$ is referred to as the **conclusion**. By rearranging and/or negating $P(x)$ and $Q(x)$, we can form various other conditional statements related to $P(x) \Rightarrow Q(x)$. You may remember doing this in a high school geometry class. Beginning with the conditional statement $P(x) \Rightarrow Q(x)$, the **converse** is $Q(x) \Rightarrow P(x)$ and the **contrapositive** is $\neg Q(x) \Rightarrow \neg P(x)$. Using the statement that we proved above as an illustration, we obtain

conditional:	if $x \geq 1$, then $x^2 \geq x$;
converse:	if $x^2 \geq x$, then $x \geq 1$;
contrapositive:	if $x^2 < x$, then $x < 1$.

There is nothing special about open statements with one variable; these terms apply to conditionals of any form, including $P \Rightarrow Q$ and $P(f, c) \Rightarrow Q(f, c)$. It is easiest to consider $P \Rightarrow Q$ and to realize that its properties extend to the other forms.

A conditional and its contrapositive always have the same truth value; they are logically equivalent:

$$(P \Rightarrow Q) \Leftrightarrow (\neg Q \Rightarrow \neg P).$$

One way to verify this is the following:

$$\begin{aligned}(P \Rightarrow Q \text{ is false}) &\Leftrightarrow (P \text{ is true and } Q \text{ is false})\\ &\Leftrightarrow (\neg P \text{ is false and } \neg Q \text{ is true})\\ &\Leftrightarrow (\neg Q \text{ is true and } \neg P \text{ is false})\\ &\Leftrightarrow (\neg Q \Rightarrow \neg P \text{ is false}).\end{aligned}$$

A truth table can also be used to prove that a conditional and its contrapositive are logically equivalent. It is important to note that the converse may or may not have the same truth value as the given conditional. The converse,

$$\text{if } x^2 \geq x, \text{ then } x \geq 1,$$

mentioned earlier in this discussion is false; the number 0 is a counterexample. It is a common mistake for students to turn theorems around without thinking much about it. Avoid this mistake by thinking carefully about the converse of each theorem.

A.5 Negation of Quantified Statements

The negation of a mathematical statement that contains quantifiers can be a little tricky and confusing at first, especially when there is more than one quantified variable. For single quantifiers,

$$\neg(\forall x)(P(x)) \Leftrightarrow (\exists x)(\neg P(x)) \quad \text{and} \quad \neg(\exists x)(P(x)) \Leftrightarrow (\forall x)(\neg P(x)).$$

In words, the first negation states that if it is not true that $P(x)$ is true for all x, then there exists at least one x for which $P(x)$ is false. The second negation states that if it is not true that $P(x)$ is true for some x, then $P(x)$ must be false for every x. The reasoning is the same if the open statement $P(x)$ is replaced by a compound expression, but the negation may then involve the use of other logically equivalent statements. For example,

$$\begin{aligned}\neg(\forall x)(P(x) \Rightarrow Q(x)) &\Leftrightarrow (\exists x)(\neg(P(x) \Rightarrow Q(x)))\\ &\Leftrightarrow (\exists x)(P(x) \wedge \neg Q(x)).\end{aligned}$$

To prove that "if $P(x)$, then $Q(x)$" is false, it is necessary to find one value of x for which $P(x)$ is true and $Q(x)$ is false. For example, the statement

if p is a prime, then $p^2 + 2$ is not a prime

is false since both 3 and $3^2 + 2 = 11$ are prime numbers.

Definitions in mathematics are biconditional in nature even though they are not always written in this form. In other words, definitions fit into the form $P \Leftrightarrow Q$. The term and its meaning are logically equivalent. To negate a definition means to write out $\neg P \Leftrightarrow \neg Q$. Since definitions often involve quantifiers, some care must be taken when doing this. Consider the following definition:

A function $f: \mathbb{R} \to \mathbb{R}$ is even if $f(-x) = f(x)$ for all $x \in \mathbb{R}$.

For example, the function $f: \mathbb{R} \to \mathbb{R}$ defined by $f(x) = x^4 + 2\cos x$ is even since

$$f(-x) = (-x)^4 + 2\cos(-x) = x^4 + 2\cos x = f(x)$$

for all $x \in \mathbb{R}$. The negation of this definition is the following.

A function $f: \mathbb{R} \to \mathbb{R}$ is not even if $f(-x) \neq f(x)$ for some $x \in \mathbb{R}$.

In symbols, we have moved from the statement $\neg(\forall x)(f(-x) = f(x))$ to the statement $(\exists x)(f(-x) \neq f(x))$. To prove that the function $g: \mathbb{R} \to \mathbb{R}$ defined by $g(x) = x^2 + x$ is not even, it is sufficient to note that $g(1) = 2$ and $g(-1) = 0$. The proof that a function is even requires a general proof (as above); testing a number of points is not sufficient.

The definition of an even function involves just one quantifier. Many definitions in real analysis involve two or more quantifiers. For example,

The sequence $\{x_n\}$ converges to L if for each $\epsilon > 0$ there exists $N > 0$ such that $|x_n - L| < \epsilon$ for all $n > N$.

The negation of this definition is the following:

The sequence $\{x_n\}$ does not converge to L if there exists $\epsilon > 0$ such that for every $N > 0$, there exists $n > N$ such that $|x_n - L| \geq \epsilon$.

It takes a moment for this negation to sink in. Each of the quantifiers has been "flipped" and the main inequality has been reversed. This negation can be carried out in symbols as follows (the symbol $x_n \to L$ means that the sequence $\{x_n\}$ converges to L):

$$\begin{aligned}
\neg\,(x_n \to L) &\Leftrightarrow \neg\,(\forall\epsilon > 0)\ (\exists N > 0)\ (\forall n > N)\ (|x_n - L| < \epsilon)\\
&\Leftrightarrow (\exists\epsilon > 0)\ (\neg\,(\exists N > 0)\ (\forall n > N)\ (|x_n - L| < \epsilon))\\
&\Leftrightarrow (\exists\epsilon > 0)\ (\forall N > 0)\ (\neg\,(\forall n > N)\ (|x_n - L| < \epsilon))\\
&\Leftrightarrow (\exists\epsilon > 0)\ (\forall N > 0)\ (\exists n > N)\ (\neg\,(|x_n - L| < \epsilon))\\
&\Leftrightarrow (\exists\epsilon > 0)\ (\forall N > 0)\ (\exists n > N)\ (|x_n - L| \geq \epsilon)
\end{aligned}$$

The last symbols then translate into the written definition given above.

Here is one more definition from calculus. We will actually write this definition as a biconditional; this is just a reminder that definitions are always of this form.

> $\lim_{x\to c} f(x) = L$ if and only if for each $\epsilon > 0$ there exists $\delta > 0$ such that $|f(x) - L| < \epsilon$ for all x that satisfy $0 < |x - c| < \delta$.

In symbols, this definition becomes

$$\lim_{x\to c} f(x) = L \Leftrightarrow (\forall\epsilon > 0)(\exists\delta > 0)(\forall x)(0 < |x - c| < \delta \Rightarrow |f(x) - L| < \epsilon).$$

The negation of this symbolized expression is

$$\begin{aligned}
&\lim_{x\to c} f(x) \neq L\\
&\quad\Leftrightarrow \neg\,(\forall\epsilon > 0)\ (\exists\delta > 0)\ (\forall x)(0 < |x - c| < \delta \Rightarrow |f(x) - L| < \epsilon)\\
&\quad\Leftrightarrow (\exists\epsilon > 0)\ (\neg\,(\exists\delta > 0)\ (\forall x)(0 < |x - c| < \delta \Rightarrow |f(x) - L| < \epsilon))\\
&\quad\Leftrightarrow (\exists\epsilon > 0)\ (\forall\delta > 0)\ (\neg\,(\forall x)(0 < |x - c| < \delta \Rightarrow |f(x) - L| < \epsilon))\\
&\quad\Leftrightarrow (\exists\epsilon > 0)\ (\forall\delta > 0)\ (\exists x)(\neg\,(0 < |x - c| < \delta \Rightarrow |f(x) - L| < \epsilon))\\
&\quad\Leftrightarrow (\exists\epsilon > 0)\ (\forall\delta > 0)\ (\exists x)\ (0 < |x - c| < \delta \wedge |f(x) - L| \geq \epsilon).
\end{aligned}$$

It may be possible to eliminate some of the intermediate steps in this symbolic negation, but it is important to exercise caution when negating complicated statements like this one. Finally, the negation of the definition in words is as follows:

> $\lim_{x\to c} f(x) \neq L$ if and only if there exists $\epsilon > 0$ such that for each $\delta > 0$, there exists an x such that $0 < |x - c| < \delta$ and $|f(x) - L| \geq \epsilon$.

It takes some practice to be able to arrive at a good wording of a negated definition. In addition, the symbolic manipulation is not a substitute for understanding. If you really understand the definition of limit, then writing its negation in words without the use of symbols should not be difficult. As a standard procedure, you should make note of the negation of every new definition.

A.6 Sample Proofs

We next present some sample proofs. The results proved in this section are not intended to be new or exciting. The purpose is to focus on the logic behind the arguments. To keep the content to a minimum, we will use the set of integers as the universe and restrict our attention to simple properties of the integers. Some familiar terms are introduced in the following definition. Remember that a definition always represents a biconditional statement even if it is not written as such.

DEFINITION A.1

a) An integer n is **even** if there exists an integer k such that $n = 2k$.

b) An integer n is **odd** if there exists an integer k such that $n = 2k + 1$.

c) An integer n is a **divisor** of m if there exists an integer j such that $m = nj$.

d) An integer n is **prime** if $n > 1$ and its only positive divisors are n and 1.

For example, the integer $18 = 2 \cdot 9$ is even and the integer $-37 = -2 \cdot 19 + 1$ is odd. Since $35 = 5 \cdot 7$, both 5 and 7 are divisors of 35. Thus, 35 is not a prime number. It is easy to verify that 1 is a divisor of every integer and that every integer is a divisor of 0. The important point to remember is that the universe is the set of integers. The only "division" that is allowed is when the quotient is also an integer and this notion is defined in terms of multiplication of integers.

The most common type of proof of a conditional such as $P(x) \Rightarrow Q(x)$ is a **direct proof**. In this type of proof, the hypothesis $P(x)$ is assumed to be true, then $P(x)$ and other known results are used to prove that $Q(x)$ is true. The first two theorems illustrate this type of proof.

THEOREM A.2 Suppose that m and n are integers. If m and n are even, then $m + n$ is even.

Proof. Let m and n be even integers. By definition, there exist integers j and k such that $m = 2j$ and $n = 2k$. Then

$$m + n = 2j + 2k = 2(j + k).$$

Let $p = j + k$. Then p is an integer and $m + n = 2p$. It follows that $m + n$ is even. ∎

In the preceding proof, the terms m and n are universally quantified variables. These represent the quantities mentioned in the hypothesis of the theorem. The terms j and k are existentially quantified variables, defined in terms of quantities already mentioned. In the same way, the integer p is an existentially quantified variable. The algebra is essentially the use of previous results and the last step is an example of the biconditional nature of definitions.

THEOREM A.3 The product of two odd integers is odd.

Proof. Let m and n be odd integers. By definition, there exist integers j and k such that $m = 2j + 1$ and $n = 2k + 1$. Since

$$mn = (2j + 1)(2k + 1) = 4jk + 2j + 2k + 1 = 2(2jk + j + k) + 1 = 2i + 1,$$

where $i = 2jk + j + k$ is an integer, the number mn is an odd integer. Consequently, the product of two odd integers is odd. ∎

Although Theorem A.3 is not written in the standard "if, then" format, it is a conditional statement. Using the set of integers as the universe, this statement would be written symbolically as

$$(\forall m)(\forall n)\big((m \text{ odd} \wedge n \text{ odd}) \Rightarrow mn \text{ odd}\big).$$

The first step in the proof is to take two arbitrary odd integers m and n. Throughout the proof, both m and n are fixed integers, but they represent generic odd integers; no special properties are assumed. Noting that $3 \cdot 5 = 15$ is an odd integer is useful, but it does not provide a general proof of the result.

Some of the theorems in this book are in the form $P(x) \Leftrightarrow Q(x)$. For example, the following theorem appears in Chapter 2.

A sequence $\{a_n\}$ is convergent if and only if it is a Cauchy sequence.

To prove a theorem of the form $(\forall x)(P(x) \Leftrightarrow Q(x))$, it is necessary to prove each of the conditionals

$$(\forall x)(P(x) \Rightarrow Q(x)) \text{ and } (\forall x)(Q(x) \Rightarrow P(x)).$$

In other words, two direct proofs are necessary. We will not present a proof of this type here, but this proof strategy for a biconditional should be kept in mind.

The next argument illustrates a proof technique known as **proof by cases**. In a direct proof of this type, the proof is split into parts using an exhaustive list. For the next theorem, the exhaustive list involves even and odd integers. Other natural choices that occur in this text include negative/zero/positive, rational/irrational, and increasing/decreasing. If the result is valid for each case of an exhaustive list, then the result is always true. This technique is illustrated in the proof of the following theorem. The theorem represents a formal "if, then" version of the fact that there is a multiple of 4 between any two consecutive perfect squares greater than 1.

THEOREM A.4 If $n > 1$ is a positive integer, then there exists a positive integer p such that $n^2 < 4p < (n+1)^2$.

Proof. Suppose first that $n > 1$ is an even integer. By definition, there exists a positive integer k such that $n = 2k$. Let $p = k^2 + k$. Then p is a positive integer and

$$n^2 = 4k^2 < 4k^2 + 4k = 4p;$$
$$4p = 4k^2 + 4k < 4k^2 + 4k + 1 = (2k+1)^2 = (n+1)^2.$$

Therefore, the result is valid for even integers.

Now suppose that $n > 1$ is an odd integer. By definition, there exists a positive integer j such that $n = 2j + 1$. Let $q = j^2 + j + 1$. Then q is a positive integer and

$$n^2 = (2j+1)^2 = 4j^2 + 4j + 1 < 4j^2 + 4j + 4 = 4q;$$
$$4q = 4j^2 + 4j + 4 < 4j^2 + 8j + 4 = (2j+2)^2 = (n+1)^2.$$

Therefore, the result is valid for odd integers as well. Since every positive integer is either even or odd, the result is true for all positive integers $n > 1$. ■

An alert reader may notice an unjustified statement in the previous proof. Although it may seem obvious that every integer is either even or odd, this statement requires proof. (The statement that an integer cannot be both even and odd also requires a proof.) This fact is actually a special case of a more general result known as the division algorithm, which is proved in Appendix C.

Since a conditional and its contrapositive are logically equivalent, a proof of $\neg Q \Rightarrow \neg P$ yields a proof of $P \Rightarrow Q$. As indicated by the next example, it is sometimes easier to prove the contrapositive than it is to prove the given conditional.

THEOREM A.5 Let $n > 1$ be a positive integer. If $2^n - 1$ is prime, then n is prime.

Proof. We will prove the contrapositive. Suppose that $n > 1$ and that n is not prime. This means that there exist integers a and b between 1 and n such that $n = ab$. The equation

$$x^b - 1 = (x - 1)\,(x^{b-1} + x^{b-2} + \cdots + 1),$$

which is valid for all integers x, is a basic factoring formula from algebra. Applying this formula, we find that

$$2^n - 1 = (2^a)^b - 1 = (2^a - 1)\left((2^a)^{b-1} + (2^a)^{b-2} + \cdots + 1\right).$$

This shows that $2^a - 1$ is a divisor of $2^n - 1$. Since $1 < 2^a - 1 < 2^n - 1$, it follows that $2^n - 1$ is not prime. ■

Another method of proof is known as **proof by contradiction**. To prove a conditional $P \Rightarrow Q$ using a proof by contradiction, begin, as usual, by assuming that P is true. In addition, assume that Q is false, then show that this leads to a contradiction. Since contradictions are not allowed to occur in mathematics, the assumption that Q is false must be incorrect. This shows that Q is true. In logical symbols, this can be expressed as

$$(\neg Q \Rightarrow (R \wedge \neg R)) \Rightarrow Q.$$

Although proof by contradiction is not necessary to prove the following result, it illustrates the technique. The proof also indicates the use of previous results.

THEOREM A.6 Suppose that m and n are integers. If m^2+n^2 is even, then $m+n$ is even.

Proof. Let m and n be integers such that $m^2 + n^2$ is even. Suppose that $m + n$ is odd. By Theorem A.3, the integer

$$(m + n)^2 = (m + n)(m + n)$$

is odd. Since $2mn$ is even and $m^2 + n^2$ is assumed to be even, Theorem A.2 implies that

$$(m + n)^2 = (m^2 + n^2) + 2mn$$

is even. Since $(m+n)^2$ cannot be both even and odd, we have reached a contradiction. It follows that $m + n$ is even. ■

When should a proof by contraposition or a proof by contradiction be attempted? There is no foolproof method for deciding when such a proof will be helpful; this is the sort of knowledge that comes with practice. However, when the hypothesis provides very little useful information, a proof by contraposition or a proof by contradiction may be helpful or necessary. The hypothesis in Theorem A.5 involves a prime number. The definition of a prime number is essentially a negative definition: a prime number does not have any positive divisors except itself and 1. By starting with the negation of the conclusion, we can work with a composite number—a number that can be factored. This factorization provides the key to the proof. These two types of proofs, which are sometimes referred to as **indirect proofs**, should be kept in mind as possible options.

Proof by contradiction makes some people uneasy; it seems a little like magic. A direct proof, or even a proof by contraposition, is often more satisfying and can provide a better indication of why a theorem is true. It is generally a good idea to avoid an indirect proof if it is possible to find a direct proof (unless a direct proof is much more difficult). However, a proof by contradiction is sometimes necessary. For instance, it would be difficult to prove the following result without a proof by contradiction. The set of real numbers will be the universe of discourse for this theorem.

THEOREM A.7 The number $\sqrt{2}+\sqrt{5}$ is an irrational number.

Proof. Suppose that $\sqrt{2}+\sqrt{5}$ is a rational number. Since the product of two rational numbers is a rational number,

$$\left(\sqrt{2}+\sqrt{5}\right)\left(\sqrt{2}+\sqrt{5}\right) = 2+2\sqrt{10}+5 = 7+\sqrt{40}$$

is a rational number. Since the sum of two rational numbers is a rational number,

$$\left(7+\sqrt{40}\right)+(-7) = \sqrt{40}$$

is a rational number. However, the number $\sqrt{40}$ is irrational since 40 is not a perfect square. This is a contradiction. We conclude that $\sqrt{2}+\sqrt{5}$ is an irrational number. ∎

This proof is more difficult than the other proofs in this appendix. The proof involves several previous results:

1. The product of two rational numbers is a rational number.
2. The sum of two rational numbers is a rational number.
3. If n is a positive integer that is not a perfect square, then $\sqrt{n}$ is an irrational number.

(Of course, knowledge of rational and irrational numbers is also assumed.) These results are either proved or given as exercises in Chapter 1. However, it is not immediately apparent from the statement of the theorem that these results will be useful. Consequently, the proof is not difficult to follow, but to come up with the proof on your own would probably take a lot of time and effort. This is what makes mathematics interesting and exciting (as well as frustrating). The above proof also indicates the advantage of knowing a number of previous results.

A.7 Some Words of Advice

We conclude this appendix with some advice on "how to do proofs". Many of the exercises in this book require a proof—usually a proof of a conditional statement. Given some hypotheses, a conclusion must be reached. In other words, the validity of some hypotheses is somehow connected to the validity of the conclusion. Your goal is to find the connection. For a simple proof, writing out what the hypotheses mean and what the conclusion means is often sufficient to see the connection (see the proofs of Theorems A.2 and A.3). Sometimes intermediate steps are involved and these usually require the use of previous results. It is therefore important to learn the theorems inside and out and to keep a list of theorems and results stored in your memory bank. There is no guaranteed method that will solve these problems; some sort of mathematical insight is required. It is possible to train your insight so that over time solving abstract problems becomes easier. One way to do this is to observe carefully the techniques used to solve other problems.

Solving a problem in this book is different from solving a problem in a calculus book. There are few models to imitate and the answers involve several sentences, not just some equations and a number. You should expect lots of trial and error, dead ends, and frustration. You should also expect some of the problems to require more than one attempt. Think about the problem, come up with ideas, experiment, scribble some notes, and ponder previous results. If you haven't solved the problem, put it aside and try it again the next day. When you believe you have a solution, write up the details very carefully and make certain that you can justify every step. The satisfaction that comes with the solution of a difficult problem justifies the effort.

It is important to read the proofs in this book very carefully. Every single step must have a reason, and you should understand each and every word in a proof. For practice, you should attempt to recreate proofs with the book closed. It is not necessary to memorize every word; determine the main ideas in the proof and learn to fill in the details. After mastering a proof, come back several days later and try the proof again without reading it first. Eventually, you should remember one or two key ideas behind most results and from these ideas write out the proof in your own words. You should adopt a similar approach with the exercises.

Learning to write proofs is hard work. It is a struggle that every person learning higher mathematics goes through. With patience, time, and effort, it is possible to acquire this skill. Working on mathematical problems requires a great deal of effort and concentration; this is true for mathematicians at every level. It is a wonderful feeling to finally solve a difficult problem. I hope that you can experience some of this joy as you work through this book.

B

Sets and Functions

B.1 SETS

As mentioned in Appendix A, some terms in mathematics must be left undefined; the concept of "set" is one such term. When a term is left undefined, some attempt must be made to explain what is meant by the term. Since everyone has some experience with sets (a set of dishes, a collection of stamps, a herd of buffalo, a pocket full of change), it is not difficult to get across the basic idea of a set. Hence, we will say that a **set** is a collection of objects. The objects in a set usually have some features in common, such as the set of real numbers or the set of continuous functions, but a set can also be any random collection of objects. (Actually, there are some restrictions on the types of objects that can be considered, but this restriction will not be important here.) The objects in a set are called **elements** or **points** in the set. If x belongs to a set S, then we write $x \in S$ and say "x is an element of S", "x belongs to the set S", or "x is a point in S". A set A is a **subset** of a set B, denoted $A \subseteq B$, if each element of A belongs to B. Two sets A and B are **equal** if and only if $A \subseteq B$ and $B \subseteq A$, that is, the sets A and B have the same elements. A set A is a **proper** subset of B if $A \subseteq B$ and $A \neq B$. A set that has no elements is called the **empty set**, denoted by the symbol $\emptyset$. In operations that involve sets (described below), this set has properties similar to those of 0 in relation to other numbers.

A set can be described in various ways. For example, each of the following represents the set of all odd positive integers (the symbol $\mathbb{Z}^+$ represents the set of positive integers):

$$\{1, 3, 5, 7, 9, \ldots\}; \quad \{n \in \mathbb{Z}^+ : 2 \text{ does not divide } n\}; \quad \{2n - 1 : n \in \mathbb{Z}^+\}.$$

The first representation is merely a listing of the elements of the set O; it is assumed that the pattern continues indefinitely. The second representation of O is read as "the set of all positive integers n such that 2 does not divide n". Finally, the third representation for O means the set of all numbers of the form $2n - 1$, where n is allowed to range over the set of positive integers.

The following operations on sets are useful. Let A and B be two sets.

a) The **intersection** of A and B, denoted $A \cap B$, is the set of all elements that belong to both A and B.

b) The **union** of A and B, denoted $A \cup B$, is the set of all elements that belong to either A or B or to both A and B.

c) The **complement** of B relative to A, denoted $A \setminus B$, is the set of all elements that belong to A but do not belong to B. This operation is sometimes referred to as **set difference** since the elements of B are removed from A.

d) The **Cartesian product** of A and B, denoted $A \times B$, is the set of all ordered pairs (a, b), where $a \in A$ and $b \in B$.

To illustrate these operations on sets, let $A = \{a, b, 2, 3\}$ and $B = \{b, c, 1, 2, 3\}$. Then

$$A \cap B = \{b, 2, 3\}; \qquad A \setminus B = \{a\};$$
$$A \cup B = \{a, b, c, 1, 2, 3\}; \qquad B \setminus A = \{c, 1\}.$$

Two sets are said to be **disjoint** if their intersection is empty, that is, the sets have no elements in common. Properties (such as $A \cup B = B \cup A$ and $A \cap \emptyset = \emptyset$) that follow automatically from the definitions will not be recorded. However, the following properties are worth noting; the proofs will be left to the reader.

1. $A \cap (B \cup C) = (A \cap B) \cup (A \cap C)$;

2. $A \cup (B \cap C) = (A \cup B) \cap (A \cup C)$;

3. $A \setminus (B \cup C) = (A \setminus B) \cap (A \setminus C)$;

4. $A \setminus (B \cap C) = (A \setminus B) \cup (A \setminus C)$.

The last two properties are known as **DeMorgan's Laws**. The interested reader might want to pursue the analogies between the operations of union and intersection and those of addition and multiplication, respectively.

The set operations defined above involve two sets, but it is easy to extend them to any finite number of sets. However, it is sometimes necessary to consider infinite collections of sets. Suppose that for each positive integer n, there is a set G_n. Then

$$\bigcap_{n=1}^{\infty} G_n = \{x : x \in G_n \text{ for every positive integer } n\};$$

$$\bigcup_{n=1}^{\infty} G_n = \{x : x \in G_n \text{ for some positive integer } n\}.$$

If the number of sets in a collection is uncountable, a different notation is needed. There are two common ways of dealing with such large collections of sets. One involves an **index set**. Let A be a set (known as the index set) and for each $\alpha \in A$, let G_α be a set. When using an index set, the symbols $\bigcap_{\alpha \in A} G_\alpha$ and $\bigcup_{\alpha \in A} G_\alpha$ represent the intersection and union, respectively, of the collection $\{G_\alpha : \alpha \in A\}$. For example, let $A = [0, 1]$ and for each $\alpha \in [0, 1]$, let G_α be the interval $[\alpha - 1, \alpha + 1]$. Then $\bigcap_{\alpha \in A} G_\alpha = [0, 1]$ and $\bigcup_{\alpha \in A} G_\alpha = [-1, 2]$. An alternate notation is to let $\mathcal{G}$ (read this as script G) denote a collection of sets. That is, $\mathcal{G}$ is a set whose elements are themselves sets. With this notation, the symbols $\bigcap_{G \in \mathcal{G}} G$ and $\bigcup_{G \in \mathcal{G}} G$ represent the intersection and union, respectively, of all the sets in the collection $\mathcal{G}$.

A collection $\mathcal{G}$ of sets is **disjoint** if no two distinct sets in $\mathcal{G}$ have elements in common. This is not the same as asserting that $\bigcap_{G \in \mathcal{G}} G = \emptyset$. For example, for each positive integer n, let E_n be the interval $[n, \infty)$. The collection $\{E_n : n \in \mathbb{Z}^+\}$ is not disjoint but $\bigcap_{n=1}^{\infty} E_n = \emptyset$. The collection of intervals $\{(n - 0.4, n + 0.3) : n \in \mathbb{Z}^+\}$ is disjoint; the intersection of any two distinct intervals in this collection is empty.

Let S be a set and let $\mathcal{G}$ be a collection of subsets of S. **DeMorgan's Laws** for arbitrary collections of sets assume the following form:

$$S \setminus \Big(\bigcap_{G \in \mathcal{G}} G\Big) = \bigcup_{G \in \mathcal{G}} (S \setminus G) \quad \text{and} \quad S \setminus \Big(\bigcup_{G \in \mathcal{G}} G\Big) = \bigcap_{G \in \mathcal{G}} (S \setminus G).$$

In words, these two formulas state that the complement of the intersection is the union of the complements and the complement of the union is the intersection of the complements.

B.2 FUNCTIONS

The notion of a function is another fundamental concept in mathematics. It is possible to define a function in terms of sets and concepts that involve sets. The advantage of defining functions in this way is that it keeps the number of undefined terms to a minimum. A disadvantage is that the notion of a function becomes more abstract. As the reader has had a great deal of experience with functions, at least real-valued functions, we will treat the term "function" as another undefined term and simply explain how the concept is used. Let A and B be two nonempty sets. A **function** $f: A \to B$ is a rule of correspondence that assigns to each element of the set A exactly one element of the set B. If f assigns the element b of B to the element a of A, then we write $f(a) = b$. We say that b is the **value** of f at a, that the **image** of a is b, and that a is a **preimage** of b. The elements of A are the **inputs** for the function f and the values $f(a)$ are the **outputs** of the function f. The set A is known as the **domain** of f, the set B is known as the **codomain** of f, and the set $\{f(a) : a \in A\}$, which is a subset of B, is known as the **range** of f. The range of f is sometimes denoted by $f(A)$.

To illustrate this concept with some simple examples, let $A = \{a, b, c\}$ and let $B = \{1, 2, 3, 4\}$. Define functions $f: A \to B$ and $g: A \to B$ by

$$\begin{aligned} f(a) &= 1; & \qquad g(a) &= 1; \\ f(b) &= 2; & \text{and} \qquad g(b) &= 2; \\ f(c) &= 4; & \qquad g(c) &= 1. \end{aligned}$$

The range of f is $\{1, 2, 4\}$ and the range of g is $\{1, 2\}$. Note that for the function g, the images of both a and c are 1. This is consistent with the definition of a function. The definition insists that each input have exactly one output, but different inputs may have the same output. In calculus and analysis, the rule of correspondence for a function is often given by an explicit formula. For example, we can define a function $h: \mathbb{R} \to \mathbb{R}$ by $h(x) = x^2$. This function assigns the real number x^2 to the real number x. However, any rule of correspondence that assigns to each element of A a unique element of B is a function, even if it does not involve a formula. As an example of this situation, for each real number x, let $\phi(x)$ be the real number for which

$$(\phi(x))^7 + (\phi(x))^5 + (\phi(x))^3 + \phi(x) + 1 = x.$$

It can be shown that this defines a function $\phi: \mathbb{R} \to \mathbb{R}$. There is no explicit formula that gives the values of this function, but it still satisfies the definition of a function; for each real number x there is exactly one real number $\phi(x)$.

Let A and B be nonempty sets. A rule of correspondence that attempts to define a function $f: A \to B$ is **well-defined** if for each $a \in A$ there is exactly one value for $f(a)$. To illustrate what is meant by this, consider the following attempt to define a function: "for each real number x, let $f(x)$ be a real number whose square is x". There are two problems with this definition. First of all, if $x < 0$, then there is no value for $f(x)$. This problem can be eliminated by writing "for each nonnegative real number x, let $f(x)$ be a real number whose square is x". However, this does not remove the second problem; for each $x > 0$, there are two real numbers whose square is x. Hence, most inputs generate two outputs, something that is not allowed in the definition of a function. The bottom line is that this rule of correspondence does not define a function. However, the function $f: [0, \infty) \to \mathbb{R}$ defined by $f(x) = \sqrt{x}$ is a valid function. (By convention, the symbol $\sqrt{x}$ means the positive square root of x.)

Let A and B be nonempty sets and let $f: A \to B$ be a function.

a) The function f is **one-to-one** if different inputs have different outputs. In other words, if $a_1, a_2 \in A$ with $a_1 \neq a_2$, then $f(a_1) \neq f(a_2)$. A one-to-one function is sometimes called an **injection** or an **injective** function.

b) The function f is **onto** if its range is the same as its codomain. In other words, for each $b \in B$, there exists $a \in A$ such that $f(a) = b$. An onto function is sometimes called a **surjection** or a **surjective** function.

c) The function f is a **one-to-one correspondence** if it is both one-to-one and onto. In other words, each element of B has exactly one preimage. A one-to-one and onto function is sometimes called a **bijection** or a **bijective** function.

In terms of preimages, a function $f: A \to B$ is one-to-one if each element of B has at most one preimage, is onto if each element of B has at least one preimage, and is a one-to-one correspondence if each element of B has exactly one preimage. Some familiar examples from calculus include the following.

1. The function $f_1: \mathbb{R} \to \mathbb{R}$ defined by the formula $f_1(x) = x^2$ is neither one-to-one nor onto.

2. The function $f_2: \mathbb{R} \to \mathbb{R}$ defined by the formula $f_2(x) = 2^x$ is one-to-one but not onto.
3. The function $f_3: \mathbb{R} \to \mathbb{R}$ defined by the formula $f_3(x) = x^3 - x$ is onto but not one-to-one.
4. The function $f_4: \mathbb{R} \to \mathbb{R}$ defined by the formula $f_4(x) = 2x - 3$ is both one-to-one and onto.

Note that the codomain of a function is an important consideration. For example, the function $f_5: (-\infty, 0] \to [0, \infty)$ defined by $f_5(x) = x^2$ is one-to-one and onto. In other words, a function really depends on the sets A and B as well as on its values.

Let A, B, and C be nonempty sets and let $f: B \to C$ and $g: A \to B$ be functions. The **composition** of f with g is the function $f \circ g: A \to C$ defined by $(f \circ g)(a) = f(g(a))$ for each $a \in A$. The following results concerning the composition of two functions such as these are all valid.

1. If f and g are one-to-one, then $f \circ g$ is one-to-one.
2. If f and g are onto, then $f \circ g$ is onto.
3. If $f \circ g$ is one-to-one, then g must be one-to-one, but f may or may not be one-to-one.
4. If $f \circ g$ is onto, then f must be onto, but g may or may not be onto.

To illustrate the proofs of these facts, assume that f and g are one-to-one. Suppose that $a_1, a_2 \in A$ with $a_1 \neq a_2$. Since g is one-to-one, we know that $g(a_1) \neq g(a_2)$. Since f is one-to-one, it follows that $f(g(a_1)) \neq f(g(a_2))$. This shows that $f \circ g$ is one-to-one, proving statement (1). Referring to the example functions defined above, note that $f_1 \circ f_2$ is one-to-one even though f_1 is not one-to-one. Hence, these functions provide an example for the second part of statement (3). Proofs of the other statements are similar.

Let A and B be two nonempty sets and let $f: A \to B$ be a one-to-one and onto function. Then for each $b \in B$ there exists a unique $a \in A$ such that $f(a) = b$. It is thus possible to define a function $g: B \to A$ as follows: for each $b \in B$, $g(b)$ is the element of A with the property that $f(g(b)) = b$. It follows that $f(g(b)) = b$ for all $b \in B$ and $g(f(a)) = a$ for all $a \in A$. The function g is called the **inverse** of f and is often denoted by f^{-1}. However, this notation can cause confusion (the -1 looks like an exponent), so in this text the inverse of f will be denoted by f_{inv}. If you think of a function f as an input/output machine, then the function f_{inv} runs the machine in reverse. The essential property that a function must possess in order to have an inverse is the one-to-one property. The domain of f_{inv} is then the range of f. As a simple example, consider the function $f: \{1, 2, 3\} \to \{a, e, i, o, u\}$ defined by $f(1) = i$, $f(2) = a$, and $f(3) = o$. Since f is a one-to-one function, it has an inverse. The domain of f_{inv} is the set $\{a, i, o\}$ and $f_{\text{inv}}(a) = 2$, $f_{\text{inv}}(i) = 1$, and $f_{\text{inv}}(o) = 3$. For more complicated functions, it may be more difficult to find an expression for the values of f_{inv}, but the concepts remain the same.

Turning to a more familiar situation from algebra, let I be an interval and let $f: I \to \mathbb{R}$ be a one-to-one function. Then f has an inverse and the range $f(I)$ of f is the domain of f_{inv}. This means that $f(f_{\text{inv}}(x)) = x$ for all $x \in f(I)$ and $f_{\text{inv}}(f(x)) = x$ for all $x \in I$. In graphical terms, if the point (a, b) is on the graph

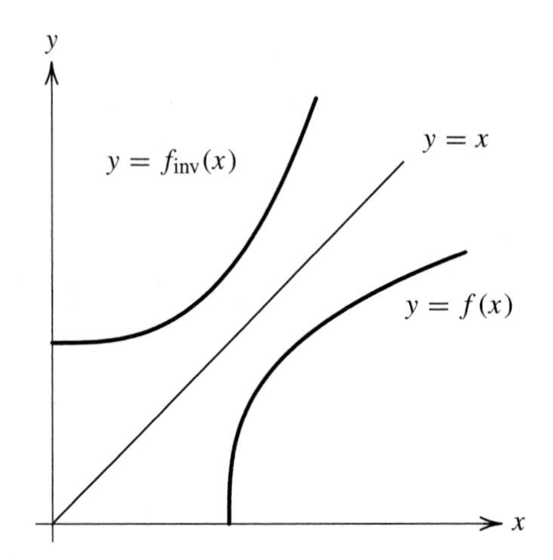

Figure B.1 The graphs of f and f_{inv} are mirror images across the line $y = x$

of f, then the point (b, a) is on the graph of f_{inv}. It follows that the graphs of f and f_{inv} are mirror images across the line $y = x$ (see Figure B.1). As simple examples, consider the functions $f\colon \mathbb{R} \to \mathbb{R}$ and $g\colon \mathbb{R} \to \mathbb{R}$ defined by $f(x) = 4 - 3x$ and $g(x) = x^3$. Each of these functions is one-to-one and onto. It is easy to verify that $f_{\text{inv}}(x) = (4 - x)/3$ and $g_{\text{inv}}(x) = \sqrt[3]{x}$ for all real numbers x.

If a function is not one-to-one, it is always possible to limit its domain so that it becomes one-to-one. For example, the function $h\colon \mathbb{R} \to \mathbb{R}$ defined by $h(x) = \sin x$ is not one-to-one. However, if we limit the domain of h to the interval $[-\pi/2, \pi/2]$, then h becomes one-to-one. On this interval, the function h has an inverse and the function $h_{\text{inv}}\colon [-1, 1] \to [-\pi/2, \pi/2]$ is usually denoted by $h_{\text{inv}}(x) = \arcsin x$.

C

Mathematical Induction

The set of integers and its properties are at the root of all mathematical disciplines. In fact, it is impossible to do mathematics without making use of the integers in some form or another. Number theory, which involves the study of the set of integers itself, is a rich and fascinating branch of mathematics. Many volumes have been written on this subject and some of the best mathematicians in history have devoted much of their time to the study of number theory. In real analysis, the integers are part of the supporting cast—important but not the primary focus. The algebraic and order properties of the integers are the properties that are most relevant when working with the set of real numbers. However, the set of integers has another property that is independent of its algebraic and order properties. This additional property of the integers, which appears frequently (both explicitly and implicitly) in this book, is the focus of this appendix.

C.1 Three Equivalent Statements

The symbols $\mathbb{Z}$ and $\mathbb{Z}^+$ will be used to represent the set of integers and the set of positive integers, respectively;

$$\mathbb{Z} = \{\ldots, -3, -2, -1, 0, 1, 2, 3, \ldots\} \quad \text{and} \quad \mathbb{Z}^+ = \{1, 2, 3, 4, 5, \ldots\}.$$

A set $A \subseteq \mathbb{Z}$ contains a least element if there exists an integer $q \in A$ such that $q \le a$ for all $a \in A$. For example, the set of prime numbers contains a least element—namely, the integer 2. The set of all even integers does not contain a least element.

Consider the following three statements about sets of positive integers.

Well-Ordering Property: Every nonempty set of positive integers contains a least element.

Principle of Mathematical Induction: If S is a set of positive integers that contains 1 and satisfies the condition "if $k \in S$, then $k+1 \in S$", then $S = \mathbb{Z}^+$.

Principle of Strong Induction: If S is a set of positive integers that contains 1 and satisfies the condition "if $1, 2, \ldots, k \in S$, then $k+1 \in S$", then $S = \mathbb{Z}^+$.

The Well-Ordering Property should make intuitive sense. Given a list of positive integers, it is always possible to find the smallest number in the list. This is not true for sets of positive real numbers. In particular, there is a smallest positive integer, but there is no smallest positive real number. The Principle of Mathematical Induction can be compared to a chain reaction. If we know that each event will set off the next (the condition in quotes) and if the first event occurs (S contains 1), then the entire chain reaction will occur. Perhaps you have seen one of those amazing domino exhibits where thousands of dominoes fall over in interesting patterns. The dominoes must be set up in such a way that each one knocks over the next, and someone must begin the process by pushing over the first domino.

Although all three statements are plausible, none of them can be proved from the algebraic or order properties of the integers. In fact, these statements must be accepted as axioms. This may come as a surprise since they seem so obvious, but it is sometimes the case that what seems obvious cannot be proved. In such situations, it is necessary to introduce an axiom. It is an interesting fact that all three of these statements effectively say the same thing, that is, that they are logically equivalent. Consequently, any one of these three statements can be taken as an axiom and the other two can be derived from it as theorems. The proof of the equivalence of these statements is given below.

THEOREM C.1 The following are equivalent:

1. Well-Ordering Property;
2. Principle of Mathematical Induction;
3. Principle of Strong Induction.

Proof. We will prove that (1) $\Rightarrow$ (2) and (3) $\Rightarrow$ (1); a proof that (2) $\Rightarrow$ (3) is very easy and will be omitted.

Suppose first that the Well-Ordering Property is true. Let S be a set of positive integers that contains 1 and satisfies the condition "if $k \in S$, then $k+1 \in S$", and let $A = \mathbb{Z}^+ \setminus S$. To prove (2), it is sufficient to show that $A = \emptyset$. We will give a proof by contradiction. Suppose that $A \neq \emptyset$. Since A is a nonempty set of positive integers, the Well-Ordering Property guarantees the existence of an integer $q \in A$ such that $q \leq a$ for all $a \in A$. Since $q \in A$, we know that $q \notin S$. It follows that $q \neq 1$, so $q - 1$ is a positive integer. Note that $q - 1 \in S$ since q is the smallest integer in A. By the properties of the set S, the integer $q = (q-1)+1$ belongs to the set S. This is a contradiction to the fact that $q \notin S$. Hence, the set A is empty.

Therefore, the Principle of Mathematical Induction follows from the Well-Ordering Principle.

Now suppose that the Principle of Strong Induction is true. Let S be the set of all positive integers n with the following property:

> Any set of positive integers that contains an integer less than or equal to n has a least element.

It is clear that $1 \in S$. Suppose that $1, 2, \ldots, k \in S$ for some positive integer k. Let A be a set of positive integers that contains an integer less than or equal to $k+1$. If A contains no integer less than $k+1$, then $k+1$ is the least element in A. If A contains an integer $a < k+1$, then A is a set of positive integers that contains an integer less than or equal to a. Since $a \in S$, the set A has a least element. This shows that every set of positive integers that contains an integer less than or equal to $k+1$ has a least element. It follows that $k+1 \in S$. By the Principle of Strong Induction, $S = \mathbb{Z}^+$. Therefore, every nonempty set of positive integers has a least element, that is, the Well-Ordering Property holds. This completes the proof. ■

As mentioned prior to the theorem, one of these statements is accepted as an axiom. Hence, all three statements are valid. The next few sections explore some of the consequences of these results.

C.2 The Principle of Mathematical Induction

Statements of the form, "for each positive integer n, something is true", occur in all branches of mathematics, not just number theory. Let $Q(n)$ be an open statement, where the variable n represents a positive integer. Using the notation discussed in Appendix A, this statement assumes the form $(\forall n)(Q(n))$. The Principle of Mathematical Induction is a useful tool for proving statements of this type. Let

$$S = \{n \in \mathbb{Z}^+ : Q(n) \text{ is a true statement}\}.$$

In order to prove the statement $(\forall n)(Q(n))$, we must show that $S = \mathbb{Z}^+$. By the Principle of Mathematical Induction, it is sufficient to prove that S contains 1 and satisfies the condition " if $k \in S$, then $k+1 \in S$". In terms of the open statement $Q(n)$, we must prove that $Q(1)$ is true and that $Q(k+1)$ is true under the assumption that $Q(k)$ is true. In almost every situation of this type, the statement $Q(1)$ is easy to prove. However, the proof of the implication $Q(k) \Rightarrow Q(k+1)$ requires more effort. For many of the induction arguments that occur in this book, the proof of this implication involves little more than algebraic manipulation. In such cases, a proof that uses the Principle of Mathematical Induction is not very difficult. However, the proof of the implication $Q(k) \Rightarrow Q(k+1)$ can be difficult in some instances.

Several examples of proofs that involve the Principle of Mathematical Induction are given below. For the first example, two complete proofs are given. The first proof uses the specific statement of the Principle of Mathematical Induction. In actual practice, this is seldom done. The second proof is the more common way of wording a proof that uses the Principle of Mathematical Induction. The remaining proofs in this appendix will be written in the more informal style.

THEOREM C.2 For each positive integer n, the formula

$$1^2 + 2^2 + 3^2 + \cdots + n^2 = \frac{n(n+1)(2n+1)}{6}$$

is valid.

Proof. We will use the Principle of Mathematical Induction. Let S be the set of all positive integers n such that

$$1^2 + 2^2 + 3^2 + \cdots + n^2 = \frac{n(n+1)(2n+1)}{6}.$$

Since $1 = (1 \cdot 2 \cdot 3)/6$, it follows that $1 \in S$. Suppose that $k \in S$ for some positive integer k. This means that

$$1^2 + 2^2 + 3^2 + \cdots + k^2 = \frac{k(k+1)(2k+1)}{6}.$$

We then have

$$\begin{aligned} 1^2 + 2^2 + 3^2 + \cdots + k^2 + (k+1)^2 &= \frac{k(k+1)(2k+1)}{6} + (k+1)^2 \\ &= \frac{k+1}{6}\big(2k^2 + k + 6k + 6\big) \\ &= \frac{k+1}{6}(k+2)(2k+3) \\ &= \frac{(k+1)\big((k+1)+1\big)\big(2(k+1)+1\big)}{6}, \end{aligned}$$

which indicates that $k+1 \in S$. We have thus shown that $k \in S \Rightarrow k+1 \in S$. By the Principle of Mathematical Induction, $S = \mathbb{Z}^+$. Hence,

$$1^2 + 2^2 + 3^2 + \cdots + n^2 = \frac{n(n+1)(2n+1)}{6}$$

for all positive integers n.

Proof. The formula is easily verified for $n = 1$. Suppose that the formula is valid for some positive integer k. Then

$$\begin{aligned} 1^2 + 2^2 + 3^2 + \cdots + k^2 + (k+1)^2 &= \frac{k(k+1)(2k+1)}{6} + (k+1)^2 \\ &= \frac{k+1}{6}\big(2k^2 + k + 6k + 6\big) \\ &= \frac{(k+1)(k+2)(2k+3)}{6}, \end{aligned}$$

so the formula is valid for $k+1$ as well. The result now follows by the Principle of Mathematical Induction. ∎

THEOREM C.3 For each positive integer n, 64 is a divisor of $9^n - 8n - 1$.

Proof. Since 64 is a divisor of 0, the statement is valid when $n = 1$. Let k be a positive integer and suppose that 64 is a divisor of $9^k - 8k - 1$. This means that there exists an integer j such that $64j = 9^k - 8k - 1$. We then have

$$9^{k+1} - 8(k+1) - 1 = 9(9^k - 8k - 1) + 64k = 64(9j + k),$$

which implies that 64 is a divisor of $9^{k+1} - 8(k+1) - 1$. The result now follows by the Principle of Mathematical Induction. ■

The purpose of the next three results is threefold. First of all, this set of results proves an important theorem known as the Binomial Theorem. Second, the proof of the middle result is another illustration of the Principle of Mathematical Induction. Finally, these three results fit the classic pattern of lemma, theorem, and corollary. A lemma is a result that is useful in the proof of another result and a corollary is a result that follows easily from a theorem (see Appendix A).

We begin with some notation. For each positive integer n, define $n!$ (read "n factorial") by $n! = n(n-1)(n-2)\cdots 3\cdot 2\cdot 1$. Even for small values of n, factorials can be very large; $70! > 10^{100}$. For sound mathematical reasons (see Exercise 32 in Section 5.6), $0!$ is defined to be 1. For a positive integer n and a nonnegative integer k such that $0 \le k \le n$, define the **binomial coefficient** $\binom{n}{k}$ by

$$\binom{n}{k} = \frac{n!}{k!\,(n-k)!} = \frac{n(n-1)\cdots(n-k+1)}{k!}.$$

To illustrate binomial coefficients, note that

$$\binom{3}{0} = \frac{3!}{0!\,3!} = 1; \qquad \binom{3}{1} = \frac{3!}{1!\,2!} = 3;$$
$$\binom{3}{2} = \frac{3!}{2!\,1!} = 3; \qquad \binom{3}{3} = \frac{3!}{3!\,0!} = 1.$$

The reader should recognize these numbers as the coefficients that appear in the expansion of $(a+b)^3$. This is no coincidence.

LEMMA C.4 If n and k are positive integers with $1 \le k \le n$, then

$$\binom{n}{k} + \binom{n}{k-1} = \binom{n+1}{k}.$$

Proof. Using the definition of binomial coefficients,

$$\begin{aligned}\binom{n}{k} + \binom{n}{k-1} &= \frac{n!}{k!\,(n-k)!} + \frac{n!}{(k-1)!\,(n-k+1)!} \\ &= \frac{(n-k+1)n!}{k!\,(n-k+1)!} + \frac{kn!}{k!\,(n-k+1)!} \\ &= \frac{(n+1)!}{k!\,(n+1-k)!} = \binom{n+1}{k}.\end{aligned}$$

This completes the proof. ■

THEOREM C.5 If x is a real number, then

$$(1+x)^n = \sum_{k=0}^{n} \binom{n}{k} x^k$$

for each positive integer n.

Proof. When $n = 1$, the formula reads

$$1 + x = \binom{1}{0} x^0 + \binom{1}{1} x^1 = 1 + x,$$

which is clearly a true statement. Suppose that the formula holds for some positive integer n. Using the previous lemma, we find that

$$\begin{aligned}
(1+x)^{n+1} &= (1+x)(1+x)^n \\
&= (1+x) \sum_{k=0}^{n} \binom{n}{k} x^k \\
&= \sum_{k=0}^{n} \binom{n}{k} x^k + \sum_{k=0}^{n} \binom{n}{k} x^{k+1} \\
&= 1 + \sum_{k=1}^{n} \binom{n}{k} x^k + \sum_{k=1}^{n} \binom{n}{k-1} x^k + x^{n+1} \\
&= \binom{n+1}{0} + \sum_{k=1}^{n} \binom{n+1}{k} x^k + \binom{n+1}{n+1} x^{n+1} = \sum_{k=0}^{n+1} \binom{n+1}{k} x^k.
\end{aligned}$$

Hence, the formula is valid for $n+1$ as well. The result now follows by the Principle of Mathematical Induction. ■

COROLLARY C.6 Binomial Theorem If a and b are real numbers, then

$$(a+b)^n = \sum_{k=0}^{n} \binom{n}{k} a^k b^{n-k}$$

for each positive integer n.

Proof. If $b = 0$, then the only nonzero term in the sum occurs when $k = n$; this yields the equation $a^n = a^n$. Assume that $b \neq 0$. Using the result in the theorem, we obtain

$$(a+b)^n = b^n \left(1 + \frac{a}{b}\right)^n = b^n \sum_{k=0}^{n} \binom{n}{k} \left(\frac{a}{b}\right)^k = \sum_{k=0}^{n} \binom{n}{k} a^k b^{n-k}.$$

This completes the proof. ■

Although the last three results illustrate the lemma, theorem, corollary sequence, it is the formula in Corollary C.6 that is known as the **Binomial Theorem**. (To a certain extent, the "title" of a result is at the discretion of the author.) The formula in Theorem C.5 is a special case of the formula in Corollary C.6. It is common in mathematics to prove a special case of a theorem first, then to show how the general case reduces to the special case.

It is not necessary that 1 be the starting point in the Principle of Mathematical Induction; any integer a will do. Let S be a set of integers that contains a and satisfies the condition "if $k \geq a$ and $k \in S$, then $k+1 \in S$", then $S = \{n \in \mathbb{Z} : n \geq a\}$. The fact that this statement is equivalent to the Principle of Mathematical Induction follows by making a simple change of variables; the details will be left to the reader. (A similar modification is also valid for the Principle of Strong Induction.) There are situations for which this slight modification to the Principle of Mathematical Induction is helpful. An example appears below.

THEOREM C.7 For each positive integer $n \geq 8$, the inequality $n < (1.3)^n$ is valid.

Proof. It is easy to verify that $8 < (1.3)^8$. Suppose that $k < (1.3)^k$ for some positive integer $k \geq 8$. Then

$$k + 1 < k + 0.3k = 1.3k < 1.3(1.3)^k = (1.3)^{k+1},$$

so the inequality is valid for $k+1$ as well. By the Principle of Mathematical Induction, the inequality is valid for all $n \geq 8$. ■

C.3 THE PRINCIPLE OF STRONG INDUCTION

What is the main difference between the Principle of Mathematical Induction and the Principle of Strong Induction? The condition "if $k \in S$, then $k+1 \in S$" is replaced by the condition "if $1, 2, \ldots, k \in S$, then $k+1 \in S$"; the hypothesis of the condition is stronger for the Principle of Strong Induction. The assumption that all of the integers $1, 2, \ldots, k$ belong to S gives more information to use in the proof that $k+1 \in S$. In terms of an open statement $Q(n)$, the difference between the two forms of induction is the following: for the Principle of Mathematical Induction, we assume that $Q(k)$ is true, whereas for the Principle of Strong Induction, we assume that $Q(1), Q(2), \ldots, Q(k)$ are true. In some cases, the stronger hypothesis is needed to prove that $Q(k+1)$ is true. An example of a proof that uses the Principle of Strong Induction is given below.

THEOREM C.8 If $a_1 = 1$, $a_2 = 2$, and $a_n = 3a_{n-1} - 2a_{n-2}$ for each positive integer $n > 2$, then $a_n = 2^{n-1}$ for every positive integer n.

Proof. By hypothesis, the formula $a_n = 2^{n-1}$ is valid for $n = 1$ and $n = 2$. Suppose that $k > 1$ is a positive integer and that the formula is valid for the integers $1, 2, \ldots, k$. Then

$$a_{k+1} = 3a_k - 2a_{k-1} = 3 \cdot 2^{k-1} - 2 \cdot 2^{k-2} = 2^{k-1}(3 - 1) = 2^k,$$

which indicates that the formula is valid for $k+1$ as well. By the Principle of Strong Induction, the formula $a_n = 2^{n-1}$ is valid for all positive integers n. ■

In the preceding proof, all we really needed was for the formula to be valid for both k and $k-1$ to prove that the formula was valid for $k+1$; the other hypotheses were simply ignored. The important point is that knowing the formula is valid only for k is not sufficient to prove that the formula is valid for $k + 1$.

By the way, the sequence $\{a_n\}$ defined in the statement of Theorem C.8 is known as a recursively defined sequence; the sequence is generated by the first few terms and a rule to determine successive terms from previous ones. The fact that a_n is defined for every positive integer n is a simple consequence of the Principle of Strong Induction. In this case, the numbers a_1 and a_2 are defined. Assuming that the numbers $a_1, a_2, \ldots, a_k$ are defined, the formula shows how to define a_{k+1}. It follows that a_n is defined for every positive integer n. (Further discussion on recursively defined sequences can be found in Chapter 2.) In some of the proofs in this book, a recursive process is set up in an informal way. The Principle of Strong Induction is generally not mentioned; instead, a comment such as "continue this process" is made. As long as the first couple of terms are defined and it is evident how to determine the next term from previous terms, the Principle of Strong Induction guarantees that there is a term defined for each positive integer.

As a second illustration of the Principle of Strong Induction, we will prove part of the Fundamental Theorem of Arithmetic. This is an important result in number theory, and it is used several times in this book. Recall that a prime number is an integer $n > 1$ such that the only positive divisors of n are n and 1 (see Appendix A).

THEOREM C.9 Fundamental Theorem of Arithmetic Every integer $n \geq 2$ is either a prime number or can be factored into a product of prime numbers. The factorization is unique except for the order in which the factors are written.

Proof. It is clear that 2 is a prime number. Suppose that each of the integers $2, 3, \ldots, k$ is either a prime number or can be factored into a product of prime numbers. Consider the integer $k + 1$. If $k + 1$ is a prime number, then the result follows. If $k + 1$ is not a prime number, then $k + 1 = ab$, where a and b are integers between 2 and k, inclusively. By the induction hypothesis, each of the integers a and b is either a prime number or can be factored into a product of prime numbers. It follows that $k + 1 = ab$ can be factored into a product of prime numbers. By the Principle of Strong Induction, every integer $n \geq 2$ is either a prime number or can be factored into a product of prime numbers.

The proof of the uniqueness part of this theorem is not difficult, but it does involve some simple facts about prime numbers that we have not discussed. The proof will therefore be omitted. (The interested reader can consult Birkhoff and MacLane [2].) ■

The first part of the Fundamental Theorem of Arithmetic is usually written as

Every integer $n \geq 2$ can be factored into a product of prime numbers.

This statement is shorter and more concise than the one stated above and the proof requires fewer words. However, the reader must make a mental adjustment by considering a single prime number, such as 2 or 3, as a product. A number by itself is not normally considered to be a product—a product requires two or more numbers. Writing $2 = 2 \cdot 1$ does not solve the problem here since 1 is not a prime number. In this instance, the single number 2 must be thought of as a product.

Simplifications and generalizations such as this occur frequently in mathematics; it is therefore necessary to learn how to make the appropriate mental adjustments.

There is a common situation in which the Principle of Strong Induction occurs in disguised form or is only mentioned as an aside. One such example from calculus is the following. After a proof of the familiar fact $(f+g)' = f' + g'$, an example such as

$$\frac{d}{dx}\left(x^3 + 2x^2 + 3x + 2\right) = \frac{d}{dx}x^3 + \frac{d}{dx}2x^2 + \frac{d}{dx}3x + \frac{d}{dx}2 = 3x^2 + 4x + 3$$

is given. What is the problem? The theorem is stated for the sum of two functions and has been applied to a sum of four functions. Some calculus texts make no mention of this; others say that it is possible to extend the result to n functions using induction. However, the proof is seldom given because it is so boring. Since it is important to see such proofs at least once, we will prove this property of derivatives. (It is assumed that the reader has some familiarity with limits and derivatives.)

THEOREM C.10 For each positive integer n, the derivative of the sum of n differentiable functions is the sum of the derivatives of the functions.

Proof. There is nothing to prove if $n = 1$, so we will first establish the result for the sum of two differentiable functions. Let f_1 and f_2 be differentiable functions and use the definition of the derivative to compute

$$\begin{aligned}
(f_1 + f_2)'(x) &= \lim_{h\to 0} \frac{\big(f_1(x+h) + f_2(x+h)\big) - \big(f_1(x) + f_2(x)\big)}{h} \\
&= \lim_{h\to 0}\left(\frac{f_1(x+h) - f_1(x)}{h} + \frac{f_2(x+h) - f_2(x)}{h}\right) \\
&= f_1'(x) + f_2'(x).
\end{aligned}$$

Hence, the derivative of $f_1 + f_2$ is $f_1' + f_2'$. Now suppose that the derivative of the sum of k or fewer differentiable functions is the sum of the derivatives of the functions and let $f_1, \ldots, f_{k+1}$ be differentiable functions. Using the induction hypothesis (both the 2 case and the k case), we obtain

$$\begin{aligned}
\big(f_1 + \cdots + f_k + f_{k+1}\big)' &= \big((f_1 + \cdots + f_k) + f_{k+1}\big)' \\
&= \big(f_1 + \cdots + f_k\big)' + f_{k+1}' \\
&= \big(f_1' + \cdots + f_k'\big) + f_{k+1}' \\
&= f_1' + \cdots + f_k' + f_{k+1}',
\end{aligned}$$

the desired result. By the Principle of Strong Induction, for each positive integer n, the derivative of the sum of n differentiable functions is the sum of the derivatives of the functions. ■

Note that the first step in the preceding proof requires the definition, but that the inductive step uses only previous results and assumptions. Since the latter part of the proof is rather routine and requires more words than thinking, it is often left out. However, it is important to know what goes on in such situations. As you read through the main body of this textbook, look for situations such as this where a result for two is extended to a result for more than two.

A cautionary word is appropriate at this point. The conclusion of Theorem C.10 can be written as

$$\frac{d}{dx}\Big(\sum_{k=1}^{n} f_k(x)\Big) = \sum_{k=1}^{n} \frac{d}{dx} f_k(x),$$

where n is any positive integer. It is not possible to conclude from this that

$$\frac{d}{dx}\Big(\sum_{k=1}^{\infty} f_k(x)\Big) = \sum_{k=1}^{\infty} \frac{d}{dx} f_k(x).$$

In fact, this result is not true in general (see Chapter 7). An induction argument only shows that a result is valid for finite sums of any size; it does not say anything about infinite sums.

C.4 The Well-Ordering Property

We have yet to mention a proof that uses the Well-Ordering Property. Since it is equivalent to the Principle of Mathematical Induction, any proof that uses the Principle of Mathematical Induction could also be done using the Well-Ordering Property. The only change is the format of the proof.

THEOREM C.11 For each positive integer n, the formula

$$1^3 + 2^3 + 3^3 + \cdots + n^3 = \Big(\frac{n(n+1)}{2}\Big)^2$$

is valid.

Proof. Let S be the set of all positive integers n for which the formula is false. We will give a proof by contradiction to show that the set S is empty. Suppose that S is a nonempty set. By the Well-Ordering Property, the set S contains a least element, call it q. It is clear that $q \neq 1$ since the formula is easily seen to be true for $n = 1$. Since $q - 1$ is a positive integer that is not in S, the formula is valid for $q - 1$. That is,

$$1^3 + 2^3 + 3^3 + \cdots + (q-1)^3 = \Big(\frac{(q-1)q}{2}\Big)^2.$$

It follows that

$$\begin{aligned} 1^3 + 2^3 + 3^3 + \cdots + (q-1)^3 + q^3 &= \Big(\frac{(q-1)q}{2}\Big)^2 + q^3 \\ &= \frac{q^2}{4}\big((q-1)^2 + 4q\big) \\ &= \Big(\frac{q(q+1)}{2}\Big)^2, \end{aligned}$$

which indicates that q is not in S, a contradiction. We conclude that S is empty. Hence, the formula is valid for all positive integers. ∎

It is probably more natural to use the Principle of Mathematical Induction rather than the Well-Ordering Property to prove the formula in Theorem C.11, but the proof does give some indication of how the Well-Ordering Principle can be used. There are some situations in which it is easier to use the Well-Ordering Property. The proof of the next result, known as the division algorithm, is such a case. Its conclusion is the rather obvious statement that when one positive integer is divided by another the result is a quotient and a remainder that is smaller than the divisor. The division algorithm is the basis for many results in number theory. The proof also provides a good example of an **existence/uniqueness proof**; a proof that establishes the existence of some "object" and shows that there is only one object with the given property.

THEOREM C.12 Division Algorithm If a and b are positive integers, then there exist unique integers q and r such that $a = bq + r$ and $0 \leq r < b$.

Proof. We first prove that integers q and r with the desired properties exist. There are two easy cases to consider.

i) If $a < b$, then $a = b \cdot 0 + a$.

ii) If $b = 1$, then $a = 1 \cdot a + 0$.

Suppose that $a \geq b > 1$. The set $C = \{k \in \mathbb{Z}^+ : bk \geq a\}$ is a nonempty set of positive integers. By the Well-Ordering Property, the set C contains a least element p. This means that $b(p-1) < a \leq bp$. Now
if $a = bp$, then let $q = p$ and $r = 0$;
if $a \neq bp$, then let $q = p - 1$ and let $r = a - b(p-1)$.
In both cases, the integers q and r have the desired properties.

To establish uniqueness, suppose that there are two representations for a:

$$a = bq_1 + r_1 \quad \text{and} \quad a = bq_2 + r_2,$$

where r_1 and r_2 are nonnegative integers less than b. By relabeling the integers if necessary, we may assume that $r_1 \geq r_2$. It follows that $0 \leq r_1 - r_2 < b$; thus

$$0 \leq r_1 - r_2 = b(q_2 - q_1) < b.$$

Since $0 \leq q_2 - q_1 < 1$ and q_1 and q_2 are integers, we find that $q_1 = q_2$ and thus $r_1 = r_2$. Therefore, there is only one representation of the desired form for a. This completes the proof. ∎

You may find some of the steps in the above proof less than obvious. (Why is the set C nonempty?) You may find some sentences require you to take out a piece of paper and do some writing. (Why do q and r have the desired properties?) You may have to think a while and/or ask for some help. The important point here is the emphasis on "you". It is important that you understand each and every step in a proof; do not be a passive reader.

C.5 Some Comments on Induction Arguments

It will be helpful to make several comments on terminology. The hypothesis "if $k \in S$" or "if $1, 2, \ldots, k \in S$" is known as the **induction hypothesis**. The part of the argument that uses this assumption to prove that $k + 1 \in S$ is called the **inductive step**. The Principle of Strong Induction is often called the Principle of Mathematical Induction since the two statements are equivalent. A quick glance at the proof will indicate which form of induction is being used. A proof that uses either the Principle of Mathematical Induction or the Principle of Strong Induction is called a **proof by induction**. In those cases in which the inductive step is easy, the proof is usually left out. A phrase such as "the result follows by induction" means that the induction argument is easy and is left to the reader.

The Principle of Mathematical Induction has two hypotheses. As we have seen, checking that the result is valid for $n = 1$ is usually very easy. This does not mean that it is not necessary. Suppose that someone claims that $n^2 + 7n - 3$ is an even number for each positive integer n. If $k^2 + 7k - 3$ is even for some positive integer k, then

$$(k + 1)^2 + 7(k + 1) - 3 = (k^2 + 7k - 3) + 2(k + 4)$$

is the sum of two even numbers and thus an even number. This establishes the condition "if $k \in S$, then $k + 1 \in S$". However, the result is false for $n = 1$ (and also false for every other positive integer n). It is generally a good idea to check a formula for several values of n before trying to find a general proof.

As a final comment, it is important to realize that not every statement that involves positive integers requires the Principle of Mathematical Induction in its proof. There may be better or easier methods to prove the result; the following result is one example.

THEOREM C.13 For each positive integer $n > 1$, the inequality $n^3 + 1 > n^2 + 2n$ is valid.

Proof. Suppose that $n > 1$. Then $n - 1 \geq 1$ and $n^2 - 2 \geq 2$. It follows that

$$0 < 2 \cdot 1 - 1 \leq (n^2 - 2)(n - 1) - 1 = n^3 - n^2 - 2n + 1.$$

Adding $n^2 + 2n$ to both sides gives the desired result. ∎